T0182046

Advances in Intelligent Systems and Computing

Volume 467

Series editor

Janusz Kacprzyk, Polish Academy of Sciences, Warsaw, Poland
e-mail: kacprzyk@ibspan.waw.pl

About this Series

The series "Advances in Intelligent Systems and Computing" contains publications on theory, applications, and design methods of Intelligent Systems and Intelligent Computing. Virtually all disciplines such as engineering, natural sciences, computer and information science, ICT, economics, business, e-commerce, environment, healthcare, life science are covered. The list of topics spans all the areas of modern intelligent systems and computing.

The publications within "Advances in Intelligent Systems and Computing" are primarily textbooks and proceedings of important conferences, symposia and congresses. They cover significant recent developments in the field, both of a foundational and applicable character. An important characteristic feature of the series is the short publication time and world-wide distribution. This permits a rapid and broad dissemination of research results.

Advisory Board

Chairman

Nikhil R. Pal, Indian Statistical Institute, Kolkata, India
e-mail: nikhil@isical.ac.in

Members

Rafael Bello, Universidad Central "Marta Abreu" de Las Villas, Santa Clara, Cuba
e-mail: rbellop@uclv.edu.cu

Emilio S. Corchado, University of Salamanca, Salamanca, Spain
e-mail: escorchado@usal.es

Hani Hagras, University of Essex, Colchester, UK
e-mail: hani@essex.ac.uk

László T. Kóczy, Széchenyi István University, Győr, Hungary
e-mail: koczy@sze.hu

Vladik Kreinovich, University of Texas at El Paso, El Paso, USA
e-mail: vladik@utep.edu

Chin-Teng Lin, National Chiao Tung University, Hsinchu, Taiwan
e-mail: ctlin@mail.nctu.edu.tw

Jie Lu, University of Technology, Sydney, Australia
e-mail: Jie.Lu@uts.edu.au

Patricia Melin, Tijuana Institute of Technology, Tijuana, Mexico
e-mail: epmelin@hafsamx.org

Nadia Nedjah, State University of Rio de Janeiro, Rio de Janeiro, Brazil
e-mail: nadia@eng.uerj.br

Ngoc Thanh Nguyen, Wroclaw University of Technology, Wroclaw, Poland
e-mail: Ngoc-Thanh.Nguyen@pwr.edu.pl

Jun Wang, The Chinese University of Hong Kong, Shatin, Hong Kong
e-mail: jwang@mae.cuhk.edu.hk

More information about this series at http://www.springer.com/series/11156

P. Deiva Sundari · Subhransu Sekhar Dash
Swagatam Das · Bijaya Ketan Panigrahi
Editors

Proceedings of 2nd International Conference on Intelligent Computing and Applications

ICICA 2015

Editors
P. Deiva Sundari
Department of Electrical and Electronics
  Engineering
KCG College of Technology
Karapakkam, Chennai, Tamil Nadu
India

Subhransu Sekhar Dash
Department of Electrical and Electronics
  Engineering
SRM University
Chennai, Tamil Nadu
India

Swagatam Das
Electronics and Communication Sciences
  Unit
Indian Statistical Institute
Kolkata, West Bengal
India

Bijaya Ketan Panigrahi
Department of Electrical and Electronics
  Engineering
Indian Institute of Technology Delhi
New Delhi
India

ISSN 2194-5357   ISSN 2194-5365   (electronic)
Advances in Intelligent Systems and Computing
ISBN 978-981-10-1644-8   ISBN 978-981-10-1645-5   (eBook)
DOI 10.1007/978-981-10-1645-5

Library of Congress Control Number: 2016944398

Printed on acid-free paper

This Springer imprint is published by Springer Nature
The registered company is Springer Science+Business Media Singapore Pte Ltd.

Preface

This AISC volume contains the papers presented at the Second International Conference on Intelligent Computing and Applications (ICICA 2015) held during February 5–6, 2016, at KCG College of Technology, Chennai, India. ICICA 2015 is the second international conference aiming at bringing together the researchers from academia and industry to report and review the latest progresses in the cutting-edge research on various research areas of electronic circuits, power systems, renewable energy applications, image processing, computer vision and pattern recognition, machine learning, data mining and computational life sciences, management of data including big data and analytics, distributed and mobile systems including grid and cloud infrastructure, information security and privacy, VLSI, antenna, computational fluid dynamics and heat transfer, intelligent manufacturing, signal processing, intelligent computing, soft computing, web security, privacy and e-commerce, e-governance, optimization, communications, smart wireless and sensor networks, networking and information security, mobile computing and applications, industrial automation and MES, cloud computing, green IT and finally to create awareness about these domains to a wider audience of practitioners.

Actually ICICA 2015 was planned during Dec 21–23, 2015. But due to recent incessant rains and floods in and around Chennai which inundated most parts of the city, it was postponed to Feb 5–6, 2016. ICICA 2015 received 173 paper submissions including two foreign countries across the globe. All the papers were peer-reviewed by the experts in the area in India and abroad, and comments have been sent to the authors of accepted papers. Finally 76 papers mainly from foreign countries like Thailand, Africa, Saudi Arabia and from various states of India were accepted for oral presentation in the conference.

This corresponds to an acceptance rate of 41.44 % and is intended to maintain the high standards of the conference proceedings. The papers included in this AISC volume cover a wide range of topics in intelligent computing and algorithms and their real-time applications in problems from diverse domains of science and engineering. The conference was inaugurated by Mr. Pashupathy Gopalan,

President of SunEdison—Asia Pacific on Feb 5, 2016. The conference featured distinguished keynote speakers as follows: Dr. Vitawat Sitakul, King Mongkut's University of Technology, Thailand; Mr. Martin Fiddler, Mr. Niko Decourt, Staffordshire University, UK; Dr. Swagatam Das, ISI, Kolkata, India; Dr. Sarathi, IIT, Madras, India; Dr. S.S. Dash, SRM University, Chennai, India; Mr. Swaminathan, Neyveli Lignite Corporation, India; and Mr. Sivakumar of NITEO Technologies, Chennai, India in various domains We take this opportunity to thank the authors of the submitted papers for their hardwork, adherence to the deadlines, and patience with the review process. The quality of a referred volume depends mainly on the expertise and dedication of the reviewers. We are indebted to the Program Committee/Technical Committee members, who produced excellent reviews in short time frames. First, we are indebted to the chairperson, director and CEO of KCG College of Technology, Chennai, India for supporting our cause and encouraging us to organize the conference there. In particular, we would like to express our heartfelt thanks for providing us with the necessary financial support and infrastructural assistance to hold the conference. Our sincere thanks to Prof. D. K. Mandal and Dr. Rajib Kar, NIT Durgapur; Dr. T. Rengaraja, principal, KCG College of Technology; Dr. Sumathi Poobal, vice principal, KCG College of Technology; Late. Dr. Amos H. Jeeva Oli; Dr. R. Senthil; Mr. S. Cloudin; Dr. J. Frank Vijay; Dr. S. Ramesh; Dr. M. Krishnamurthy; Dr. Raja Paul Perinbam; Dr. P.S. Mayurappriyan; Mr. J. Arun Venkatesh for their continuous support. We thank the International Advisory Committee members for providing valuable guidelines and inspiration to overcome various difficulties in the process of organizing this conference. We would also like to thank the participants of this conference. The members of faculty and students of KCG College of Technology deserve special thanks because without their involvement, we would not have been able to face the challenges of our responsibilities. Finally, we thank all the volunteers who made great efforts in meeting the deadlines and arranging every detail to make sure that the conference could run smoothly. We hope the readers of these proceedings find the papers inspiring and enjoyable.

Chennai, India	P. Deiva Sundari
Chennai, India	Subhransu Sekhar Dash
Kolkata, India	Swagatam Das
New Delhi, India	Bijaya Ketan Panigrahi
February 2016	

Contents

About the Editors

Dr. P. Deiva Sundari is Professor and Head of Department of Electrical and Electronics Engineering, KCG College of Technology, Chennai, India. She received her Ph.D. degree from College of Engineering, Guindy, Anna University, Chennai, India. She has published papers in many reputed international journals such as IET, Springer, and World Scientific journals. She has received many awards such as "IET CLN Young Women Engineer Award, Best Teacher Award, and Best Academic Researcher." Her research interest focuses on power electronics, solar PV systems, converters, and application of artificial intelligence for electrical drives.

Dr. Subhransu Sekhar Dash is presently Professor in the Department of Electrical and Electronics Engineering, SRM Engineering College, SRM University, Chennai, India. He received his Ph.D. degree from College of Engineering, Guindy, Anna University, Chennai, India. He has more than seventeen years of research and teaching experience. His research areas are power electronics and drives, modeling of FACTS controller, power quality, power system stability, and smart grid. He is a visiting professor at Francois Rabelais University, POLYTECH, France. He is the chief editor of International Journal of Advanced Electrical and Computer Engineering.

Dr. Swagatam Das received the B.E. Tel.E., M.E. Tel.E (control engineering specialization), and Ph. D. degrees, all from Jadavpur University, India, in 2003, 2005, and 2009, respectively. He is currently serving as assistant professor at the Electronics and Communication Sciences Unit of the Indian Statistical Institute, Kolkata, India. His research interests include evolutionary computing, pattern recognition, multi-agent systems, and wireless communication. He has published one research monograph, one edited volume, and more than 200 research articles in peer reviewed journals and international conferences.

Dr. Bijaya Ketan Panigrahi is associate professor of Electrical and Electronics Engineering Department in Indian Institute of Technology, Delhi, India. He received his Ph.D. degree from Sambalpur University. He is a chief editor in the International Journal of Power and Energy Conversion. His interests focus on power quality, FACTS devices, power system protection, and AI application to power system.

Time-Domain Analytical Modeling of Current-Mode Signaling Bundled Single-Wall Carbon Nanotube Interconnects

Yash Agrawal, M. Girish and Rajeevan Chandel

Abstract Bundled single-wall carbon nanotube (SWCNT) interconnects have been investigated as one of the prominent replacements to copper interconnects in nano-scale regime. This paper investigates the performance of SWCNT bundle interconnect using current-mode signaling (CMS) scheme. Time-domain analytical model is derived for CMS SWCNT bundle interconnects using finite-difference time-domain (FDTD) technique. This model for the first time accurately considers the CMOS gate driver in its analytical FDTD formulation for CMS SWCNT bundle interconnects. The performance of CMS SWCNT bundle interconnect is compared with the conventional CMS copper interconnects. It is investigated that CMS SWCNT bundle interconnects have lower propagation delay than CMS copper interconnects.

Keywords Bundled single-wall carbon nanotube (SWCNT) · Current-mode signaling (CMS) · Finite-difference time-domain (FDTD)

1 Introduction

The integrated circuit technology has advanced from micron to nano-scale dimensions [1]. The conventional copper and aluminum on-chip interconnects possess several fabrication and performance limitations due to increased electro-migration, hillock formation at miniaturized dimensions [2]. As a substitute to these interconnect materials, graphene based materials such as carbon nanotube (CNT) and graphene nano-ribbon (GNR) have been reported by researchers in

Y. Agrawal (✉) · M. Girish · R. Chandel
Department of Electronics and Communication Engineering, National Institute of Technology, Hamirpur 177 005, Himachal Pradesh, India
e-mail: mr.yashagrawal@gmail.com

M. Girish
e-mail: giri.frds@gmail.com

R. Chandel
e-mail: rchandel@nith.ac.in

© Springer Science+Business Media Singapore 2017
P. Deiva Sundari et al. (eds.), *Proceedings of 2nd International Conference on Intelligent Computing and Applications*, Advances in Intelligent Systems and Computing 467, DOI 10.1007/978-981-10-1645-5_1

literature [1–3]. CNTs can be classified as single-wall CNTs (SWCNTs) and multi-wall CNTs (MWCNTs) [2]. Bundled structure of SWCNT possess high mean free path (MPF) [4], have ballistic transport [2] and are easy to fabricate [1]. These improves the overall performance of the interconnect system.

Signals in on-chip interconnects can be transmitted by various signaling schemes such as voltage-mode signaling (VMS) [1–5] or current-mode signaling (CMS) [6–9]. VMS has full voltage swing while CMS scheme has reduced voltage swing over the interconnects. The CMS scheme has advantage of lower latency and higher bandwidth over VMS scheme [7]. The reduced voltage swing in CMS scheme is achieved by using specialized current-mode low input impedance receiver circuits [7, 8]. In [6–9], CMS scheme has been explored for copper interconnects. However, till now, no significant work has been reported for SWCNT bundle interconnects using CMS scheme. Thus, this paper analyzes the SWCNT bundle interconnect using CMS scheme and proposes a novel time-domain analytical model using finite-difference time-domain (FDTD) technique. Moreover, earlier analytical FDTD models for CNT based interconnects have approximated CMOS gate driver as lumped resistor and capacitor elements [3]. This approximation however leads to inaccurate results. This is because, CMOS gate driver cannot be approximated by single lumped value in different regions of operation of PMOS and NMOS transistors. In order to remove this limitation, the present model accurately incorporates driver as CMOS gate instead of approximating it as lumped parameters.

The paper is organized as follows. Section 2 presents the model for CMS interconnect system. In Sect. 3, analytical model using FDTD technique is formulated. The performance analysis of CMS copper and SWCNT bundle interconnects using FDTD and SPICE is made in Sect. 4. Finally, conclusion is drawn in Sect. 5.

2 Model of Current Mode Signaling Interconnect System

The current mode signaling interconnect system consists of driver, interconnect and receiver subsystems and is illustrated in Fig. 1a. The driver is implemented using CMOS gate. The interconnect is represented by equivalent single conductor

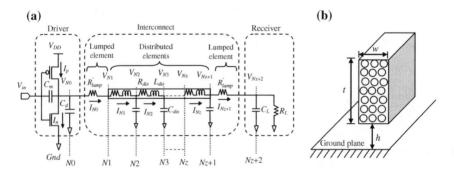

Fig. 1 a The current mode interconnect system. **b** Schematic of SWCNT bundle placed above a ground plane

(ESC) model [1, 3]. It consists of lumped and distributed interconnect parasitics. The current mode receiver provides low impedance termination [7, 8]. It is equivalently modeled using parallel combination of resistance (R_L) and capacitance (C_L) [6, 9]. $N0$, $N1$, ..., Nz, $Nz+1$, $Nz+2$ represent nodes along the interconnect line in Fig. 1a. The schematic of SWCNT bundle interconnect is shown in Fig. 1b. w and t represent the width and height of SWCNT bundle interconnect. h is the distance between the SWCNT bundle and ground plane.

3 Analytical Model Formulation

The time-domain analytical modeling for CMS copper and SWCNT bundle interconnect is formulated using FDTD technique. The distributed section of interconnect can be modeled by Telegraph's equations as

$$\frac{\partial V(z,t)}{\partial z} + L_{dis}\frac{\partial I(z,t)}{\partial t} + R_{dis}I(z,t) = 0 \tag{1}$$

$$\frac{\partial I(z,t)}{\partial z} + C_{dis}\frac{\partial V(z,t)}{\partial t} = 0 \tag{2}$$

where $V(z,t)$ and $I(z,t)$ represent voltage and current variables. z and t define the spatial and temporal position along the interconnect line. R_{dis}, C_{dis}, and L_{dis} are interconnect parasitics. The voltage and current variables are interlaced in time and space [9]. For FDTD formulation, (1) and (2) are discretized in time and space and are presented by (3) and (4) respectively.

$$I_k^{n+3} = BDI_k^{n+1} + B\left(V_k^{n+2} - V_{k+1}^{n+2}\right) \tag{3}$$

where $B = \left(\frac{\Delta z}{\Delta t}L_{dis} + \frac{\Delta z}{2}R_{dis}\right)^{-1}$, $D = \left(\frac{\Delta z}{\Delta t}L_{dis} - \frac{\Delta z}{2}R_{dis}\right)$, $k = N1, N2, N3, \ldots,$ Nz and $n = 0, 2, 4, \ldots, Tm$.

$$V_k^{n+2} = V_k^n + A\left(I_{k-1}^{n+1} - I_k^{n+1}\right) \tag{4}$$

where $A = \left(\frac{\Delta t}{\Delta z} \cdot \frac{1}{C_{dis}}\right)$, $k = N2, N3, N4, \ldots, Nz$ and $n = 0, 2, 4, \ldots, Tm$.

For correct output realization, near- and far-end boundary conditions need to be specified precisely.

The near-end boundary condition is defined at node $N0$ in the current mode interconnect system as shown in Fig. 1a.

Voltage at node $N1$ is evaluated by putting $k = 1$ in (4). The distance between current at node $N0$ and $N1$ is $\Delta z/2$. Hence, Δz is replaced as $\Delta z/2$ in (4) [9].

$$V_{N1}^{n+2} = V_{N1}^n + 2A\left(\frac{I_{N0}^n + I_{N0}^{n+2}}{2} - I_{N1}^{n+1}\right) \tag{5}$$

Current I_{N0} is evaluated by applying KCL at node $N0$ and is given as

$$I_{N0} = I_p - I_n + C_m\frac{d(V_{in} - V_{N0})}{dt} - C_d\frac{dV_{N0}}{dt} \tag{6}$$

The PMOS and NMOS transistor currents in (6) are evaluated using nth power law model [9, 10].

Using KVL, V_{N0} for varying time instant n is derived as

$$V_{N0}^{n+2} = V_{N1}^{n+2} + R'_{lump} \cdot I_{N0}^{n+2} \tag{7}$$

Using (7) in (6), the discretized branch current I_{N0}^{n+2} is evaluated as

$$I_{N0}^{n+2} = H\left(\begin{array}{c} I_p^{n+2} - I_n^{n+2} + C_m\left(\frac{V_{in}^{n+2} - V_{in}^n}{\Delta t}\right) - (C_m + C_d)\left(\frac{V_{N1}^{n+2} - V_{N1}^n}{\Delta t}\right) \\ + R'_{lump}(C_m + C_d)\left(\frac{I_{N0}^n}{\Delta t}\right) \end{array}\right) \tag{8}$$

Substituting (8) in (5), gives

$$V_{N1}^{n+2} = \left(\begin{array}{c} V_{N1}^n + EAH\left[C_m\left(\frac{V_{in}^{n+2} - V_{in}^n}{\Delta t}\right) + \left(\frac{1}{H} + \frac{R'_{lump}(C_m + C_d)}{\Delta t}\right)I_{N0}^n\right] \\ -2EAI_{N1}^{n+1} + EAH\left(I_p^{n+2} - I_n^{n+2}\right) \end{array}\right) \tag{9}$$

where H and E in (8) and (9) are defined as

$$H = \left(1 + \frac{R'_{lump}(C_m + C_d)}{\Delta t}\right)^{-1} \quad \text{and} \quad E = \left(1 + \frac{AH(C_m + C_d)}{\Delta t}\right)^{-1} \tag{10}$$

The far-end boundary condition is defined at node $Nz + 2$ in Fig. 1a. Using Ohm's law, it is derived as

$$V_{Nz+2}^{n+2} = V_{Nz+1}^{n+2} - R'_{lump} \cdot I_{Nz+1}^{n+2} \tag{11}$$

The nodal voltage V_{Nz+1}^{n+2} is computed by putting $k = Nz + 1$ in (4). Here also, Δz is replaced as $\Delta z/2$ in (4).

$$V_{Nz+1}^{n+2} = V_{Nz+1}^n + 2A\left(I_{Nz}^{n+1} - I_{Nz+1}^{n+1}\right) \tag{12}$$

Applying KCL at node $Nz + 2$, results

$$I_{Nz+1} = C_L \frac{dV_{Nz+2}}{dt} + \frac{V_{Nz+2}}{R_L} \tag{13}$$

Using (11) and (13), I_{Nz+1}^{n+2} is given as

$$I_{Nz+1}^{n+2} = J\left(\left(\frac{R'_{lump}C_L}{\Delta t}\right)I_{Nz+1}^n + \left(\frac{C_L}{\Delta t} + \frac{1}{R_L}\right)\left(V_{Nz+1}^{n+2}\right) - \left(\frac{C_L}{\Delta t}\right)\left(V_{Nz+1}^n\right)\right) \tag{14}$$

The nodal voltage V_{Nz+1}^{n+2} is computed using (12) and (14) as

$$V_{Nz+1}^{n+2} = FGV_{Nz+1}^n + 2FAI_{Nz}^{n+1} - FA\left(1 + \frac{JR'_{lump}C_L}{\Delta t}\right)I_{Nz+1}^n \tag{15}$$

where constants in (14)–(15) are defined as

$$J = \left(1 + \frac{R'_{lump}C_L}{\Delta t} + \frac{R'_{lump}}{R_L}\right)^{-1}, \quad F = \left(1 + \frac{AJC_L}{\Delta t} + \frac{AJ}{R_L}\right)^{-1} \quad \text{and} \tag{16}$$
$$G = \left(1 + \frac{AJC_L}{\Delta t}\right)$$

Finally, far-end output voltage V_{Nz+2}^{n+2} is evaluated using (11), (14) and (15).

4 Results and Discussion

The CMS SWCNT bundle interconnect is analyzed for 32 nm technology node. The performance of CMS SWCNT bundle interconnect is compared with conventional copper interconnects. Further, propagation delay for varying diameters of individual SWCNTs is presented in this section. The interconnect parasitics are computed from ITRS [11]. The width of NMOS and PMOS transistors are 1 and 2 μm respectively. The diameter of SWCNT in a bundle interconnect is 1.5 nm [1]. The distributed parameters namely R_{dis}, L_{dis} and C_{dis} for SWCNT bundle interconnects are 5.62 Ω/μm, 5.27 pH/μm and 11.89 aF/μm respectively. These values for copper interconnects are 3.18 Ω/μm, 1.48 pH/μm and 21.8 aF/μm respectively [12]. The lumped resistance $\left(R'_{lump}\right)$ is 6.18 Ω for SWCNT bundle and zero for copper interconnects. The receiver is equivalently model by load resistance (R_L) and load capacitance (C_L). R_L and C_L are 1 KΩ and 0.5 fF respectively [9]. The various analyses are performed using analytical FDTD based model and its validation is ascertained using SPICE [13] simulation model.

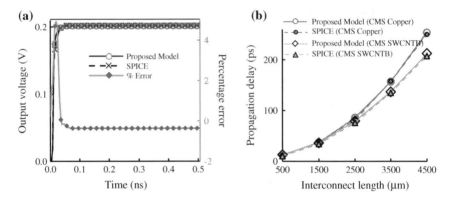

Fig. 2 **a** Transient response of CMS SWCNT bundle interconnects. **b** Propagation delay in CMS copper and SWCNT bundle interconnects with variation in interconnect length

4.1 Transient Analysis

The transient analysis of CMS SWCNT bundle interconnect is presented in Fig. 2a. Input is a ramp signal with transition period of 5 ps. The interconnect length considered is 1000 μm. From the figure, it is observed that the analytical FDTD based model and SPICE simulation results match closely. The average percentage error between these two models is 0.126 %.

4.2 Propagation Delay Evaluation for CMS SWCNT Bundle and Copper Interconnects

The propagation delay with variation in interconnect length in CMS copper and SWCNT bundle interconnects is shown in Fig. 2b. The interconnect length is varied from 500 to 4500 μm. Irrespective to interconnect materials, it is analyzed that propagation delay increases with increase in interconnect length. This is because of higher interconnect parasitics for longer wire lengths. From the figure it is also analyzed that propagation delay in CMS SWCNT bundle is lesser as compared to CMS copper interconnects. The former has about 16.27 % lesser propagation delay than the later at interconnect length of 4500 μm. This justifies the efficacy of CMS SWCNT bundle over CMS copper interconnects. It is observed from the figure that FDTD and SPICE simulation results are in close agreement. The percentage variations in between FDTD and SPICE model results at 4500 μm for CMS copper and SWCNT bundle interconnects are 1.53 and 2.37 % respectively.

In Fig. 3a, propagation delay for varying signal transition period is presented. The signal transition period is varied from 20 to 100 ps. The interconnect length is

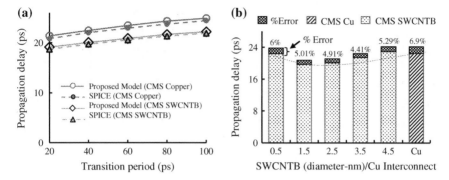

Fig. 3 **a** Propagation delay in CMS copper and SWCNT bundle interconnects with variation in signal transition period. **b** Propagation delay for copper and varying SWCNT diameters

kept constant at 1000 μm. It is seen form the figure that propagation delay varies nominally with signal transition period for both the interconnect materials. Also, it is observed that CMS SWCNT bundle have higher edge over CMS copper interconnects since these result in lower propagation delay in the circuit.

4.3 Performance Assessment with Variation in Diameter of Individual SWCNTs in a Bundle Interconnect

The diameter of individual SWCNTs in a bundle interconnect may vary due to lack of control over chirality and growth process. In the present analysis, the diameter of individual SWCNT is varied from 0.5 to 4.5 nm. The variation of propagation delay for different diameters of individual SWCNTs is presented in Fig. 3b. With variation in diameter of individual SWCNTs in a bundle interconnect, the number of SWCNTs also change. These parameters along with interconnect parasitics for copper and SWCNT bundle interconnect for 32 nm technology node is illustrated in Table 1.

It is analyzed that propagation delay has about parabolic variation effect as SWCNT diameter changes. The least propagation delay is observed for SWCNT diameter of 1.5 nm. However, propagation delay increases below and after this point. The propagation delay is high at low SWCNT diameters due to high parasitic capacitance. On the other hand, for large SWCNT diameters, parasitic interconnect resistance is high. Both increased resistance and capacitance cause higher propagation delay in the circuit. It is also seen that propagation delay in CMS SWCNT bundle interconnect is lesser as compared to CMS copper interconnects for all SWCNT diameters except at 4.5 nm. This suggests that diameter of individual SWCNT in a bundle interconnect should be carefully monitored and controlled to achieve the best results.

Table 1 Interconnect parasitics for SWCNT and copper interconnects

Interconnect dimensions	SWCNT bundle					Cu
Diameter of SWCNT (nm)	0.5	1.5	2.5	3.5	4.5	–
No. of SWCNTs (Metallic nature)	3795	765	324	179	108	–
R'_{lump} (Ω)	1.24	6.18	14.59	26.40	43.76	0
R_{dis} ($\Omega/\mu m$)	3.40	5.62	7.97	10.30	13.28	3.18
L_{dis} (pH/μm)	1.06	5.27	12.45	22.54	37.35	1.48
C_{dis} (aF/μm)	20.88	11.89	8.90	7.52	6.28	21.8

5 Conclusion

The present paper analyzes the performance of SWCNT bundle interconnects using current mode signaling scheme. The analytical model is proposed using FDTD technique. The proposed model is validated using SPICE simulations. The interconnect is driven by practical CMOS gate. The performance of CMS SWCNT bundle interconnects is compared with the conventional CMS copper interconnects. It is analyzed that CMS SWCNT bundle interconnects lead to lesser propagation delay with variation in both interconnect lengths and transition periods. Also, the effect of variation of individual SWCNT diameters in a bundle interconnect is analyzed. It is investigated that SWCNTs with diameter of 1.5 for 32 nm technology nodes result in least circuit propagation delay. The various analyzes in the present research work suggest that CMS SWCNT interconnects have enhanced performance and are better than conventional CMS copper interconnects for future technology nodes.

References

1. Rai, M.K., Khanna, R., Sarkar, S.: Crosstalk analysis in CNT bundle interconnects for VLSI applications. IEEJ Trans. Electr. Electron. Eng. **9**(4), 391–397 (2014)
2. Srivastava, A., Marulanda, J.M., Xu, Y., Sharma, A.K.: Carbon-Based Electronics: Transistors and Interconnects at the Nanoscale. Pan Stanford Publishing, Singapore (2015)
3. Duksh, Y.S., Kaushik, B.K., Agarwal, R.P.: FDTD technique based crosstalk analysis of bundled SWCNT interconnects. J. Semicond. **36**(5), 055002–055009 (2015)
4. Sahoo, M., Ghosal, P., Rahaman, H.: Performance modeling and analysis of carbon nanotube bundles for future VLSI circuit applications. J. Comput. Electron. **13**(3), 673–688 (2014)
5. Chandel, R., Sarkar, S., Agarwal, R.P.: An analysis of interconnect delay minimization by low-voltage repeater insertion. Microelectron. J. **38**(4–5), 649–655 (2007)
6. Bashirullah, R., Liu, W., Cavin, R.K.: Current-mode signaling in deep submicrometer global interconnects. IEEE Trans. Very Large Scale Integr. Syst. **11**(3), 406–417 (2003)

7. Agrawal, Y., Chandel, R., Dhiman, R.: High performance current mode receiver design for on-chip VLSI interconnects. Chapter 54. In: Springer Proceedings of the International Conference on ICA. Series: Advances in Intelligent Systems and Computing, vol. 343, pp. 527–536 (2014)
8. Dave, M., Jain, M., Baghini, M.S., Sharma, D.: A variation tolerant current-mode signaling scheme for on-chip interconnects. IEEE Trans. Very Large Scale Integr. Syst. 21(2), 342–353 (2013)
9. Agrawal, Y., Chandel, R.: Crosstalk analysis of current-mode signalling-coupled RLC interconnects using FDTD technique. IETE Tech Rev 33(2), 148–159 (2016)
10. Sakurai, T., Newton, A.R.: A simple MOSFET model for circuit analysis. IEEE Trans. Electron Devices 38(4), 887–894 (1991)
11. International technology roadmap for semiconductors (ITRS). http://public.itrs.net
12. Predictive technology model (PTM). http://ptm.asu.edu
13. Tanner EDA tool. www.tannereda.com

Stability Analysis of Carbon Nanotube Interconnects

Mekala Girish Kumar, Yash Agrawal and Rajeevan Chandel

Abstract This paper deals with frequency and stability response of single wall carbon nanotube bundle (SWB) and multiwall carbon nanotube bundle (MWB) at global interconnect lengths. The performance of SWB and MWB interconnects are analyzed using driver-interconnect-load system. It is analyzed that MWB interconnects are more stable than SWB interconnects. It is illustrated that stability of both SWB and MWB interconnects increases with increase in interconnect length. The analytical model for stability and frequency response using *ABCD* matrix has been formulated. Using frequency response, it is observed that the bandwidth of SWB and MWB interconnects are 7.94 and 22.2 GHz respectively for an interconnect length of 500 μm. The results are verified using SPICE simulations. The time delay analysis has been performed for different interconnect lengths. Further, it is investigated that delay reduces with increasing number of shells in MWB interconnect.

Keywords Carbon nanotubes (CNT) · Kinetic inductance · MWCNT · Quantum resistance · Quantum capacitance · SWCNT

1 Introduction

In VLSI technology, the chip performance and signal integrity is dependent on interconnect delay in deep submicron technology (DSM) [1]. As technology scales down interconnects are a major concern as these degrades system performance and

M.G. Kumar (✉) · Y. Agrawal · R. Chandel
Electronics and Communication Engineering, National Institute of Technology Hamirpur,
Hamirpur, Himachal Pradesh, India
e-mail: giri.frds@gmail.com

Y. Agrawal
e-mail: mr.yashagrawal@gmail.com

R. Chandel
e-mail: rchandel@nith.ac.in

© Springer Science+Business Media Singapore 2017
P. Deiva Sundari et al. (eds.), *Proceedings of 2nd International Conference on Intelligent Computing and Applications*, Advances in Intelligent Systems and Computing 467, DOI 10.1007/978-981-10-1645-5_2

11

causes reliability problems. With scaling technology, grain boundary and surface scattering phenomenon increases in copper interconnects [2]. At high frequencies, issues like skin effect, operational bandwidth and stability response affect the performance of copper interconnects [3]. To overcome these limitations, the carbon nanotubes (CNT) have been proposed as potential materials for interconnect applications as these possess extremely long mean free path (MFP), large current capability [4]. CNTs are made by graphene sheet that are rolled up into cylindrical structure. CNTs are classified into single wall carbon nanotubes (SWCNTs) and multiwall carbon nanotubes (MWCNTs). SWCNTs are either metallic or semiconducting depending on their chirality [5]. But, MWCNTs are always metallic in nature [6]. Also, MWCNTs are easier to fabricate as compared to SWCNTs due to their growth process. The most promising material for global interconnects is MWCNTs due to its high current carrying capability than SWCNTs.

To analyze the performance of SWB and MWB interconnects for on-chip applications, frequency and stability response have been performed. It is important to note that stability and the frequency responses are affected by interconnect parameters, such as resistances, capacitances, and inductances.

Li et al. [7] have analyzed on-chip inductor design of CNTs at high frequencies. Nasiri et al. [8] derived input-output transfer function of a MWCNT using transmission line model. The stability and Nyquist plot for CNT bundle interconnect have been presented in Fathi et al. [9]. It is assumed that the driver parasitic and contact resistances are only for local interconnects.

This paper deals with stability and frequency response of CNT based interconnects. Rest of the paper is organized as follows. A brief description of SWB and MWB interconnect is presented in Sect. 2. In Sect. 3 transfer function and its analytical formulation is using driver-interconnect-load (DIL) system is presented. Stability and frequency domain analysis have been performed in Sect. 4. Finally, conclusions are drawn in Sect. 5.

2 Interconnect Model

This section presents a comprehensive analytical model for SWB and MWB interconnects. SWB and MWB of width (w), diameter (d) and thickness (t) is placed above the ground plane at a distance (h) as sown in Fig. 1.

The number of SWCNTs in a bundle as shown in Fig. 1a along x and y directions are expressed as [10]:

$$n_x = \frac{w - d}{sp} \tag{1}$$

$$n_y = \frac{2(t - d)}{\sqrt{3}sp} + 1 \tag{2}$$

where sp is inter-CNT distance of interconnect.

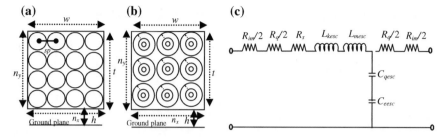

Fig. 1 a Cross sectional view of SWB. **b** MWB. **c** Equivalent single conductor model for SWB/MWB interconnect

Using (1) and (2), the total number of SWCNTs in bundle are given as

$$N_S = n_x n_y - \frac{n_y}{2} \quad \text{if } n_y \text{ is even} \tag{3a}$$

$$N_S = n_x n_y - \frac{(n_y - 1)}{2} \quad \text{if } n_y \text{ is odd} \tag{3b}$$

The number of MWCNTs in a bundle as shown in Fig. 1b along x and y directions are given as:

$$n_x = \frac{w - d_N}{d_N + \delta} + 1 \tag{4a}$$

$$n_y = \frac{2(t - d_N)}{\sqrt{3}(d_N + \delta)} + 1 \tag{4b}$$

Using (1) and (2), the total number of MWCNTs in bundle are given as

$$N_{MWCNT} = n_x n_y - \frac{n_y}{2} \quad \text{if } n_y \text{ is even} \tag{5a}$$

$$N_{MWCNT} = n_x n_y - \frac{(n_y - 1)}{2} \quad \text{if } n_y \text{ is odd} \tag{5b}$$

Figure 1c represents the equivalent single conductor model of SWB/MWB interconnects. The *RLC* parasitic values in the figure are calculated from Das and Rahaman [11].

3 Stability Analysis of SWB and MWB Interconnects

The transfer function (*TF*) of the SWB and MWB interconnect distributed line is derived to get the frequency and stability response at global interconnect lengths using DIL system.

The *TF* accurately considered the driver parasitics i.e. resistance (R_d) and capacitance (C_d) as shown in Fig. 2a. To obtain overall gain (V_o/V_i), the DIL system is represented as a cascaded connection of two-port networks as shown in Fig. 2b. Here, the two-port system is represented by *ABCD* parameters. The *ABCD* parameters of each of the two-port networks are denoted as f_1, f_2, f_3, f_4 and f_5. The Telegrapher's equation of distributed interconnect line is used to obtain *ABCD* matrix parameters f_4. Using the method, voltage and current at any point x of the DIL are expressed as [12, 13]

$$\frac{dV}{dx} = -(R_{esc} + sL_{esc})I(x) \tag{6}$$

$$\frac{dI}{dx} = -sC_{esc}.V(x) \tag{7}$$

To obtain overall gain, it is needed to evaluate transmission matrix (T_{final}) from Fig. 2b.

$$T_{final} = \begin{bmatrix} 1 & R_d \\ 0 & 1 \end{bmatrix} \begin{bmatrix} 1 & 0 \\ sC_d & 1 \end{bmatrix} \begin{bmatrix} 1 & R_{lump} \\ 0 & 1 \end{bmatrix}$$
$$\times \begin{bmatrix} \cosh(\gamma px) & Z_0 \sinh(\gamma px) \\ 1/Z_0 \sinh(\gamma px) & \cosh(\gamma px) \end{bmatrix} \begin{bmatrix} 1 & R_{lump} \\ 0 & 1 \end{bmatrix} = \begin{bmatrix} A & B \\ C & D \end{bmatrix} \tag{8}$$

The resultant transmission matrix of (8) can be obtained in terms of input and output parameters as

$$\begin{bmatrix} V_i \\ I_i \end{bmatrix} = \begin{bmatrix} A & B \\ C & D \end{bmatrix} \begin{bmatrix} V_0 \\ I_0 \end{bmatrix} \tag{9}$$

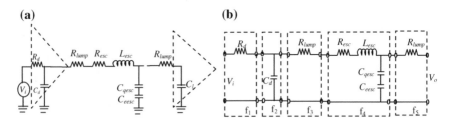

Fig. 2 a A driver-interconnect-load system. **b** Cascaded connections of the DIL of (**a**)

Thus, the *TF* is given as

$$TF = \frac{V_o}{V_i} = \frac{1}{A + sBC_l} \tag{10}$$

The parameters A and B of (10) are detailed in Appendix.

4 Results and Discussion

The Nyquist stability and frequency responses for SWB and MWB interconnects are evaluated using (10). The cross-sectional dimensions considered for interconnects are $w = 48$ nm and $t = 144$ nm. The parameters for 32 nm technology node are taken as per international technology road map (ITRS) [14]. The *RLC* values for SWB interconnect are 4.4 Ω/μm, 2.75 pH/μm, 14.97 aF/μm. The parasitic values for MWB interconnect are 1.6 Ω/μm, 2.17 pH/μm, 13.37 aF/μm. The driver resistance and capacitances are 13.8 KΩ and 0.07 fF respectively. The load capacitance (C_l) is 1 fF.

4.1 Stability Analysis

The Nyquist stability response of MWB for 500, 1000 and 2000 μm interconnect lengths is shown in Fig. 3a.

It is observed that for global interconnect lengths, the encirclement moves away from point $(-1, 0)$ that interprets to higher system stability [15]. At the same time the Nyquist diagram imaginary values decrease leading to lower fluctuations in the system.

The Nyquist plot of SWB and MWB for an interconnect length of 1000 μm is shown in Fig. 3b. It is noticed that encirclement moves far away from point $(-1, 0)$ for MWB interconnect as compared to SWB interconnect. This clearly states that MWB interconnects is more stable than SWB interconnects.

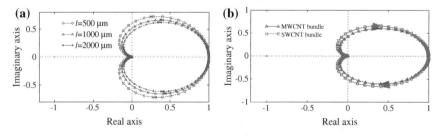

Fig. 3 **a** Nyquist plot for MWB interconnect at different interconnect lengths. **b** Nyquist plot for SWB and MWB interconnects

Fig. 4 Nyquist plot for
MWB interconnect with
different number of shells

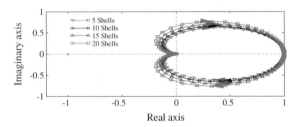

The Nyquist response of MWB interconnects for 5, 10, 15 and 20 shells is shown in Fig. 4. As number of shells increases in MWCNT, the point $(-1, 0)$ moves inside the encirclement. It is due to the number of shells increase in MWB interconnect, the damping coefficient (ξ) decreases. Higher quantitative value of ξ makes system becomes more stable.

4.2 Frequency Analysis

The open loop TF of (10) is used to analyze the frequency response of SWB and MWB interconnects. It is observed from Fig. 5a that MWB interconnect offers higher operating frequency as compared to SWB interconnect.

The cut off frequency for SWB and MWB interconnects are 7.8 and 22.2 GHz at interconnect length of 500 μm respectively.

Figure 5b plots the frequency response of MWB for different global interconnect lengths ranging from 500 to 2000 μm. It is noticed that operating frequency reduces for longer interconnects. It is observed from Fig. 5b, that SPICE simulation and analytical results are good in agreement. The average percentage error between SPICE and analytical results is 3.79 %.

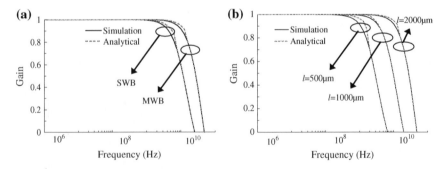

Fig. 5 **a** Frequency response of SWB and MWB interconnects. **b** Frequency response of MWB at different interconnect lengths

4.3 Delay Analysis

The time delay of SWB and MWB interconnects for different global interconnect lengths is shown in Fig. 6a. It is known that 50 % time delay (τ) for SWB and MWB interconnect can be obtained as the time it takes to reach 50 % of output voltage steady value [16],

$$\tau_{50\%} = \left(1.48 + e^{-2.9\xi^{1.35}}\right)\sqrt{L_{esc}l(C_{esc}l + C_l)} \tag{11}$$

where

$$
\xi = 0.5\left(1 + \frac{C_l}{C_{esc}l}\right)^{-0.5}\left[\left(0.5R_{esc}l + 2R_{lump} + R_d\right)\sqrt{\frac{C_{esc}}{L_{esc}}}\right.
$$
$$
\left. + \left(R_{esc}l + 2R_{lump} + R_d\right)\sqrt{\frac{C_l^2}{L_{esc}C_{esc}l^2}}\right] \tag{12}
$$

It is observed from Fig. 6a, that 50 % time delay (τ) of the DIL system of SWB and MWB at interconnect length 500 µm is 51.77 and 49.38 ps respectively. The delay calculated from SPICE simulation for SWB and MWB interconnects are 54.25 and 51.72 ps respectively. Thus the average error between analytical and simulation is 2.86 %.

Figure 6b shows that the delay of MWB with different number shells at global interconnect length ranging from 500 to 2500 µm. It is noticed that the delay reduces significantly with increasing number of shells.

The delays for 5 shells and 20 shells MWB interconnect for a length of 1000 µm are 96.1 and 98.4 ps respectively. The average error between simulative and analytic is 1.67 % and 2.34 % for 5 shells and 20 shells respectively.

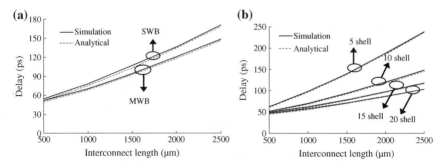

Fig. 6 a Time delay analysis of SWB and MWB at different interconnect lengths. **b** Time delay analysis of MWB with different number of shells

5 Conclusion

The present paper analyzes frequency and stability response of SWB and MWB interconnects. The driver interconnect load is considered. The interconnect is represented by equivalent ESC model for SWB/MWB interconnects. The transfer function is derived using *ABCD* model. At different global interconnect lengths, operating frequency and the 50 % time delay are computed for SWB/MWB interconnects. It is found that the MWB interconnect exhibits lesser time delay as compared to SWB interconnects. Further, it is observed that as the number of shells increases in MWB interconnect, the delay reduces substantially.

Appendix

$$
A = 1 + s \left[\frac{R_{esc} C_{esc} (px)^2}{2} + R_d C_d + C_{esc}(px)\left(R_{lump} + R_d\right) \right] + s^2 \left[\begin{array}{l} \frac{L_{esc} C_{esc}(px)^2}{2} + \frac{R_{esc}^2 C_{esc}^2 (px)^4}{24} \\ + \frac{R_{esc} R_d C_{esc} C_d (px)^2}{2} \\ + \frac{R_{esc} C_{esc}^2 (px)^3 \left(R_{lump} + R_d\right)}{6} + R_{lump} R_d C_d C_{esc}(px) \end{array} \right]
$$

$$(A1)$$

$$
B = \left(2R_{lump} + R_d + R_{esc}(px)\right) + s\left[\frac{R_{esc} C_{esc}(px)^2}{2}\left(2R_{lump} + R_d\right) + 2R_{lump} R_d C_d + \frac{R_{esc} C_{esc}(px)^3}{6} \right.
$$
$$
\left. + L_{esc}(px) + R_{lump}^2 C_{esc}(px) + R_{esc} R_d C_d(px) + R_{lump} R_d C_{esc}(px) \right]
$$
$$
+ s^2 \left[\left(2R_{lump} + R_d\right)\left(\frac{L_{esc} C_{esc}(px)^2}{2} + \frac{R_{esc}^2 C_{esc}^2 (px)^4}{24} \right) + R_{lump} R_d R_{esc} C_{esc} C_d (px)^2 \right.
$$
$$
+ \frac{2R_{esc} L_{esc} C_{esc}(px)^3}{6} + \frac{R_{esc}^3 C_{esc}^2 (px)^5}{120} + \frac{R_{lump} R_{esc} C_{esc}^2 (px)^3}{6}
$$
$$
\left. + \frac{R_{esc}^2 R_d C_d C_{esc}(px)^3}{6} + R_d C_d \left(L_{esc} + R_{lump}^2 C_{esc}\right)(px) \right]
$$

$$(A2)$$

References

1. Chandel, R., Sarkar, S., Agarwal, R.P.: An analysis of interconnect delay minimization by low-voltage repeater insertion. Microelectron. J. **38**, 649–655 (2007)
2. Wen, W., Brongersma, S.H., Hove, M.V., Maex, K.: Influence of surface and grain-boundary scattering on the resistivity of copper in reduced dimensions. Appl. Phys. Lett. **84**, 3238–3240 (2004)
3. Goel, A.K.: High-speed VLSI Interconnections. Wiley-IEEE Press, New York (2007)
4. Li, H., Xu, C., Srivastava, N., Banerjee, K.: Carbon nanomaterials for next-generation interconnects and passives: physics, status, and prospects. IEEE Trans. Electron Devices **56**, 1799–1821 (2009)

5. McEuen, P.L., Fuhrer, M.S., Park, H.: Single-walled carbon nanotube electronics. IEEE Trans. Nanotechnol. **1**, 78–85 (2002)
6. Naeemi, A., Meindl, J.: Compact physical models for mutliwall carbon-nanotube interconnects. IEEE Electron Device Lett. **27**, 338–340 (2006)
7. Li, H., Banerjee, K.: High-frequency analysis of carbon nanotube interconnects and implications for on-chip inductor design. IEEE Trans. Electron Devices **56**, 2202–2214 (2009)
8. Nasiri, H., Rahim, F., Farshi, M.K.M.: Stability analysis in multiwall carbon nanotube bundle interconnects. Microelectron. Reliab. **52**, 3026–3034 (2012)
9. Fathi, D., Forouzandeh, B., Mohajerzadeh, S., Sarvari, R.: Accurate analysis of carbon nanotube interconnects using transmission line model. Micro Nano Lett. IET **4**, 116–121 (2009)
10. Li, H., Yin, W.Y., Banerjee, K., Mao, J.F.: Circuit modeling and performance analysis of multi-walled carbon nanotube interconnects. IEEE Trans. Electron Devices **55**, 1328–1337 (2008)
11. Das, D., Rahaman, H.: Analysis of crosstalk in single-and multiwall carbon nanotube interconnects and its impact on gate oxide reliability. IEEE Trans. Nanotechnol. **10**, 1362–1370 (2011)
12. Majumder, M.K., Narasimha Reddy, K., Kaushik, B.K.: Frequency response and bandwidth analysis of multi-layer graphene nanoribbon and multi-walled carbon nanotube interconnects. Micro Nano Lett. IET **9**, 557–560 (2014)
13. Fathi, D., Forouzandeh, B.: A novel approach for stability analysis in carbon nanotube interconnects. IEEE Electron Device Lett. **30**, 475–477 (2009)
14. International Technology Roadmap for Semiconductors, 2012. [Online]. Available: http://public.itrs.net/
15. Nagrath, I.J., Gopal, M.: Control Systems Engineering. Halsted Press, Sydney (1977)
16. Amore, D., Marcello Sarto, M.S., Tamburrano, A.: Fast transient analysis of next-generation interconnects based on carbon nanotubes. IEEE Trans. Electromagn. Comp. **52**, 496–503 (2010)

Multi-robot Assembling Along a Boundary of a Given Region in Presence of Opaque Line Obstacles

Deepanwita Das, Srabani Mukhopadhyaya and Debashis Nandi

Abstract This paper presents a distributed algorithm for assembling a swarm of autonomous mobile robots on a common boundary of a given polygonal region in presence of opaque horizontal line obstacles. Robots and obstacles are initially scattered in an unknown environment and they do not have direct communication among themselves. The algorithm guarantees successful assembling of all the robots on the left boundary of the given region within finite amount of time and without facing any collision during their movement. The intermediate distances among the assembled robots are not fixed. In this proposed algorithm, the robots follow the basic *Wait-Observe-Compute-Move* model together with the *Full-Compass* and *Synchronous/Semi-synchronous* timing models.

Keywords Robot swarm · Assembling · Passive communication · Distributed algorithm

1 Introduction

Research works on swarm robots consider a group of relatively simple robots in various environments and coordinate the robots to perform desired task by using only local information. The main advantage of swarm is that, it can perform tasks that are beyond the capabilities of the individuals. For example, the ants move

D. Das (✉) · D. Nandi
National Institute of Technology, Mahatma Gandhi Avenue,
Durgapur 713209, West Bengal, India
e-mail: deepanwita.das@it.nitdgp.ac.in

D. Nandi
e-mail: debashis.nandi@it.nitdgp.ac.in

S. Mukhopadhyaya
Kolkata Extension Centre, Birla Institute of Technology,
Kolkata 700107, West Bengal, India
e-mail: smukhopadhyaya@bitmesra.ac.in

© Springer Science+Business Media Singapore 2017
P. Deiva Sundari et al. (eds.), *Proceedings of 2nd International Conference on Intelligent Computing and Applications*, Advances in Intelligent Systems and Computing 467, DOI 10.1007/978-981-10-1645-5_3

together in some particular direction in search of any food source and then they fetch the food portions in a coordinated manner [6] which could have been an impossible mission for a single ant. Inspired by this type of basic behavior researchers are working on designing algorithms to solve computational problems related to swarm robots, like geometric pattern formation, flocking, coverage, partitioning, spreading, searching, gathering, convergence etc.

Problem of assembling a swarm of robots on a given line or shape is an interesting and very relevant problem in the area of swarm research. Assembling of robots on a common line may also be considered as one of the basic processing steps in solving different complex problems. Area partitioning may be mentioned as one such problem in which assembling of robots on a line has significant contribution. Area partitioning has several applications like, scanning or coverage of a free space [2, 3], lawn mowing and milling, sweeping, search and rescue of victims [8], space explorations, terrain mapping etc.

In the assembling problem, the objective is to assemble all the robots on a common given line, when initially the robots are randomly scattered over a plane. In this paper, we assume that initially all the robots are randomly deployed within a given rectangular area and the robots are required to assemble along a boundary of that area. Moreover, during the movement robots should not collide with each other.

In *spreading problem* [1], the objective is to uniformly spread a group of robots over a line, when initially the robots are randomly placed over it. However, deploying a group of robots over a single line is not only a difficult task but impractical also. Usual trend is to deploy the whole swarm over a region in a random manner. Here comes the utility of the assembling problem, when after deploying the robots over a plane, a situation may demand uniform assembling of the robots along a given line within the region. Assembling followed by spreading can be a solution to this problem.

This paper presents a distributed algorithm for assembling a swarm of mobile robots along a boundary of a given rectangular area in presence of opaque horizontal line obstacles. We have assumed that a swarm of autonomous mobile robots are randomly deployed within the region and several opaque horizontal line obstacles are scattered inside the region. It is also assumed that the internal environment, like, the total number and positions of the robots, as well as obstacles are unknown to the robots. Moreover, the obstacles are assumed to be opaque horizontal lines with negligible width. During execution of the proposed algorithm the robots are guaranteed to follow collision free paths.

The research on swarm robotics mainly concern about the robots' characteristics and capabilities to solve any task. The main characteristics of the robots are with respect to their orientation, visibility capacity, memory and mode of communication. In the literature, several computational models are proposed; among which the most commonly known model is *CORDA* model. The algorithm proposed here executes the computation by following the *CORDA* model. The robots are assumed to have total agreement in direction and orientation of their local coordinate systems. The proposed algorithm may execute the cycles in both synchronous and semi-synchronous [4, 7] manner. The proposed algorithm also assumes unlimited visibility radii for the robots. However, the view of any robot can be restricted due to the presence of the obstacle(s) on the line of sight of any robot. In this paper, the left boundary of the given region is assumed to be the target line of assembling. It is assumed that initially any two objects (robots and obstacles) are separated by a minimum distance δ. The positions of the obstacles are fixed but unknown to the robots. The proposed algorithm does not require any direct communication (message passing) among the robots and the robots can carry on their task without depending on others. Moreover, the robots are assumed to be oblivious in a sense that they can retain only $O(1)$ amount of information during execution.

The organization of the paper is as follows. In Sect. 2, the models and characteristics of the robots are discussed; Sect. 3 discusses the algorithm *ASSEMBLE*. Finally, Sect. 4 concludes the paper, along with the discussion on future scope of work.

2 Problem Definition, Assumptions and Models

Let us assume that the swarm of robots is randomly deployed over a bounded rectangular area such that no two robots occupy the same position. The rectangular area consists of some opaque horizontal line obstacles. It is assumed that length of the obstacles can not exceed a given length, say l, which is less than the length of the rectangular region and the obstacles are separated by a minimum distance δ, both horizontally and vertically. This assumption is made for robots also. In the *ASSEMBLE* problem the robots are required to assemble on the left boundary of the rectangular area. Each robot has its local coordinate system. We further assume that the entire environment is completely unknown to the robots. By environment we mean the positions of the robots and obstacles, number of robots and/or obstacles etc. To summarize, the characteristics of the robots and the models assumed in the paper are listed below:

(a) Robots are *identical and homogeneous* with respect to their computational power. They are *autonomous* in the sense that there is no central control. (b) All the robots execute the same algorithm independently of the others so that the goal is achieved in a completely distributed way. (c) There is no direct communication (message passing) among the robots. However, there is a passive mode of communication in the sense that a robot observes the position of other robots and accordingly take the decision for further action. (d) Robots are having *unlimited* visibility. Each robot can view almost everything unless its view is restricted by the position of any obstacle(s) and/or any other robot. (e) *Full-compass* [4, 7]: Each robot has its own local co-ordinate system in which they are placed at the origin. The robots agree on directions and orientations of both axes. (f) The robots are assumed to follow *CORDA* model [5]. In *CORDA* model, a *computational cycle* is defined as a sequence of three steps, *look, compute* and *move*. Each of the robots executes same instructions in all the computational cycles. A robot takes a look of its surroundings in the *look* step; based on the observations made in the *look* step, it computes a destination point in the *compute* step and then it moves to the destination in the *move* step. In some situations, an observation might lead a robot not to change its position in *move* step. In such cases the robot seems to be idle, though it is actually executing all the three steps. Once a computational cycle is over a robot starts the next computational cycle. Thus, a sequence of computational cycles are carried out continuously throughout the entire execution of the algorithm. (g) Robots are assumed to be oblivious or memoryless. That is, robots can retain only the snapshot obtained during the *look* step of the current computational cycle. (h) The robots are capable to move freely on a plane. In this paper, we assume rigid motion of the robots. By rigid motion we mean that in the *move* step of a computational cycle a robot reaches the destination as computed in the current *compute* step without any halt in-between. (i) The robots can have two different states, *active* state and *sleep* state. In the *active* state the robots execute the computational cycles continuously and actively. In the *sleep* state, robot remain idle which is similar to the *power off* state. However, a robot cannot *sleep* infinite amount of time and becomes active within a finite time interval. (j) *Fully Synchronous/Semi-Synchronous Model* [4, 7]: The proposed algorithm works in both synchronous and semi-synchronous model. In synchronous model, all the robots follow a common notion of time. They execute different phases of the computational cycles synchronously. For example, in the *look* step, all the robots take the snapshot of their surroundings at the same instant of time. After computing the destination they all move to their destinations at the same time. In the synchronous model, robots are always active. Whereas, in the semi-synchronous model, not all robots are active always, only a finite subset of robots are active at a time. However, all active robots are fully synchronous.

3   Assembling Algorithm

The first part of this section describes the proposed algorithm. The correctness of the algorithm is proved in the next section.

3.1   Algorithm ASSEMBLE

The objective of the problem is to assemble all the randomly deployed robots on a boundary of a rectangular area. The rectangular region consists of several horizontal line obstacles of different lengths. The length of the longest obstacles, say l, is assumed to be less than L, the length of the rectangular area. Moreover, it is assumed that any two objects (obstacle or robot) are separated both vertically and horizontally, by a minimum distance δ. It is assumed that the dimensions of the rectangular area are known to all the robots. The robots are having unlimited view of the environment. However, robots' view may get obstructed due to the presence of obstacle(s). Based on the snapshot of the environment taken in the *look* phase, the robots compute a path to assemble on the left boundary of the region. Since, we assume *full-compass* model, by *left* we mean the direction of x-negative of the local coordinate system, which is same for all the robots.

If the robot R is initially on the left boundary it will not move to anywhere. Otherwise, the robot R first checks whether it can see the left boundary directly looking towards left horizontally. If so, the robot moves directly towards left (following a horizontal movement) to reach the left boundary. If the direct view is obstructed due to the presence of obstacle(s) and/or any robot(s) (say, R') on the same horizontal line, then robot R first checks whether it is on the right boundary or not. If it is on the right boundary, it first changes its position, moving through a very small distance, say ε, towards left along the x-axis. A robot can always do so because of the assumption that two objects are separated by a minimum distance δ. For all other positions of the robots the final destination on the left boundary is computed in the following manner.

According to the local co-ordinate system of R, let $A(x_{left}, 0)$ and $B(x_{right}, 0)$ be the points on the left and right boundaries of the region respectively, at which the horizontal line passing through $R(0,0)$ (x-axis of the local coordinate systems) intersects the left and the right boundary of the region respectively. In case, R is not located on the left boundary initially, i.e., $x_{left} \neq 0$, let $P(\alpha, \beta)$ be a point on an obstacle or a robot which is immediately above R. That is, P is either a robot or a point on an obstacle above R such that neither any robot nor an obstacle is present in between P and R. However, more than one such points (as P) can be identified. That is, several robots as well as obstacles may be present at the same vertical distance immediately above R. In that case, the robot R will choose one of these robots as P. However, in absence of any robot, P is taken as a point on one of these obstacles. If the robot R is itself on the top boundary of the region, it will find a similar point $P(\alpha, \beta)$ in the downward direction.

Depending on the position of R and P, what would be the final destination of a robot is described below. It may be noted here that these final destinations may not be computed by the robots in a single computational cycle. However, at the end of the execution, the robots ultimately reach their destinations. The following cases may be considered separately.

Case I: Neither R nor P are on the top boundary of the region.

Let $D(x_{left}, \beta)$ and $C(x_{right}, \beta)$ be two points where the horizontal line passing through P intersects the left and right boundary of the region respectively. This scenario is illustrated in Fig. 1a. Let $Q(0, Y_1)$ be the point where the diagonal AC of the rectangle ABCD intersects the vertical line passing through R. The final destination of R on the left boundary as computed in the algorithm is $S(x_{left}, Y_1)$ where $Y_1 = \frac{-\beta \cdot x_{left}}{x_{right} - x_{left}}$.

Case II: P is a point on an obstacle which is lying along the top boundary of the region.

In this case also, the final destination of R is computed as in Case I.

Case III: P is a robot which is located on the top boundary of the region.

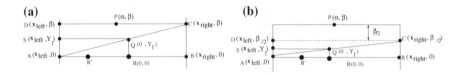

Fig. 1 Destination of $R(0, 0)$ **a** in Case I when both $R(0, 0)$ and $P(\alpha, \beta)$ are not on the top boundary, **b** in Case III when $P(\alpha, \beta)$ is on the top boundary

In this case, the robot R computes its destination as shown in Fig. 1b. Let $D\left(x_{left}, \frac{\beta}{2}\right)$ and $C\left(x_{left}, \frac{\beta}{2}\right)$ be two points on the left and right boundary of the region respectively. Unlike case I and II, in this case the line CD is not passing through P. Let $Q(0, Y_1)$ be the point where the diagonal AC of the rectangle $ABCD$ intersects the vertical line passing through R. The final destination of R on the left boundary is $S(x_{left}, Y_1)$, where $Y_1 = \frac{-\beta.x_{left}}{2(x_{right}-x_{left})}$.

In all the above three cases, R follows a path to the final destination as $(0, 0)$ to Q $(0, Y_1)$ in vertically upward direction and then Q to S horizontally towards left. As already indicated, robots may require more than one computational cycles to reach the destination.

Case IV: The robot R is on the top boundary of the region and $P(\alpha, \beta)$ is a point on an obstacle below it. Let $D(x_{left}, \beta)$ and $C(x_{right}, \beta)$ be two points where the horizontal line passing through P intersects the left and the right boundary of the region respectively. Let $Q(0, Y_1)$ be the point where the diagonal AC of the rectangle $ABCD$ intersects the vertical line passing through R. The final destination on the left boundary is $S(x_{left}, Y_1)$, where $Y_1 = \frac{-\beta.x_{left}}{2(x_{right}-x_{left})}$ as shown in Fig. 2a.

Case V: The robot R is on the top boundary of the region and $P(\alpha, \beta)$ is a robot below R which is vertically closest to R.

The scenario is shown in Fig. 2b. Let $Q(0, Y_1)$ be the point where the diagonal AC of the rectangle $ABCD$ intersects the vertical line passing through R, where. $D\left(x_{left}, \frac{\beta}{2}\right)$ and $C\left(x_{left}, \frac{\beta}{2}\right)$ be two points on the left and right boundary of the region respectively. Here, the final destination on the left boundary is $S(x_{left}, Y_1)$, where $Y_1 = \frac{-\beta.x_{left}}{2(x_{right}-x_{left})}$.

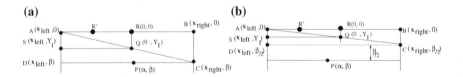

Fig. 2 Destination of $R(0, 0)$ when R is on the top boundary and **a** $P(\alpha, \beta)$ is an obstacle below R, **b** $P(\alpha, \beta)$ is a robot below R

In the cases IV and V, R follows its path to the final destination as $(0, 0)$ to $Q(0, Y_1)$ in vertically downward direction and then Q to S horizontally towards left.

When a robot R is located on the top boundary it needs to move in the downward direction. However, if the vertically closest point below R is another robot P, then that robot P is required to go in the upward direction. So to avoid the conflict, the cases III and V are treated specially. Before formally describing the algorithm some of the terminologies are defined as follows:

- $R(0,0)$ be the robot which is currently executing the algorithm. All coordinates mentioned are taken with respect to the local co-ordinate system of R.
- $ROBO = \{(x_i, y_i), 0 \leq i \leq k\}$ be the set of all robots visible to R including itself and $k \leq N$, where, N is the total number of robots.
- OBS be the set of the co-ordinates of all end points of the obstacles visible to R.
- $(x_{left}, 0)$ and $(x_{right}, 0)$ be two points on the X-axis (w.r.t., R) where this axis intersects the left and right boundary of the region. These points are not visible to R if there exists any obstacle/robot on this horizontal line in between the corresponding boundary and R. In case, the left boundary is not visible to the robot R, it would rotate its sensor through an angle $\theta\left(-\frac{\pi}{2} < \theta < \frac{\pi}{2}\right)$ to observe a point (x', y') on the left boundary. R can always observe such a point because of the assumption that any two objects are separated by at least a small distance δ and length of an obstacle is always less than that of the region. R then compute x_{left} as $d \cos\theta$, where $d = \sqrt{(x')^2 + (y')^2}$ as shown in Fig. 3. It is to be noted here that once R is able to compute $(x_{left}, 0)$, then it can compute $(x_{right}, 0)$ directly from the relation $|x_{left}| + |x_{right}| = L$, where L is the length of the area, known to R.
- $TOP = \{(0, y_{top})\}$, where $(0, y_{top})$ is the point where a vertical line through R intersects the top boundary. If this point is not visible to R, then $TOP = \phi$.
- $L_{visible} = ROBO \cup OBS \cup TOP$, the set of all objects (including robots and obstacles) visible to R.

This algorithm $ASSEMBLE$ (Fig. 4) is executed by all the robots in each computational cycle until they reach their final destination on the left boundary. Within

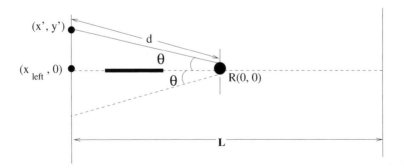

Fig. 3 Calculation of the distance of the left boundary from R

finite number of computational cycles, the algorithm successfully assembles all the robots along the left boundary of the given region in presence of horizontal line obstacles, under synchronous and semi-synchronous timing models provided the robots are assumed to have rigid motion.

Algorithm 1: ASSEMBLE

State *Look*
R takes a snapshot of the positions of all the visible robots, obstacles (end points) and boundaries of the region, according to the local coordinate system of R, where R occupies the origin $(0,0)$.

State *Compute* (returns destination point (X_d, Y_d))

if *(R is on left boundary)* then
 | $X_d = 0$; $Y_d = 0$; and the algorithm terminates
else
 if *(R does not find any other robot R' or any obstacle to the left of it on the x-axis)* then
  | $X_d = x_{left}$; $Y_d = 0$;
 else
  if *(R finds itself on the right boundary)* then
   | $X_d = -\epsilon$; $Y_d = 0$;
  else
   R identifies a point $P(\alpha, \beta)$ (either robot/obstacle) on top (below) of it which is vertically closest to R. This can be obtained by sorting all the element of $L_{visible}$ according to their y-coordinate and identifying the one having smallest positive (largest negative) non-zero value. If R is on the top boundary then it identifies $P(\alpha, \beta)$ as the closest object below it.

   Case I: R and P both are not on the top boundary of the region.
   $X_d = 0$; $Y_d = \frac{-\beta . x_{left}}{x_{right} - x_{left}}$;

   Case II: P is on the top boundary of the region.
   if P *is a point on an obstacle* then
    | $X_d = 0$; $Y_d = \frac{-\beta . x_{left}}{x_{right} - x_{left}}$;
   else
    | $X_d = 0$; $Y_d = \frac{-\beta . x_{left}}{2(x_{right} - x_{left})}$;

   Case III: R is on the top boundary of the region.
   if P *is a point on an obstacle below R* then
    | $X_d = 0$; $Y_d = \frac{-\beta . x_{left}}{x_{right} - x_{left}}$;
   else
    | $X_d = 0$; $Y_d = \frac{-\beta . x_{left}}{2(x_{right} - x_{left})}$;

State *Move*
Move to (X_d, Y_d);

Fig. 4 Algorithm *ASSEMBLE*

4 Conclusion and Future Work

The algorithm is based on *full-compass* and *synchronous/Semi-synchronous* model. The robots do not have direct message exchange during the process. The algorithm guarantees assembling of the robots on the left boundary within finite time. In addition to that, the path followed by the robots are all disjoint so that the robots do not collide with each other or with any obstacle while moving. The proposed algorithm can be extended for any convex region also. The work can be extended for limited visibility. Considering variation in speed, various types and shapes of the obstacles (static or mobile, convex or concave), one can make the problem more pragmatic.

References

1. Cohen, R., Peleg, D.: Local spreading algorithms for autonomous robot systems. In: Theoretical Computer Science, vol. 399(12), pp. 71–82. Elsevier, Amsterdam (2008)
2. Das, D., Mukhopadhyaya, S.: Painting an area by swarm of mobile robots with limited visibility. In: International Conference on Information Processing ICIP 2012, CCIS 292, pp. 446–455. Springer, Heidelberg (2012)
3. Das, D., Mukhopadhyaya, S.: An algorithm for painting an area by swarm of mobile robots. Int. J. Inf. Process. 7(3), 1–15 (2013)
4. Efrima, A., Peleg, D.: Distributed algorithms for partitioning a swarm of autonomous mobile robots. Technical Report MCS06-08, The Weizmann Institute of Science (2006)
5. Flochinni, P., Prencipe, G., Santoro, N., Widmayer, P.: Distributed coordination of a set of autonomous mobile robots. In: Proceedings of the IEEE Intelligent Vehicles Symposium, pp. 480–485 (2000)
6. Hoff III, N.R., Sagoff, A., Wood, R.J., Nagpal, R.: Two foraging algorithms for robot swarms using only local communication. In: IEEE International Conference on Robotics and Bioinformatics, pp. 123–130 (2010)
7. Pagli, L., Principe, G., Viglietra, G.: Getting close without touching: near-gathering for autonomous mobile robots. In: International Journal of Distributed Computing, vol. 28(5), pp . 333-349. Springer (2015)
8. Stormont, P.D.: Autonomous rescue robot swarms for first responders. In Pro-ceedings of IEEE International Conference on Computational Intelligence for Home-land Security and Personal Safety, pp. 151-157. IEEE (2005)

An Approach to Identify Data Leakage in Secure Communication

Suhasini Sodagudi and Rajasekhara Rao Kurra

Abstract In the field of wireless networks, Mobile Ad hoc Network (MANET) is one of the most unique applications due to mobility and scalability. As it is vulnerable to malicious attackers with the open medium and wide distribution, network performance is degraded with the behavior attacks in MANET's, which makes protection most important. In any ad hoc networks, one of the situations is that, when a data distributor possesses sensitive data destined to a set of supposedly trusted parties, some data can be leaked at an authorized place. With the introduction of fake objects the data allocation strategies gives a path in identifying leakages ensuring more security with encryption. In this context, light weight system is proposed to characterize the data loss and to minimize the performance degradation in such networks. The light weight system includes the concept of cryptography and routing protocol implementation at different stages of data transfer. A social network is taken as a challenging issue to achieve the proposed scheme.

Keywords Trusted parties · Encryption · Fake object · Data leakage · Light weight cryptography · Data loss · Routing

1 Introduction

Now a day's the data leakage or loss is major problem being faced in the outside world specifically in the field of communication technology. At a high end of communication, like a wireless network, Mobile Ad hoc Network (MANET) is one of the most modest unique applications due to mobility and scalability [1]. The

S. Sodagudi (✉)
Department of Information Technology, VR Siddhartha Engineering College,
Kanuru, Vijayawada, Andra Pradesh 520007, India
e-mail: Ssuhasini09@gmail.com

R.R. Kurra
Sri Prakash College of Engineering, Tuni, Andra Pradesh, India
e-mail: Krr_it@yahoo.co.in

© Springer Science+Business Media Singapore 2017
P. Deiva Sundari et al. (eds.), *Proceedings of 2nd International Conference
on Intelligent Computing and Applications*, Advances in Intelligent Systems
and Computing 467, DOI 10.1007/978-981-10-1645-5_4

medium of communication is completely open and hence it is vulnerable to malicious attackers with wide distribution of communication [2]. The target of the malicious attacker is to degrade the communication technology with the introduction of their misbehavior [3]. For instance, when a data distributor in a social networking site possess sensitive data (i.e. which is not to be shared with everyone) but must be distributed to a set of supposedly trusted third parties, then there is a scope for some part of the sensitive data be leaked at an unauthorized location. To handle such situation, an alternative mechanism is proposed in usage of fake objects [4]. By introducing fake objects, maintained as a set of sequence characters by data allocation strategies in sharing the data to private or trusted parties, the identification of leakages is possible that ensures more security [5].

This paper covers the major aspects of data loss with security proven mechanisms. Section 2 introduces and elaborates the data leakage concept. Section 3 describes the literature used towards the study of the problem and its related analysis of existing approaches. Section 4 covers the proposed model to detect the data loss. Section 5 illustrates the experimental analysis with results depiction and in Sect. 6, conclusion is mentioned. The major possibilities of data leakage with real time applications are discussed in the paper, along with illustration of facebook application taken in this work schedule.

2 Data Leakage Overview

Data leakage is the unauthorized transfer of classified data from one computer or datacenter to the other system or to outside world. In the proposed task of Data leakage recognition and prevention process, it is suggested to learn the data leakage found to occur mainly at the trusted parties either intentionally or by mistake. The unauthorized transfer of classified data from one computer or datacenter to the other system or to outside world is said to be the Data Leakage [1]. Such data can come in the form of private or company information, intellectual property (IP), financial or patient information, credit-card data, and other information depending on the business and the industry. Czerwinski et al. [6], discussed watermarking schemes for digital music distribution and audio media. Papadimitriou and Garcia-Molina [5] presented and extended a section on Data Leakage Detection. They initiated a model in 2011 for data leakage detection and proposed data allocation strategies will improve the probability of identifying leakages. The proposed techniques implemented data distribution strategies that had improved the probability of identifying leakages.

In this work, facebook social network application is considered to observe the data leakage. As far as to this application, the owner of the data (facebook) is known as the Distributor and other companies (amazon…) are supposedly known as trusted third parties who are referred as Agents. In this context, a sample scenario is shown in Fig. 1. Sometimes a data distributor gives away sensitive data to a set of third parties. Sometime later, the distributor observes that some of the data is

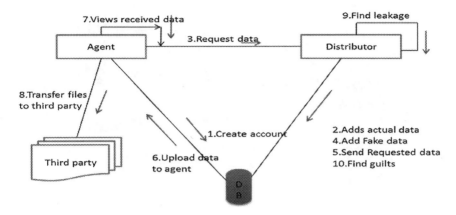

Fig. 1 Data leakage scenario

overseen at an unauthorized place (e.g., on the web or on a user's laptop). Then the distributor starts to investigate that the data sent was leaked at some point by one or more third parties, or if it was independently gathered by some means.

2.1 Facebook Application

In the data leakage detection process [5], if either sensitive data is disclosed to unauthorized personnel either by malicious agent or by mistake, or if data is found at an unauthorized place, then, it is declared as data loss. In the data leakage detection process, if either sensitive data is disclosed to unauthorized personnel either by malicious agent or by mistake, or if data is found at an unauthorized place, then, it is declared as data loss [7]. These Applications are so called as data loss prevention (DLP) Applications. In the application considered "Facebook", the distributor is said to be facebook, one who has all the user information. Agent is one, who requests the distributor to provide the user information so that the agent can share this data with the outside users. Figure 2 as shown depicts a sample facebook application where users are U1–U4 at some instance.

At a current instance of time, suppose there are four Facebook profiles available at the distributor (facebook). At this point, say U1, U2, U3, U4 are the four agents that request the profile information of the users. Let each agent has a request of two profiles information from the distributor. Say user information is sent to the agents through their requests R1, R2, R3, R4. Assume that the information 'S' is shared by the distributor through the agents U1, U3 is the "leaked set of data". Now, there is a need to conclude to state that this is a real "Data Leakage". It is required to calculate the data leakage with probability function. To illustrate this guilt detection process, consider that a distributor has an information set T = {t1, t2, t3} of valuable objects. Suppose say he wants to share some set to agents U1, U2, U3 and U4, but

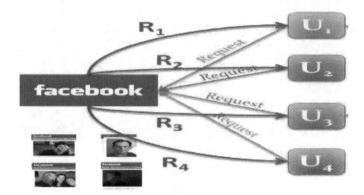

Fig. 2 Sample facebook application

does not expect the objects to be leaked. Let these sets be $R1 = t1, t2$ $R2 = t1, t3$ and Target set $= t1, t2, t3$. Now it is necessary to determine whether S is leaked data or not. An agent receives a subset of objects

$$Ri \subset Ti \tag{1}$$

If suppose say S is leaked. Then there is a need to calculate the probability of guilt between U1 and U3 agents which is said to be "Pr{G|S}". Here "guilt" means the user/agent who leaked the data.

$$Pr\,G1\,S \gg Pr\,G2\,S \tag{2}$$

$$Pr\,G1\,S \gg Pr\,G4\,S \tag{3}$$

By knowing the probability values of the U1, U2, U3, U4 the identification of guilt's can be done along with tracing the leaked data. To achieve this objective the distributor has to distribute sets Ri, ..., Rn that minimize as:

$$\sum_i \frac{1}{|R_i|} \sum_{j \neq i} |R_i \cap R_j|, \quad i,j = 1,\dots,n$$

3 Background Study

In every field of business or personal, data leak is the biggest problem being faced in the outside world. As the data is stored in a PC, it is found that digital information is being difficult to protect, and it is now the primary source of data theft and trade secret theft [8]. Data Leakage mainly is performed under malicious activities or unauthorized transfer of data by the trusted third parties [6]. If to distribute

(a)

Poor				
U_1	✓	✓		
U_2	✓	✓		
U_3	✓	✓		
U_4	✓	✓		

(b)

Minimize $\sum_{i \neq j}|R_i \cap R_j|$				
U_1	✓	✓		
U_2	✓	✓		
U_3			✓	✓
U_4			✓	✓

(c)

Accurate				
U_1	✓	✓		
U_2			✓	✓
U_3			✓	✓
U_4	✓			✓

Fig. 3 **a** Poor strategy, **b** minimum strategy, **c** accurate strategy

datasets Ri, ..., Rn to the agents U1..., then the minimized data sharing among agents makes leaked data reveal the guilt agents. Here is an illustration of data distribution strategies that provide ways to decrease the data leakage possibilities. The columns in Fig. 3a–c represent data of a user from facebook application which has to be shared among third parties and rows represent the agents who require the user details. In Fig. 3a, the distribution is said to be "poor strategy" since two user's information (column) is shared with 4 agents (row). Similarly, the second form of strategy shown in Fig. 3b is said to be of "minimum strategy".

This strategy of distribution reflects to be "minimum" because every two agents were sent a minimum of two user profiles information and the request of all the agents is satisfied. Equivalently, it is seen in Fig. 3c that this strategy of distribution is said to be "accurate" strategy since here it is assumed that every agent must be distributed equally as per their request without any overlap and must be able to receive the data. But unlike as minimum strategy, every two agents will not receive the same set of data.

4 Proposed Approach

In most of the organizations, the data protection programs are emphasized in protecting sensitive data from external malicious attacks with the dependence on technical controls like perimeter security, network/wireless surveillance, application and point security management, and user awareness and education [4]. The problem aimed in this context is to include prevention and detection scheme, when the distributor's sensitive data has been leaked by agents, and to identify the agent who leaked the data in any organization. The proposed system DLDR (Data Leakage detection and Reduction using Light weight cryptography) consists of a distributor with one or more agents. Various distribution strategies are discussed in Sect. 3. Now, the distributor functioning is to distribute data to agents using any of the data allocation strategies, through request-response method. Initially agent sends a request to the distributor for files of user related data. Distributor sends the files based upon the request. Using optimized allocation strategies, the distributor sends the fake objects from the fake database, so that if a particular data is found to be leaked, the distributor can assess the guilt (who leaked) using S-Max algorithm. From this, it is possible to predict that data is leaked.

The aim of the proposed investigation is to include secure enhanced application on to the hosts in the network so that the data exchange will ensure data availability, data confidentiality and data integrity services. Indeed, the problem described is illustrated in Fig. 4 and designed in various modules like: Handshake, Lightweight cryptography, Behavior analysis and Data leakage detection. Before communication begins, handshake module is designed to establish the cooperative agreement between the source and the destination in the network [9]. Simple Mail Transfer Protocol (SMTP) is considered in the implementation. Through mail server, the source and destination perform handshaking to send and receive mail messages. For sending, SMTP is used and for receiving, client applications normally use Post Office Protocol (POP3) or Internet Message Access Protocol (IMAP). Further light weight encryption scheme is implemented and data confidentiality is maintained in data exchange. During the data in transmission, the data leak is detected and the user that performs the data loss activity is also traced in this module [2]. In this view, the proposed method is called as "Data Leakage Detection and Reduction using Lightweight Cryptography (DLDR)" is discussed further.

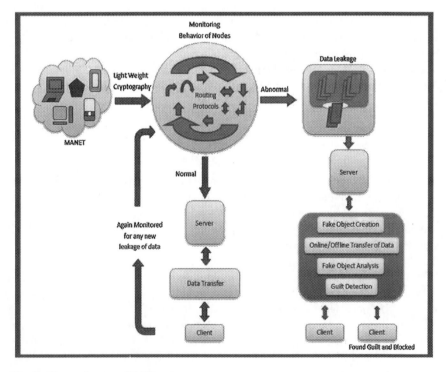

Fig. 4 Proposed system DLDR

4.1 Handshake Mechanism

This module is coded in such a way that any mail server like yahoo, gmail, rediff, aol or hotmail can send mail with proper credentials. To assess the guilt, a model is developed as an application where it can be used by any distributor to estimate the agent's behavior [5, 6]. As email systems are inbuilt with security mechanisms like SSL, it is considered to use e-mail systems that are used to send mail over the Internet with SMTP to send messages from source to destination. These messages are retrieved with the client mail by POP or IMAP [9]. The motivation to include this mechanism is the necessity to configure both the sender and destination. Moreover, webmail systems like gmail, yahoo, hotmail, aol use their own non-standard protocols to access mail messages on their own mail servers and all use SMTP when sending or receiving email from outside their own systems. The number 25 in the code indicates the default TCP port 25 and 587 is the mail submission protocol. The port 465 is the default port for SMTPS.

4.2 Light Weight Cryptography (LWC)

Lightweight cryptography is a branch of the modern cryptography, which covers cryptographic algorithms intended for use in devices with low or extremely low resources [10]. The motivation towards the usage of Light weight cryptography in this work is to introduce security for sensitive data that is being distributed across the web. As the medium of transfer is an open medium, data confidentiality is addressed. The main aim included here is to provide encryption only for sensitive part of the data before distribution. To meet this, data is summarized as sensitive and non-sensitive parts. Of course this summarization varies according to the application standards [11]. But in this work, the information possessing sensitive data in the facebook application is treated as username, user-email and user-contact-number. So such sensitive data is not seen by the agent and thus cannot leak to outside world. In this regard, light weight cryptography has given a scope to address the guilt agent. To generate the round keys, as per Present algorithm, it takes keys of either 80 or 128 bits [10]. However in this investigation, the key is taken as 80 bits. The key is stored initially in a key register K and is denoted as $k79$, $k78$, ..., $k0$. Lightweight is suitable for small devices like hand held mobile gadgets. In this work, the performance of this encryption technique is defined in terms of data privacy without any lags and compromised nature [12]. Lightweight cryptography does not determine strict criteria for classifying a cryptographic algorithm as lightweight, but the common features of lightweight algorithms are extremely low requirements to essential resources of target devices. Each of the 31 rounds as specified, consists of an XOR operation. The round key Ki for $1 \leq i$ 32, consists of a linear bitwise permutation and a non-linear substitution (Sbox) layer. The non-linear layer uses a single 4-bit S-box and is applied 16 times in parallel in each round since the plaintext is 64 bits. Light weight encrypted bytes are sensitive and are transferred through the file stream, the code bytes are being added. For this purpose, it was necessary to encode these bytes with Base64, which provides more security to the content.

4.3 Fake Objects

A fake object returns realistic looking results, which are simpler objects, implementing the same interface as the real object that they represent and return pre-arranged responses. Proposing the "realistic, but fake data" along with the original data, the information can be further tested for data escape [10]. While developing, it is implemented in a class with a method that return a fixed value or values which can either be hardcoded or set programmatically. In general, fake object can be anything and in this work, fake object is taken as a .txt file, created by an agent system with a real file existing at a distributor system. In this system, to generate fake objects, there is a provenance through user interface to create fake object as input which can be saved as fake object name in "fakeobject" database. The constraint assumed here is the length must be greater than or equal to 4 alpha-numeric/alpha/numeric characters. The property for the fake object included in this task is a file of byte stream content. Example for Input: ABCD, 12345, ZXcvb@1, etc. By introducing fake objects it was easy to verify user behavior also in the system against genuinely [10].

4.4 Agent Guilt Model

In this model, S-Max approach is applied that assigns fake objects to agents and determines the overlap in the agents. This yields the minimum increase of the maximum relative overlap among any pair of agents. The relative overlap gives the details whether same number of objects is sent between any pair of agents. At this point, fake objects are sent to an agent using S-Max algorithm and thus determined the possibility of an overlap between pair of agents. If the relative probability is 0.5 or 1.0 for any pair, then there is a need to send fake object to any of the pair. Figure 5 describes the S-Max function. There are many variables shown in the function S-max as of Fig. 5. U_i represents the agents and t is the object. Here the distributor's objective is to give each agent a unique subset of T of size m. The S-Max function is used to allocate an agent the data record that yields the minimum increase of the maximum relative overlap among any pair of agents.

Function SELECTOBJECT (i:R$_1$;......R$_n$; m1;.....m$_n$)

Step 1 – Initialize min_overlap as 1, which is the minimum out of the maximum relative overlaps that the allocations of different objects to users U$_i$

Step 2 – For K ∈ {k | t$_k$ ∈ R$_i$} do Initialize max_relov as 0 and it is the maximum relative overlap between R$_i$ and any set Rj that the allocation of t$_k$ to U$_i$

Step 3 – For j=1 to n; j=i and tk ∈ R$_j$ do
  Calculate absolute overlap as abs_ov=|Ri ∩Rj| +1
  Calculate relative overlap as relov = abs_ov /min (m$_i$, m$_j$)

Step 4 – Find maximum relative.max_relov = MAX (max_relov, relov)
  If max_relov <= min_overlap then
  min_overlap = max_relov

Step 5 – Return ret_k = k

Fig. 5   S-Max function for object selection

5   Results and Discussion

To accomplish the task, real time data is taken from facebook.com. An interface is designed where a data distributor and agents are provisioned to get register. Also each distributor maintains a set of agents in the backend. Moreover, an agent can also move from distributor to another distributor. The first module proposed in the investigation is Handshaking module. Either the Distributor or agent will initiate this process. It is been implemented such that the distributor initiates with hand-shake process. The distributor sends a mail to its agents before sending the actual information that the agent seek. All the agents automatically receive the message from their distributor. The agent who ever needs to seek, acknowledges the distributor with its details and the information it wants the distributor to distribute. This provisions both the parties ready for communication. Now the distributor selects the agent and decides to send fake object or not and enters fake object minimum of 4 size in length, and selects information file that the agent requested to send. Now the distributor encrypts the file in Fig. 6. The data in the file is divided into 64-bit blocks with key size of 80 bits and 31 rounds. The key for encryption is included in two ways. One is manual approach in which the key can be entered manually by the admin or any distributor. Second approach is the automatic key generation.

Figure 7 is to be understood in many views. It consists of send data, send file information, file allocation and fake record information operations. Here the service data availability is to be provided at any point of time but if the agent is an authorized agent. The other view shows the client details including their host and username. The distributor sends the file requested by the client. Depending on the requirements, fake object is also sent along with the data file and the file is also sent in cipher format by applying the present light weight algorithm. Key used in this context is also shown and it is randomly generated. Similarly, an agent is ready on other side. In order to provide agent services, agent can select Receive/Request Files option and then can choose the path and name to saves the file. Now the client

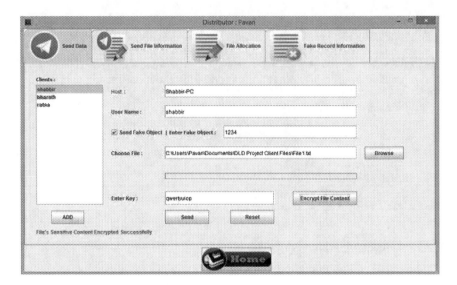

Fig. 6 Light weight encryption

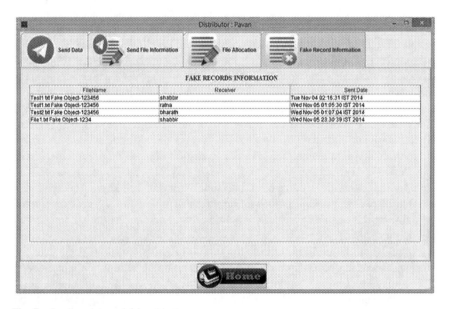

Fig. 7 Sending data and fake object

or the agent is able to receive the data and can ensure the distributor or server that data has been received with a confirmation message. It also exhibits that data is received successfully by the client. The received file is File1.txt, which consists of cipher. Decrypt File content is an operation to be selected by the agent to convert the received cipher into plaintext format. Since encrypted data is sent, the agent receives the cipher and performs decryption to get the original data file with the key provided by distributor during handshake procedure.

Now the Agent/Distributor can send data requests to each other as if needed. Distributor can see the all data sent to the agents. Now there exists a scope for data leakage to happen. Hence Agent guilt model has to be further proceeded. Using S-Max scheme, the relative overlap between the agents whom the files are sent is determined. This leads to identify the behavior of agents and it is considered as data leakage in this application. Here the calculation of the probability with the common files shared between the agents as Relative Overlap is done. This will help the distributor to identify the guilt parties. This is obtained from the execution of guilt detection module and selects that particular file with the operation "FIND GUILT". Whether the agent is guilt or not, can be concluded with complete details of the agent. The agent with high probability of leaking is shown as the result and it is observed in Fig. 8 with 66.666664 % of guilt is identified from the agent "shabbir" in this case on the File1.txt, being distributed. Face book datasets of 2014 are taken for the evaluation. Nearly 25 text files were collected of 71 KB up to 238 KB. These are taken from face book social media and for this module, 71 KB data is taken to determine the guilt users. "Analyze" button in Fig. 8 is used for plotting a graph with respect to agents on X-axis and obtained probabilities on Y-axis.

Figure 8 shows a graph that is been plotted automatically between agents with respect to their probability of guiltiness. X-axis is shown with three agents,

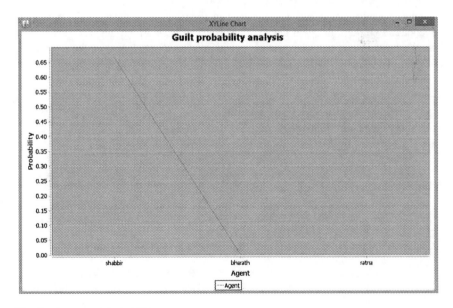

Fig. 8 Agents and probabilities

"shabbier", "bharath" and "ratna". Y-axis shows the probability values. The graph shows that agent "shabbier" is guilt with a probability of 0.6666.

6   Discussion and Future Work

A new scheme "Data Leakage Detection and Reduction with Light weight cryptography (DLDR)" is modeled to identify and reduce data leakage across the Internet. This developed application acts as a resource for an end user to include this system to test against data loss. Two approaches are included to deal the detection and prevention schemes of data leakage. For this to address, initially, it is emphasized to learn data leakage strategies as these could be the probable resources of data loss. Fake objects are introduced and steered towards the third parties to perform the detection and prevention process obsolete. Thus identified that any data leakage is happening at the point of its distribution, unauthorized communication parties, malicious activities etc. Hence in this direction, in order to lessen the rate of data loss, data distribution strategies are implemented with S-Max algorithm and thus defined the probability of leakage with respect to the data distribution strategies. If these strategies are followed by the data distributor, then there would be less scope for data leakage to happen.

The developed scenario supports data confidentiality through the implementation of cryptography. In this direction, the proposed work can include extensions of various allocation strategies to handle data distribution requests with known in advance the vulnerabilities of data loss. In spite of these difficulties, it is possible to assess the likelihood that an agent is responsible for a leak, based on the overlap of his data with the leaked data and the data of other agents, and based on the probability that objects can be "guessed" by other means. The proposed model is relatively simple, and it captures the essential trade-offs. The algorithms have presented implement a variety of data distribution strategies that can improve the distributor's chances of identifying a leaker.

References

1. Royer, E.M., Toh, C.K.: A review of current routing protocols for ad-hoc mobile wireless networks. IEEE Pers. Commun. (1999)
2. Renu, B., Mandoria H.L., Pranavi, T.: Routing Protocols in Mobile Ad-Hoc Network: A Review, pp. 52–60. Springer, LNCS (2013)
3. Vulimiri, A., Gupta, A., Roy, P., Muthaiah, S.N., Kherani, A.A.: Application of secondary information for misbehavior detection in VANETs. In: IFIP, vol. 6091, pp. 385–396. Springer, LNCS, Berlin (2010)
4. Cayrichi, E., Rong, C.: Security in Wireless Ad Hoc and Sensor Networks. Wiley, New York (2009)

5. Papadimitriou, P., Garcia-Molina, H.: Data leakage detection. IEEE Trans. Knowl. Data Eng **23**, 51–63 (2011)
6. Czerwinski, S., Fromm, R., Hodes, T.: Digital Music Distribution and Audio Watermarking. UCB IS 219, Technical report (2010)
7. Singh, A.: Identifying malicious code through reverse engineering. In: Advances in Information Security, p. 115. Springer, Berlin (2009)
8. Lu, S., Li, J., Liu, Z., Cui, G.: A secure data transmission protocol for mobile ad hoc networks. In: Chapter Intelligence and Security Informatics, vol. 4430, pp. 184–195. Springer, LNCS (2007)
9. Youssef, M.W., El-Gendy, H.: Securing authentication of TCP/IP layer two by modifying challenge-handshake authentication protocol. Adv. Comput. Int. J. (ACIJ) **3** (2012)
10. Bogdanov, A., Knudsen, L.R., Leander, G., Paar, C., Poschmann, A., Robshaw, M.J.B., Seurin, Y., Vikkelsoe, C.: PRESENT: an ultra-lightweight block cipher. In: Paillier, P., Verbauwhede, I. (eds.) Cryptographic Hardware and Embedded Systems—CHES 2007. Proceedings of 9th International Workshop, Vienna, Austria, vol. 4727, pp. 450–466. Springer, LNCS (2007)
11. Augot, D., Finiasz, M.: Direct construction of recursive MDS diffusion layers using shortened BCH codes. In: Cid, C., Rechberger, C. (eds.) Fast Software Encryption—21st International Workshop, FSE 2014, London, UK, March 3–5, 2014. Revised Selected Papers. Lecture Notes in Computer Science, vol. 8540, pp. 3–17. Springer (2014)
12. Eisenbarth, T., Paar, C., Poschmann, A., Kumar, S., Uhsadel, L.: A survey of lightweight cryptography implementations. IEEE Design Test Comput. (2007)

Quality of Service of Traffic Prediction Mechanism in WiMAX Network Using RBFNN and RSM

J. Sangeetha, G. Harish Kumar and Abhishek Jindal

Abstract In this fast growing world, there is a great demand for multimedia applications in WiMAX networks. To fulfill this demand, the network should be capable of handling QoS of traffic. Hence, we are predicting the QoS of traffic in the network. This is a prediction based problem. To solve this problem, we have applied two algorithms, namely, Response Surface Methodology (RSM) and Radial Basis Function Neural Network (RBFNN) on two different applications, namely, CBR based traffic and file transfer applications. From the experiment, we have observed that RBFNN performs better than RSM.

Keywords Wimax · Quality of service · Response surface methodology · Radial basis function neural network

1 Introduction

IEEE 802.16 or WiMAX is a wireless version of Ethernet that is considered as an alternative to wired technologies such as Cable Modems, DSL and T1/E1 links to provide broadband access. WiMAX aims to provide broadband wireless data access over long distances and at high speed. The bandwidth and the range of WiMAX

J. Sangeetha (✉) · G. Harish Kumar
Department of Information Science and Engineering, PES Institute of Technology,
Bengaluru, India
e-mail: sangeethaj@pes.edu

G. Harish Kumar
e-mail: gharishkumar.2023@gmail.com

A. Jindal
Department of Computer Science and Engineering, PES Institute of Technology,
Bengaluru, India
e-mail: abhishekjaipurian@gmail.com

© Springer Science+Business Media Singapore 2017   45
P. Deiva Sundari et al. (eds.), *Proceedings of 2nd International Conference on Intelligent Computing and Applications*, Advances in Intelligent Systems and Computing 467, DOI 10.1007/978-981-10-1645-5_5

makes it suitable for providing the mobile broadband connectivity across cities. WiMAX can be used to connect to WLAN hotspots (i.e. Wi-Fi). A QoS mechanism [1, 2] is included in MAC layer [3] architecture. The QoS parameters are bandwidth, average jitter, average end-to-end delay, packet loss ratio, etc. [3, 4].

In this paper, two algorithms, RSM [5–7] and RBFNN [8–10] are used to predict the QoS of traffic in WiMAX network by training it with help of QoS parameters such as bandwidth, average jitter, average end-to-end delay and throughput. In this paper, we are analyzing the throughput performance of WiMAX network using the aforementioned algorithms for CBR based traffic and file transfer application. From the result, we found that RBFNN performs better than RSM.

Section 2 of this paper deals with Problem Formulation followed by Methodology in Sect. 3, where the RBFNN and RSM are discussed in detail. Section 4 deals with the Results and Discussion. Finally in Sect. 5 we conclude our result.

2 Problem Formulation

We have considered WiMAX network scenario and given different values for input parameters such as Bandwidth (BW), average end-to-end Delay (D), average Jitter (J) and recorded its corresponding output parameter (such as throughput) for both, CBR based traffic and file transfer applications. Training datasets are used for training purpose and obtaining system models. Testing datasets are used for validating the obtained system models. RSM and RBFNN are applied for predicting the QoS of traffic in the WiMAX network. These algorithms are used for formulating the mathematical equations for the output parameter in terms of input parameters. In our study, we can establish the general relation between input and output parameters by the equation given below:

$$\text{Throughput} = f(\text{BW}, \text{D}, \text{J}) \tag{1}$$

The function 'f' changes according to the algorithm used. The RBFNN is used to generate system models for throughput with different set of combinations of hidden layer neurons and spread constant. On the other hand, RSM is a polynomial regression algorithm which is used for curve fitting. The RSM is implemented using second order model. The results obtained by two algorithms are then compared.

The throughput of network depends on all input parameters. The WiMAX network is then trained so that network understands the mathematical relation between the given input and output parameters. Hence, the network can predict the throughput value given input parameters are known. These predicted values can be helpful for network engineers in decision making process.

3 Methodology

3.1 Response Surface Methodology

RSM is a linear or polynomial regression model used for optimization of the response curve. The response curve is influenced by many factors depending on the application and use. The RSM model is designed using training dataset. The unknown function 'f' in (1) is derived by either first-order factorial model or second-order factorial model [5]. In this study, we have considered second order factorial model to get better results (lower error value).

The general second-order factorial model [5] is:

$$y = \beta_0 + \sum_{j=1}^{k} \beta_j x_j + \sum_{j}^{k} \beta_{jj} x_j^2 + \sum_{i<}^{k} \sum_{j=2}^{k} \beta_{ij} x_i x_j + \varepsilon \qquad (2)$$

where k is number of input variables, x_k is an input variable, β_k is model coefficient and ε is a statistical error. With respect to our problem 'k' corresponds to three input variables and x_1, x_2, x_3 corresponds to BW, D and J respectively. From the training dataset we calculate the value of β and predict the throughput of the testing dataset.

3.2 Radial Basis Function Neural Network

A supervised RBFNN [11] has a characteristic feature that causes it's response to decrease monotonically with the distance from the center. Although there are many kinds of RBF, in our study we have considered Gaussian RBF [9].

$$\varphi(r) = e^{\left(\frac{-(x-c)^2}{r^2}\right)} \qquad (3)$$

where c is the center vector of the function which is chosen from the training dataset, x is the input training and testing vector in the input layer, and r is the Euclidean distance between the input vector and the center. RBFNN consists of three layers which are discussed below:

3.2.1 Input Layer

A vector with n dimensions is given as input. Hence n neurons in this layer will be used to represent n dimensions of the input. In our study, the three input neurons are input parameters such as BW, D and J.

3.2.2 Hidden Layer

This is where the processing happens. The activation value [12] of each hidden neuron is determined by Gaussian RBF. Each RBF neuron compares the input vector to its center value, and generates a value between 0 and 1. If the input is equal to the center, the output of that hidden neuron will be 1, otherwise the response curve falls off exponentially towards 0 [13, 14].

3.2.3 Output Layer

This layer consists of only one neuron which gives a weighted sum of the outputs of individual hidden layer neurons. The weighted sum is a value given by output node, which associates a weight value with each of the RBF neurons, and multiplies the neuron's activation by this weighted sum before adding it to the total response [9, 10, 14].

$$y_m = \sum_{m=1}^{N} w_m e^{\left(\frac{-\|x_n - x_m\|^2}{2\sigma^2}\right)} \tag{4}$$

where W_m is the weight value associated with the output node and mth hidden neuron

$$\beta = \frac{1}{2\sigma^2} \tag{5}$$

In this study, the weighted sum is equivalent to throughput. We try to estimate the throughput with a Mean Square Error (MSE) goal [15] of 0.01 on training dataset. At each iteration, a hidden layer neuron is added until either the MSE goal is achieved or the hidden layer neurons reach the maximum limit [16]. The center vectors are determined from the training dataset that maximizes the correlation between hidden nodes and output layer targets [17]. The equivalent matrix [18] of (4) is given as:

$$
\begin{pmatrix}
e^{\left(\frac{-\|x_1 - x_1\|^2}{2\sigma^2}\right)} & \cdots\cdots & e^{\left(\frac{-\|x_1 - x_N\|^2}{2\sigma^2}\right)} \\
e^{\left(\frac{-\|x_2 - x_1\|^2}{2\sigma^2}\right)} & & e^{\left(\frac{-\|x_2 - x_N\|^2}{2\sigma^2}\right)} \\
\vdots & \ddots & \vdots \\
e^{\left(\frac{-\|x_{N-1} - x_1\|^2}{2\sigma^2}\right)} & & e^{\left(\frac{-\|x_{N-1} - x_N\|^2}{2\sigma^2}\right)} \\
e^{\left(\frac{-\|x_N - x_1\|^2}{2\sigma^2}\right)} & \cdots\cdots & e^{\left(\frac{-\|x_N - x_N\|^2}{2\sigma^2}\right)}
\end{pmatrix}
\times
\begin{pmatrix}
w_1 \\
w_2 \\
\vdots \\
\vdots \\
w_N
\end{pmatrix}
=
\begin{pmatrix}
y_1 \\
y_2 \\
\vdots \\
\vdots \\
y_N
\end{pmatrix}
\tag{6}
$$

The $N \times N$ matrix is denoted by $\emptyset$, the $N \times 1$ matrix is denoted by W_n and the output matrix by Y_n. Hence, the final equation [18] can be written as:

$$\emptyset W = Y \qquad (7)$$

4 Results and Discussion

In this section, we are analyzing the throughput performance of the WiMAX network based on QoS of traffic prediction mechanism. Here, we have considered two applications, namely, CBR based traffic and file transfer applications. The considered QoS parameters are Bandwidth (BW), average end-to-end Delay (D) and average Jitter (J) as input parameters and throughput as output parameter. In order to analyze the throughput performance of the network, we have applied Radial Basis Function Neural Network (RBFNN) and Response Surface Methodology (RSM). The results obtained from the RBFNN were compared with RSM.

4.1 Dataset Generation

The datasets were generated using QualNet^{TM} 7.1 network simulator. The tabulated values were classified into training and testing datasets. There were a total of 140 samples out of which 69 samples were used for training phase and 71 samples were used for testing phase. This is done for both, CBR based traffic and file transfer applications.

4.2 CBR Based Traffic

4.2.1 Response Surface Methodology

For the second order mathematical equation with output as throughput and inputs as BW, D and J, have been formulated. The model coefficients of the second order factorial model, as shown in (2) are obtained by applying regression on the input parameters of training dataset. The testing dataset is used to validate the efficiency of model using the mean and standard deviation. The model coefficients (i.e. β values) are shown in Table 1.

From Fig. 1a, it has been observed that values obtained using RSM model and target values did not closely match with each other.

Table 1 Values of β coefficients obtained for CBR based traffic using RSM

β_0	3.77e+03	β_5	−1.17e+03
β_1	455.53	β_6	−1.22e+03
β_2	−1.56e+03	β_7	−99.18
β_3	7.31e+03	β_8	302.31
β_4	419.32	β_9	−8.04e+03

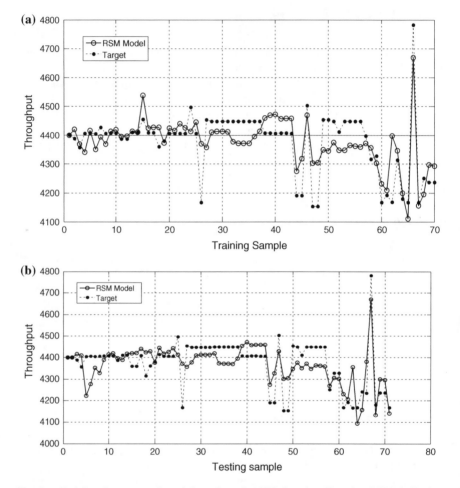

Fig. 1 **a** Training dataset samples of throughput for CBR based traffic using RSM. **b** Testing dataset samples of throughput for CBR based traffic using RSM

The β values obtained from Table 1 are used for fitting the curve during the testing phase as shown in Fig. 1b. The mean and standard deviation of error values corresponding to Fig. 1a, b are shown in Table 2. After n number of iterations, it is observed that the mean and standard deviation values of both training and testing datasets did not vary.

Table 2 Mean and standard deviation of error values for CBR based traffic using RSM	Training dataset		Testing dataset	
	Mean	Standard deviation	Mean	Standard deviation
	55.7397	44.8104	64.2077	45.4096

4.2.2 Radial Basis Function Neural Network

In this neural network, there are three input layer neurons which are dedicated for three input parameters, namely, BW, J and D, for both, training and testing datasets. One output neuron is used to generate the throughput value. The algorithm is now trained with the training dataset using different combinations of maximum number of hidden layer neurons and spread constant ranging from 10–50 and 1–150 respectively in the intervals of 1. During the training phase, the standard deviation corresponding to each combination (i.e. number of hidden layer neurons and spread constant) is recorded and a surface graph is plotted as shown in Fig. 2a, b. The Fig. 2b is a bottom view of Fig. 2a which gives a better view of the regions.

The darker regions in both the above surface plots signify the standard deviation produced using RSM, while the lighter portions signify that using RBFNN. Similarly another surface is plotted during the testing phase using the testing dataset which is shown in Fig. 3a, b.

The above surfaces are an effective way of showing parameter values for RBFNN. After having conducted this experiment for different trials, the mean and standard deviation values have been recorded in Table 3. From Table 3, it is observed that the 1st trial gives the minimum error value.

Comparing the error values of RSM with RBFNN (i.e. Tables 2 and 3), we observe that RBFNN gives a lower error value compared that given by RSM. Hence, the least values of standard deviation clearly depicts that RBFNN gives a better performance than RSM.

4.3 File Transfer Applications

4.3.1 Response Surface Methodology

Like CBR based traffic in file transfer application, the experiment has been carried out using training and testing datasets. Using second order model, the obtained β values are shown in Table 4.

Figure 4a shows the curve fitting during training phase. The β values obtained from Table 4 are used for fitting of curve during the testing phase which is shown in Fig. 4b. The mean and standard deviation of error values corresponding to Fig. 4a, b

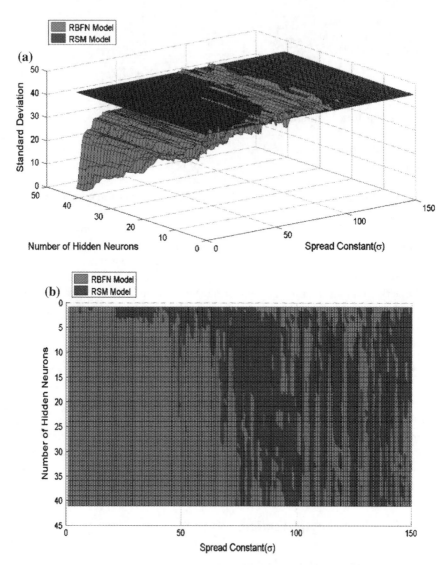

Fig. 2 **a** Surface plot for comparison of RSM and RBFNN under training dataset. **b** Bottom view of **a**

is shown in Table 5. Based on many number of trials, we have observed that the mean and standard deviation error values of training and testing datasets were same. From Fig. 4a, it has been observed that model values and target values were fitting better than that in CBR based traffic.

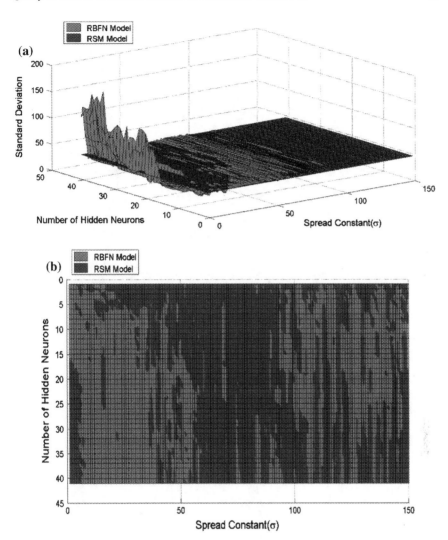

Fig. 3 **a** Surface plot for comparison of RSM and RBFNN Model under testing dataset. **b** Bottom view of **a**

Table 3 Mean and standard deviation of error values for CBR based traffic using RBFNN

Trail no.	Training dataset		Testing dataset	
	Mean	Standard deviation	Mean	Standard deviation
1	**26.6289**	**30.9817**	**31.7065**	**32.9866**
2	28.1573	28.3411	36.0666	31.9629
3	28.3266	28.2994	37.0048	34.2159
4	53.7442	45.8812	61.6282	46.8829
5	28.5112	27.7034	39.4544	34.9565

Bold values indicates the best results obtained from the experiments conducted

Table 4 Values of β coefficients obtained for file transfer application using RSM

β_0	8.20e+05	β_5	6.61e+06
β_1	−2.58e+04	β_6	7.62e+07
β_2	5.54e+05	β_7	2.94e+03
β_3	−5.08e+07	β_8	−1.17e+06
β_4	−1.90e+05	β_9	−7.76e+08

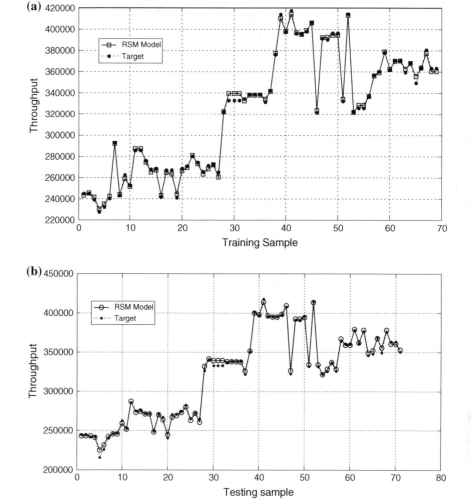

Fig. 4 a Training dataset samples of throughput for file transfer application using RSM. **b** Testing dataset samples of throughput for file transfer application using RSM

Table 5 Mean and standard deviation error values for file transfer application using RSM

Training dataset		Testing dataset	
Mean	Standard deviation	Mean	Standard deviation
1.9351e+03	1.5497e+03	2.1815e+03	1.8663e+03

4.3.2 Radial Basis Function Neural Network

Like CBR based traffic, here also the surface plots obtained during training phase are shown in Fig. 5a, b, where Fig. 5b is the bottom view of Fig. 5a.
The surface plots obtained during testing phase are shown in Fig. 6a, b.

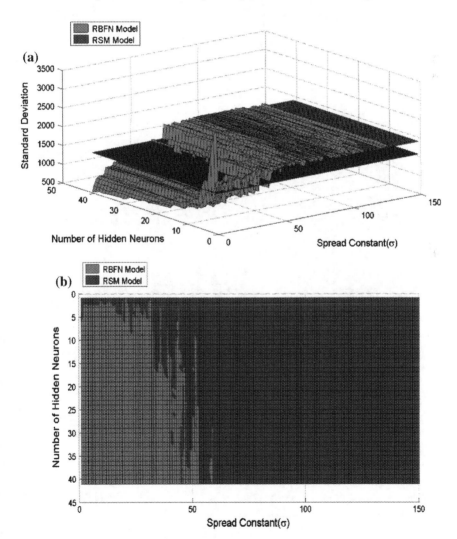

Fig. 5 a Surface plot for comparison of RSM and RBFNN Model. **b** Bottom view of **a**

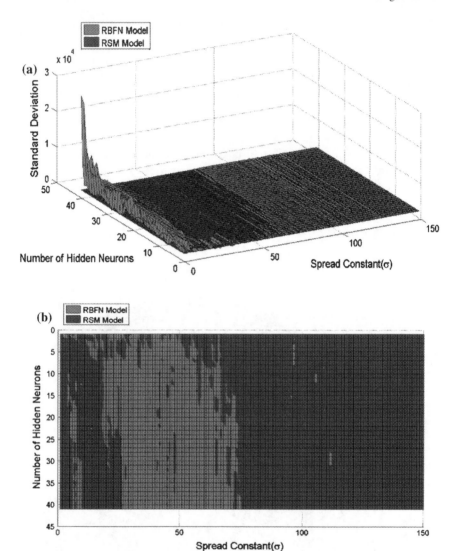

Fig. 6 **a** Surface plot for comparison of RSM and RBFNN Model. **b** Bottom view of **a**

From Figs. 5b and 6b, a clear boundary is visible between the lighter and darker regions. From this we can interpret that, in case of file transfer applications, the region for better performance is more localized than that in CBR based traffic. Different trials for the experiment have been conducted and random values of spread constant and hidden layer neurons were selected from Figs. 5b and 6b. The mean and standard deviation corresponding to these values are calculated and are shown in Table 6. From Table 6, it is observed that 3rd trial gives a minimum error value.

Table 6 Mean and standard deviation of error values using RBFNN (file transfer application)

Trail no.	Training dataset		Testing dataset	
	Mean	Standard deviation	Mean	Standard deviation
1	1.4669e+03	1.1581e+03	2.4396e+03	3.6950e+03
2	1.1088e+03	930.5172	1.4049e+03	1.2080e+03
3	**1.2808e+03**	**868.4987**	**1.6319e+03**	**1.4485e+03**
4	1.8052e+03	1.9464e+03	1.9979e+03	1.9236e+03
5	1.4074e+03	1.0902e+03	2.1510e+03	3.3171e+03

Bold values indicates the best results obtained from the experiments conducted

The least value of standard deviation clearly depicts that RBFNN gives better performance than RSM in this case too.

It is important to note that Gaussian RBF results in a lower standard deviation than that given by RSM, if the number of hidden layer neurons and spread constant of RBFNN are properly chosen. A proper choice of spread constant is needed because a very low value may result in RBF to decrease very rapidly hence lowering the interpolation between the functions. It may also lead to an untidy curve fitting because of several sharp edges in the resultant model. On the contrary a large spread constant may result in RBF to decrease very gradually hence again lowering the interpolation and accuracy due to the occurrence of blunt peaks in the resultant model.

5   Conclusion

In this paper, two algorithms, namely, RBFNN and RSM were used for analyzing the throughput performance of the WiMAX network. Here, we are predicting QoS of traffic in the network with respect to the CBR based traffic and file transfer applications. From the results, it is observed that number of hidden layer neurons and spread constant of Gaussian RBF plays a major role in performance of the RBF neural network. Hence, a proper selection of these parameters is needed. From the experiment, we observe the regions where RBFNN gives an optimal solution compared to RSM. This helps to predict a range of values of spread constant and number of hidden layer neurons that can be used to set up an ideal network for a particular WiMAX application. Finally, we conclude that the RBFNN performs better than RSM.

Acknowledgment   The authors would like to thank PES Institute of Technology, Bangalore, for providing the infrastructure and resources in completing this work successfully.

References

1. Ma, M., Lu, J.: QoS Provision Mechanisms in WiMax. Current Technology Developments of WiMax Systems, pp. 85–114. Springer, Netherlands (2009)
2. Vijayalakshmy, G., Shivaradje, G.: WiMAX and WiFi Convergence Architecture to Achieve QoS, Global Trends in Computing and Communication Systems, pp. 512—521. Springer (2012)
3. Talwalkar, R.A., Ilyas, M.: Analysis of quality of service (QoS) in WiMAX networks. In: 16th IEEE International Conference on Networks, 2008, ICON 2008, pp. 1–8 (2008)
4. Sangeetha, J., Balipadi, P.G, Ismail, K.H, Murthy, K.N.B, Rustagi, R.P.: An analysis and comparison of different routing algorithms in WiMAX networks. In: IACC, Bangalore, pp. 734–737 (2015)
5. Carley, K.M., Kamneva, N.Y., Reminga, J.: Response Surface Methodology. CMU-ISRI-04-136 (2004)
6. Montgomery, D.C., Peck, E.A., Vining, G.G.: Introduction to Linear Regression Analysis, 3rd edn. Willey, New York (2001)
7. Myers, R.H.: Classical and Modern Regression with Applications, 2nd edn. Duxbury Press, Boston (1990)
8. Romyaldy Jr., M.A.: Observations and guidelines on interpolation with radial basis function network for one dimensional approximation problem. In: 26th Annual Conference of the IEEE Industrial Electronics Society, 2000. IECON 2000, vol. 3, pp. 2129–2134 (2000)
9. Wang Jr., H., Xinai, X.: Determination of spread constant in RBF neural network by genetic algorithm. Int. J. Adv. Comput. Technol. (IJACT) 5(9), 719–726 (2013)
10. Baxter, B.J.C.: The Interpolations Theory of Radial Basis Functions. PhD Thesis, Cambridge University, UK (2010)
11. Deshmukh, S.C., Senthilnath, J., Dixit, R.M, Malik, S.N., Pandey, R.A., Vaidya, A.N., Omkar, S.N., Mudliar, S.N.: Comparison of Radial Basis Function Neural Network and Response Surface Methodology for Predicting Performance of Biofilter Treating Toluene. Scientific Research Publishing (2012)
12. The Machine Learning Dictionary [Online]. http://www.cse.unsw.edu.au/~billw/mldict.html#activnfn
13. Dong, C.: Neural Network Application in Matlab. National Defense Industry Publisher, China (2005)
14. RBFN Tutorial by Chris McCormick. https://chrisjmccormick.wordpress.com/2013/08/15/radial-basis-function-network-rbfn-tutorial/
15. Allen, D.M.: Mean square error of prediction as a criterion for selecting variables. Technometrics 13(3), 469–475 (1971)
16. Mathworks Inc., Matlab User Guide version 3.0, Mathworks (1998)
17. Wen, X.-L., Wang, h.-T., Wang, H.: Prediction model of flow boiling heat transfer for R407C inside horizontal smooth tubes based on RBF neural network. Int. Conf. Adv. Comput. Model. Simul. 31, 233–239 (2012)
18. Watkins, D.S.: Fundamentals of Matrix Computations. Wiley, Washington (1991)

Dissimilar Regulatory Actions Between Neurodegenerative Disease Pairs Through Probablistic Differential Correlation

Aurpan Majumder and Mrityunjay Sarkar

Abstract Rigorous research on gene expression analysis comparing the expression samples over normal and diseased states has helped us in understanding the pattern of various diseases. To gain better insight some state of the art methodologies goes a step beyond through exploration of differential co-expression patterns, where the co-expression level between genes alters across different states. In our work we have utilized a recently published concept which frames up a probabilistic score for differential correlation to detect differentially co-expressed genes modules. Our contribution lies in implementing the probabilistic differential correlation score for the identification of significantly paired gene networks possessing extremely dissimilar regulatory actions across multiple diseases. This has been tested upon differential gene expression profiles of genes present in a set of neurodegenerative diseases. Our investigation gets more interesting testing the impact of not only common (among multiple disease sets) differentially expressed genes but the revealing role of mutually exclusive disease specific gene sets in the spreading of diseases. Hence, we can explore through the fruitful aspects of our work scrutinizing the disease specific problem.

Keywords Differential correlation (DC) · KEGG pathway · Log Likelihood ratio (LLR) · Neurodegenerative disease · T score

1 Introduction

Gene expression analysis using microarray technology has now become a pivotal zone in biomedical research [1]. It gets primarily involved in understanding different diseases which further helps in showing the right track in the diagnosis of the

A. Majumder (✉)
Department of ECE, N.I.T Durgapur, Durgapur, India
e-mail: aurpan.nitd@gmail.com

M. Sarkar
Department of ECE, D.I.A.T.M, Durgapur, India
e-mail: mrityu1488@gmail.com

© Springer Science+Business Media Singapore 2017
P. Deiva Sundari et al. (eds.), *Proceedings of 2nd International Conference on Intelligent Computing and Applications*, Advances in Intelligent Systems and Computing 467, DOI 10.1007/978-981-10-1645-5_6

corresponding disease. However, understanding a particular form of disease requires the identification of differences between normal and affected tissues. A fundamental method is to infer through co-expressed gene sets, which assumes that the expression patterns of the gene set are correlated under all conditions [2]. Another approach is to discover genes having different expression levels across varied conditions known to be differentially expressed genes (henceforth will be represented by DE genes) [3].

Some recent methodologies have also explored the concept of differential connectivity of gene modules in terms of Topological Overlap (TO) to understand the behaviour of genes across different conditions [4]. However in the recent past analysis have gone beyond simple co-expression or differential-expression or differential connectivity measures to instigate through differential co-expression (also called differential correlation, henceforth will be abbreviated as DC) to unveil the differential genetic regulation [5]. In DC system the co-expression between a gene pair changes around different conditions. This phenomenon is especially prominent in gene regulatory network, assuming that genes co-expressed in any condition are controlled by a common regulator. Thus major change in co-expression patterns between conditions may be attributed towards the changes in regulation. As an example we can take a Transcription Factor (TF)-target gene (specifically DE) network, where tailored regulation of target genes by TFs cause different kinds of diseases, although the mean expression levels of TFs hardly change between conditions [6]. To extend further evidences from literature prove the importance of DC not only in direct physical gene interaction models [7] but also shows equal involvement in preparing significant clustering architectures [8].

In our work we have experimented upon the integration of DE and DC approaches. Our motto is to discover the gene sets having extreme differential regulation between dissimilar disease pairs. To unveil the DC characteristic we have taken the help from an existing architecture in which we find a probabilistic score to detect differentially co-expressed gene modules. Basically, here we have gone for a two tier analysis where at first we have applied the probabilistic score [9] to select gene pairs possessing significant DC in at least one condition. As a next step we have applied a kind of DC approach in search of genes associated with the pair having differential regulation across dissimilar conditions.

We have applied our proposed algorithm over human genome wide gene expression data containing expression values from six kinds of neurodegenerative diseases [10], namely Alzheimer's disease (AD), Amyotrophic lateral sclerosis (ALS), Huntington's disease (HD), Multiple sclerosis (MS), Schizophrenia (SCZ), and Parkinson's disease (PD). Then based upon their relationship [11–17] we have paired up some diseases. Thereafter the disease pairs are again combined in pairs in order to fit our two tier DC structure. It also helps us to understand those gene pairs having significant DC in at least one disease pair. The application of second step (non probabilistic DC structure) is done on both common and uncommon (disease pair specific) DE genes. Finally, a complete biological significance analysis reveals the pivotal role of the uncommon DE genes to illustrate the dynamic association of different neurodegenerative diseases.

2 Methodology

Multiple neurodegenerative diseases as mentioned above have gene expression values over normal and disease conditions. In this section first of all we would discuss on procedures to form different disease pairs. Then stepwise we would divulge into other subsequent aspects of our analysis, such as formation of disease sets through further pairing up of different disease pairs, calculation of T and LLR scores, concluding with unearthing of common and uncommon DE genes having extreme differential T score across conditions.

Algorithm: Differentially co-expressed gene set selection based on the combination of probabilistic and non-probabilistic frameworks.

---

1. Find DE genes between the conditions *control* and *disease* individually for different diseases.
2. Formation of multiple disease pairs based upon the proximity of different diseases.
3. Further combination of disease pairs in order to form a disease set d_i, yielding up a total of D disease sets.
4. while(each and every d_i is considered; i=1,2......,D)
    begin
    ➤ Evaluation of T score for a particular pair by (1)
    ➤ Disease pair specific T_{OVA} calculation by comparison of T scores using (2)
    ➤ Computation of LLR score, having both positive (L_P) and negative (L_N) values, from T_{OVA} using (3).
    while(all possible $L_P\_L_P$, $L_P\_L_N$ and $L_N\_L_P$ combinations across disease pairs in a disease set d_i are taken into consideration)
        begin
        ➤ Extraction of significant gene pairs, in terms of possessing extreme LLR values, across both disease pairs.
        ➤ Filtering of common genes, showing dysfunctional regulation with that gene pair across disease pairs.
        ➤ Further extraction of disease pair specific exclusive genes showing dissimilar regulations.
        end
    end

2.1 Formation of Disease Pairs

Let us have the expression values of two ailments A and B distributed over control and diseased states. The diseased states being mutually exclusive, the control states are only adjoined on putting together A and B.

Let us assume to have m and n ($n \leq m$) number of samples (i.e. expression levels) individually across A and B for the control state. Now in order to find the inter disease proximity we have computed the correlation values for all possible sample pairs (i.e. correlation between the 1st sample of A to all n samples of B continued up to the mth sample of A to all n samples of B). Accordingly, we obtain a m × n correlation matrix and form a threshold value equal to the average of the matrix. This is followed by the selection of the top n samples from pairs which exceed the threshold to yield up a common control state. However, if we do not get at least n samples then that particular disease pair is not been considered for further calculations.

2.2   Disease Set Formation and LLR Calculation

On completion of the previous step performed over those 6 neurodegenerative diseases gives us 5 possible disease pairs. Next we have independently combined those pairs and got 10 disease sets. The purpose of forming these sets is to study the strong differential regulation (up/down) of genes across different disease pairs. This regulation can easily be understood by a special metric, known as LLR (Log Likelihood Ratio) [9].

In this work the prerequisite for LLR is the T score. For a gene pair u and v acting under common control state C_1 and diseased states C_{21} and C_{22}, the T score is given as

$$T_{C_1,C_2}^{u,v} = \frac{\left(R_{C_2}^{u,v} - R_{C_1}^{u,v}\right) - (\mu_2 - \mu_1)}{\sqrt{\sigma_2^2 + \sigma_1^2}} \qquad (1)$$

The condition specific parameters $(\mu_1, \mu_2, \sigma_1, \sigma_2)$ are calculated directly from the expression values. For each pair of genes u and v the T scores $T_{C_1,C_2}^{u,v}$ between C_1 and the other diseased conditions are checked to see whether it possesses the same sign or not. If the sign is uniform with respect to both the conditions C_{21} and C_{22}, then the aggregated T score is given as

$$T_{OVA}(u,v) = sign\left(T_{C_1,C_2}^{u,v}\right) \min\left|T_{C_1,C_2}^{u,v}\right| \qquad (2)$$

For inconsistent sign the T_{OVA} value is set to 0. Under this assignment positive T_{OVA} score denotes the situation where the correlation in the disease states is greater than healthy (i.e. control) state. This is up-correlation. However, when the T_{OVA} score is consistently negative, we call it down-correlation.

Before proceeding further we would like to mention that the implementation of T_{OVA} is done over two set of genes. First on common DE genes and then on uncommon DE genes, both of which are calculated across two disease pairs of a set.

Now with the T_{OVA} values in hand we have taken help of the framework given in [9], where it compares T_{OVA} (u,v) scores on real and random data sets. Assuming both the distributions (real and random) are normal (henceforth will be represented as f_0 and f_1), the probability that a T score belongs to the first distribution is p. Given the mean and standard deviation of f_0 and f_1 are μ_0, μ_1 and σ_0, σ_1, the LLR score for a T score (x) can be written as

$$LLR_{1,0}(x) = \log \frac{pf(x/\mu_1,\sigma_1)}{(1-p)f(x/\mu_0,\sigma_0)} = \log \frac{p\sigma_0}{(1-p)\sigma_1} + \frac{(x-\mu_0)^2}{2\sigma_0^2} - \frac{(x-\mu_1)^2}{2\sigma_1^2}$$

(3)

The p value can be computed using the following formula.

$$p = Pr_{T_{OVA}^{real}} \left(x \geq \mu_{T_{OVA}^{random}} + k\sigma_{T_{OVA}^{random}} \right)$$

(4)

For $k \geq 0$. Throughout our calculation we have taken k = 2.

Following the idea given in [9] we can say that a positive LLR value represents significant DC and vice versa.

2.3 Evaluation of Maximally Differential Regulated Genes

At this stage we unravel the crux part of our algorithm. The necessary aspects are highlighted below:

- At first for a particular disease set we have to discover gene pairs having extreme LLR scores (maximum positive or minimum negative) across disease pairs. Gene pairs having extreme negative LLR across both disease pairs are ignored.
- Next we have searched for those disease set specific common DE genes having differential regulation (in terms of T score) with the gene pairs across the disease pairs.
- For uncommon DE genes our focus is to discover the disjoint gene sets over different diseases. Dissimilar regulation based upon the sign of T_{OVA} score can be treated to be a solution in this context.

3 Results and Discussion

Now let us have a quick look on the data set over which we have tested our algorithm. This is a genome wide gene expression data [10] (Accession No. GSE 26927) for Alzheimer's disease (AD), Amyotrophic lateral sclerosis (ALS), Huntington's disease (HD), Multiple sclerosis (MS), Schizophrenia (SCHIZ), and Parkinson's disease (PD) sampled from an extensive cohort of well characterized post-mortem CNS tissues. In all the above diseases there are 20,859 genes each comprising of 118 samples distributed across normal and diseased states.

At the very beginning we have searched for DE genes exclusively for different diseases using DEGseq [18]. It is an R package to find the differentially expressed genes from the genome wide gene expression value. Here DE genes are computed using the expression values of genes at different time instants (specifically between conditions) and by setting a particular threshold p value/z score/q value. As per the aforementioned order of the diseases there are 6236, 10831, 9382, 9384, 9695, and 8892 DE genes respectively. Next through an exhaustive clinical survey [11–17] and following our algorithm we are able to pair up some diseases based upon their proximity. through common control expression values. Disease pairs thus formed are AD_HD, AD_SCHIZ, ALS_MS, MS_SCHIZ, and PD_SCHIZ, with 2794, 1937, 3749, 3403, and 2556 DE genes.

With these disease pairs in hand next we have concentrated to evaluate the T_{OVA} and LLR values using the expression values from the pool of common control and diseased states. After this our quest lies in locating that paired up gene for a particular gene possessing extreme LLR value (maximum positive/minimum negative). Eventually we generate two sets of gene pairs containing positive and negative LLR values, L_P and L_N. Here, L_P and L_N also denote significant and non-significant set of DC pairs.

Concurrently we have also formed 10 disease sets by adjoining every possible disease pairs. The corresponding disease sets are AD_HD_AD_SCHIZ, AD_HD_ALS_MS, AD_HD_MS_SCHIZ, AD_HD_PD_SCHIZ, AD_SCHIZ_ALS_MS, AD_SCHIZ_MS_SCHIZ, AD_SCHIZ_PD_SCHIZ, ALS_MS_MS_SCHIZ, ALS_MS_PD_SCHIZ and MS_SCHIZ_PD_SCHIZ with 1061, 856, 1023, 872, 1139, 785, 1147, 804, 1027 and 1291 DE genes respectively. Next for a particular disease set say Z if the disease pairs are X and Y possessing fundamental diseases A, B and C, D respectively, then we have searched for common gene pairs between L_P of X and L_N of Y or L_N of X and L_P of Y or L_P of both X and Y. However, we have refrained ourselves from taking L_N for both X and Y, because it will give us non-significant DC pairs in both the diseased pairs.

Let us assume that U-V is a particular gene pair found to be common using the above mentioned procedure. To elucidate further we compare L_P and L_N, i.e. in the first disease pair the DC is significant and for the second disease pair it stands insignificant. Next, we have considered those genes having positive T_{OVA} score with U and negative T_{OVA} score with V in X, and negative T_{OVA} score with U and positive T_{OVA} score with V in Y. Finally, for common DE genes between X and Y, we have searched for the intersection of genes showing positive T_{OVA} score with U in X and negative T_{OVA} score with U in Y, and vice versa for V across X and Y. Without loss of generality we can claim that these sets of intersected genes maintain extreme differential regulation across the disease pairs. Let us denote this combination as p1pn-n2np. Here p1 and n2 stands for positive and negative LLRs with pn and np implying sign of T_{OVA} scores for the corresponding pair U-V in the disease pairs. For the same L_P and L_N another combination is p1np-n2pn. In a similar way the other possible combinations that turn up are n1pn-p2np, n1np-p2pn, p1pn-p2np, and p1np-p2pn.

The idea is to consider the interaction of uncommon genes with those gene pairs found common between disease pairs (U-V in our case). Now as for uncommon DE genes we are subtracting disease set common DE genes from the disease pair specific DE genes, thus unlike common genes in uncommon genes we obtain four disjoint sets, with two each for U and V respectively for a particular disease set. Now in order to check the impact of common and uncommon DE genes on the spreading/association of different diseases we have in total 6 sets of genes (2 sets for common with 1 each for U and V and the remaining 4 sets for uncommon genes) for every common pair between disease pairs in a disease set.

This is followed by pathway enrichment analysis for the combined gene set. A particular disease set combination yields up a number of common genes pairs with varied LLR and T_{OVA} score characteristics. To get a complete picture we have analyzed all such disease set combinations using KEGG [19] pathway analysis.

In Table 1 we have mentioned some significant pathways hence obtained, especially with the number of common DE genes between disease pairs. This particular information is given in the last column of the table.

For a given pathway, let's say *Parkinson's disease* in AD_HD_AD_SCHIZ the LLR and T metric combination of n1np_p2pn with a common gene index pair as 1, the last column shows $1 \rightarrow 5$ and $2 \rightarrow 3$. Here $1 \rightarrow 5$ means among 41 participating genes, there are 5 genes common between total down regulated genes (n) across first disease pair X and up regulated genes (p) across second disease pair Y associated with the first gene of the given gene pair. Accordingly, $2 \rightarrow 3$ means 3 genes are found common between total up regulated genes (p) across first disease pair X and down regulated genes (n) across second disease pair Y associated with the second gene of the given gene pair. A close look into these pathways reveals that the association of uncommon DE genes in any pathway is much higher compared to the common counterpart, irrespective of gene pairs, LLR and T score combinations or disease combinations. Theoretically too this seems to be quite obvious because it is the disease specific genes (uncommon DE genes) which show disjoint interactive patterns between similar types of neurodegenerative disorders.

In order to gain a better insight on the functionalities of the common DE genes we have gone for significance testing of the common genes (present in a pathway) individually across the disease pairs (genes having p to n, and n to p transitions from first to second pair). Significance testing is conducted via random shuffling of the sample labels between the control and diseased states followed by computation and comparison of T_{OVA} scores. Table 2 enlists a detailed view of significance testing of some selected pathways. In this table we have mentioned the effect of removing significant genes over the enrichment of a pathway. In such cases where we do not get any significant gene, we retain the original enrichment score shown italicized. For an example let us take the earlier case of *Parkinson's disease*. From

Table 1 Some significant pathways in terms of enrichment score/number of participating genes, along with the specific disease combination set, LLR and T metric combination, common gene pair index (among multiple gene pairs found common between two LLR combinations), and common DE genes association

Disease set	LLR and T metric combination	Common gene pair index	Pathway	Enrichment _FDR score	Total number of participating genes	Participating number of common DE genes
AD_HD_AD_SCHIZ	n1np-p2pn	1	Parkinson's disease	8.42E-07	41	1 → 5, 2 → 3
			Huntington's disease	1.62E-04	46	1 → 4, 2 → 2
	p1np-p2pn	1	Oxidative phosphorylation	6.13E-04	36	1 → 2, 2 → 4
			Alzheimer's disease	1.33E-03	41	1 → 3, 2 → 3
	p1pn-p2np	1	Parkinson's disease	5.25E-03	36	1 → 2, 2 → 3
			Oxidative phosphorylation	7.7E-03	36	1 → 3, 2 → 2
AD_HD_ALS_MS	n1np-p2pn	3	Huntington's disease	3.5E-02	36	1 → 12, 2 → 6
	n1pn-p2np	1	Ribosome	3.24E-02	21	1 → 11, 2 → 3
	p1pn-p2np	2	Ribosome	5.3E-03	22	1 → 8
AD_HD_PD_SCHIZ	n1pn-p2np	1	Huntington's disease	3.9E-02	35	1 → 2, 2 → 4
			Oxidative phosphorylation	4.7E-02	28	1 → 2, 2 → 3
			Parkinson's disease	9.7E-02	27	1 → 3, 2 → 2

(continued)

Table 1 (continued)

Disease set	LLR and T metric combination	Common gene pair index	Pathway	Enrichment _FDR score	Total number of participating genes	Participating number of common DE genes
AD_SCHIZ_MS_SCHIZ	p1np-n2pn	2	Oxidative phosphorylation	1.67E−02	32	1 → 4, 2 → 9
			Alzheimer's disease	5.3E−02	36	1 → 3, 2 → 10
AD_SCHIZ_PD_SCHIZ	n1np-p2pn	1	Ribosome	4.43E−03	26	1 → 5, 2 → 3
		3	Ribosome	1.3E−03	21	1 → 4, 2 → 3
	n1pn-p2pn	3	Ribosome	5.63E−05	28	1 → 4, 2 → 4
	p1np-n2pn	2	Ribosome	3.4E−03	24	1 → 4, 2 → 3
		7	Ribosome	8.45E−08	33	1 → 4, 2 → 4
	p1pn-n2np	4	Ribosome	7.31E−02	21	1 → 2, 2 → 5
		5	Ribosome	1.53E−02	24	1 → 3, 2 → 4
		6	Ribosome	5.37E−05	28	1 → 3, 2 → 4
		7	Ribosome	3.27E−02	23	1 → 4, 2 → 3
	p1pn-p2np	1	Ribosome	1.7E−04	26	1 → 5, 2 → 4
		2	Ribosome	7.5E−06	28	1 → 5, 2 → 3
		3	Ribosome	1.73E−04	27	1 → 3, 2 → 5
		4	Ribosome	2.4E−06	23	1 → 3, 2 → 5
		5	Ribosome	2.26E−04	27	1 → 6, 2 → 3
		6	Ribosome	9.91E−08	32	1 → 3, 2 → 7
ALS_MS_MS_SCHIZ	p1np-p2pn	8	Long-term potentiation	3.13E−03	22	1 → 1, 2 → 2

Table 2 Enrichment analysis of certain pathways without the significant common genes in first and second condition independently across gene pairs (here C1 stands for condition 1 and C2 stands for condition 2)

Pathways	Disease set with LLR, T metric combination for a gene pair index (mentioned in brackets)	No. of significant genes				Enrichment score excluding significant genes			
		1st gene specific		2nd gene specific		1st gene specific		2nd gene specific	
		C1	C2	C1	C2	C1	C2	C1	C2
Parkinson's disease	AD_HD_AD_SCHIZ n1np-p2pn (1)	Nil	3	1	Nil	$8.42E{-}07$	$3.4E{-}05$	$3E{-}06$	$8.42E{-}07$
Huntington's disease	AD_HD_ALS_MS n1np-p2pn (3)	3	4	2	2	2.8	5.12	0.77	0.77
Ribosome	AD_HD_ALS_MS n1pn-p2np (1)	7	4	2	Nil	6.4	1.16	0.22	$3.24E{-}02$
Oxidative phosphorylation	AD_SCHIZ_MS_SCHIZ p1np-n2pn (2)	1	1	4	3	$4E{-}02$	$4E{-}02$	0.61	0.46
Alzheimer's disease	AD_SCHIZ_MS_SCHIZ p1np-n2pn (2)	Nil	Nil	2	4	$5.3E{-}02$	$5.3E{-}02$	0.26	1.2
Ribosome	AD_SCHIZ_PD_SCHIZ n1np-p2pn (3)	1	Nil	Nil	Nil	$2.3E{-}04$	$5.63E{-}05$	$5.63E{-}05$	$5.63E{-}05$
Ribosome	AD_SCHIZ_PD_SCHIZ p1np-n2pn (2)	3	3	Nil	Nil	0.123	0.123	$3.4E{-}03$	$3.4E{-}03$
Ribosome	AD_SCHIZ_PD_SCHIZ p1np-n2pn (7)	1	Nil	3	Nil	$4.1E{-}07$	$8.45E{-}08$	$8E{-}06$	$8.45E{-}08$
Ribosome	AD_SCHIZ_PD_SCHIZ p1pn-p2np (6)	Nil	Nil	3	Nil	$9.91E{-}08$	$9.91E{-}08$	$2.3E{-}06$	$9.91E{-}08$

Table 1 we have got 5 and 3 genes commonly associated with gene 1 and 2 of the concerned pair. Now from Table 2 we can see that out of 5 common genes none has been found significant with the first gene of the gene pair across first condition (C1), where as in second condition (C2) 3 genes happen to be significant. Removal of these 3 common genes worsens the FDR score to 3.4E−05. On the other hand across 3 common genes 1 gene has been found significant with the second gene of the gene pair in C1, but none has been found to be significant in C2. Here removal of the significant gene has given the enrichment score 3E−06 (here first and second condition suggests first and second disease pair of the disease set).

We have got varieties of disease pathways by feeding the gene set into DAVID [20, 21], as listed in Tables 1 and 2. The tables also reveal that most of these pathways are nothing but neurodegenerative diseases only. Only the three pathways, which do not bear any disease name, are *Oxidative phosphorylation*, *Ribosome*, and *Long term potentiation*. In order to check whether any/all of them have any role in such diseases we have gone through an exhaustive clinical survey. Our survey discloses the crucial role played by these three above mentioned pathways in neurodegenerative diseases.

Let us first check the role of *Oxidative phosphorylation*. Basically it regulates the neuronal actions via the help of Mitochondria (abbreviated as Mt). As given in [22] oxygen takes part in glucose break down in Mt through oxidative phosphorylation and generates ATP, which works as energy currency of the cell. Now any sort of mutation of Mt DNA (works as molecular machinery) leads to impaired ATP generation and perturbed oxidative phosphorylation cascade that may further lock the neuronal function, which specifically leads to AD [22, 23]. Now talking about *Ribosome*, we have found that the absence of some binding partners (for an example GTPBP2), of the ribosome recycling protein may cause ribosome stalling and widespread neuro-degeneration [24]. Finally, when we consider the role of *Long term potentiation*, then recent literatures suggests that it causes different neurodegenerative disorders due to synaptic dysfunctions [25].

Looking into Table 2 it is apparent that here the pathways are listed based upon the inclusion of high amount of significant genes across both conditions. This enables us to predict the importance of these genes portraying significant impact over a disease set. Hereby, we can also claim bridging effect of these pathways over different diseases of a corresponding set. To support our thoughts we have gone for exhaustive literature mining which has given us ample evidences in this context.

At the very beginning while searching for the association between PD and AD, HD, SCHIZ we have found that it is the TRANSGLUTAMINE (TG) kind of enzymes which affect (act as a common factor) different neurodegenerative disorders like PD and AD/HD [26]. As given in [26] different kinds of TGs which are activated in AD and HD CSF (Cerebrospinal Fluid) also contribute to the formation of proteinaceous deposits in PD. Both PD and SCHIZ have got a common originating link. As stated in [27] these are the results of redox process (i.e. joint activity of Reactive Oxygen Species (ROS) and Oxidative Stress (OS)). This redox process works as a common link between HD and AD, ALS, MS. In [27] we also find common redox association between AD and MS. Again, apart from having a common chemically reactive baseline all these different neurodegenerative diseases happen to be a subset of Neurodegenerative Misfolding Diseases (NMD) triggered by the misfolding of one or two proteins and their accumulation in the aggregated species toxic to neurons [28], especially due to the effect of protein disulfide isomers (PDI).

Next we move on to assess the exclusive biological significance of the significant genes obtained via permutation test. In this connection at first we re-checked the enrichment of the pathways in which different significant genes participate, but this time formed by those significant genes only across different conditions. In the following section we have discussed on those pathways where this enrichment is far better compared to the scenario having all other genes excluding significant participants (given in Table 2).

First case where we have encountered this kind of scenario is Huntington's disease. There corresponding to the 1st gene in the second condition we have obtained 4 significant genes. They are ILMN_4450, ILMN_10087, ILMN_13178, and ILMN_16327. Taken together they have formed the same pathway with enrichment score of 3.03E−02 against 5.12 as listed in Table 2. We observed a similar situation in Ribosome (obtained from AD_HD_ALS_MS), where corresponding to the 1st gene in the first condition there are 7 significant genes. These are ILMN_138835, ILMN_137528, ILMN_10289, ILMN_137046, ILMN_138635, ILMN_138392, and ILMN_13487. A combination of these genes has enriched the same pathway with a score of 7.99E−06 compared to 6.4, given in Table 2. For the 1st gene only with the same pathway in second condition there are 4 significant genes, which are ILMN_137046, ILMN_138635, ILMN_139337 and ILMN_137876. Taken together have given an FDR score of 4.84E−04 much better than 1.16, given in Table 2. Similar observations on Oxidative Phosphorylation where enrichments corresponding to the 2nd gene in first and second condition based significant genes, which are ILMN_20286, ILMN_2295, ILMN_19166, ILMN_137342 and ILMN_20286, ILMN_19166, ILMN_16064 respectively, are far better compared to the enrichments obtained excluding them. From Table 2 we have got the FDR score excluding those significant genes as 0.61 and 0.46, whereas simple combination of significant genes yields FDR scores equal to 8.42E−03 and 0.28 respectively. Significant genes obtained from Alzheimer's disease in connection to second condition of 2nd gene are ILMN_1351, ILMN_20286, ILMN_5679, and ILMN_19166. Collectively only these 4 genes have given an enrichment score of 3.1E−02 as compared to 1.2 excluding them. Finally, in the Ribosome pathway obtained from AD_SCHIZ_PD_SCHIZ with a LLR combination of p1np-n2pn across 1st gene we have got 3 significant genes in both conditions. These are ILMN_21554, ILMN_16945, and ILMN_16298 respectively. Combination of these genes only has given an enrichment score of 2.89E−02 compared to 0.123, except those.

At the second level we find the significance individually assessing the biological contribution of a gene in the corresponding pathway. We have applied this strategy especially for those cases where our first phase of exclusive significance assessment does not turn fruitful. From Table 2 in Parkinson's disease the 1st gene specific significant genes under second condition are ILMN_11281, ILMN_1167 and ILMN_17626. As given in [29] first 2 genes are involved/activated in encoding of ubiquitin activated enzyme E1, and NADH dehydrogenase (ubiquinone) Fe-S protein 4, whereas the third gene is a cytochrome c oxidase subunit VIIc (COX7C) one. Involvement of them in activation/spreading of Parkinson's disease in different organisms are given in [30–32]. In the same disease across 2nd gene under first condition we have got only one significant gene ILMN_10929 known to be involved in encoding of ubiquinol-cytochrome c reductase core protein II (UQCRC2). Now as given in [33] this protein actively participates in this diseased pathway. Again from Table 2 for Huntington's disease across 1st gene in the first condition we have got 3 significant genes which are respectively ILMN_22085, ILMN_10087, and ILMN_1167. Among them we have got evidence of active participation in this disease for the latter two genes. As given in [34, 35] they participate in this diseased pathway via the encoding of cytochrome c, somatic (CYCS), nuclear gene encoding mitochondrial protein and Homo sapiens NADH dehydrogenase (ubiquinone) Fe-S protein 4. This is the same way through which ILMN_20348 (found across 2nd gene in both conditions) participates in this pathway [35]. Another gene found here is ILMN_138125, but surprisingly no existing literature describes its role in Huntington's disease. In Ribosome (by AD_HD_ALS_MS shown in Table 2) we got 2 significant genes across 2nd gene in condition 1, ILMN_137810 and ILMN_138613. Involvement of them in encoding of different ribosomal proteins is given in [29]. In Oxidative Phosphorylation (OP) as per Table 2 we have got ILMN_17626 as the significant gene present across both conditions of the 1st gene. ILMN_17626 is a cytochrome c oxidase subunit VIIc (COX7C) gene, whose effect on OP has been given in [36]. From Table 2 only in Alzheimer's disease (AD) across first condition of 2nd gene we have got two significant genes, ILMN_1351, and ILMN_20286. Individual assessment of them over AD via mitogen-activated protein kinase 1 (MAPK1) and NADH dehydrogenase (ubiquinone) 1 beta sub complex encoding is described in [37–39]. ILMN_21554 is the only significant gene found in Ribosome (obtained from AD_SCHIZ_PD_SCHIZ with a LLR combination of n1pn_p2np and third gene pair specific), and its role in ribosomal protein is given in [29]. Whereas there are all total 3 significant genes obtained from another Ribosome pathway (this time it is from the same disease set and same LLR combination as the previous one but from 7th gene pair). These are ILMN_2271 (found with both the genes in the pair), ILMN_2500, and ILMN_1815 (last two are found exclusively with the 2nd gene). Ref. [29] has listed their involvements in activation of different ribosomal proteins. Finally, the last combination is the group of 3 significant genes ILMN_11712, ILMN_15150, and ILMN_138613 (this time also the pathway is Ribosome, with the same disease set combination having the LLR combination p1pn_p2np, with sixth gene pair). Ref. [29] also enlists their functionalities over the encoding of different ribosomal proteins.

4 Conclusion and Future Work

In this work we have tried to frame an algorithm in order to select genes having extreme dysfunctional activities across differential conditions. Our algorithm has started from the estimation of DE genes individually across different diseases, accomplished by the comparison of expression levels across two conditions: (healthy) control and disease. Parallely we have completed another significant segment by pairing up some diseases based upon their proximity to one another. Finally, every possible disease pairs are again combined in order to form different disease sets.

Upon the completion of the above framework next we have headed towards the estimation of differentially co-expressed (DC) gene pairs. In order to estimate DC pairs we have taken help of a probabilistic disease pair specific measure, known as Log Likelihood Ratio (LLR). Thus from our prior knowledge we can say that for each set there are two LLR metrics. Then we have searched for those common gene pairs possessing extreme LLR values (maximum positive or minimum negative) across both disease pairs. However as negative LLR value signifies insignificant DC thus we have refrained taking negative LLR across both disease pairs.

We have also taken help of the DC score, known as T score (considered to be the basic building block of LLR metric) in order to search for common set of genes having dissimilar dysfunctional regulatory activities with those common gene pairs across different conditions. In our context these are the common DE genes between the disease pairs of a given set. For uncommon DE genes across each disease pair there will be separate two gene sets having alternate regulations [upward (p)/downward (n)] with each gene of the common gene pairs.

Now in order to check the impact of common as well as uncommon genes over the regulation of different disease pairs we have combined different common and uncommon gene sets to form an adjoined set for biological enrichment. From Table 1 we can conclude on the dominating impact of uncommon DE genes over the common DE genes across various enriched pathways.

We have also gone for significance analysis of common DE genes individually across conditions. In order to assess the importance of the significant genes we have removed the same from the total set and re-evaluated the biological enrichment.

In future we would like to explore the analogy into Transcriptional Factor (TF) based regulatory network. This procedure will initiate through the discovery of the TF pairs, having DC association across conditions. Next in order to find the best suitable targets the TFs will be fitted in an existing procedure [40, 41], where based upon the strong dissimilar association between TF pairs and targets, targets are chosen. This LLR metric is applicable over TF_DE interaction patterns where for a particular DE gene we can search for regulatory pathway via some other TFs. One can also apply this analogy in the concept given in [42] in order to find TF_bridged_DCLs.

References

1. Schulze, A., Downward, J.: Navigating gene expression using microarrays: a technology review. Nat. Cell Biol. **3**, E190–E195 (2001)
2. Spellman, P.T., et al.: Comprehensive identification of cell cycle-regulated genes of the yeast *Saccharomyces cerevisiae* by microarray hybridization. Mol. Biol. Cell **9**(12), 3273–3297 (1998)
3. Sarkar, M., Majumder, A.: Quantitative trait specific differential expression (qtDE). Procedia Comput. Sci. **46**, 706–718 (2015)
4. Sarkar, M., Majumder, A.: TOP: an algorithm in search of biologically enriched differentially connective gene networks. In: Proceedings of the 5th Annual International Conference on Advances in Biotechnology. Kanpur (2015)
5. de la Fuente, A.: From 'differential expression' to 'differential networking' identification of dysfunctional regulatory networks in diseases. Trends Genet. **26**, 326–333 (2010)
6. Carter, S.L., Brechbuhler, C.M., Griffin, M., Bond, A.T.: Gene co-expression network topology provides a framework for molecular characterization of cellular state. Bioinformatics **20**(14), 2242–2250 (2004)
7. Altay, G., Asim, M., Markowetz, F., Neal, D.E.: Differential C3NET reveals disease networks of direct physical interactions. BMC Bioinform. **12**, 296 (2011)
8. Tesson, B.M., Breitling, R., Jansen, R.C.: DiffCoEx: a simple and sensitive method to find differentially coexpressed gene modules. BMC Bioinform. **11**, 497 (2010)
9. Amar, D., Safer, H., Shamir, R.: Dissection of regulatory networks that are altered in disease via differential co-expression. PLoS Comput. Biol. **9**(3), e1002955 (2013)
10. Durrenberger, P.F., et al.: Selection of novel reference genes for use in the human central nervous system: a BrainNet Europe Study. Acta Neuropathol. **124**(6), 893–903 (2012)
11. Palmer, B.W., et al.: Assessment of capacity to consent to research among older persons with schizophrenia, Alzheimer disease, or diabetes mellitus. Arch. Gen. Psychiatry **62**(7), 726–733 (2005)
12. Sutherlad, M.K., Somerville, M.J., Yoong, L.K.K., Bergeron, C., Haussler, M.R., McLachlan, D.R.C.: Reduction of vitamin D hormone receptor mRNA levels in Alzheimer as compared to Huntington hippocampus: correlation with calbindin-28k mRNA levels. Mol. Brain Res. **13**(3), 239–250 (1992)
13. Rao, S.M., Huber, S.J., Bornstein, R.A.: Emotional changes with multiple sclerosis and Parkinson's disease. J. Consult. Clin. Psychol. **60**(3), 369–378 (1992)
14. Healthline, http://www.healthline.com
15. Frisoni, G.B., Filippi, M.: Multiple sclerosis and Alzheimer disease through the looking glass of MR imaging. AJNR Am. J. Neuroradiol. **26**, 2488–2491 (2005)
16. Andreassen, O.A., et al.: Genetic pleiotropy between multiple sclerosis and schizophrenia but not bipolar disorder: differential involvement of immune-related gene loci. Mol. Psychiatry **20**(2), 1–8 (2014)
17. Ghanemi, A.: Schizophrenia and Parkinson's disease: selected therapeutic advances beyond the dopaminergic etiologies. Alex. J. Med. **49**, 287–291 (2013)
18. Wang, L., Fenq, Z., Wang, X., Wang, X., Zhang, X.: DEGseq: an R package for identifying differentially expressed genes from RNAseq date. Bioinformatics **26**(1), 136–144 (2010)
19. Kanehisa, M., Goto, S.: KEGG: kyoto encyclopedia of genes and genomes. Nucl. Acids Res. **28**, 27–30 (2000)
20. Huang, D.W., Sherman, B.T., Lempicki, R.M.: Systematic and integrative analysis of large gene list using DAVID bioinformatics resources. Nat. Protoc. **4**(1), 44–57 (2008)
21. Huang, D.W., Sherman, B.T., Lempicki, R.M.: Bioinformatics enrichment tools: paths towards the comprehensive functional analysis of large gene lists. Nucl. Acids Res. **37**(1), 1–13 (2013)

22. Uttara, B., Singh, A.V., Zamboni, P., Mahajan, R.T.: Oxidative stress and neurodegenerative diseases: a review of upstreamand downstream antioxidant therapeutic options. Curr. Europharmacol. 7(1), 65–74 (2009)
23. Hroudová, J., Singh, N., Fišar, Z.: Mitochondrial dysfunctions in neurodegenerative diseases: relevance to Alzheimer's disease. BioMed. Res. Int. 2014, Article ID 175062 (2014)
24. Ishimura, R., et al.: Ribosome stalling induced by mutation of a CNS-specific tRNA causes neurodegeneration. Science 6195, 455–459 (2014)
25. Marttinen, M., Kurkinen, K.M.A., Soinien, H., Haapasalo, A., Hiltunen, M.: Synaptic dysfunction and septin protein family members in neurodegenerative diseases. Mol. Neurodegener. 10, 16 (2015)
26. Martin, A., Vivo, G.D., Ventile, G.: Possible role of the transglutaminases in the pathogenesis of Alzheimer's disease and other neurodegenerative diseases. Int. J. Alzheimer's Dis. 2011 (2011)
27. Kovacic, P., Somanathan, R.: Redox processes in neurodegenerative disease involving reactive oxygen species. Curr. Neuropharmacol. 10, 289–302 (2012)
28. Mossuto, M.F.: Disulfide bonding in neurodegenerative misfolding diseases. Int. J. Cell Biol. 2013 (2013)
29. https://www.ebi.ac.uk/arrayexpress/files/A-GEOD-6171/A-GEOD-6171_comments.txt
30. Viquez, O.M., Caito, S.W., McDonald, W.H., Friedman, D.B., Valentine, W.M.: Electrphilic adduction of ubiquitin activating enzyme E1 by N,N-diethyldithiocarbamate inhibits ubquitin activation and is accompanied by striatal injury in the rat. Chem. Res. Toxicol. 25(11), 2310–2321 (2012)
31. Genetics Home Reference, ghr.nlm.nih.gov/gene/NDUFS1
32. Arnold, S.: Cytochrome c oxidase and its role in neurodegeneration and neuroprotection. Adv. Exp. Med. Biol. 748, 305–339 (2012)
33. NCBI database, https://www.ncbi.nlm.nih.gov (Gene ID: 7385)
34. Wang, X., et al.: Inhibitors of cytochrome c release with therapeutic potential for Huntingtons disease. J. Neurosci. 28(38), 9473–9485 (2008)
35. Human Metabolome Database. http://www.hmdb.ca/proteins/HMDBP00180
36. GeneCards. http://www.genecards.org/cgi-bin/carddisp.pl?gene=COX7C
37. Zhu, X., Lee, H.G., Raina, A.K., Perry, G., Smith, M.A.: The role of mitogen-activated protein kinase pathways in Alzheimer's disease. Neurosignals 11(5), 270–281 (2002)
38. NCBI database. https://www.ncbi.nlm.nih.gov (Gene ID: 4729)
39. Kim, S.H., Vlkolinsky, R., Crains, N., Fountoulakis, M., Lubec, G.: The reduction of NADH: Ubiquinone oxidoreductase 24- and 75-kDa subunits in brains of patients with Down syndrome and Alzheimer's disease. Life Sci. 68(24), 2741–2750 (2001)
40. Majumder, A., Sarkar, M.: Simple transcriptional networks for differentially expressed genes. In: International Conference on Signal Propagation and Computer Technology (ICSPCT), pp. 642–647 (2014)
41. Majumder, A., Sarkar, M.: Paired transcriptional regulatory system for differentially expressed genes. In: Lecture Notes on Information Theory, vol 2(3) (2014)
42. Wang, J., et al.: DCGL v2.0: an R package for unveiling differential regulation from differential co-expression. PLoS One 8(11), e79729 (2013)

Multiobjective Ranked Selection of Differentially Expressed Genes

Mrityunjay Sarkar and Aurpan Majumder

Abstract The regulatory action of a gene in a complex network is guided by the differential functionalities of the gene acting under varied conditions. Many methodologies have been proposed in recent years to unveil this regulation. However in this context the gene ranking obtained via separate methodologies according to their significance is quite dissimilar to one another making regulatory assessment of genes very difficult. In this paper, we have developed a novel procedure to compute significant genes using more than one ranking strategy. Accordingly, we have explored this idea applying the concept of non-dominated set of solutions residing in different Pareto optimal fronts. Our main objective is to find a set of non-dominated genes in the primary Pareto front each of which having an optimal combination of significant ranking across different ranking algorithms. In the results we have shown that most of the KEGG pathways formed from the set of DE genes contain at most two genes from the non-dominated set. This helps us to understand the independent regulatory function of a gene from the non-dominated set with the set of dominated genes. In other words, the existence of enriched control pathways with significant ranked genes non-dominant to one another is almost absent.

Keywords Activity score · Dominance · Gene set enrichment analysis (GSEA) · Rank sum test (RS) · TOP

M. Sarkar (✉)
Department of ECE, D.I.A.T.M. Durgapur, Durgapur, India
e-mail: mrityu1488@gmail.com

A. Majumder
Department of ECE, N.I.T. Durgapur, Durgapur, India
e-mail: aurpan.nitd@gmail.com

© Springer Science+Business Media Singapore 2017
P. Deiva Sundari et al. (eds.), *Proceedings of 2nd International Conference on Intelligent Computing and Applications*, Advances in Intelligent Systems and Computing 467, DOI 10.1007/978-981-10-1645-5_7

1 Introduction

Gene differential expression analysis using microarrays has been the fundamental strategy to explore the differential functionalities of genes across different conditions. Based on this strategy many approaches had been framed to prioritize the disease specific genes. Amongst these algorithms some access the significance observing the change in the expression level only (differential expression) [1], whereas others investigate the change in the connectivity with neighbour genes across conditions (differential connectivity) [2]. Again there are recent literatures which try to integrate the gene-gene association with the change in expression level (i.e. an amalgamation of differential expression with differential connectivity) [3].

The implementations of the different algorithms give us statistically significant unique gene combinations. The confusion comes to the front after ranking them according to their significance. Overlapped portions from the top ranked genes amongst these methodologies are quite low which signify different methods infer separate set of genes to be significant. Hence, it gets difficult in determining the contributory role of genes to differential functionality of a complex gene regulatory network.

In order to solve the above problem we have adopted a simple approach through finding a non-dominated solution set (gene-set) utilizing multiple objective functions (ranking methodologies). In this connection a set of genes is designated non-dominated provided they possess low ranking in all objectives with non-dominance to one another in the set across objectives and positively not being dominated by any other gene outside the referred set.

Application of multiobjective evolutionary algorithms has got an extensive research in the domain of gene classification [4] by clustering [5–8]. In this context, most of the works try to optimize clustering indices in a multiobjective paradigm. The commonly used algorithms have been GA-II, NSGA-II, PESA-II [9], to name a few. Extensively it focuses on methods to improve clustering via number of clusters present in a chromosome, intra-cluster compactness, inter-cluster separation, and cluster size [10]. However, one of the major drawbacks of clustering approach is time complexity, irrespective of optimization techniques such as GA [11], PSO [12], BPNN [13], ACO [14]. This aspect occurs mainly due to clustering with generation and validation of new populations in different iterations [7, 8]. This time challenging constraint does not arise in our approach as we attack the problem with the parent population of genes only. Our intention is to find the best possible genes from the existing ones on the basis of ranking and not to generate any new set of solutions with predicted gene expression levels.

Let us assume having expression levels of m genes with n objectives. Henceforth the different ranking algorithms based on gene significance would be mentioned as objective functions. We take into consideration a minimization problem for all the objectives. Thus a feasible gene x is said to dominate another feasible gene y if $z_i(x) \le z_i(y)$ for $i = 1, 2, …,$ n with $z_j(x) < z_j(y)$ for at least one objective function j [9]. If any gene is not dominated by others in the assumed gene space then such a gene is said to be a *Pareto optimal gene*. The set of all possible non dominated genes in m is defined as the *Pareto optimal set* [9].

In our work, we have tested the above mentioned concept utilizing 4 objective functions. They are rank sum statistics [15], gene significance based enhancement analysis (GSEA) [16], activity score (AS) [17], and a recently published TOP based gene significance [18]. To explore this thought with four objective functions a gene is dominated or regulated if and only if from the entire gene set forming a pair of genes with the concerned gene we obtain a higher rank or worse score for it in one, or two, or three, or four objectives with equal score in the other three, two, one, and zero objectives respectively compared to a paired gene. Accordingly, the genes which do not meet any of the above criteria are the non-dominated genes. The scores from different algorithms (ranksum, AS, Z, and TOP) have been used as the objective functions.

These measures either highlight a maximization or minimization problem depending on the individual algorithms. However, for our computation we have designed minimized objectives for all the four cases.

After obtaining the non-dominated gene set we have gone for KEGG pathway analysis to verify the presence of more than one gene in any pathway from the non-dominated set.

In any pathway amongst other genes if only one gene belongs from the non-dominated set then this validates our finding. However participation of more than one gene from the non-dominated set in any pathway suggests that though computationally they fall on the non-dominant Pareto optimal front but from a biological view one is dominated or regulated by the other.

The rest of the paper is as follows. In next section we have discussed on the methodology. In the results we are able to give a detailed view of the problem and its implementation on a dataset. To validate our findings brief descriptions on some important pathways found in the analysis have also been given. We conclude with applying this idea to build up of transcription factor regulatory networks.

2 Methodology

Assuming a gene expression profile of a biological dataset distributed over a minimum of two conditions, our target is to first find the genes having significant differential contribution across conditions using different strategies and finally to discover the set of non-dominated genes amongst these strategies. The flowchart of the entire operation is given Fig. 1.

As discussed earlier the best way to find genes having strong differential role is to find genes having altered expression level (differentially expressed) [1] as well as possessing differential connectivity with neighbours across conditions. In this context estimation of DE genes have been done using DEGseq [19] and qtDE [20]. As a next step we have computed the generalized topological measure (GTOM)

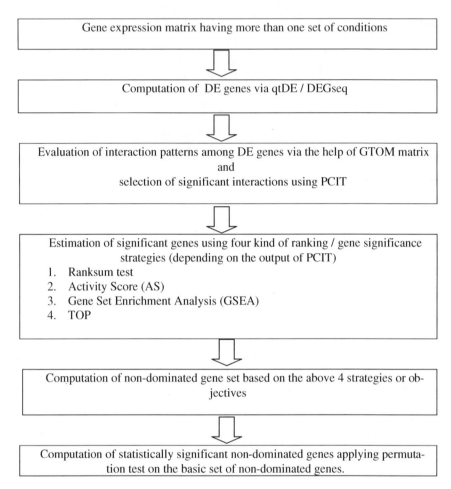

Fig. 1 Flowchart for evaluating significantly non-dominated differentially expressed (DE) genes

[21] among the DE genes at different conditions. In short, result of GTOM depends upon a connectivity measure between every pair of genes directly as well as via all other genes. Now as discussed in [18] it would emphasize those gene pairs having strong connectivity between themselves via indirect gene regulations, giving us genes having highly connected neighbourhood in both conditions.

Using these GTOM matrices we have searched for densely connected modules across samples using PCIT [22]. Further calculations are conducted on the filtered GTOM matrices (coming out to be significant after the application of PCIT). This operation is followed by the estimation of differentially significant genes based on the filtered GTOM matrices involving the four types of ranking strategies. Brief descriptions of these ranking techniques are given below:

2.1 Ranksum Test

In our work we have applied Wilcoxon rank sum test [15]. It is a non-parametric test of the null hypothesis where the two populations are same against an alternative hypothesis, where a particular population puts up a skewed effect to the combined distribution.

Let us assume to have two populations A and B with independent random samples $a_1, a_2, ..., a_m$ and $b_1, b_2, ..., b_n$ of sizes m and n respectively.

As per the algorithm we merge the data and rank the measurements from lowest to highest values. Here the mean (μ) and standard deviation (s) of the merged data is

$$\mu = m(m+n+1)/2 \tag{1}$$

$$s = \sqrt{[mn(m+n+1)/12]} \tag{2}$$

Now based on the merged data two kinds of hypothesis are checked. First one being the null hypothesis H_0: A = B, which means the distribution of X measurements in population A is same as that of B or in other words the ranked distribution pattern in the merged data happens to be an association of samples taken at random from the individual distributions. In this case we cannot predict the differential pattern of the data vectors (in our case the filtered GTOM vectors for each gene).

In this context the alternative hypothesis is of two kinds. First one is H_1: A > B, which means in the ranked merged distribution samples from A is right shifted compared to B, and the second one is H_2: A < B, suggesting the samples from A is shifted to the left of B. In both of the cases we are able to ascertain the differential pattern of the filtered GTOM vectors highlighting the differential contribution of the concerned gene.

As per our thought the application of PCIT [22] yields filtered GTOM matrices. In this connection we get two matrices (for each condition) with entries 1 and 0. As a next step we have searched for common set of interactions for every gene across both conditions. This means searching for those entries possessing 1 at the same location in both matrices for each individual gene. Finally, we replace these 1's by the original filtered GTOM values in both conditions and other entries (uncommon 1's as well as 0's) are made equal to 0. At this stage we have applied the Wilcoxon-rank-sum test over these modified versions of the filtered weighted GTOM matrices A and B.

2.2 Activity Score (AS)

According to [17] this score is used for prioritizing deregulated genes and also to group them into modules.

Like the rank sum test here also at first we take the common set of interactions between two conditions. Next we have replaced the (common) 1's by the filtered GTOM values in both conditions and other fields by 0.

Next as given in [17] AS score is calculated by the following set of equations

$$AS = (-1)^\alpha \times \max_{i \in neighbour} \left(\frac{w_i - u_i}{\sigma_i} \times \frac{w_i - u_i}{w_{\max} - u_i} \right) \qquad (3)$$

where $w_i = \sum_{j \in X_i} rank(z_j)$, $u_i = \frac{i(N+1)}{2}$ and

$$\sigma_i = \sqrt{\frac{i(N-i)(N+1)}{12}} \qquad (4)$$

Here corresponding to any row (i.e. a gene) of the filtered GTOM matrix, X_i means the genes having a non-zero entry. Z_j is the Z score of the differential expression of any such gene using rank-sum test. Further as given in (4) we calculate w_i, u_i, and σ_i where N is the total number of genes used in the analysis. Thus utilising the values produced by (4) we are able to compute the AS score as given in (3), provided $\alpha = 0$ if $\frac{w_i - u_i}{\sigma_i}$ is >0, otherwise $\alpha = 1$.

Finally the genes are ranked according to their AS scores.

2.3   Gene Set Enrichment Analysis (GSEA)

Starting with the filtered GTOM matrices obtained via PCIT we build up a gene matrix maintaining the non-zero entries just for the real values of significant gene to gene interactions in both matrices. Further computations are performed on this gene matrix.

As given in [16], enrichment score is calculated by the following set of equations:

$$P_{hit}(S, i) = \sum_{E_j \in S, j \leq i} \frac{r_j^P}{N_R} \qquad (5)$$

where

$$N_R = \sum_{E_j \in S} r_j^P \qquad (6)$$

and

$$P_{miss}(S, i) = \sum_{E_j \in S, j \leq i} \frac{1}{N - N_H} \tag{7}$$

At first, corresponding to every gene, separately in each condition, we have ranked the gene interaction values from lowest to highest level and accordingly assigned rank labels for every significant interaction. This operation is followed by the generation of a random vector for every gene comprising of values meant to index locations of the significant GTOM matrix. In each case we check the rank of the indexed gene interaction following which we add up the weight (filtered GTOM value) of the genes having less rank than the indexed gene interaction. This idea is framed up in Eqs. (5–7), where 'i' is the indexed rank and 'j' are those genes possessing lower rank with 'S' being the neighbourhood of significant interactions for a gene.

The term N in Eq. (6) represents the total number of genes and N_H is the number of significant gene interactions corresponding to a gene. Hence the ES score for every indexed gene interaction is calculated as

$$ES(S, i) = P_{hit}(S, i) - P_{miss}(S, i) \tag{8}$$

Using Eq. (8) we have an ES matrix in each condition. The differential attitude of the ES vectors thus formed is calculated via rank-sum test which yields us the rES vector.

As a next step we compute the permuted ES and rES scores considering 200 cases of random shuffling of the parent GTOM matrices. For each gene we have calculated the average $(\overline{rES})$ as well as the standard (S') deviation of the permuted rES scores. Finally, we compute the Z score shown below and rank the genes.

$$Z = \frac{rES - \overline{rES}}{S'} \tag{9}$$

2.4   TOP Based Analysis

This analysis can be performed using two approaches: unweighted and weighted [23]. However as discussed in [18] results obtained by unweighted measure are far better than weighted counterpart, thus in our work we have proceed via unweighted measure only.

Basically this is a combination of TO value [23] with its significance using t test [24]. Let A is a matrix whose entries represents significant interactions. Application of PCIT reveals that the interaction between i and j gene is significant. So we have made the corresponding entry in A as $1(A_{ij} = 1)$ else we have made $A_{ij} = 0$. Thus the continuation of this process across two conditions will give us two matrices with entries as 1 and 0. Again using the concept given in [23] we are assuming that for gene i in condition $1X$ no. of interaction(s) is/are significant, and in condition 2 it is Y. So like [23] the TO of gene i between two networks can be defined as

$$T \cdot O_i = (X \cap Y)/\max(X_i, Y_i) \qquad (10)$$

Next as in [18] we have calculated the average of TO measure and the p value (using permutation/t test [24]) and termed it as *TOPavg*. Mathematically this can be written as

$$TOPavg_i = (TO_i + p\ value_i)/2 \qquad (11)$$

2.5   Non-dominated Gene Set

Here the quest lies in finding genes non-dominant to one another based on the four above mentioned objectives.

In this connection we discover genes to be on the primary non-dominated front which yields better ranking in at least one, two, or three objective functions out of four compared to the all other genes. According to this the number of possible combinations to look through in order to satisfy the above strategy will be $\binom{4}{1} + \binom{4}{2} + \binom{4}{3} = 14$. The scores from different algorithms (ranksum, AS, Z, and TOPavg) have been used as objective functions. Here a superior ranked gene is obtained depending on the minimized value for the individual objectives.

2.6   Significant Non-dominated Gene Set

At this stage we focus on the computation of significant non-dominated genes conducting permutation test [24] on the filtered GTOM matrices. In this context we are able to validate the non-dominance of the set of genes obtained above. Thereafter genes showing a p value <0.2 are considered to be significant.

Next we performed KEGG pathway analysis on the entire set of genes. This biological validation helps us to understand the importance of the significantly non-dominated participating genes. In any such pathway we are interested to find as minimum as possible genes from the non-dominated set associated with other dominated genes. This leads us to interpret the regulatory action of non-dominated genes over the dominated ones. However, being non-dominant to one another ideally there should be one such significant gene in any noteworthy pathway indicating non-regulatory action amongst these genes.

3 Results

Our proposed algorithm has been tested on an open access mice data set [25]. We perform the computation of DE genes using two different approaches. The data-set contains expression levels of male and female mice over four tissues: adipose, brain, liver and muscle. Details on these data sets, microarray analysis, and data reduction using pre-processing have been discussed in [25], and [26].

As given in [20] qtDE genes from the mice data have been computed using three techniques which are correlation, mutual information and polynomial regression respectively. The qtDE genes in adipose, brain, liver, and muscle using correlative measure are 856, 579, 837 and 1132. Using mutual information there happens to be 1236, 1499, 1479, and 2503 qtDE genes respectively, whereas polynomial regression gives us 938, 675, 1395 and 1163 qtDE genes respectively. On the other hand traditional DEGseq [19] gives us 732, 373, 424 and 301 DE genes.

As per the methodology we have gone through the four kinds of ranking/significance measures to generate the set of Pareto optimal DE genes. Non-dominated qtDE genes across adipose, brain, liver and muscle using correlation are 77, 105, 8, and 114 respectively, whereas mutual information and polynomial regression yields us 28, 21, 8, 99 and 20, 112, 33, 160 qtDE genes respectively. Using DEGseq we have 9, 21, 7 and 3 DE genes respectively.

3.1 Significant Non-dominated Gene Set

In Tables 1, 2, 3, and 4 we have given a detailed pathway analysis on the total DE genes detected via qtDE and DEGseq. Genes from basic non-dominated set are highlighted in bold characters, and the significantly non-dominated ones are accentuated through bold and italics:

3.1.1 Correlation

The number of non-dominated qtDE genes obtained across adipose, brain, liver and muscle via linear correlative measure are 77, 105, 8, and 114 where 12, 91, 6, and 99 genes are found to be significantly non-dominated in these four tissues.

3.1.2 Polynomial Regression

Applying this non-linear based approach we have got 20, 112, 33, and 160 non-dominated qtDE genes in adipose, brain, liver and muscle. Here 7, 67, 9, and 124 genes come out to be significant in the corresponding tissues.

Table 1 Significant pathways by correlation

Pathways	p values	Genes
1.1 Adipose		
Olfactory transduction	3.49E−07	**Olfr584**, Olfr599, Olfr957, Clca1
Leishmaniasis	5.51E−03	Jun, **H2-Aa**, Mapk1, **Prkcb**, Jak2, Tgfb3, Il1b, H2-DMa, H2-Ab1
ErbB signalling pathway	8.31E−02	Jun, Rps6kb2, Mapk1, Stat5a, Erbb3, **Prkcb**, Cdkn1b, Areg
Graft versus host disease	8.31E−02	**H2-Aa**, H2-Q8, Il1b, Cd86, H2-DMa, H2-T10, H2-Ab1
Malaria	8.31E−02	Itgal, Tgfb3, Il1b, **Ccl2**, Hgf, Vcam1
Type I diabetes mellitus	8.31E−02	**H2-Aa**, H2-Q8, Il1b, Cd86, H2-DMa, H2-T10, H2-Ab1
Viral myocarditis	8.31E−02	Rac2, H2-Aa, H2-Q8, Itgal, Casp3, Cd86, **H2-DMa**, H2-T10, H2-Ab1
Cell adhesion molecules (CAMs)	8.31E−02	**H2-Aa**, H2-Q8, Itgal, Cdh2, Cntnap2, Cd86, H2-DMa, H2-T10, Jam2, H2-Ab1, Vcam1
Ribosome	0.10	Rpl15, Rps3a, Rps3, Rps8, Rpl3 l, Rpl29, Rpl6, Rps14, **Rpl35**
1.2 Brain		
Olfactory transduction	6.69E−04	Olfr1226, **Olfr206**, Olfr380, Olfr599, Olfr1234
Leukocyte transendothelial migration	4.03E−02	*Txk*, Jam2, Itgal, Vegfb
Galactose metabolism	8.79E−02	*Gck*, Galt, **Gaa**, Ugp2
1.3 Liver		
Olfactory transduction	4.27E−07	Clca1, Olfr535, Olfr380, **Olfr599**, Olfr1234, Clca2
Drug metabolism— cytochrome P450	3.59E−02	Fmo3, *Cyp2d22*, Gsta2, Cyp2c40, Gstm2, Mgst2, Cyp2d10, Ugt1a9, Mgst3
Glutathione metabolism	3.59E−02	Ggt1, Gsta2, Gstm2, Mgst2, *Pgd*, Gclm, Mgst3
Chagas disease	6.39E−02	*C1qb*, Cd3g, Car, Jun, Cd3d, Tgfb3, Il1b, Pik3r1, Tlr2, Tnf
1.4 Muscle		
Olfactory transduction	6.96E−09	*Olfr571*, **Olfr957**, Olfr380, Clca2, Olfr584, Olfr1234, Olfr535, *Clca1*
Asthma	4.3E−02	H2-Eb1, **H2-Aa**, H2-DMa, H2-DMb1, H2-Ab1, Tnf
Cell adhesion molecules (CAMs)	4.3E−02	Pvrl2, Jam2, H2-Eb1, Cd86, Cd22, Cntnap2, **H2-Aa**, H2-Q8, H2-DMa, H2-DMb1, H2-Ab1, Cdh2, Itga8, Ptprc, Itgal
Complement and coagulation cascades	4.3E−02	Vwf, F11, Masp1, C1qa, Thbd, Cd59a, Serpine1, **C8b**, F5
Focal adhesion	4.3E−02	*Parva*, Vwf, Chad, Rac2, Rap1a, Pdgfrb, Col5a3, Vegfb, Pik3r1, Vtn, Hgf, Ccnd2, Fyn, Myl2, Itga8, Pak7, Kdr, Flnb

(continued)

Table 1 (continued)

Pathways	p values	Genes
Glutathione metabolism	4.3E−02	Gstm1, Gclm, **Pgd**, Gstm2, Gpx7, Ggt1, Gpx4, Gsta2
Intestinal immune network for IgA production	4.3E−02	H2-Eb1, Cd86, **H2-Aa**, H2-DMa, Pigr, H2-DMb1, H2-Ab1, Tnfsf13b
Renal cell carcinoma	4.3E−02	Slc2a1, Rap1a, ***Tceb1***, Vegfb, Pik3r1, Hgf, Pak7, Ets1, ***Cul2***
Type I diabetes mellitus	4.3E−02	H2-Eb1, Cd86, **H2-Aa**, H2-Q8, H2-DMa, H2-DMb1, H2-Ab1, Tnf, Ins1, Hspd1
Viral myocarditis	5.8E−02	Rac2, H2-Eb1, Cd86, **H2-Aa**, H2-Q8, Myh7, H2-DMa, H2-DMb1, H2-Ab1, Fyn, Itgal
Allograft rejection	6.3E−02	H2-Eb1, Cd86, **H2-Aa**, H2-Q8, H2-DMa, H2-DMb1, H2-Ab1, Tnf
Graft-versus-host disease	6.3E−02	H2-Eb1, Cd86, **H2-Aa**, H2-Q8, H2-DMa, H2-DMb1, H2-Ab1, Tnf

3.1.3 Mutual Information

This non-linear based approach has given us 28, 21, 8 and 99 non-dominated genes in which 6, 5, 2, and 33 genes respectively are significantly qtDE.

In brain we did not get any significant KEGG pathway from the obtained qtDE set.

3.1.4 DEGseq

Application of DEGseq has given us 9, 21, 7 and 3 non-dominated DE genes in adipose, brain, liver and muscle.

Significant non-dominance analysis yields us 3, 9, and 3 genes in adipose, brain and liver respectively. Unfortunately in muscle out of the 3 genes none happens to be significant, thus in muscle we have not gone for any pathway analysis.

Apart from the listed pathways (in Tables 1, 2, 3, and 4) we have also got some other significant pathways having null contribution of non-dominated genes.

Using qtDE linear correlative measure in brain we have got *Leishmaniasis* (3.07E−02); in liver the null contributed pathways are *Leishmaniasis* (2.83E−02), *Focal Adhesion* (3.59E−02), *Hematopoietic cell lineage* (3.59E−02), and *Asthma* (7.27E−02). We have got *Complement and coagulation cascades* (2.73E−02), *Asthma* (4.17E−02), *Hematopoietic cell lineage* (4.17E−02), and *Intestinal immune network for IgA production* (8.9E−02) showing null contribution of non-dominated genes in adipose by non-linear polynomial regression based method, whereas using the same method in brain retrieved *Fc gamma R-mediated phagocytosis* (9.11E −02); in liver we have got *Cytokine-cytokine receptor interaction* (3.07E−03), *Complement and coagulation cascades* (3.06E−02), *Maturity onset diabetes of the*

Table 2 Significant pathways by polynomial regression

Pathways	p values	Genes
2.1 Adipose		
Olfactory transduction	7.23E−11	Olfr584, Olfr380, Olfr1234, Clca1, **Olfr957**
Cytokine-cytokine receptor interaction	2.88E−03	*Ifng*, Bmp2, Tnfsf13b, Hgf, Ccl5, Ccl2, Ccr5, Ccl7, Il7r, Csf3r, Cxcl14, Cxcl5, Tnfrsf18, Tgfb3, Tnf, Kitl, Il10rb, Tnfsf8, Inhbb, Il22ra2, Tnfrsf12a, Csf2rb2, Ltbr, Kdr
TGF-beta signaling pathway	2.73E−02	*Ifng*, Bmp2, Rbl1, Bmp5, Dcn, Tgfb3, Rps6kb2, Tnf, Inhbb, Bmp8b, Ltbp1
Malaria	4.17E−02	*Ifng*, Hgf, Ccl2, Lrp1, Itgal, Tgfb3, Tnf, Tlr2
Leishmaniasis	7.55E−02	*Ifng*, Tgfb3, H2-DMa, Tnf, H2-DMb1, H2-Aa, H2-Eb1, Tlr2
Allograft rejection	7.9E−02	*Ifng*, H2-T10, H2-DMa, Tnf, H2-DMb1, Cd86, H2-Aa, H2-Eb1
Graft versus host disease	7.9E−02	*Ifng*, H2-T10, H2-DMa, Tnf, H2-DMb1, Cd86, H2-Aa, H2-Eb1
Amino sugar and nucleotide sugar metabolism	9.19E−02	Hexa, **Gnpnat1**, Gne, Gck, Galt, Hexb
2.2 Brain		
Olfactory transduction	3.21E−07	Olfr1226, **Olfr1234**, Olfr894, Olfr380
Steroid biosynthesis	5.73E−02	*Sc4mol*, Dhcr7, Sqle, *Nsdhl*
Glycerolipid metabolism	7.31E−02	Akr1b8, Ppap2b, **Gpam**, Lpl, Agpat2, Gyk
2.3 Liver		
Olfactory transduction	6.46E−15	Olfr1234, Olfr599, Olfr584, Clca1, **Olfr535**, Olfr380, Olfr957, Olfr894
Leishmaniasis	3.07E−03	Tlr2, Ncf2, H2-Ab1, Il1b, H2-Eb1, Jun, Ifng, H2-Aa, Tgfb3, *Prkcb*, **Tnf**, Jak2, H2-DMa
Focal adhesion	3.06E−02	Vtn, Rac2, Jun, Tnc, Col1a2, Pdgfc, Chad, Fyn, Rap1a, Pik3r1, Kdr, Pdgfrb, Parva, Col5a3, Vwf, Itga8, Lamb3, Prkca, Mylpf, Flnb, Actn2, *Prkcb*, Hgf
Amoebiasis	4.3E−02	Tlr2, Il1b, Serpinb6a, Ifng, Col1a2, Casp3, Pik3r1, C8b, Col5a3, Tgfb3, Lamb3, Prkca, Actn2, *Prkcb*, **Tnf**
Hematopoietic cell lineage	4.3E−02	Il1b, H2-Eb1, Cd2, Cd22, Il7r, Csf3r, Cd59a, Mme, Kitl, Cd3d, Kit, **Tnf**
Type I diabetes mellitus	4.3E−02	Cd86, H2-Ab1, Ins1, Il1b, H2-Eb1, Ifng, H2-Q8, H2-Aa, **Tnf**, H2-DMa, H2-T10
Graft-versus-host disease	5.61E−02	Cd86, H2-Ab1, Il1b, H2-Eb1, Ifng, H2-Q8, H2-Aa, **Tnf**, H2-DMa, H2-T10
Malaria	5.61E−02	Tlr2, Il1b, Ifng, Itgal, Lrp1, Tgfb3, **Tnf**, Hgf, Ccl2
TGF-beta signaling pathway	5.61E−02	Bmp8b, Rps6kb2, Rbl1, Ifng, Inhbb, Bmp2, Bmp5, Dcn, Ltbp1, Tgfb3, **Tnf**, Gdf5
Galactose metabolism	9.14E−02	Pfkp, Gaa, Galt, **Akr1b8**, Gck

(continued)

Table 2 (continued)

Pathways	p values	Genes
2.4 Muscle		
Olfactory transduction	1.76E−04	Olfr957, Olfr894, Olfr535, Olfr571, *Olfr599*, Clca1, Clca2, Olfr893, Olfr584
Cardiac muscle contraction	3.86E−02	**Cox7b**, Myh7, **Tpm3**, Cacnb1, Myl3, *Slc8a1*, Cox6a2, Actc1, Cacna2d1, Tpm1, Cox7a1, Cox7a2
Cytokine-cytokine receptor interaction	3.86E−02	Kdr, Tnf, Cxcl14, Il10ra, Bmp2, *Vegfb*, Csf1r, Il7r, Flt4, Inhbb, Tnfrsf12a, Tnfsf13b, Gdf5, Ifng, Cxcr3, Egfr, Tnfrsf21, Kitl, Tnfrsf18, Ccl6, Ccr5, Il1b, Ccl4, Cxcl1
Hematopoietic cell lineage	5.87E−02	*Cd22*, Tnf, H2-Eb1, Csf1r, Il7r, Cd14, Mme, Kitl, Cd2, Il1b, Cd34
Hypertrophic cardiomyopathy (HCM)	5.87E−02	Itga8, Myh7, Tnf, **Tpm3**, Cacnb1, Myl3, *Slc8a1*, Actc1, Cacna2d1, Tpm1, Actb
Dilated cardiomyopathy	9.92E−02	Itga8, Myh7, Tnf, **Tpm3**, Cacnb1, Myl3, *Slc8a1*, Actc1, Cacna2d1, Tpm1, Actb

Table 3 Significant pathways by mutual information

Pathways	p values	Genes
3.1 Adipose		
Selenoamino acid metabolism	3.76E−02	Ahcy, *Mat1a*
Cysteine and methionine metabolism	4.46E−02	Ahcy, *Mat1a*
Intestinal immune network for IgA production	4.46E−02	Pigr, *H2-Aa*
Ribosome	8.14E−02	Rpl3 1, **Rps3a**
3.2 Liver		
Ubiquitin mediated proteolysis	2.37E−03	Tceb1, Ube2m, *Aco2*
3.3 Muscle		
Olfactory transductions	1.838E −06	Olfr837, Olfr918, *Olfr1450*, *Olfr307*

young (4.3E−02), and *Cell adhesion molecules* (CAMs) (5.61E−02). Using non-linear mutual information based method *Axon guidance* (4.46E−02), and *Cytosolic DNA-sensing pathway* (4.46E−02), present in adipose possess null contribution.

Table 1, 2, 3, and 4 show that the pathways have one or at most two genes from the non-dominated set. From the previous discussions we can conclude on better significance of these non-dominated genes in all four ranking algorithms, which in turn suggest the enriched regulatory action on the dominated set of genes.

These tables also show that there is a very rare probability of grouping more than one significant non-dominated gene in a single pathway. Exceptions are *Olfactory Transductions*, *Renal cell carcinoma* (both present in liver via linear correlation),

Table 4 Significant pathways by DEGseq

Pathways	p values	Genes
4.1 Adipose		
Amoebiasis	2.75E−02	Serpinb6a, Prkcb, **Serpinb9e**, Tlr4, Casp3, Rab5a, Tfgb3, Adcy1, Il1b, Prkca
Leishmaniasis	3.59E−02	Mapk1, Tlr4, H2-DMa, Tgfb3, Il1b, **Prkcb**, Jak2
ErbB signaling pathway	3.68E−02	Mapk1, Nck1, Crk, **Prkcb**, Rps6kb1, Areg, Prkca, Pak7
Fc gamma R-mediated phagocytosis	5.34E−02	Mapk1, Arpc3, Myo10, Crk, **Prkcb**, Rps6kb1, Prkca, Pla2g6
Amino sugar and nucleotide sugar metabolism	7.2E−02	**Hk1**, Ugdh, Gpi1, Ugp2, Galt
Renal cell carcinoma	9.12E−02	Mapk1, **Rap1a**, Tgfb3, Crk, Tceb1, Pak7
4.2 Brain		
Tight junction	3.57E−03	Myh7, *Jam2*, Myh2, Mylpf, Jam3, Prkcd, Ash1 l, Csnk2a1
4.3 Liver		
Galactose metabolism	3.07E−03	Hk2, Gaa, Galt, **Pfkp**, Akr1b8

Steroid Biosynthesis (present in brain using polynomial regression), *Olfactory Transduction* (present in muscle using non-linear mutual information).

However in these cases apart from a single gene the other genes present in the significant non-dominated set possess high p value very close to the cut-off level 0.2. From table 1.4. we can see that for *Olfactory Transduction* the significantly non-dominated genes are Olfr571, and Clca1. They are having p values 0.07, and 0.188; for *Renal Cell Carcinoma* the significant non-dominant genes are Tceb1 and Cul2 with a p value of 0.084 and 0.147. This shows us an appreciable difference between the p values of Olfr571 and Clca1 in one case and between Tceb1 and Cul2 in the other. In both the cases based on our threshold selection the second gene is on the verge of elimination from the significantly non-dominated set. We observe a similar pattern for *Steroid Biosynthesis* and in *Olfactory Transduction* (obtained via mutual information in qtDE approach). The participating genes from the Pareto front are Sc4 mol, and Nsdhlin the previous one and Olfr1450 and Olfr307 in the latter. Analysis shows that they are having p values (0.185, 0.084) and (0.108, 0.179) respectively.

4   Discussion

In this section we discuss about the involvement of some significant pathways towards differential evolution of mice. One pathway which is common in almost all tables is *Olfactory Transduction*. As discussed in [27] this pathway has a significant level of association with the development of obesity in both adipose and muscle tissues. This pathway is also having a role in connection with functioning of olfactory sensory neurons (OSN) in the septal tissue [28] and in [27] as functioning of rodent olfactory epithelium on live.

Some other significant pathways are *Leishmaniasis* (present in adipose and liver via correlative qtDE measure as well as via polynomial regression qtDE measure; in adipose via DEGseq), *Cell adhesion molecules* (CAMs) (present in adipose and muscle through correlative qtDE). Effects of these pathways in different developmental stages are discussed in [29, 30]. Ref. [30] gives us a detailed description on the involvement of CAMs in the development of embryonic stem cell (ESC) markers. Effect of *Graft versus Host disease* (shown significant in adipose and muscle by correlation qtDE method; in adipose and liver using polynomial regression qtDE) and *Alograft rejection* (present in muscle via correlative qtDE, in adipose via polynomial regression qtDE) pathways in early embryonic development and fatal growth across different sex is given in [31]. Tables 1, 2, 3, and 4 give us such significant pathways functional across various organs. *Hematopoietic cell lineage* (present in liver and muscle via polynomial regression qtDE), *TGF-beta signalling pathway* (find in adipose and liver using polynomial regression qtDE), *Complement and coagulation cascade* (present in muscle via correlative qtDE), *Cytokine cytokine receptor interaction* (present in adipose and muscle via polynomial regression qtDE), *Ubiquitin mediated proteolysis* (found in liver via mutual information qtDE) and *Fc gamma R-mediated phagocytesis* (present in adipose via DEGseq) are to name a few. Involvement of these pathways over differential development like gastrulation, axis symmetry of the body, organ morphogenesis, liver development and tissue homeostasis in adults of mice/other mammals across different sex are discussed in [32–37].

5   Conclusion and Future Work

Differential association of a gene in a complex network is responsible for the evolution of different species/sex, and now days in general spreading of different complex diseases. Different methodologies are available to assess the gene significance working across different conditions. Although all methods work in near about same direction but the ranking of the genes (via gene significance) hence obtained are different. In this paper we have explored in search of such optimal genes non-dominant to one another. As a next step using permutation test we have evaluated the significance of non-dominated genes.

In this work we have searched for the most optimal genes from the existing ones on the basis of gene ranking. Though it is feasible to generate a new set of differential gene expression levels using various evolutionary computing techniques which may contribute to the true non-dominated set in the long run, we have not generated any such predicted gene expression level. The reason behind this being the final Pareto set, hence obtained, might not contain any physical gene name, making it difficult to interpret the biological independence of such genes. In order to support our statement we had to search for physical genes which correlate with the predicted genes. So here we have two possibilities in hand. Either the physically correlated gene happens to be a DE gene or a non-DE gene. But our problem being guided through the selection of DE genes using the expression level only, there is virtually no possibility of the corresponding physical gene to be in the non-DE set. It is always a part of the DE gene set. As the new genes are checked for differential expression thus the expression levels of the qualifying genes should correlate with differentially expressed genes only. Hence, we can perform our analysis without the generation of new set of gene expression levels.

While validating we have shown in the pathways formed by DE genes, the number of participating genes from the non-dominated set is at most two. This suggests the dominance over the other participating genes with very low chance of having more than one significant non-dominant gene in any such pathway. Also the pathways thus formed have an active role in the development of organs across different sex.

In future we would like to extend this work on Transcription Factor (TF) based network. Here we can rank the TFs based on TF to DE gene interaction and from the enrichment analysis assess the significant TFs working in tandem to differential regulatory functions. This will help us to promote a general view on the regulatory effect of one or more TFs on different sets of DE genes.

References

1. Allison, D.B., Cui, X., Page, G.P., Sabripour, M.: Microarray data analysis: from disarray to consolidation and consensus. Nat. Rev. Genet. **7**, 55–65 (2006)
2. Lai, Y., Wu, B., Chen, L., Zhao, H.: A statistical method for identifying differential gene-gene co-expression patterns. Bioinformatics **20**(17), 3146–3155 (2004)
3. Bockmayr, M., Klauschen, F., Györffy, B., Denkert, C., Budczies, J.: New network topology approaches reveal differential correlation patterns in breast cancer. BMC Syst. Biol. **7**, 78 (2013)
4. Handl, J., Knowles, J.: On semi-supervised clustering via multiobjective optimization. In: Proceedings of the 8th Annual Conference on Genetic and Evolutionary Computation, GECCO'06, pp. 1465–1472. ACM, New York (2006)
5. Mitra, P., Murthy, C.A., Pal, S.K.: Unsupervised feature selection using feature similarity. IEEE Trans. Pattern Anal. Mach. Intell. **24**(3), 301–312 (2002)
6. Maulik, U., Mukhopadhyay, A., Bandyopadhyay, S.: Combining pareto-optimal clusters using supervised learning for identifying co-expressed genes. BMC Bioinformatics **10**, 27 (2009)
7. Bandyopadhyay, S., Mukhopadhyay, A., Maulik, U.: An improved algorithm for clustering gene expression data. Bioinformatics **3**(21), 2859–2865 (2007)
8. Mukhopadhyay, A., Maulik, U., Bandyopadhyay, S.: A novel biclustering approach to association rule mining for predicting HIV-1–human protein interactions. PLoS ONE **7**(4), e32289 (2012)
9. Konak, A., Coit, D.W., Smith, A.E.: Multi-objective optimization using genetic algorithms: a tutorial. Reliab. Eng. Syst. Saf. **91**, 992–1007 (2006)

10. Saha, S., Ekbal, A., Alok, A.K., Spandana, R.: Feature selection and semi-supervised clustering using multiobjective optimization. Springer Plus **3**, 465 (2014)
11. Goldberg, D.E.: Genetic Algorithms in Search, Optimization, and Machine Learning. Addison-Wesley, Boston (1989)
12. Khanesar, A.M., Teshnehlab, M., Shoorehdeli, M.A.: A novel binary particle swarm optimization. In: Proceedings of the 15th Mediterranean Conference on Control and Automation, Athens-Greece (2007)
13. Specht, D.F.: A general regression neural network. IEEE Trans. Neural Netw. **2**(6), 568–576 (1991)
14. Dorigo, M., Blum, C.: Ant colony optimization theory: a survey. Theoret. Comput. Sci. **344**, 243–278 (2005)
15. Wilcoxon, F.: Individual comparisons by ranking methods. Biometrics Bull. **1**(6), 80–83 (1945)
16. Subramanian, A., et al.: Gene set enrichment analysis: a knowledge-based approach for interpreting genome-wide expression profiles. Proc. Natl. Acad. Sci. USA **102**(43), 15545–15550 (2005)
17. Wu, C., Zhu, C., Zhang, X.: Network-based differential gene expression analysis suggests cell cycle related genes regulated by E2F1 underlie the molecular difference between smoker and non-smoker lung adenocarcinoma. BMC Bioinformatics **14**, 365 (2013)
18. Sarkar, M., Majumder, A.: TOP: an algorithm in search of biologically enriched differentially connective gene networks. In: 5th Annual International Conference on Advances in Biotechnology, Kanpur, India (2015)
19. Wang, L., Fenq, Z., Wang, X., Wang, X., Zhang, X.: DEGseq: an R package for identifying differentially expressed genes from RNAseq date. Bioinformatics **26**(1), 136–144 (2010)
20. Sarkar, M., Majumder, A.: Quantitative trait specific differential expression (qtDE). Procedia Comput. Sci. **46**, 706–718 (2015)
21. Yip, A.M., Horvath, S.: Gene network interconnectedness and the generalized topological overlap measure. BMC Bioinformatics **8**, 22 (2007)
22. Revarter, A., Chan, E.K.: Combining partial correlation and an information theory approach to the reversed engineering of gene co expression networks. Bioinformatics **24**(21), 2491–2497 (2008)
23. Majumder, A., Sarkar, M.: Exploring different stages of Alzheimer's disease through topological analysis of differentially expressed genetic networks. Int. J. Comput. Theory Eng. **6**(5), 386–391 (2014)
24. Wolfram Mathworld. http://www.mathworld.wolfram.com
25. Ghazalpour, A., et al.: Integrating genetic and network analysis to characterize genes related to mouse weight. PLoS Genet. **2**(8), e130 (2006)
26. http://www.genetics.ucla.edu/labs/horvath/CoexpressionNetwork/MouseWeight/
27. Choi, Y., Hur, C.-G., Park, T.: Induction of olfaction and cancer-related genes in mice fed a high-fat diet as assessed through the mode of action by network identification analysis. PLoS ONE **8**(3), e56610 (2013)
28. Oshimoto, A., et al.: Potential role of transient receptor potential channel M5 in sensing putative pheromones in mouse olfactory sensory neurons. PLoS ONE **8**(4), e61990 (2013)
29. Cruz, A., Nieto, J., Moreno, J., Canavate, C., Desjeux, P., Alvar, J.: HIV co-infections in the second decade. Indian J. Med. Res. **123**, 357–388 (2006)
30. Zhao, W., Ji, X., Zhang, F., Li, L., Ma, L.: Embryonic stem cell markers. Molecules **17**, 6196–6236 (2012)
31. Baker, D.G.: Natural pathogens of laboratory mice, rats, and rabbits and their effects on research. Clin. Microbiol. Rev. **11**, 231–266 (1998)
32. Jaffredo, T., Yvernogeau, L.: How the avian model has pioneered the field of hematopoietic development. Exp. Hematol. **42**(8), 661–668 (2014)
33. Bandyopadhyay, A., Tsuji, K., Cox, K., Harfe, B.D., Rosen, V., Tabin, C.J.: Genetic analysis of the roles of BMP2, BMP4, and BMP7 in limb patterning and skeletogenesis. PLoS Genet. **2**(12), e216 (2006)

34. Wynn, J.L., Wong, H.R.: Pathophysiology and treatment of septic shock in neonates. Clin. Perinatol. Natl. Inst. Health **37**(2), 439–479 (2010)
35. Patil, A., Kumaga, Y., Liang, K.-C., Suzuki, Y., Nakai, K.: Linking transcriptional changes over time in stimulated dendritic cells to identify gene networks activated during the innate immune response. PLoS Comput. Biol. **9**(11), e1003323 (2013)
36. Hamazaki, J., Sasaki, K., Kawahara, H., Hisanaga, S.-I., Tanaka, K., Murata, S.: Rpn10-mediated degradation of ubiquitinated proteins is essential for mouse development. Mol. Cell. Biol. **27**(19), 6629–6638 (2007)
37. Mencheet, J., et al.: A diVIsive shuffling approach (VIStA) for gene expression analysis to identify subtypes in chronic obstructive pulmonary disease. BMC Syst. Biol. **8**(Suppl 2), S8 (2014)

A Rule Based Approach for Connective in Malayalam Language

S. Kumari Sheeja, S. Lakshmi and Lalitha Devi Sobha

Abstract Discourse connectives signal the relationship between two coherent spans of text. Arguments of connective are the text spans in discourse. Discourse relations link clauses in text and compose overall text structure. Discourse connectives play an important role for modeling the Malayalam discourse and its structure. We present our work on rule based approach for identifying the discourse connective in Malayalam language. Discourse connectives may or may not be explicitly present in the relation. In our work, we have focused on the rule based approach for the identification of particular connective in Malayalam text and showed encouraging results.

Keywords Discourse connectives · Rule based approach · Malayalam discourse · Arguments

1 Introduction

Discourse connectives connect sentences and clauses in the discourse and creates the overall structure of the text. In Natural Language Processing, Discourse analysis is concerned with analyzing how sentence or clause level units of discourse are related to each other within a larger unit of discourse. Discourse markers and their arguments are the two basic units of discourse relations. The discourse markers in

S. Kumari Sheeja · S. Lakshmi · L.D. Sobha (✉)
AU-KBC Research Centre, Anna University, Chrompet, Chennai, India
e-mail: sobha@au-kbc.org

S. Kumari Sheeja
e-mail: sheeja@kcgcollege.com

S. Lakshmi
e-mail: slakshmi@au-kbc.org

S. Kumari Sheeja
KCG College of Technology, Karapakkam, Chennai, India

© Springer Science+Business Media Singapore 2017
P. Deiva Sundari et al. (eds.), *Proceedings of 2nd International Conference on Intelligent Computing and Applications*, Advances in Intelligent Systems and Computing 467, DOI 10.1007/978-981-10-1645-5_8

93

text are the phrases or words which connect two clauses or sentences and establish a relation between two discourse units in NLP.

Kamala went to hospital but doctor was not there.

In the above given example, the conjunction "but" makes a connection between two clauses or sentences and makes coherent text. Discourse connectives are an important part of NLP applications and it is essential for discourse analysis. Identification of discourse relation in natural language processing is the most challenging task in NLP. Discourse connectives are acting as a conjunction along with their common function of connecting the contents of two different clauses or sentences [1]. So it is a difficult process to differentiate discourse and non-discourse markers in text. The identification of connectives and argument boundaries in text is very difficult process in large text. Malayalam is a South Indian or Dravidian language and also the language with free word order but maintains the verb in final position. Discourse connectives are important for producing or interpreting text in malayalam language. The various sections of this paper is arranged as follows. The literature study related to this paper is described in Sect. 2. Discourse connectives and its types are explained in Sect. 3. The fourth section describes the rule based approach of connectives in Malayalam Language and also the paper ends with the conclusion of the rule based approach for connective in Malayalam Language.

2   Related Work

The literature study of discourse connectives their arguments have been used in natural languages such as Turkish [2], Arabic [3], English [4], etc. Penn Discourse Tree Bank follows the lexically grounded approach and it is very special in adopting a theory-neutral approach for annotation of discourse connectives. PDTB gives argument structure and sense labels of discourse relations in text which follows hierarchical classification scheme. Elwell et al. [5] worked using maximum entropy rankers and identified the arguments of discourse connectives. Tagging of German discourse connectives and arguments using English training data and a German_ English parallel corpus have done by Versley [1]. Versely also worked for transferring a tagger for English discourse connectives. He worked with annotation of connectives and arguments by annotation projection with a freely accessible list of connectives. The F-score for the identification of discourse connectives of Versley's work is 68.7 %. Identification of arguments of explicit discourse connectives have done by Ghosh using data driven approach in the PDTB corpus [6]. In Arabic language, Al Saif used machine learning algorithms for automatically identifying explicit discourse connectives and its arguments [7]. Wang et al.'s paper stated that significant improvement in identifying arguments, explicit and implicit discourse relations used sub-trees as features for connectives and their arguments. Annotation of discourse connectives and their arguments in Indian languages are available for Tamil, Hindi, and Malayalam by Sobha et al. [8]. They have also worked on automatic identification of Discourse connectives for the three Indian

Languages [9] using CRFs technique. Other published works in Indian languages are in Hindi [4, 10] and Tamil [11]. In this paper, we have explored various Discourse connectives and rule based approach for particular connective in Malayalam language.

3 Discourse Connectives in Malayalam

Malayalam is a free-word order language and words are seen agglutinated, hence most of the connectives are seen in agglutinated form. The discourse relation in Malayalam language can be syntactic (a suffix) or lexical [9]. It can be within a clause, inter-clausal or inter-sentential. Discourse connectives are an important part of modeling discourse structure. In this paper, we now describe various connectives present in Malayalam language and a rule based approach to figure out the connective "pakshe" (But).

3.1 Discourse Relation Categorization

The discourse markers can be realized in any of the following ways. There are two major category of relations. They are Explicit and Implicit relations. In this paper, we observed the different types of relations in discourse.

3.2 Explicit Relations

The explicit relations in text are morphemes or free words that trigger discourse connectives in Malayalam language. Explicit connectives signal the presence of discourse connectives between sentences or clauses [12]. The relations in text can occur at the final, medial and initial position of arguments in Malayalam language [2]. Below are the examples for explicit connectives in malayalam language.

[prameham   oru   nishabdha   kolayaaLiyaaN.]/arg1

diabetes   one   silent   killer

ennaal [niyanthrichu   nirthiyaal   kuzhappamilla]/

but   control   kept   if no problem

(Diabetes is a silent killer. But it is not a problem when kept in control.)

In the above example, "ennaal" is the connective which occurs inter sententially by connecting the two sentences in Malayalam Language. Connective occurs at the initial position in the second argument. We see that the connectives are explicitly realizing relations between two arguments. We have observed four types of explicit connectives in discourse.

3.3   Explicit Connective Types

Subordinate Conjunctions. This type of conjunction connects the main clause with the adverbial clause, noun or an adjectival clause. Most commonly observed subordinate conjunctions in natural languages are since, because and when. Consider the following examples which give the distribution of subordinate conjunctions in malayalam language.

> [pachakkarikaL vevichu
>
> Vegetables   boil
>
> kazhikk**umpoL**]/arg1
>
> when eat
>
> [athiluLLa poshakam nashtamaakum]/arg2
>
> In that   nutrients   loss
>
> (When vegetables are boiled and consumed, the nutrients in it are lost)

In the above example " ümpoL" connects the above Malayalam text which act as connective.

Co-ordinate Conjunctions. This conjunction gives equal emphasis for two clauses or phrases in text. They connect two clauses, words and phrases. The most commonly used co-ordinate conjunction in the corpus are "and" and "but". The conjunction "but" in Malayalam language is "pakshe" which is a co-ordinate conjunction. The intra sentential coordinating conjunction can occur between the clauses in discourse.

Conjunct Adverbs. Conjunct Adverb is another type of connective to modify the clauses or sentences in text which they occur. This type of conjunction joins independent clauses of Malayalam discourse together. These are special type of conjunctions as they are part of adverbs and conjunction. Example of Conjunct adverb is given below.

> [geetha   nannAyi   pATum.]/arg1   **athinAl**
>
> Geetha   well   sing   therefore
>
> [skooL   yuvajanOlsavaththil   onnAmatheththi.]/arg2
>
> School   Youthfestival   first prize
>
> (Geetha sings well. **Therefore** she secured first prize in Youth festival.)

In the above example "athinAl" is the adverbial conjunction which actually shows a cause and effect relationship where arg1 is effect and arg2 is the cause.

Correlative conjunction. This is another type of conjunction which occurs in simple pair. This type of conjunction is used in a sentence to join different words or group of words in text. This conjunction is not used to connect the sentences themselves. But they link two or more clauses or clauses of same importance within a sentence itself. Important feature of correlative conjunction is that it occurs within a sentence.

[indyayennaal innu   sachin

india means   today sachin

maathramalla,]/arg1 **[pakshe** innum

not only   but also today

Sachinillaathe   indyaye

sachin without   india

sankalppikkaan prayaasam.]/arg2

think   cannot

(Today India means not only Sachin, but also   cannot think of an India without Sachin.)

Here "maathramalla-pakshe" is the correlative connective. But the "pakshe" is even said to be dropped in certain cases.

Complementizer clause. This clause is also considered as a connective and this type of conjunction marks a complement clause in Malayalam Language.

[avare vila kalppikkunnilla]/arg1 **ennu** [nethaakkal

they   value   not given   that   leaders

abhinayichu]/arg2

pretend

(The leaders pretended that they were not given a value.)

3.4 Implicit Connectives

Implicit relation is the second major category of connectives. An implicit relation can be identified if there exist a relationship between the adjacent pair of sentences. An explicit connective is not present in the text. We used the label "IMPLICIT" when identified an implicit relation in Malayalam discourse [2].

(7) [pilkaalath niravadhi svadeshikal bekkarute

    later   many   people   bekkar's

    paatha pinthutarnnu.]/arg1 IMPLICIT [mattu

    way   followed   some

    chilaraakatte kaayalil   svadesheeyamaaya

    People   backwater   traditional

    Reethiyil kayal   nikathi krishi bhoomi

    style   backwater filled   farm land

    uNdaakkiyetuthu.]/arg2

    made

(Later many people followed bekkar's path. Some people in their traditional style filled up back waters and made their farm land.)

In the above example, the two given sentences are not explicitly connected but a relationship between the sentences can be inferred implicitly in Malayalam Language.

4 Rule Based Approach

Malayalam is a language of the Dravidian family and words are seen agglutinated. In this work, we have collected Malayalam sentences from websites and the document consists of 3000 sentences. The sentences were tagged for the connectives and their arguments (Fig. 1).

4.1 Our Approach

Explicit connectives are identified and labelled as arg1, arg2 and CONN.

Fig. 1 Tags used for annotation

Malayalam Corpus	Start Tag	End Tag
Arguments	<ARG1>	</ARG1>
	<ARG2>	</ARG2>
Explicit Connectives	<CONN>	</CONN>

We have analysed the rule based approach for most frequently occuring connective "pakshe" (but) in malayalam. "Pakshe" is the co-ordinating connective which connect words, phrases or clauses which are of equal importance.

In our approach we have done a syntactic based tagging. Discourse connectives can occur within a Malayalam sentence or between the sentences. In Malayalam, inter sentence connectives are said to occupy sentence initial position.

The connective "pakshe" is seen occuring as a inter sentence connective and always in the sentence initial position. The patterns and rule of the connective "pakshe" in Malayalam language is described below.

Rule 1. Neither arg1 or arg2 describes a situation that is asserted on the basis of other one to indicate contrast.

[Raamu skoolil poyi]/arg1 pakshe

Ramu   school   went   but

[Teecchar klaasil vannilla]/arg2

Teacher class   didn't come

(Ramu went to school but teacher didn't come to class)

Rule 2. When one argument specify a fact which creates an expectation and the other argument is denying it.

[raamu eppozhum ente veetil vararuNtu.]/arg1

ramu   always   my house come

pakshe [ithuvare ammaye koNtu vannittilla.]/arg2

but   till now mother   bring   not

(Ramu always comes to my house but till now he   has not brought his mother)

Rule 3. Both the arguments indicate alternate values and connect the arguments using "Pakshe".

[ammu pareekshayil 80 mArk vAngngi]/arg1 pakshe

Ammu examination   80 mArk scored   but

[anu pareekshayil 90 mark vAngngi]/arg2

Anu examination 90 mark scored

(Ammu scored 80 marks in the examination but Anu scored 90 mark in the examination)

Rule 4. When connective indicates a contrast between one of the arguments and the inference can be drawn from the other argument.

[innu lOkaththil akramam kooTi varukayAN]/arg1.

Today world   violence   more   coming

pakshe [AlkAr ithinekuRichu mumpE bOthavAnmArANu]/arg2

but people   this about   before   conscious

(Today violence are increasing in the world. But people are more conscious about this)

5 Conclusion

We have presented a detailed description of the discourse relation existing in Malayalam language. The syntactic pattern of the discourse relation has been explained. We have presented our work on rule based approach for the identification of connective in Malayalam language. Our future work will be developing automatic identification system for discourse relation and their arguments using machine learning techniques.

References

1. Versley, Y.: Discovery of ambiguous and unambiguous discourse connectives via annotation projection. In: Proceedings of Workshop on Annotation and Exploitation of Parallel Corpora (AEPC), pp. 82–83 (2010)
2. Wang, X., Li, S., Li, J., Li, W.: Implicit discourse relation recognition by selecting typical training examples. In: Proceedings of International Conference on Computational Linguistics, Mumbai, India, pp. 2757—2772 (2012)
3. Al-Saif, A., Markert, K.: The Leeds Arabic discourse treebank: annotating discourse connectives for Arabic. In: Proceedings of Language Resources and Evaluation Conference. Valletta, Malta (2010)
4. Prasad, R., Dinesh, N., Lee, A., Miltsakaki, E., Robaldo, L., Joshi, A.K., Webber, B.L.: The Penn discourse TreeBank 2.0. In: Proceedings of Language Resources and Evaluation Conference. Marrakech, Morocco (2008)
5. Elwell, R., Baldridge, J.: Discourse connective argument identification with connective specific rankers. In: Proceedings of the International Conference on Semantic Computing. Santa Clara, CA (2008)
6. Ghosh, S.: End-to-End discourse parsing with cascaded structured prediction. Doctoral dissertation, University of Trento (2012)
7. AlSaif, A.: Human and Automatic Annotation of Discourse Relations for Arabic. Ph.D. thesis, University of Leeds (2012)
8. Devi, S.L., Lakshmi, S., Gopalan, S.: Discourse tagging for Indian languages. Computational Linguistics and Intelligent Text Processing, pp. 469–480. Springer, Berlin (2014)
9. Devi, S.L., Gopalan, S., Lakshmi, S.: Automatic identification of discourse relations in Indian languages. In: Proceedings of 2nd Workshop on Indian Language Data: Resources and Evaluation, Organized under LREC2014, Reykjavik, Iceland (2014)

10. Kolachina, S., Prasad, R., Sharma, D.M., Joshi, A.K. Evaluation of discourse relation annotation in the hindi discourse relation bank. In: Proceedings of Language Resources and Evaluation Conference, pp. 823–828 (2012)
11. Rachakonda, R.T., Sharma, D.M.: Creating an annotated tamil corpus as a discourse resource. In: Proceedings of the 5th Linguistic Annotation Workshop, pp. 119–123 (2011)
12. Faiz, S.I., Mercer, R.E.: Identifying Explicit Discourse Connectives in Text, pp. 64–76. Springer, Berlin (2013)

Dynamic Scheduling of Machines Towards the Vision of Industry 4.0 Studio—A Case Study

Nagarajan Ayvarnam and P.S. Mayurappriyan

Abstract Today, consistent data exchange between engineering applications such as special purpose machines, Manufacturing Execution Systems (MES) and Enterprise Resource Planning (ERP) systems is indispensable for efficient, error free planning and operation of plant and equipment. The approach towards Industry 4.0 Studio (I4.0)—an integration project, integrates value creation chains horizontally and processes and systems vertically. Hence the customers are indirectly benefitted with standardized and reliable better quality products within specified time at affordable cost. This paper illustrates the development of a MES system through high-level understanding process with the aid of concrete examples of functioning automation and IT delivery teams together to ensure success for the approach towards smart factory or I4.0.

Keywords Dynamic scheduling · Manufacturing execution systems · Enterprise resource planning · Smart factory

1 Introduction

Integration of smart mechanical systems with information technology has pushed the industrial controls and factory automation market to the new heights. These systems are now being used in the industries to build standardized and reliable production activities. The scope of the industrial controls and factory automation is not limited to the production floor, but it is extended to the overall business processes. The software systems like Manufacturing Execution Systems (MES) and

N. Ayvarnam (✉)
Robert Bosch Engineering and Business Solutions Private Limited,
Bangalore, India
e-mail: nagaraj.eee.eng@gmail.com

P.S. Mayurappriyan
KCG College of Technology, Chennai, India
e-mail: mayurpriya@yahoo.com

© Springer Science+Business Media Singapore 2017
P. Deiva Sundari et al. (eds.), *Proceedings of 2nd International Conference on Intelligent Computing and Applications*, Advances in Intelligent Systems and Computing 467, DOI 10.1007/978-981-10-1645-5_9

Enterprise Resource Planning (ERP) are helping the market to extend their service offerings [1]. Even though MES exist for several years, many companies didn't recognize its critical role in lean manufacturing operations, control, and performance. This paper focuses on the role of MES in the area of integrating shop floor management to top management.

2 Need of MES System

Over the past two decades, discrete manufacturing and assembly operations have seen significant advances in the automation of industrial processes and interaction with supply chain activities. Many of these advances were driven through investment in the automation of the production equipment/assembly lines and through the implementation of ERP systems. It soon became apparent that to continue the natural progression of improvement, advances were necessary between the shop floor (equipment) and the high level business system (ERP) [2].

MES includes applications to manage plant schedules, often to a more detailed level than ERP. MES systems often connect directly to machine controls and collect information like status, rates, piece counts, parametric data as well as managing instructions and programs from engineering in coordination with the schedules developed by ERP and supervised within MES [3].

Thus a MES is an information system that drives the execution of manufacturing operations. It does so by providing a set of functions that manage production operations from the point an order is released to manufacturing to the point a product is delivered as a finished good [2]. In 2004, the global MES market crossed the billion-dollar mark, demonstrating the escalation in significance of MES to modern manufacturing operations [4].

3 MES of the Future

New standards and functions are required to cover future production processes, which will especially affect MES applications. The MES of future covers the following topics:

- Interoperability and Flexibility
- Horizontal integration
- Online capability and Integrative data management
- Unified Shop Floor Connectivity

All the above objectives will be covered through various modules of MES as per the customer requirement. The fragmentation of manufacturing systems will increase further. This does not necessarily mean that the factory of the future will be smaller, as the production systems will be controlled in detail. From the point of view of production planning and control, the advantage of smaller units is a greater degree of flexibility and easier local optimization of production [5].

4 Production Control with Up-to-Date Information

To achieve production control that is as realistic as possible, constantly up-to-date information is required regarding the actual situation in production with respect to the existing resources, the current properties of these resources and the present and future order situations.

In order to do this, the physical production system and the automation technology must be deeply integrated with information technology [2].

The order management software provides the platform for integrating the production planning from Top floor (Management) to Shop floor (Machines). The architecture of production planner module is portrayed in Fig. 1.

For ease, the communication between the MES and the machine is through XML Telegrams. The traceability of the part becomes easy with the database maintained in the server.

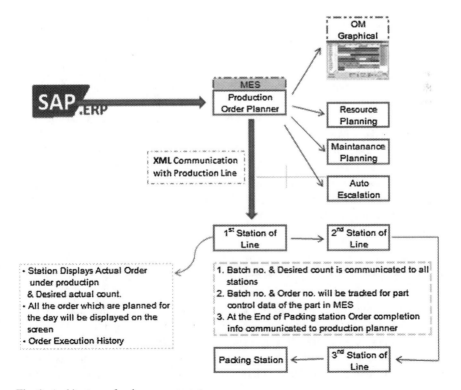

Fig. 1 Architecture of order management

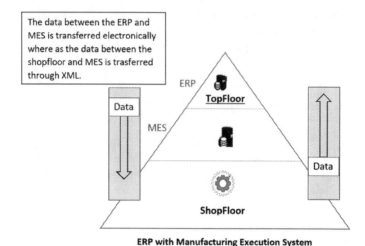

Fig. 2 Pyramid view of Shop floor to Top floor (*Source* Computing and control engineering, volume 17, issue 1, February 2006)

It is not displayed in the above picture that the individual station is communicating with MES for traceability of parts. Thus the real time data of the part is stored in the server.

OM Graphical is the display available in the production line which shows the order received from MES and real time monitoring of data. Sample pictures of display in the running machine are shown in Fig. 5. Resource Planning will assign the required operator for the stations. Now the operator can measure his performance lively in the machine. The productivity will improve with this methodology of working. The maintenance module raise the request for maintenance activity needs to be carried out [6]. The maintenance procedure can be accessed from screens as video or document.

To connect the Shop floor to Top floor in the existing factories, special programming needs to be done in the existing/additional hardware to deploy the above framework for dynamic scheduling of machines to implement the orders, as depicted in Fig. 2. In the real world the customer orders are different which depends on various social, technological and environmental factors.

5   Case Study of Agriculture Pump Manufacturing Unit

TEXMO is a manufacturing unit in Coimbatore, Tamil Nadu which manufactures the pumps required for agricultural sectors. They produce approximately 3000 variants of pumps annually which is a great challenge. In their business, the greatest challenge was improving the forecasting accuracy. Because the business is highly volatile, this depends on factors like monsoon, rainfall, water level and other environmental factors. Forecasting the business with these variables is a huge difficult task. Industrial automation and MES enhance the way to communicate and make decisions.

Earlier followed process was fully manual which uses Kanban cards, Marker boards and usual system entry using Microsoft Excel tool to keep the records of the production machines. Earlier in TEXMO past/shift data is recorded manually and sent to the management. Then the skilled person records the data and analyse it using Microsoft Office Excel tool and decide how much production that is required for the successive years. Since the analyzing methodology is not so powerful, the production or business forecast to meet the demand in time is not so accurate. Also the quality and accuracy involved in analyzing the data and forecasting the results for 3000 variants are highly questionable.

Here a Multinational German company which have its software branch in India pitched in, made a detailed technical study and implemented the MES solution for providing the better business forecast. The key technologies used are as follows:

1. Open Platform Communications (OPC) for collecting the data from running machines [1]
2. MES server for traceability of parts, order management and escalations
3. ANDON display to show the real time data from Shop floor to Top floor

The functionalities are detailed as follows:

5.1   Data Collection

An additional hardware is used in individual machines to collect the real time data from running machines through OPC. This controller handles the connections between MES and machine. The collected information are recorded in MES through XML concept. For example, the following information are recorded in the MES: Type no, Type Variant, Batch no., Quality parameters of the process being carried out like Welding, Vision, measurement etc., and many others like this which depends on the customer requirement. All the above information will be collected and stored in the server through MES Telegrams. MES involves mutual communication between machine and server through XML telegrams.

After the part is processed in specific position, machine sends the result to server in an XML format to have the information database. After receiving it, MES send the acknowledgement telegram to machine. Through this the traceability of parts can be achieved.

5.2 Andon Display

ANDON display is used in individual machines to convey the item needs to be produced for the present shift. The sample ANDON screen for Order to be executed in a machine is shown in Table 1. It has separate Button to show the raw materials required for the production of above quantity. Also, the mail will be triggered to the supervisors involved, if the machine has shortage of raw materials. In special cases, mobile devices have been used to quickly react for escalations. The ANDON screen as shown in Fig. 3, displays the machine setter details if it involves the authorized access.

The screen shot of horizontal integration view in ANDON display is shown in Fig. 3. One of the characteristics of the smart factory is the ability to reconfigure the production system. The boundaries between ERP and MES systems will become increasingly blurred in the context of Industry 4.0 are expected.

Vertical integration involves the integration of sensors, actuators, field devices like Programmable Logic Controllers (PLC), MES, as well as business applications like ERP systems [7]. Seamless integration of information and messages were implemented to connect the Shop floor to Top floor [4].

Then with the special software, the rough business forecast is portrayed by the MES module. The special module is production planning module, where it does the Overall Equipment Effectiveness (OEE) and other necessary calculations.

Table 1 Sample ANDON screen

Order ID	Type no.	Type variant	Batch	Target count	Good part	Bad part
16470	254,854	A002	100	150	0	0

Fig. 3 Horizontal integration view in ANDON display

5.3 OEE Measurement

OEE is the key indicator of how effectively the production machines and time is used for intended purpose. It uses simple mathematic methodology to categorize and evaluate all the losses in production.

The overview of OEE is detailed in Fig. 4.

Total Productivity (TP)

$$TP = OEE * Planned Production factor$$

The effects of accurate OEE measurement are:

- Cost reduction
- Increased output with Less Input (Efficient handling of Inputs)
- Expanding the line investments with advanced technologies

Other benefits include [8]:

- Accurate production planning
- Understanding the production process
- Cooperation and team work

5.4 Forecasting

Production Planner Module (PPM) in MES provides all the relevant data required to make most of the real time production/process decisions.

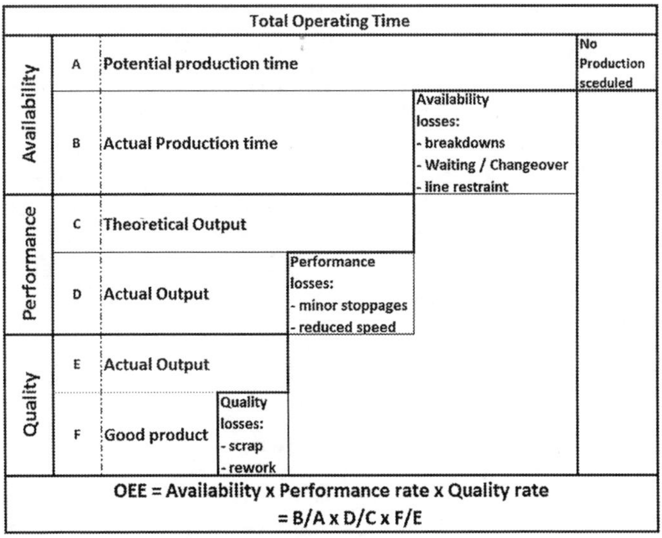

		Total Operating Time		
Availability	A	Potential production time		No Production sceduled
	B	Actual Production time	Availability losses: - breakdowns - Waiting / Changeover - line restraint	
Performance	C	Theoretical Output		
	D	Actual Output	Performance losses: - minor stoppages - reduced speed	
Quality	E	Actual Output		
	F	Good product	Quality losses: - scrap - rework	
		OEE = Availability x Performance rate x Quality rate = B/A x D/C x F/E		

Fig. 4 Overview of OEE (*Source*: "OEE for the production Team" by Arno Koch)

- Intelligent algorithms transforms the raw machine signals into Understandable user interface charts/various screens
- Existing systems can be automatically challenged with variances and alarms activated and forwarded to the operator and/or escalated to management by email, pager or text messages [8]
- Powerful data historians allow for centralisation of all data in an MES database and for detailed analysis of data by shift, batch, product, operator etc. [8]

Flexible processors allow the operators to collect data to suit their own shift patterns. PPM comes in a range of variants with an upgraded path as the customer requirements expand [8]. Thus the PPM module includes OEE calculation and report templates that can be customised to different customer requirements. The productivity of the machine can be improved with the OEE calculation.

On successful implementation of production planner in TEXMO, the productivity has been increased from 62.9 to 75.3 %. MES schedules the machines to execute the orders which need to be carried out in the respective shift. ANDON displays the order needs to be executed with respect to business forecast/MES scheduling. With the conventional market forecast or emergency delivery, the system has the flexibility to edit the orders manually in the machine as shown in Fig. 5.

Hence better business forecast can be done by deploying the above framework and dynamic scheduling can be done with the availability of machines. By this framework, the error from conventional business forecast is avoided and the lean manufacturing is implemented in TEXMO where the delivery time to reach the customer is very less. Now the rainfall/fluctuations in customer demand to meet the delivery time is under control with the implemented solution.

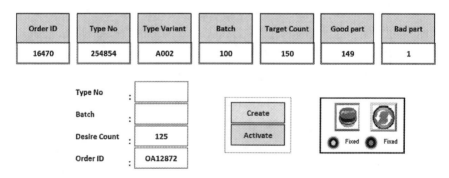

Order ID	Type No	Type Variant	Batch	Target Count	Good part	Bad part
16470	254854	A002	100	150	149	1

Fig. 5 Manual order creation screen in ANDON

6 Conclusion

In this paper, the requirements of future manufacturing systems which have to be adaptable to face turbulent markets are described with a case study. There are many ways to raise the OEE to meet the demands of market in ON-time. Some of these improvements may require substantial investments. With good OEE measurement, it is possible to select the project with quickest returns [8]. With the framework mentioned in this paper smart manufacturing solutions can be achieved on integrating the 3 M's of Man, Machine and Material. Seamless Integration of Information and messages were implemented to connect the shop floor to top floor through MES, dashboards and ANDON displays. From networking point of view, vertical integration between the manufacturing and office level networks can be best achieved by means of gateway, which translates between different protocols and offers unified representation of various field bus systems and resources [9]. MES plays the major role in this methodology.

With the above architecture the vision towards I4.0 is achieved to enable the small factory to SMART factory.

References

1. AIS, webpage. aispro.com
2. Pyramid solutions webpage. mfg.pyramidsolutions.com/importance-of-implementing-a-mes-system
3. Whitepaper by Infor. www.gobiztech.com/docs/infor-importance-of-execution-lean-mes.pdf, p. 2
4. ROCKWELL AUTOMATION: Manufacturing execution systems for sustainability. In: Extending the Scope of MES to Achieve Energy Efficiency and Sustainability Goals, Paper SUST-WP001A-EN-P, May 2009
5. Loskyll, M., Heck, I., Schlick, J., Schwarz, M.: Context-based orchestration for control of resource-efficient manufacturing processes. In: Future Internet (2012)
6. Ramya, L.N.: A case study on effective consumption of energy using sensor based switching system. In: International Conference on Innovations in Information Embedded and Communication Systems, (ICIIECS) (2015)
7. PSIPENTA Software Systems GmbH, a white paper on ERP/MES integration in the age of Industry 4.0
8. Global Pharma Networks, Presentation about OEE on 08 Dec 2013. www.globalnetworksgroup.com/Documents/OEE%20presentation.pps, slide no. 8(25)
9. Lobashov, M.: Vertical integration in distributed automation environment. In: E&I Elektrotechnik and Informationstechnik (2006)

Enhancing Trust of Cloud Services and Federation of Multi Cloud Infrastructures for Provisioning Reliable Resources

L. Pavithra and M. Azhagiri

Abstract Cloud computing is a new methodology that is used in several organizations and enterprises. It supports immediate provisioning of network access and also shares data to a pool of computing resources on the basis of paying for resource usage. In today trend, providing trust resources has become a problem but it is not a technical issue. Perhaps the technology can enhance trust, reliability, credibility and surety in internet services. In order to increase the use of Web and cloud services, cloud service providers must first enhance trust and security among the enormous amount of users. These issues are addressed by a reputation-based trust-management scheme augmented with trust broker, hybrid and adaptive credibility model, maximizing deviation method, cloud security framework. Trust broker serves as middleware which lies between operating system and applications and it is user effectively matching trusted service resources. Using hybrid trust computing model can calculate the trust degree of service resources and adaptive credibility model that differentiates between credible trust feedbacks and malicious and vulnerable feedbacks. Maximizing deviation method can gain users experiences on using the services by direct interaction and networking risk can be reduced rigorously and efficiency of the system can be improved using light weight mechanism.

Keywords Cloud computing · Trust in cloud · Credibility model · Multi cloud environment · Cloud security · Cloud broker

L. Pavithra (✉) · M. Azhagiri
Department of CSE, Kingston Engineering College, Vellore, India
e-mail: pavi26992@gmail.com

M. Azhagiri
e-mail: azhagiri1687@kingston.ac.in

© Springer Science+Business Media Singapore 2017
P. Deiva Sundari et al. (eds.), *Proceedings of 2nd International Conference on Intelligent Computing and Applications*, Advances in Intelligent Systems and Computing 467, DOI 10.1007/978-981-10-1645-5_10

113

1 Introduction

Cloud Computing

Cloud computing is a technique for provisioning internet access to a shared infrastructure of computing service resources. Cloud computing a new computing methodology for provisioning on demand self services and dynamically provisioning computing resources to end-users through internet applications. Cloud federation is a process of connecting two or more cloud service providers.

Types of Cloud

Private cloud: A cloud infrastructure or environment that is operated for a single organization or enterprises for security.
Public cloud: Services that are delivered through a network that is operated mainly for the purpose of usage of public entities.
Community cloud: Different organization form a separate community and it is operated for several uses.
Hybrid cloud: It is a combination of two or multiple clouds with same entities.
Distributed cloud: It is a different set of distributed machines that is being operated at various locations.

Service Models of Cloud

Infrastructure as a Service (IaaS): A service that is provided to the user for data processing, storing, and other cloud resources where the user could deploy and even compute an arbitrary software, which includes operating system, applications, software (Fig. 1).
Platform as a Service (PaaS): A service provided to the user for deploying onto the consumer-created cloud platform or retrieved applications are developed using programming paradigm, libraries, services (Fig. 1).
Software as a Service (SaaS): Service provided to the user for utilizing the cloud infrastructure environment. The services offered over applications can be accessed through various devices such as interfaces or browsing applications such as internet technologies or any programming interfaces (Fig. 1).

Fig. 1 Service models

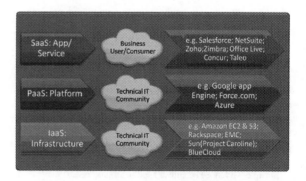

Issues in Cloud Computing

Security, data, Load Balancing, Performance, lack of trust, Processing data, Sharing data, etc. are the issues in cloud computing.

2 Cloud Brokering Environment

Cloud brokers provide intermediation and aggregation capabilities between cloud users and cloud service providers. The future of cloud computing will be established only with the frequent emergence of many cloud brokers. Some of the cloud brokering systems which provide resources to users are Aeolus, Right scale, PCMONS, Spot cloud, Reservoir, Optimis (Fig. 2).

Spot cloud is a secure central platform for buying or selling resources and it also increases utilization and drives new revenue.

Right scale is a web based cloud computing managing tool for managing cloud infrastructures from multiple cloud providers. This enables an organization to easily manage and deploy business applications across all types of cloud.

Aeolus is open source cloud management software which runs on Linux systems. It eases the burden of managing enormous number of clouds and also ensures cloud consumers can use large number of cloud. The components of Aeolus are Conductor-the application used for interacting by users and administrators, Orchestrator-where number of applications is considered as one machine, Composer-an application used for maintaining and building images.

Optimis can be used to identify and modify the optimized cloud resource that depicts risk, trust and efficiency of available services.

Most of these brokers do not provide trust management capabilities for multiple cloud computing services, so as to improve trust among users several trust attributes has been used.

Cloud Services Brokerage (CSB)

"The future of cloud computing will be permeated with the notion of brokers negotiating relationships between providers of cloud services and the service customers."

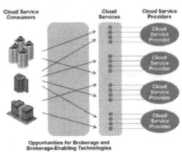

Fig. 2 Cloud broker

3 Trust Attributes in Cloud

The lack of trust between cloud users and providers has become a major impact in recent trends. In order to enhance trust among cloud users four key factors have been identified which is control, security, prevention, ownership [1]. These key factors enhance trust among cloud users, cloud brokers, cloud providers. The above key factors may also diminish control and security capabilities among different cloud providers.

The control of cloud services has to be promulgated with the authorized cloud provider if provided the user may be facilitated with service resources [2]. The service resources provisioned by cloud service provider must have a copyright of data and secured storage of data and proper protection to data. The frequency, capacity and quality service has to be enormously incremented [3].

In today's networked environment with heterogeneous infrastructure and platforms enormous number of software's and application of software are running on different platforms so it has become an important requirement of providing trust among cloud service resources. Cloud computing platform usage has increased rigorously among the networked entities and it can run different applications together at the same time. The utilization rate has also increased so there is a possibility for vulnerability of security and hacking of data stored. Nowadays attacks are becoming much sophisticated and they also have predefined ready-made software for exploiting security vulnerabilities in cloud services. Multiple users are using the homogeneous platforms so there is a possibility for vulnerability. Therefore an authentication and authorization mechanism has to be introduced among cloud services to ensure whether legitimate users are accessing the data resources. The providers must ensure the integrity, identity, security, notarization, confidentiality of a platform in a multi user distributed cloud environment.

Limitation

In existing service provisioning cloud providers do not consider feedback as an important requirement for integrity of services in a virtualized environment. Without collecting feedback from consumers or users trustworthy services may not be established and it can also not control capabilities and distribute best service resources. The efficiency and performance capability may also seem to be iteratively decreased among multiple services.

4 Cloud Broker Computation

As identified in previous studies no cloud broker presents users a trusted service resources. Those cloud brokers as described in Sect. 2 indicates that they do not consider user feedback as an important requirement and there are no control capabilities on services. But the modified cloud system architecture approach focuses mainly on user feedback. Enormous amount of users are accessing or using

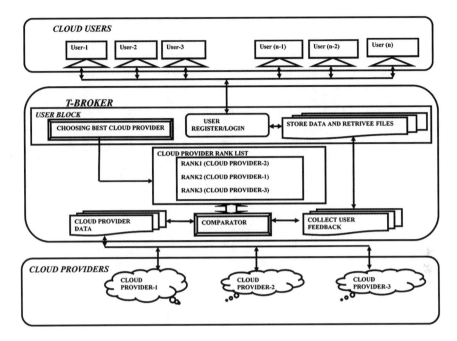

Fig. 3 Cloud system architecture

services in a day. There may also be infinite number cloud providers and cloud brokers. But not all cloud providers provide trusted resources to users (Fig. 3).

The modified cloud system architecture resolves all these requirements and provides trusted resources. Cloud users perhaps first login into their account if an existing user else registers a new account. Then requests for cloud services to trusted broker. Further the trust broker check for availability of services and produces a list of service resources to cloud users based on different providers and also shows the feedback of previous users of the same resources. Trust broker uses a comparator tool for distinguishing feedbacks of cloud users. Furthermore to ensure secured, controlled, ownership of service resources trust broker uses several techniques and methodologies to resolve the issues [4]. They are described as follows in a detailed way.

Distributed Soft Sensor Monitoring

This technique is used to monitor service behavior of provided resources. Distributed soft sensor monitoring is a process of gaining the state information of service resources. This can be used to authorize Service level agreement with cloud users. This also measures the utilization rate of service resources and also guarantees quality of service levels. Some of the parameters that are measured are CPU frequency, latency, throughput, bandwidth, response time, resource utilization ratio, task completion and also ensures whether authorized users are using services.

Virtual Resource Control Center

Every cloud providers has virtual machines configured as default parameter. This focuses on collecting and indexing all the resource information of a cloud provider [5]. Gathers data from each cloud provider and maintains repository that can store all the resource information.

Service Agreement and Resource Matchmaking

Each cloud broker offers Service level agreement with users an important requirement for multiple collaborative services. Service level agreement is a negotiation between consumers and providers and brokers [6]. Using this trust broker can provide efficient trusted resources to users [7]. A contract is signed between cloud brokers and providers.

Hybrid and Adaptive Trust Computation

Security controls are designed based on trust on services. Hybrid trust computation is a technique based on first hand trust that is directly monitoring service behavior of resources and second hand trust based on gathering data directly from cloud users. Adaptive trust computation is used to distinguish between trust feedback and malicious feedbacks [8]. This computation is used to compute entire trust rate of service resources. This model supports users to determine secured and trusted services. This can generate high performance of service resources. This computation allows users to provide their opinions, suggestions, feedback of resources. Those requirements will be satisfied by computation.

Data normalization is a process that is used within the network entity. The mail goal of normalization is to decrease or eliminate redundant data in a database so that the data storage or repository space might be increased and can store more amounts of data. This feedback normalization can be performed through vector normalization or linear transformation methodologies. This can also compute all the quality of service level factors and its parameters.

Service Response Clustering

All network entities have a built in relationship with all types of systems. Consider a large computing environment in which there may be multiple numbers of machines which are hosted and produces millions of instructions per second. Since many machines are running at a same time there may be delay in response time. This method resolves those delay problems and can also increase efficiency of the system and decrease the risk occurrence in the system services.

The feedback is ranked based on priority and a list is prepared for showing it to cloud users. Enormous feedback may be collected and aggregated by using certain fixed value. If the rating of specified service exceeds the fixed value then it may be categorized as positive or negative feedback. The comparator hosted in broker service does all jobs of categorizing and reducing risks involved in processes.

Cost Based Aggregation

The goal of cloud providers is to provide resources and increase their revenue but this has a drawback of affecting users of multi cloud environment. A pricing model is introduced which calculates cost of service resources based on the response time of the cloud resource [9]. And also depends on efficiency of the product and provisioned services.

5 Comparison on Cloud Brokers

The below depicted table illustrates the security and control capabilities and trust management capabilities among the existing brokers and proposed modified trust brokering system for multiple cloud computing environment.

Cloud brokers	Security	Control capabilities	Trust management
Spot cloud	Yes	No	No
Right scale	Yes	No	Yes
Optimis	Yes	No	No
Aeolus	Yes	No	Yes
Reservoir	No	No	Yes
Pcmons	Yes	No	No
Trust broker	YES	YES	YES

6 Protection of Data in Cloud

The protection of data in cloud plays a crucial role in current networking environment. There exist many encryption techniques for cloud some of them are Diffie-Hellman key exchange, Advanced encryption standard, Data encryption standard, Kerberos, RSA, Elliptic curve cryptography, Elgamel public key crypto systems, and so on. To ensure security to data RSA algorithm can be used. RSA is a public key crypto system which is a block cipher where the plain text and cipher text are of integers from 0 to n − 1. It is an authentication algorithm that is used to verify whether legitimate users are accessing data from cloud [10]. Also ensures integrity of data services [11]. The operations involved are encryption, key generation, and decryption. This is a multiplication of two large prime numbers and generates public key and private key. These keys are used for encrypting and decrypting of text or message or data.

Step 1: Take any two prime values assigned to a, b where a $\neq$ b
Step 2: Compute n = a*b
Step 3: Compute g(n) = (a − 1) × (b − 1)
Step 4: Select e which is a prime value to n and less than g(n)
Step 5: Compute d $\equiv$ e^{-1} mod g(n)
Step 6: Encryption c = Message^{e}mod n
Step 7: Decryption p = Cipher^{d}mod n

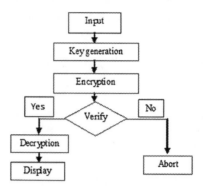

Thus the data in cloud can be protected and secured by using this authentication algorithm. Only legitimate users can gain access to the cloud system and use the service resources [12]. Among all other encryption and decryption algorithms this algorithm ensures ultimate protection and security to data access. Thus RSA authentication algorithm has an edge over all other algorithms and an efficient technique.

7 Conclusion

Cloud brokers are used for satisfying the request of the user. Thus the modified approach of brokering systems that is trust brokers provide users secured and best service among all other brokering system. This trust broker mainly considers user feedback and aggregates them to offer enhanced service resources. Therefore multiple data can be distributed or provided to various users and utilization rate is increased.

References

1. Habib, S.M., Varadharajan, V., Muhlhanser, M.: A trust-aware framework for evaluating security controls of service providers in cloud marketplace. In: 12th International Conference on Trust, Security and Privacy in Computing and Communications, pp. 459–469 (2013)
2. Muhlhauser, M., Habib, S.M., Hauke, S., Ries, S.: Trust as a facilitator in cloud computing: a survey. J. Cloud Comput. Adv. Syst. Appl. **1**, 9 (2012)
3. Hwang, K., Kulkareni, S., Hu, Y.: Cloud security with virtualized defense and reputation-based trust mangement. In: Eighth IEEE International Conference on Dependable, Autonomic and Secure Computing IEEE Computer Society, pp. 717–722 (2009)
4. David, M.N., Jingwei, H.: Trust mechanisms for cloud computing. J. Cloud Comput. Adv. Syst. Appl. **2**, 9 (2013)

5. Abawajy, J.: Establishing trust in hybrid cloud computing environments. In: Proceedings of the 2011 IEEE 10th International Conference on Trust, Security and Privacy in Computing and Communications, pp. 118–126 (2011)
6. Zhou, F., Ma, H., Yao, W., Li, X.: T-broker: a trust-aware service brokering scheme for multiple cloud collaborative services. IEEE Trans. Inf. Forensics Secur. 10(7), 1402–1415 (2015)
7. Gui, X., Li, X., Ma, H., Zhou, F.: Service operator-aware trust scheme for resource matchmaking across multiple clouds. In: IEEE Transactions on Parallel and Distributed Systems, vol. 26, no. 5, pp. 1419–1430 (2015)
8. Li, X., Du, J.: An adaptive and attribute-based trust model for SLA guarantee in cloud computing. Inf. Secur. 7(1), 39–50 (2013)
9. Mohammadi, H.F., Prodan, R., Fahringer, T.: A truthful dynamic workflow scheduling mechanism for commercial multi-cloud environments. In: IEEE Transactions on Parallel and Distributed Systems, vol. 24(6), pp. 1203–1213 (2012)
10. Jamuna, P., Sudha, I.: Data coloring by cloud watermarking using RSA for periodic authentication. Int. J. Adv. Res. Comput. Sci. Softw. Eng. 3(11), 627–630 (2013)
11. Miers, C., Redigolo, F., Simplicio, M., Naslund, M., Pourzandi, M., Gonzalez, N., Carvalho, T.: A quantitative analysis of current security concerns and solutions for cloud computing. J. Cloud Comput. Adv. Syst. Appl. 1, 11 (2012)
12. Alfazi, A., Noor, T.H., Sheng, Q.Z.: Reputation attack detection for effective trust assessment among cloud services. In: 12th International Conference on Trust, Security and Privacy in Computing and Communications, pp. 469–477 (2013)
13. Jain, P., Patidar, S., Rane, D.: A novel cloud bursting brokerage and aggregation(CBBA) algorithm for multi-cloud environment. In: Second International Conference on Advanced Computing & Communication Technologies (ACCT), pp. 383:387 (2012)

Effective Use of GIS Based Spatial Pattern Technology for Urban Greenery Space Planning: A Case Study for *Ganesh Nagar* Area of Nanded City

Govind Kulkarni, Deshmukh Nilesh, Bhalchandra Parag, Pawan Wasnik, Kailas Hambarde, Preetam Tamsekar, Vijendra Kamble and Vijay Bahuguna

Abstract The aim of this study is to find out the conceptual framework for urban greenery planning. The present investigation proposes a GIS based model which will summarizes a strategy for supplying the ecosystem services for urban area through the required planning process. The proposed model acts as a decision support system for local government by providing suggestion of best suitable tree on particular location with their description to help the environment and human health. The model will also allow easier adaptation for Local Government and for other similar communities wishing to implement standard policies for Green City/Green India programs. The study area chosen is Ganesh Nagar of Nanded city.

Keywords GIS · Urban green space · Spatial patterns · Correlations

---

G. Kulkarni (✉) · D. Nilesh · B. Parag · P. Wasnik · K. Hambarde · P. Tamsekar · V. Kamble
School of Computational Sciecnes, S.R.T.M. University, Nanded 431606, MS, India
e-mail: govindcoolkarni@gmail.com

D. Nilesh
e-mail: nileshkd@yahoo.com

B. Parag
e-mail: srtmun.parag@gmail.com

P. Wasnik
e-mail: pawan_wasnik@yahoo.com

K. Hambarde
e-mail: kailas.srt@gmail.com

P. Tamsekar
e-mail: pritam.tamsekar@gmail.com

V. Kamble
e-mail: vijendrakamble5@gmail.com

V. Bahuguna
Department of Geography, DBS PG College, Dehradun 248001, UK, India
e-mail: vijaybahugunadbs@gmail.com

© Springer Science+Business Media Singapore 2017
P. Deiva Sundari et al. (eds.), *Proceedings of 2nd International Conference on Intelligent Computing and Applications*, Advances in Intelligent Systems and Computing 467, DOI 10.1007/978-981-10-1645-5_11

1 Introduction

In Today's world, entire human race is facing tremendous problems caused by rapid urbanization scenario, global climate change, explosion in population and developmental activities. These activities disturb natural flora and puts pressures on environment & ecosystem. To maintain eco balance collaborative, long-term and strategic management of urban green space planning is necessary. This must support proactive management approaches. Further, its results must have increased operational efficiency, risk reduction, increased urban forest canopy and leaf area. This has enforced us to create greenery spaces with urbanization and developmental activities. Various planning processes have been introduced in past [1] with a synergistic planning approach acting as a key factor for reducing urban heat and its associated effects. We have also understood that while planning the urban environment, it is very important understand the fundamental needs of urban area by having concise understanding for how to preserve the green infrastructure, how long it required the planning, how to implement such plans such questions should be focused while developing any plan for urban green space. The creation of greenery space is a complex decision making process. It needs to consider all variables related to planning, selection, plantation and cultivation of plants. It demands several inputs about soil, air, water, fertilizers, diseases, preventive measures etc. We felt that, this decision making for can be supported by applying Geographical Information System (GIS). The GIS is an emerging technology for developing a spatial analysis of data using geographical information and highly generated computerized maps as an interface based on the location [2]. GIS software is the natural evolution of the modern database, which combine the functionality of a database management system (DBMS) [2]. The GIS also provide the platform to perform the spatial analysis and allow displaying the data visually. Various studies have been introduced in the GIS research area which has shown how a GIS based analysis can support integrated distribution of population, manpower, material resources and financial resources, city greening planning design, construction, transportation cost and the maintenance management and other comprehensive factors which explores the importance of urban green space [1, 3]. On the backdrop of this discussion, the main goal of this research is to create a Geographical Information System model for decision making for Greenery Space within a limited study area. The suburbs of Nanded City, Maharashtra State, India were considered as research area. The outcome of this research will be totally based on the data criteria and their parameters. This system model will guide us for determining the area for the urban greenery space planning implementation. The result of this study will be represented as the tools for education of the general public on green infrastructure, its concepts, and implementation. The model will also allow easier adaptation for Municipal Corporation and for other similar communities wishing to implement standard policies.

2 Proposed Model

The main focus of this study is to establish the technical needs of green infrastructure for urban areas. Before collecting the actual data we have performed various necessary actions on data collection protocols which result efficient process for data collection. The first-hand data gives us primary source. Primary data are created by spectators or recorders who went through the events or conditions being documented. The strength of a GIS is to handle the sophisticated functions which consist of numerous tasks for utilizing both spatial and attribute information stored within them. Tree spots are assigned by putting 10 × 10 m and 5 × 5 m distance according to space requirement of selected trees. The Poly lines are created when more than two lines are joined to form lines of line. The representations of buffer zone in GIS application are always in vector polygon enclosing other polygon, line or point feature. We have used the Polygon feature to represent the commercial areas, industrial area, open space, etc. The Import to Layers is the mechanism which used to display geo datasets in GIS tool. Each layer addresses a dataset and specifies how dataset is depicted using symbols and text labels. Layers can be displayed in a particular order and can be seen in content table of the map. The above mentioned features process are merged in layer to create the geo-database. The component Geo-database is a collection of various types of geo-graphical data sets stored in common file system. While creating a datasets it is important to design and build a geo-database. Geo-database represents the spatial relationship between each data set. We can define geo-database as a spatial relational database management system which has storage of geographical data stored in central location for access, retrieve and management. The Map is the visual representation of information. These maps are main source of data for the GIS tool which has roots in the analysis of information to overcome the limitations of manual analysis. Our model also needs Secondary data to provide the base for the comparison of data to understand the research problem. In this study the secondary data is collected by applying field work method which includes collection of soil samples, air pollution reports, ground water test reports. Various statistical analysis methods are used for interpretation of data which is a major component of data management. Data Elucidation is a fundamental success plan for the projects. The data interpretation technique is used to reach on important decisions. There is also an attachment of Elucidated data to GIS: Attaching the data to a GIS is a management technique which provides the accuracy in results.

3 Experimentations, Results and Discussions

There is need to keep some efforts for urban tree management which will be beneficial for those trees which regrettably die due to various imbalances including water scarcity. It is very important to understand type of tree to be planted with respected to chemical composition of soil. By putting focus on theses point and discussion with expertise following trees are suggested for Nanded city with their availability and geographical nature. We choose following six trees as per our survey of planted trees in Nanded city. We have chosen most common Ashoka tree {Saraca asoca}, the Banyan Tree {Ficus Benghalensis}, the Gulmohar tree {Delonix Regia}, the Neem {Azadirachta Iindica}, the Pimple Tree {Ficus Religiosa} and the Subabul Tree {Leucaena Leucocephala}.

We have chosen some basic chemical terms like pH, EC, Alkanity, Chloride, Magnesium, Calcium, Sodium, Potassium, Moisture Contents, Organic Matter, Phosphors. The selected soils samples are represented in Tables 1 and 2. In the present research the collected soil data exposed that there were considerable variations in the quality with respect to their physicochemical characteristics. A composite surface soil sample from 0 to 20 cm depth was collected from the experimental area before initiating the experiment and was analyzed for physical and chemical properties. The Table 3 represents the correlation analysis which describes the relationship of each parameter with each other. The correlation analysis defines the dependence of each parameter with each other. The positive and negative results have perspective values. The negative range with different values defines variations in results. Correlation analysis also defines the strength and directions of relationship. After examine and discussion with the expertise we concluded that the selected soil samples consist of rich calcium and magnesium. The soil is black cotton soil which has high range of water holding capacity. It is very important to analysis the ground water quality due to increased substantially and suitability for consumption, irrigation and industrial activities. As per the expert opinion the ground water quality is very unhealthy. The ground water of this area is

Table 1 Physical and chemical properties samples-2 of soil

Sample no.	pH	EC	Alkanity	Chlorides (Cho)	Magnesium (Mg)	Calcium (Ca)
1	8.43	440	12.00	0.07	0.42	1.15
2	8.08	280	5.50	0.09	0.59	0.65
3	7.89	300	6.00	0.09	0.25	0.94
4	8.14	220	7.00	0.09	0.65	1.07
5	8.07	260	5.00	0.08	0.70	0.24
6	7.82	300	5.00	0.04	0.20	0.65
7	8.51	440	10.37	0.05	0.79	0.93
8	8.73	470	8.40	0.05	0.63	0.36
9	8.48	480	11.50	0.09	0.52	1.25
10	8.92	390	9.50	0.09	0.58	0.98
11	8.13	280	8.50	0.07	0.75	0.74
12	8.18	320	11.00	0.11	0.41	0.90
AVG	8.28	348	8.31	0.07	0.54	0.82
MAX	8.92	480	12.00	0.11	0.79	1.25
MINI	7.82	220	5.00	0.04	0.20	0.24

Table 2 Physical and chemical properties of soil samples-1

Sample no.	Sodium (Na)	Potasium (K)	Moisture contents	Organic matter	Phophors
1	0.06	0.02	3.07	0.49	0.02
2	0.00	0.01	5.28	0.49	0.04
3	0.00	0.01	2.11	0.44	0.05
4	0.03	0.01	4.42	0.46	0.01
5	0.02	0.01	1.89	0.50	0.07
6	0.01	0.01	3.58	0.48	0.04
7	0.01	0.02	2.77	0.48	0.01
8	0.02	0.01	4.81	0.49	0.04
9	0.02	0.01	1.79	0.49	0.02
10	0.02	0.01	0.71	0.42	0.01
11	0.02	0.02	3.91	0.46	0.05
12	0.01	0.01	1.06	0.49	0.03
AVG	0.02	0.01	2.95	0.47	0.033
MAX	0.06	0.02	5.28	0.50	0.070
MINI	0.00	0.01	0.71	0.42	0.010

Table 3 Correlation analysis of soil samples

Sr. no.	pH	EC	Al	Cho	Mg	Ca	Na	K	MC	O.M	P
pH	1	0.74	0.63	**−0.62**	0.30	0.34	0.28	0.33	0.36	0.49	**−0.56**
EC		1	0.72	**−0.64**	**−0.29**	**−0.23**	0.27	0.18	0.09	0.60	**−0.43**
Al			1	**−0.36**	**−0.11**	**−0.09**	0.46	0.26	0.17	0.70	**−0.63**
Cho				1	0.03	**−0.01**	0.14	**−0.25**	**−0.19**	**−0.33**	0.21
Mg					1	1.00	0.19	0.47	0.53	**−0.39**	0.06
Ca						1	0.19	0.49	0.53	**−0.37**	0.07
Na							1	0.49	**−0.27**	0.36	**−0.32**
K								1	0.08	0.01	**−0.11**
MC									1	−0.30	0.19
O.M										1	−0.64
P											1

The bold values are used to represent the negative correlation

not suitable for human health as it not meets the drinking water standard. The average calculation of NO_3 and total hardness had crossed the maximum standard which represents the unhealthy quality of ground water. As per the expert opinion the ground water quality has failed to meets the bacteriological quality parameters. The Table 4 represents ground water sample of study area. The Table 5 represents correlation analysis of ground water parameters. Figure 1 illustrates it graphically. Based on above experimentation, tables and analysis, we have selected only those trees which satisfy the correlations. The tree properties and their pollution

Table 4 Ground water chemical parameters

Sr. no.	Ganeshnagar	Fl	Ca	Ph	No_3	Ir	Tds	Tur	TH	AL
1	Labour colony	0.18	188	7.68	45.76	0.10	659	0.82	266	262
2	Nagsennagar	0.68	117	7.76	8.47	0.13	610	0.75	300	224
3	Kabaranagar	0.36	97	7.03	8.51	0.01	532	0.46	218	320
4	Yashvantnagar extension	0.38	201	8.21	75.73	0.05	536	0.48	224	210
5	Laxminagar	0.67	102	7.32	7.23	0.05	956	0.15	422	324
6	Maroti Tample Ganesh Nagar	0.23	108	7.39	9.63	0.54	548	4.23	232	222
7	Vijaynagar	0.20	227	7.42	223.00	0.30	110	0.06	482	356
8	Ravindra nagar	1.10	254	7.35	55.00	0.08	885	0.11	382	216
9	Fire Briga de Office	0.74	117	7.61	19.06	0.05	751	0.51	230	344
10	Vaibhavnagar	0.66	235	7.17	140.30	0.09	992	0.88	258	264
11	Parimalnagar	0.33	77	7.46	23.05	0.10	512	0.88	244	246
12	Ambikanagar	0.25	164	1.48	23.05	0.08	522	0.28	152	248
	Avg.	0.48	157	6.99	53.24	0.12	717	0.80	284	270

Table 5 Correlation analysis

Sr. no.	Fl	Cl	pH	No3	Ir	Tds	Tur	Total hardness	Alkialirity
FL	1	0.23	**−0.12**	**−0.18**	**−0.37**	0.13	**−0.33669**	**−0.02261**	**−0.34**
CL		1	**−0.01**	0.72	**−0.10**	0.83	**−0.35157**	0.619243	0.160781
pH			1	0.13	0.09	0.31	0.119797	0.491463	0.157315
NO3				1	0.034	0.87	**−0.28321**	0.702332	0.492179
IR					1	−0.08	0.900543	0.024826	**−0.24698**
TDS						1	−0.34882	0.805138	0.50306
TUR							1	**−0.30766**	**−0.36283**
TH								1	0.385267
ALK									1

The bold values are used to represent the negative correlation

Fig. 1 Graphical representation of ground water elements

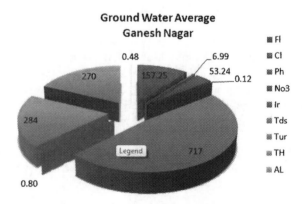

Ground Water Average Ganesh Nagar

0.48   6.99
270   157.25   53.24
284   0.12
717
0.80

Legend: Fl, Cl, Ph, No3, Ir, Tds, Tur, TH, AL

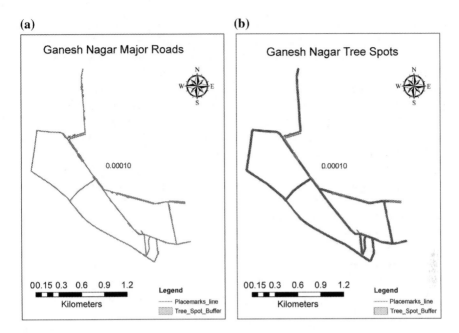

Fig. 2 a Study area (Ganesh nagar) map. b Spotted tree points in GIS

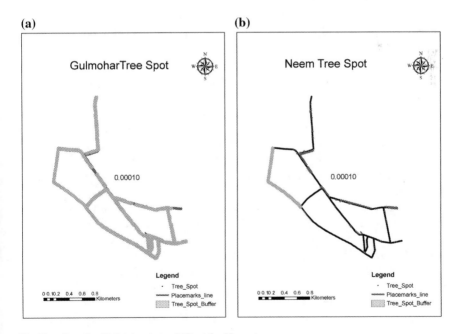

Fig. 3 a Spot for Gulmohar trees. b Spot for Neem tree

controlling aspects have taken into consideration. These values have added to Arc GIS 10.1 software (trial and legal version valid for 2 months). The results are shown in Figs. 2 and 3. The selected study area comes under the new Nanded city. Ganesh Nagar area is situated in north region of Nanded city. The Ganesh Nagar is well known residential area in Nanded. Various Government offices are located in Ganesh Nagar. The selected study area consists of some standard residential societies. The southwest part of this study region consists of slum area with poverty and inferior living conditions. It causes soil and water pollution in the region. The study area consists of well-known major road through which we can reach to other part of the city. The study area also have complex road network which causes traffic problems. This leads to vehicle pollution in high amount which is a major problem in this area. We have selected major road network for the study purpose. The road network covers various residential areas like Snehnagar, Labourcolony, pushpa Nagar, Naagsen Nagar, Ambika Nagar, Kabara Nagar and Chatrapatichowk. 12 km of road network was calculated for plantation purpose. Total 1331 points are selected by applying 10 × 10 m distance for tree cultivation purpose. Location which having tree at the site are shown by using buffer zone in the map.

We have selected various points with their latitude and longitude which commonly known as coordinate system. Some tree spots are manually assigned to place where there is a free space is available in between two different locations. The locations for tree cultivations are selected as per the NWMC (Nanded Waghala Municipal Corporation) guideline. Tree spots are plotted at the both end of road network. The present map depicts the zoom in view of Ganesh Nagar study area. The map also shows the assigned tree spots along polygon feature. The green line which is displayed in map along with road network indicates the spots where trees are already presents.

The map also shows the GIS features which are used in the Map. It is highly recommended that trees should not be planted at the location where road is having large curve to avoid the accident because most of the accident are happened at the spots where road having large curve sites. The map also shows the assigned tree spots along polygon feature. The green line which is displayed in map along with road network indicates the spots where trees are already presents. The map also shows the GIS features which are used in the Map. It is highly recommended that trees should not be planted at the location where road is having large curve to avoid the accident because most of the accident are happened at the spots where road having large curve sites. The Ashoka tree is exalted evergreen tree which founded in Indian subcontinent. The reason behind planting Ashoka tree is its effectiveness in assuaging noise pollution, dust particles absorption. The Ashoka trees have pyramidal growth with height of 12 m and above. It required near about 2 m of space for its circumferences. Due to its pyramidal structure the leaves are long narrow with undulate margin. The Ashoka tree is a rainforest tree. The recommended space for planting Ashoka tree is 5 × 5, 7 × 7. Our study suggests planting Ashoka tree towards the area with both side of road network, because Ashoka tree have high capacity to control noise pollution as well as it will absorb dust particles in higher amount. The air pollution record of Ganesh Nagar area various chemical gases are entered in environment with more laid-back noise and scatter dust particles. Ashoka tree will absorb such pollutant elements and helps to protect the human health with environment. The Banyan tree is an evergreen tree which grows in any region across the world with independent growth. The Banyan can grow to an average height of around 100 feet (approximately 30.5 m). The trunk of banyan tree can covers around 200 m of size which have large amount of spread out branches. The Banyan tree can group very fatly. The leaves are broad, oval and glossy. The Banyan tree is a long-lived tree with high resistance to wind dark-gray and rough bark. The Banyan tree canopy can cover near about 450 m of area. Our model recommends planting the banyan tree where there is a less amount of residential site, industrial area is present. While planting Banyan tree we should focus on amount of open space is available. The Gulmohar tree is evergreen tree which is useful for air pollution control. The maximum height of Gulmohar is 12 m, standard height is 5 m. The height of this tree is depending up on the soil type. This tree can grow up on any type of soil. Tree trunk is tall and un branched. This tree consists of large no. of leaves with 10–12 pairs of pinnate. Each pinnate consist of 30–60 opposite leaflets. This tree can cover the width around 6–12 m. The recommended space for planting for Gulmohar is 5×5 or 10×10. The Gulmohar tree looks like an umbrella which provide high amount of shadow. The Gulmohar tree is useful for increasing rainfall level in the area. Small road network are recommended for planting Gulmohar tree which also provide attractiveness to area. The Neem tree is evergreen and fast growing tree. It has a diameter of 60 cm at top and 2 m diameter at the base. This tree can grow up very fatly in black cotton soil. Neem tree can improve the water holding capacity of soil and soil fertility. Neem tree is a natural chemist which has various chemical in its body. This tree can grow up widely with soil pH range up to 10 which make it more versatile and important tree in sub-continent. The recommended space for planting Neem tree is 5×5 or 10×10. Long road network are more suitable for Neem tree plantation. Pimple tree have some sort of same characteristics like Banyan tree have. Pimple tree is large dry season-deciduous evergreen tree which can grow up to 30 (close to 98 ft.) m in height and 3 ms (near about 10 ft.) of trunks diameter. The tree leaves are cordite in shape. The leaves are 10–18 cm long with 8–12 cm broad. The Pimple is a large in size whose roots travel very long distances and whose branches spread expansively; it can give a wide area of shade. Well known fact about pimple tree is that it utilizes Co_2 for photosynthesis and generates the oxygen during daytime. Our model suggests planting Pimple tree at the end of road network with small residential area with large amount free open space. Most commonly planted in urban forestry is Subabul tree which is fast growing evergreen species. The Subabul tree is best suitable in warm region where temperature is 30 °C and above. It is fastest growing tree which is popular in farm forestry. At worm location the Subabul tree are more suitable as it has strong and deep root system. This tree can grow under a wide range of conditions as a range plant, roadside plant, in pastures, etc. The Subabul tree can adopt 1.27 m $\times$ 1.27 m (50″ $\times$ 50″) of area. The recommended space for planting Subabul tree is 3×3 m, 5×5 m. The Subabul tree is highly suitable for green-manuring and composting which fixes the flow of nitrogen gas in environment. Our model has suggested a site for Subabul tree at residential area along with road network. Based on above all research and investigations, our model recommends planting two types of trees and the Neem tree. Our model did not consider the remaining trees as they do not fit the selection criteria.

4 Conclusion

The presented research work has proposed a GIS based model as a strategy for urban planning process. The proposed model acts as a decision support system for local government by providing suggestion of best suitable tree on particular location with their description to help the environment and human health. An attempt has been made to represents the systematic spatial analysis of urban green space planning for selected work area—Ganesh Nagar area of Nanded city. Our model has demonstrated suitability of only two trees out of considered six trees. Our model can be used as role model for greenery planning of any area of any city.

References

1. Luley, C.J., Jerry, B., and Cooperating Agencies.: A Plan to Integrate Management of Urban Trees into Air Quality Planning. Davey Resource Group, Naples (2002)
2. Anna, C. et al.: Urban Tree Mapping. PhD diss., Worcester Polytechnic Institute (2010)
3. Willoughby, R.: GIS-Based Land Use Suitability Modeling for Open Space Preservation in the Tijuana River Watershed. Diss. San Diego State University (2005)

Automated Hand Radiograph Segmentation, Feature Extraction and Classification Using Feed Forward BPN Network in Assessment of Rheumatoid Arthritis

U. Snekhalatha and M. Anburajan

Abstract Rheumatoid arthritis is a chronic autoimmune disorder which affects the multiple joints especially small joints in hands and feet of both the sides of the body. Even though modern imaging techniques such as MRI, PET, SPECT, color Doppler ultrasound was used to quantify the inflammatory conditions of RA, radiograph was considered as the gold standard technique in evaluation of RA. The aim and objective of the study were given as follows (i) to perform an automated segmentation of hand from the radiograph using DTCWT based watershed algorithm; (ii) to compare the measured statistical features of the joint space of the hand using GLCM method, and (iii) to perform feed forward back propagation network for classification of subjects into RA and normal with high accuracy. The dual tree complex wavelet transform based watershed algorithm was found to be satisfactory in segmenting the hand automatically using the digitized radiograph. The feed forward BPN network was utilized for the classification of subjects into normal and RA patients from statistical features (total inputs, n = 45) extracted from hand X-ray image. The calculated accuracy of the classifier was found to be 95.7 %, whereas its sensitivity and specificity was found to be 88.2 and 98.1 % respectively. When both the statistical features as well as geometric features (total inputs, n = 53) were used in the feed forward BPN network, it was found that accuracy was 92.9 %, whereas its sensitivity and specificity were found to be 82.4 and 96.2 % respectively.

U. Snekhalatha (✉) · M. Anburajan
Department of Biomedical Engineering, SRM University, Kattankulathur,
Chennai 603203, Tamil Nadu, India
e-mail: sneha_samuma@yahoo.co.in

M. Anburajan
e-mail: hod.biomedi@ktr.srmuniv.ac.in

© Springer Science+Business Media Singapore 2017
P. Deiva Sundari et al. (eds.), *Proceedings of 2nd International Conference
on Intelligent Computing and Applications*, Advances in Intelligent Systems
and Computing 467, DOI 10.1007/978-981-10-1645-5_12

Keywords Dual tree complex wavelet transforms · Feature extraction · Feed forward back propagation network · Rheumatoid arthritis · Watershed algorithm

1 Introduction

'Rheumatoid arthritis (RA)' is a systemic disease that affects the joint connective tissues, muscles, tendons and fibrous tissue and it is a chronic disabling condition often leads to pain and joint deformity and progressive physical disability [1]. It is an autoimmune disease which can progress to very rapidly cause swelling and damaging cartilage and bone around the joints [2]. It can affect any joint but it is common in the wrist and fingers. RA affects 0.5–1 % of adults in developed countries, and the disease is three times more frequent in women than men [3]. The high prevalence rate of RA of about 7.1 % had been reported in American Indian tribes in the year 2005 [4]. The lower prevalence rate of about 0.2 and 0.19 % has been reported in countries like China and France in the year 2005 [5, 6]. Hand radiograph is a standard technique and frequently used imaging modality to monitor the progression of joint damage and bone erosion in RA patients [7]. Valid visual scoring methods are available to study the inflammatory activity in the hand joints of RA [8]. The simple manual scoring methods available to evaluate the radiograph of the hand are Larsen score and Genant/sharp score. The Larsen method is based on a global score of each joint ranges from 0 to 5. The Genant/Sharp scoring method provide a separate score for bone erosions (0–3.5) and joint space narrowing (0–4) in hand joints. The Vander Heijde scores ranged from 0 to 5 in the evaluation of erosions and scores (0–4) for joint space narrowing. It is very difficult and time consuming task to perform the long term assessment of bone erosions and joint space narrowing using various visual scoring methods due to inter and intra reader variations [9, 10]. In order to overcome these issues, it is necessary to introduce the computer based automated measurements for quantifying the bone erosions and joint damage in hand radiograph of RA patients. The computer assisted analysis will provide better disease treatment if the assessment procedure was faster, accurate and reproducible.

Hand bone segmentation is a challenging task due to its anatomical structure and variation in size from person to person. Also its bone structure presents great variation in the image intensities. In some cases, the background, flesh and the bone represents the same gray level value at some points. Hence it is very much essential to propose an automated image segmentation algorithm which segments the hand bone region in order to quantify the arthritis parameters such as bone erosion and joint space width in the assessment of RA. Several researchers used different types of automated segmentation methods such as otsu thresholding techniques [11], canny edge based segmentation methods [12], active contour models [13]. Point distribution and geometric based active shape models [14], live wire algorithm [15], Bayesian inference approach [16], graph based approach [17] for automated segmentation of hand radiograph in evaluation of RA patient. The limitations of these methods were as follows: (i) fails to segment the metacarpal bones located inside the palm of the hand especially thumbs and little finger; (ii) these methods eliminates the joints that are too impaired and images with poor quality; (iii) finds challenging to overcome the hand tissue muscle attenuation in the segmented region. The aim of the study were given as follows (i) to perform an automated segmentation of hand from the radiograph using dual tree complex wavelet transform (DTCWT) based watershed algorithm; (ii) to correlate the statistical features extracted from the joint space of the hand using GLCM method, and (iii) to perform feed forward back propagation network for classification of subjects into RA and normal with high accuracy.

2 Methodology

2.1 Subjects

The study was approved by ethical committee of SRM University, Kattankulathur, Chennai, Tamil Nadu, India and obtained ethical clearance number 35/iec/2010. A medical camp was conducted for screening the RA patients in the suburban South Indian population at SRM Hospital and Research Centre, Kattankulathur, Tamil Nadu, India on 3rd–5th of August 2010. The inclusion criteria were patients who were suffering from joint pain, joint swelling, inflammation and arthritis. The exclusion criteria were patients who had undergone joint surgery or previous old bone fractures. In this study, the following criteria for diagnosing RA, outlined by the IRA consensus report 2008 [18] was used: (i) individual with persistent inflammatory arthritis with more than four hand joints; (ii) high ESR, (iii) high CRP values; and (iv) Positive Igm RF. A total number of 100 subjects, age ranged from 30 to 75 years were registered in the camp. According to the diagnostic criteria of the study, 50 subjects found to have RA and another 20 were age and-sex-matched normal. The remaining 30 subjects were excluded according to the exclusion criteria of the study. The study group were organized as follows:

Group I: RA (n = 50, M/F ratio: 1:3, age = 46.5 ± 12.2 years, disease duration: 4.6 ± 1.5 years)
Group II: Age- and sex-matched normal (n = 20, M/F ratio: 1:2, age = 46.2 ± 12.2 years).

2.2 Hand Radiograph Acquisition and Measurements

A standard PA view hand radiograph was obtained in each subject. From the radiograph, the whole hand was segmented automatically using DTCWT based watershed algorithm. The statistical features of the joints in MCP1-MCP5 were extracted using GLCM method. The geometric measurements such as cortical thickness and joint space width were measured using mimics software. The DTCWT based watershed algorithm was used for segmentation of the hand radiograph of each subject [19, 20]. The rectangular region of interest (ROI) of size 9 mm × 10 mm was marked semi-automatically near the centre of 1st–5th MCP joints in the final segmented image. Features such as mean, standard deviation, variance, entropy, normalized mean, mean angle, and autocorrelation were extracted from the segmented image of the total studied population using the gray level co occurrence matrix (GLCM) method. The feed forward back propagation network was used to classify the subjects into RA and normal.

2.3 Block Diagram of Automated Segmentation Algorithm

The hand X-ray image was preprocessed using Gaussian filter for smoothening and reduces the noise. The DTCWT calculates the complex transform of a signal using two separate DWT decompositions. The filters used in one are specifically designed DWT to produce the real coefficients and the other the imaginary. From the complex magnitude obtained from the DTCWT, the directional median filtering of order 9 was applied to remove the noise and double edge effect. The texture gradient was obtained by applying morphological erosion operator and interpolation on the subbands of the filtered image. On the other hand, modulated gradient was achieved by applying the Gaussian gradient operator and interpolation on the filtered image. The total gradient was obtained from the aggregation of texture and modulated gradient. Here the Watershed Algorithm is applied to detect bones from the surrounding tissues of the hand radiographs. The two most prominent families of the WT algorithm are rain-falling and immersion simulations. Both simulations are to regard the gradient image as a topographic surface. The initial segmented was obtained for different threshold T < 10. Then for threshold T = 10, segmented image is obtained. Then segmented region contours are superimposed on the original image. Then the ROI was marked semiautomatically on the MCP joints of both hands, from which features are extracted using GLCM algorithm (Fig. 1).

Fig. 1 Flow chart of the automated image segmentation using DTCWT based watershed algorithm and feature extraction

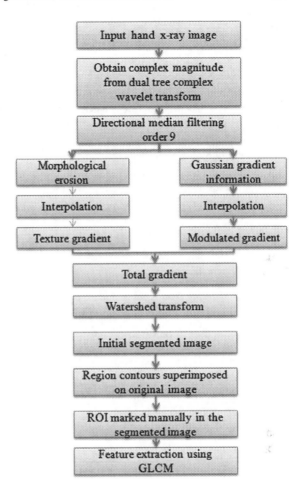

2.4   *Feature Extraction*

The GLCM method is used to distinguish the texture of the image based on pixel intensity distribution and its relative position in the image. It is the statistical method of examining the texture features from the second order statistics as follows: The mean is defined as the average gray level values in the image. The standard deviation is a measure of dispersion or variation exists from the average value. A low standard deviation represents the data points to be very close to the mean value and a high standard deviation represents the data points are spread out over the large range of values [21]. The variance is the square of the standard deviation. The small variance indicates data points are close to the mean, whereas the large variance represents the data points are spread out from the mean. Entropy is a degree of the randomness present in the intensity distribution. It is the significant measure of information provided by the image. If the pixels have same gray level in the image, then minimum entropy is achieved, whereas for the pixels with uniform distribution of gray levels or histogram equalized gray level image, maximum entropy is achieved. The normalized mean is scaling the mean of the image to the normalized value. The mean angle will calculate the set of angles in degrees based on polar considerations in gray levels [22]. The autocorrelation texture measures the linear dependency of gray levels on those of neighboring pixels. Covariance is the measure of how two random variables vary together with respect to their mean. It provides the measure of correlation between the two or more sets of random variables [23].

The fundamental equations of nine features obtained from GLCM method [23] are given as follows:

(i) Mean

$$\text{mean} = \Sigma i \, \Sigma j \, i \, P(j, i) \tag{1}$$

(ii) Standard deviation

$$\sigma = \left(\sum_{i=0}^{N-1} \sum_{j=0}^{N-1} (i - \mu)^2 p(i,j) \right)^{1/2} \tag{2}$$

(iii) Variance

$$\text{Var} = \Sigma i \Sigma j (i - \mu)^2 P(i, j) \tag{3}$$

(iv) Autocorrelation

$$\text{Auto corr} = \sum_{i=0}^{N-1} \sum_{j=0}^{N-1} p(i,j) \frac{(i - \mu)(j - \mu)}{\sigma^2} \tag{4}$$

(v) Entropy

$$\text{Ent} = \Sigma i \Sigma j P(i, j) \log(P(i, j)) \tag{5}$$

(vi) Normalized mean

$$\text{Norm mean} = 1/N \, i\Sigma j \, i \, P(i, j) \tag{6}$$

(vii) Exponential component

$$y = exp^{-[p(i,j)]^2} \tag{7}$$

(viii)   Mean angle

$$\bar{\alpha} = a\ tan\ 2 \left(\frac{1}{n} \cdot \sum_{j=1}^{n} sin\ \alpha_j, \ \frac{1}{n} \cdot \sum_{j=1}^{n} cos\ \alpha_j \right) \qquad (8)$$

(ix)   Covariance

$$C_{x=1/N}\ \Sigma i\ P(i,j)\ P(i,j)^T - \mu\,\mu^T \qquad (9)$$

2.5   Feed Forward Back Propagation Method

Feed forward neural networks are one class of neural network used to solve complex problems by modeling complex input-output relationships [24]. The back propagation learning algorithm is a widely used method for feed forward neural networks in many medical applications [25]. The feed forward back propagation network consists of three layers namely input layer, hidden layer and output layer. It is trained by adjusting the weights in order to perform perfect classification using the back propagation learning algorithm. The feed forward back propagation network make use of binary sigmoid activation function to scale the hidden layer and output layer of neural network. The input features are normalized between 0 and 1 before feeding to the network. The desired output is indicated as '0' for normal and '1' for abnormal. The classification process is categorized into four phase as follows: (i) training phase (ii) testing phase (iii) validation phase and (iv) resultant phase. By using the input patterns, learning repetitions is performed with different training and validation data sets. The ROC plots are used to determine the accuracy of the classification method obtained from the network.

2.6   Statistical Analysis

Data were analyzed using SPSS software package version 19.0 (SPSS Inc., Chicago, USA). The measured feature extracted parameters such as mean, standard deviation, variance, entropy, mean angle and exponential component in RA group and normal were compared using a student's t-test.

3 Results

The DTCWT based watershed algorithm was found to be satisfactory in segmenting the hand automatically using the digitized radiograph (Fig. 2a–d). Figure 2a Indicates the hand X-ray (PA view) of normal individual, Fig. 2b depicts the segmented image of normal subject. Figure 2c shows the hand X-ray (PA view) of RA patient. The arrow marks indicate erosion at the 1st and 2nd MCP joint regions. Figure 2d Indicates an automated segmented whole hand X-ray of RA patient using DTCWT based watershed algorithm. The ROI marked in MCP joints (1–5) to extract statistical features using GLCM method.

A comparison study is made to find the statistical features of ROI in MCP1-MCP5 joints before segmentation and after segmentation in both RA patients and normal. Hence the results are tabulated below as Tables 1, 2, 3, 4 and 5. It was found that, there was no statistically significant difference between statistical features derived from the image just before and after segmentation. But the segmentation process is useful for the following reasons: (i) to minimize the

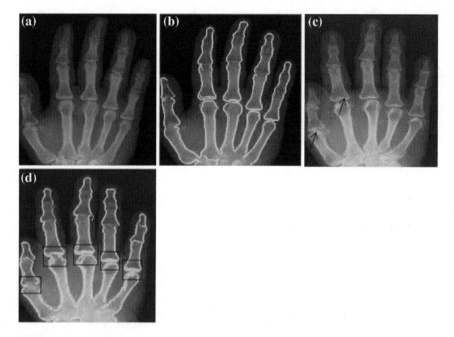

Fig. 2 Input and segmented output image for normal (**a, b**) and RA (**c, d**)

Table 1 Feature extraction of hand MCP1 joints of RA patients and normal before segmentation and after segmentation

Features	MCP1						
	RA (N = 50) mean ± SD			Normal (N = 20) mean ± SD			
	Before segmentation	After segmentation	Statistical significance (p)	Before segmentation	After segmentation	Statistical significance (p)	
Mean	0.250 ± 0.1	0.257 ± 0.1	0.75	0.159 ± 0.04	0.173 ± 0.03	0.16	
Std. deviation	0.076 ± 0.01	0.079 ± 0.01	0.9	0.032 ± 0.01	0.03 ± 0.01	0.09	
Variance	0.0001 ± 0.001	0.0001 ± 0.001	0.96	0.000001 ± 1.25E−06	0.000001 ± 1.27E−06	0.39	
Autocorrelation	0.0017 ± 0.001	0.001 ± 0.001	0.94	0.0009 ± 0.0002	0.0009 ± 0.0002	0.33	
Entropy	0.0003 ± 0.0007	0.0003 ± 0.0005	0.91	0.0001 ± 0.00005	0.0001 ± 0.00005	0.64	
Norm mean	0.0001 ± 0.0001	0.0001 ± 0.0001	0.95	0.00007 ± 2.46E−05	0.00007 ± 0.00002	0.4	
Exp. comp.	0.0002 ± 0.0002	0.0002 ± 0.0002	0.97	0.0001 ± 0.00004	0.0001 ± 0.00003	0.33	
Mean angle	0.0018 ± 0.002	0.001 ± 0.001	0.96	0.001 ± 0.0003	0.001 ± 0.0002	0.75	
Covariance	0.0001 ± 0.0001	0.0001 ± 0.0001	0.97	0.00005 ± 0.00001	0.00006 ± 0.00001	0.22	

Table 2 Feature extraction of hand MCP2 joints of RA patients and normal before segmentation and after segmentation

Features	MCP2 RA (N = 50) mean ± SD			Normal (N = 20) mean ± SD		
	Before segmentation	After segmentation	Statistical significance (p)	Before segmentation	After segmentation	Statistical significance (p)
Mean	0.252 ± 0.2	0.257 ± 0.2	0.92	0.136 ± 0.03	0.142 ± 0.02	0.28
Std. deviation	0.086 ± 0.1	0.09 ± 0.1	0.9	0.032 ± 0.01	0.03 ± 0.007	0.37
Variance	0.0004 ± 0.002	0.0005 ± 0.003	0.86	0.0002 ± 0.0002	0.000001 ± 6.9E–07	0.34
Autocorrelation	0.006 ± 0.03	0.007 ± 0.02	0.93	0.0008 ± 0.0002	0.0008 ± 0.0002	0.38
Entropy	0.0003 ± 0.0005	0.0003 ± 0.0006	0.95	0.0001 ± 0.00005	0.0001 ± 0.00003	0.51
Norm mean	0.0001 ± 0.0002	0.0001 ± 0.0002	0.94	0.00007 ± 2.7E–05	0.00006 ± 0.00001	0.15
Exp comp	0.0002 ± 0.0004	0.0002 ± 0.0004	0.97	0.0001 ± 4.9E–05	0.0001 ± 0.00002	0.12
Mean angle	0.002 ± 0.003	0.002 ± 0.003	0.96	0.0009 ± 0.0003	0.0009 ± 0.0001	0.1
Covariance	0.0001 ± 0.0001	0.0001 ± 0.0001	0.87	0.00005 ± 1.46E–05	0.00004 ± 0.00001	0.14

Table 3 Feature extraction of hand MCP3 joints of RA patients and normal before segmentation and after segmentation

Features	MCP3 RA (N = 50) mean ± SD			Normal (N = 20) mean ± SD		
	Before segmentation	After segmentation	Statistical significance (p)	Before segmentation	After segmentation	Statistical significance (p)
Mean	0.27 ± 0.2	0.264 ± 0.2	0.89	0.133 ± 0.04	0.139 ± 0.05	0.57
Std. deviation	0.09 ± 0.1	0.08 ± 0.1	0.7	0.024 ± 0.01	0.03 ± 0.01	0.77
Variance	0.0004 ± 0.003	0.0004 ± 0.002	0.98	0.0000008 ± 1.48E−06	0.000001 ± 2.08E−06	0.18
Autocorrelation	0.001 ± 0.003	0.001 ± 0.002	0.78	0.0008 ± 0.0002	0.0008 ± 0.0002	0.25
Entropy	0.0004 ± 0.0008	0.0003 ± 0.0005	0.64	0.0001 ± 0.00008	0.0001 ± 0.00006	0.12
Norm mean	0.0001 ± 0.0002	0.0001 ± 0.0001	0.67	0.00007 ± 3.3E−05	0.00006 ± 0.00001	0.34
Exp comp	0.0002 ± 0.0005	0.0002 ± 0.0003	0.65	0.0001 ± 6.09E−05	0.0001 ± 0.00003	0.45
Mean angle	0.002 ± 0.003	0.001 ± 0.002	0.58	0.001 ± 0.0004	0.0009 ± 0.0002	0.48
Covariance	0.01 ± 0.0002	0.0001 ± 0.0003	0.32	0.00004 ± 1.62E−05	0.00003 ± 0.00001	0.61

Table 4 Feature extraction of hand MCP4 joints of RA patients and normal before segmentation and after segmentation

Features	MCP4					
	RA (N = 50) mean ± SD			Normal (N = 20) mean ± SD		
	Before segmentation	After segmentation	Statistical significance (p)	Before segmentation	After segmentation	Statistical significance (p)
Mean	0.29 ± 0.1	0.31 ± 0.2	0.69	0.156 ± 0.06	0.16 ± 0.04	0.33
Std. deviation	0.09 ± 0.1	0.09 ± 0.1	0.95	0.027 ± 0.02	0.04 ± 0.01	0.31
Variance	0.02 ± 0.01	0.0009 ± 0.006	0.3	0.00002 ± 0.0001	0.000002 ± 2.14E−06	0.37
Autocorrelation	0.003 ± 0.001	0.002 ± 0.003	0.3	0.0009 ± 0.0003	0.0008 ± 0.0002	0.51
Entropy	0.0003 ± 0.0002	0.0003 ± 0.0005	0.35	0.0001 ± 0.00007	0.0001 ± 0.00005	0.49
Norm mean	0.0001 ± 0.0001	0.0001 ± 0.0002	0.43	0.0001 ± 5.57E−05	0.00007 ± 0.00001	0.47
Exp comp	0.0003 ± 0.0001	0.0003 ± 0.0004	0.68	0.0001 ± 6.61E−05	0.0001 ± 0.00003	0.46
Mean angle	0.002 ± 0.004	0.002 ± 0.003	0.73	0.001 ± 0.0004	0.001 ± 0.0002	0.51
Covariance	0.0001 ± 0.0002	0.0001 ± 0.0001	0.74	0.00005 ± 2.04E−05	0.00005 ± 0.00001	0.34

Table 5 Feature extraction of hand MCP5 joints of RA patients and normal before segmentation and after segmentation

Features	MCP5 RA (N = 50) Mean ± SD			Normal (N = 20) Mean ± SD		
	Before segmentation	After segmentation	Statistical significance (p)	Before segmentation	After segmentation	Statistical significance (p)
Mean	0.45 ± 0.3	0.478 ± 0.2	0.81	0.159 ± 0.05	0.169 ± 0.03	0.85
Std. deviation	0.169 ± 0.2	0.175 ± 0.1	0.93	0.0304 ± 0.01	0.04 ± 0.009	0.46
Variance	0.005 ± 0.03	0.005 ± 0.02	0.96	0.000001 ± 1.83E−06	0.000002 ± 1.37E−06	0.29
Autocorrelation	0.005 ± 0.001	0.003 ± 0.001	0.41	0.001 ± 0.0003	0.001 ± 0.0002	0.25
Entropy	0.0005 ± 0.0001	0.0006 ± 0.0001	0.63	0.0001 ± 0.0001	0.0001 ± 0.00004	0.97
Norm mean	0.0004 ± 0.0001	0.0002 ± 0.0002	0.8	0.00009 ± 0.00004	0.00007 ± 0.00002	0.87
Exp comp	0.0005 ± 0.001	0.0005 ± 0.001	0.91	0.0001 ± 8.3E−05	0.0001 ± 0.00003	0.78
Mean angle	0.006 ± 0.01	0.004 ± 0.007	0.3	0.001 ± 0.0006	0.0009 ± 0.0004	0.81
Covariance	0.0002 ± 0.0007	0.0002 ± 0.0004	0.68	0.00006 ± 2.11E−05	0.00006 ± 0.00001	0.23

subjective error in keeping ROI at desired region and its feature measurement; (ii) It enhance the reproducibility of the feature extraction measurements.

3.1   Feed Forward Back Propagation Network

The feed forward Back propagation (BPN) network has two different approaches for determining the input features in the input layer of the network. The input features were given as follows: (i) only X-ray image features (n = 45) and (ii) both X-ray image features and hand geometry variables (n = 53). The selection of number of hidden neurons for the hidden layer for two different input features was determined by performing the analysis as given in Table 6 and Fig. 1. Hence at

Table 6 Results of feed forward BPN network illustrating the percentage accuracy for different number of hidden units-input X-ray image features (n = 45) and input [combination of X-ray image and geometric features (n = 53)].

No. of units in the hidden layer	Accuracy (%) (n = 45)	Accuracy % (n = 53)
1	74.3	91.4
2	72.9	84.3
3	72.9	81.4
4	94.3	90.9
5	75.7	90.3
6	72.9	90
7	75.7	91.4
8	92.9	88.6
9	75.7	72.9
10	**95.7**	**92.9**
11	92.9	81.4
12	94.3	88.6
13	72.9	87.1
14	84.3	75.1
15	92.9	74.3
16	92.7	85.7
17	92.7	84.3
18	87.1	74.3
19	92.9	87.1
20	92.9	87.1

At hidden unit 10, the maximum accuracy of 95.7 % and 92.9 % was achieved and hence highlighted with bold

10th hidden neurons, maximum accuracy was achieved in both the approaches of input features. The percentage accuracy obtained from feed forward BPN network for different number of hidden units for the input features such as image features (n = 45) and combination of image features and geometric features (n = 53) was illustrated in Table 6. The line graph illustrating the selection of hidden neurons based on the maximum accuracy for input X-ray image features (n = 45 and n = 53) was depicted in Fig. 3a, b. Hence at hidden unit 10, maximum accuracy of 95.7 and 92.9 % was obtained.

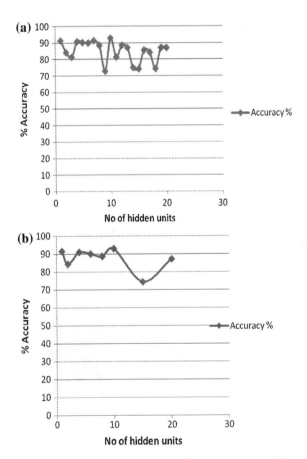

Fig. 3 a, b % Accuracy for different number of hidden units for input X-ray image features (n = 45) and (n = 53)

Figure 4a shows the architecture of feed forward BPN network with input as image features. According to Fig. 4a, the number of input units used in input layer was 45, hidden units used in hidden layer was 10 and one output unit which classifies the RA group and normal. Similarly Fig. 4b depicts the architecture of feed forward BPN network with input as combination of geometric and image features. The number of units used in input layer was 53, hidden units 10 and one output unit. Figure 5a, b depicts the confusion matrix for the feed forward BPN network with input as image features for different phases such as training phase, validation phase, testing phase and overall resultant phase. From the confusion matrix, the sensitivity of 88.2 %, specificity of 98.1 %, PPV of 93.8 %, NPV of 96.3 % and accuracy of 95.7 % was achieved for input features n = 45. The sensitivity of 82.4 %, specificity of 96.2 %, PPV of 87.5 %, NPV of 94.4 % and accuracy of 92.9 % was obtained for input features n = 53. The ROC curves were shown in Fig. 6a, b. The percentage of correct classification measured by the area under the ROC curve was 95.7 % for image feature as input and 92.9 % for the combination of geometric and image feature as input.

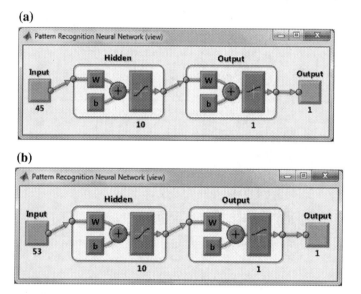

Fig. 4 **a, b** Architecture of feed forward BPN network for input image features (n = 45 and 53)

(a)

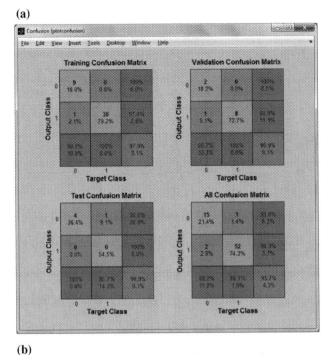

(b)

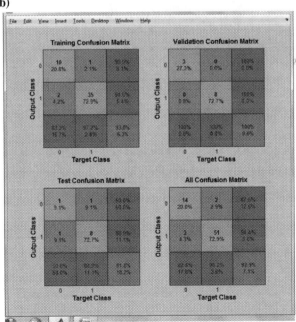

Fig. 5 **a, b** Confusion matrix for feed forward BPN network for input as image features (n = 45 and n = 53)

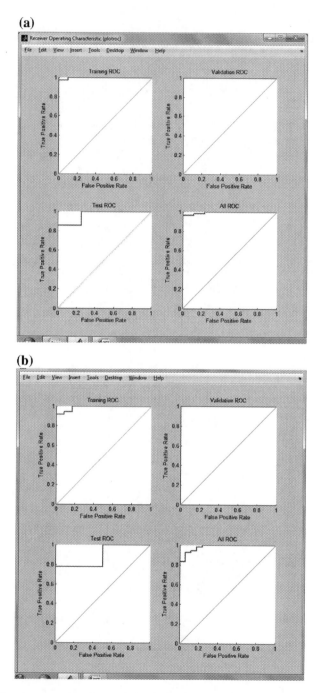

Fig. 6 a, b Receiver operating characteristic plot for feed forward BPN network for input image features (n = 45 and n = 53)

4 Discussions

The results of our experiment depicted that DTCWT based watershed algorithm gives accurate segmentation of hand X-ray images. Based on statistical features and geometric features extracted from the segmented output image, the feed forward BPN network effectively classified the normal and RA images with good sensitivity and specificity with better accuracy. Pfeil et al. [26] performed the semi-automated measurements of joint space distance at the finger articulations based on hand radiographs. They quantified the severity of joint space narrowing by sharp score and erosions by using z score based on joint space width (JSW) differentiated for each peripheral finger joint in RA patients. The results from their data revealed that the high sensitivity (85.4 %) and low specificity (55.2 %) were obtained for MCP-JSW differentiated for the detection of bone erosions [26]. Vera [11] developed an automated method for joint detection and joint space characterization for evaluation of RA. The image processing techniques like binarization and thinning method were used to analyze the skeleton image on the hand radiograph of RA patients. The algorithm provided the global detection rate of JSW above 95 % [11]. Zielinski [27] applied fully automated algorithm to detect the joint cavity width in the MCP and proximal inter phalangeal (PIP) joints for the evaluation of RA in hand radiographs. The image processing techniques used were binarization and thinning methods. It was reported that algorithm provided satisfactory and accurate result (81 %) in the joint cavity width analysis [27]. Hence in our study, DTCWT based watershed algorithm produced better segmentation results in hand bone X-ray images irrespective of size and damaged joints. Scheel et al. [28] in their study utilized multilayer perceptron classifier to detect the inflammatory changes in affected hand joints of RA patients. They found the inflammatory status of RA joints examined correctly by laser examination and joint circumference determination. They obtained the sensitivity of 80 %, specificity of 89 % and accuracy of 83 % in detecting the inflammatory changes of the affected hand joints [28]. In our study, feed forward back propagation network was used for the classification of subjects into RA and normal using the statistical extracted features of hand segmented image and hand geometry variables as input features. The accuracy of 95.7 and 92.9 % was obtained for input feature as image features (n = 45) and combination of image and geometric features (n = 53) respectively.

The limitations of this study was described as follows: (i) It is known that Mimics software is used to measure geometry from the image with high accuracy. The reason for not using an image processing algorithm in the measurement of both JSW and CCT are given as follows (a) In this study a conventional standard radiograph was taken due to non-availability of digital radiograph. After that, it was digitized with a depth of 12 bit per pixel. This poor resolution of image limits the utilization of an automated image processing algorithm; (b) According to the designed protocol of this study, both JSW and CCT were measured at multiple skeletal sites and bone joints. This multiple measurements at different regions limit the utilization of an automated image processing algorithm. (ii) Fully automated method may not be possible due to the following reason: (a) In this study, conventional radiograph was used for image processing. The image was digitized using digitizer (LASER film digitizer, Model: 2905 Array Corporation Netherlands, Europe). But, it had poor resolution. Further, preprocessing tools were used to enhance the image quality. Hence semi-automated method was attempted and subsequently features were extracted; (b) there was a inter-technological persons variations in positioning of the hand while obtaining the image which was really unavoidable. But, we followed standard X-ray tube parameter for obtaining the image. Hence the first two points limits the fully automated image processing. The following are referred in the literature in connection with limitation of image processing in the evaluation of Kauffman et al. [29] reported in his study that the fully automated method subjects to particular conditions and tasks for which they have been designed. Hence in cases of abnormal and severe deviations, it may results in false measurements. They also recommended in their study that a standardized radiographic acquisition protocol is required to minimize the positioning variability in order to improve the segmentation results and selection of required ROI [29]. Due to the different input sources and different age groups of the subjects in radiographs, there is some variation in terms of illumination, size of the bones and locations of the bone. Hence these variations affect the performance and quality of the segmentation techniques [30].

5 Conclusion

The DTCWT based watershed algorithm provided effective segmentation in the digitized hand radiograph. The feed forward BPN network was utilized for the classification of normal and RA patients from statistical features (total inputs, n = 45) extracted from hand X-ray image. The calculated accuracy of the classifier was found to be 95.7 %, whereas its sensitivity and specificity was found to be 88.2 and 98.1 % respectively. When both the statistical features as well as geometric features (total inputs, n = 53) were used in the feed forward BPN network, it was found that accuracy was 92.9 %, whereas its sensitivity and specificity were found to be 82.4 and 96.2 % respectively.

References

1. www.who.int/chp/topics/rheumatic/en/. Accessed on 20th June 2015
2. www.nras.org.uk/about_rheumatoid_arhtritis/whatisra/. Accessed on 20th June 2015
3. Scott, D.L., Wolfe, F., Huizinga, T.W.: Rheumatoid arthritis. Lancet 376, 1094–1108 (2010)
4. Ferucci, E.D., Templin, D.W., Lanier, A.P.: Rheumatoid arthritis in American Indians and Alaska natives: a review of the literature. Semin. Arthritis Rheum. 34(4), 662–667 (2005)
5. Zeng, Q.Y., Chen, R., Darmawan, J., Xiao, Z.Y., Chen, S.B., Wigley, R., Chen, S.L., Zhang, N.Z.: Rheumatic disease in China. Arthritis Res. Ther. 10(1), R17 (2008)

6. Biver, E., Beague, V., Verloop, D., Mollet, D., Lajugie, D., Baudens, G., Neirinck, P., Flipo, R.M.: Low and stable prevalence of rheumatoid arthritis in northern France. Joint Bone Spine **76**(5), 497–500 (2009)
7. Van der Heijde, D.M.F.M.: Radiographic imaging: the gold standard for assessment of disease progression in rheumatoid arthritis. Rheumatology **39**(1), 9–16 (2000)
8. Landewe, R., Vander Heijde, D.: Radiographic progression in rheumatoid arthritis. Clin. Exp. Rheumatol. **23**(39), S63–S68 (2005)
9. Larsen, A., Dale, K., Eek, M.: Radiographic evaluation of rheumatoid arthritis and related conditions by standard reference films. Acta Radiol. **18**, 481–491 (1977)
10. Arbillaga, H.O., Montgomery, G.P., Cabarrus, L.P., Watson, M.M., Martin, Liam, Edworthy, S.M.: Internet hand X-rays: a comparison of joint space narrowing and erosion scores (Sharp/Genant) of plain versus digitized X-rays in rheumatoid arthritis patients. BMC Musculoskelet. Disord. **3**, 13 (2002). doi:10.1186/1471-2474-3-13
11. Vera, S.: Finger joint modeling from hand X-ray images for assessing rheumatoid arthritis. In: Center de Visio per computador CVC Technical report, #164, pp. 1–30 (2010)
12. Hsien, C.W., Jong, T.L., Tiu, C.M.: Bone age estimation based on phalanx information with fuzzy constrain of carpals. Med. Biol. Eng. Comput. **45**(3), 283–295 (2007)
13. De Luis-Garcia, R., Martín-Fernández, M., Arribas, J.I., Alberola-López, C.: A fully automatic algorithm for contour detection of bones in hand radiographs using active contours. In: Proceedings of the IEEE International Conference on Image Processing, pp. 421–424 (2003)
14. Michael, D.J., Nelson, A.C.: HANDX: a model-based system for automatic segmentation of bones from digital hand radiographs. IEEE Trans. Med. Imaging **8**(1), 191–193 (1990)
15. Peloshek, P., Langs, G., Weber, J., Sailer, M., Reisegger, H., Imhof, H., Bischof, H., Kainberger, F.: An automatic model-based system for joint space measurements on hand radiographs: initial experience. Radiology **245**(3), 855–862 (2007)
16. Levit, T.S., Hedgcock, M.W., Dye, J.W., Johnston, S.E., Shadle, V.M., Vosky, D.: Bayesian inference for model based segmentation of computed radiographs of the hand. Artif. Intell. Med. **5**(4), 365–387 (1993)
17. Sotaca, J.M., Inesta, J.M., Belmonte, M.A.: Hand bone segmentation in radioabsorptiometry images for computerized bone mass assessment. Comput. Med. Imaging Gr. **27**, 459–467 (2003)
18. Mishra, R., Sharma, B.L., Gupta, R., Pandya, S., Agarwal, S., Agarwal, P., Grover, S., Sarma, P., Wanjam, K.: Indian Rheumatology Association consensus statement on the management of adults with rheumatoid arthritis. Indian J. Rheumatol. **3**(3), S1–S16 (2008)
19. Selesnick, I.W., Baraniuk, R.G., Kingsbury, N.G.: The dual tree wavelet transform. IEEE Signal Process. Mag. **22**(6), 123–151 (2005)
20. Beucher, S., Meyer, F.: The mathematical approach to segmentation: the watershed transform (Chap. 12). In: Dougherty, E.R. (ed.) Mathematical Morphology in Image Processing, pp. 433–481. Marcel Dekker, New York (1993)
21. Haralick, R.M., Shanmugam, K., Dinstein, I.: Texture features for image classification. IEEE Trans. Syst. Man Cybern. **SMC-3**, 610–621 (1973)
22. Snekhalatha, U., Anburajan, M.: Computer based measurements of joint space analysis at metacarpal morphometry in hand radiograph for evaluation of rheumatoid arthritis. J. Rheum. Dis., Int (2015). doi:10.1111/1756-185X.12559
23. He, D.C., Wang, L., Juibert, J.: Texture feature extraction. Pattern Recognit. Lett. **6**, 269–273 (1987)
24. Choi, B., Lee, J.H., Kim, D.H.: Solving local mimima problem wit hlarge number of hidden nodes on two layered feed-forward artificial neural network. Neurocomputing. **71**, 3640–3643 (2008)
25. Pratiwi, D., Santika, D.D., Paradamean, B.: An application of back propagation artificial neural network for measuring the security of osteoarthrtitis. Int. J. Eng. Technol. **11**(3), 102–105 (2011)
26. Pfeil, A., Oelzner, P., Bornholdt, K., Hansch, A., Lehmann, G., Renz, D.M., Wolf, G., Bottcher, J.: Joint damage in rheumatoid arthritis: assessment of a new scoring method. Arthritis Res Ther 15, R27 (2013). doi:10.1186/ar4163

27. Zielinski, B.: Hand radiograph analysis and joint space locations improvement for image interpretation. Schedae Inform. **17**, 45–61 (2009)
28. Scheel, A.K., Krause, A., Rheinbaben, I.M., Metzger, G., Rost, H., Tresp, V., Mayer, P., Borst, M.R., Muller, G.A.: Assessment of proximal finger joint inflammation in patients with rheumatoid arthritis using a novel laser-based imaging technique. Arthritis Rheum. **46**(5), 1177–1184 (2002)
29. Kauffman, J.A.: Automated radiographic assessment of hands in rheumatoid arthritis. Ph.D. thesis, doi:10.3990/1.9789036528306. http://doc.utwente.nl/61094/1/thesis_J_Kauffman.pdf (2009)
30. Hum, Y.C., Lai, K.W., Utama, N.P., Salim, M.I.M., Myint, M.: Review on segmentation of computer aided skeletal maturity assessment. In: Advances in Medical Diagnostic Technology. Lecture Notes in Bioengineering, pp. 23–38. Springer, Berlin. doi:10.1007/978-981-4585-72-9-2 (2014)

Comfort Sensor Using Fuzzy Logic and Arduino

S. Sharanya and Samuel John

Abstract Automation has become an important part of our life. It has been used to control home entertainment systems and household activities. One of the main parameters to control in a smart home is the atmospheric comfort. Atmospheric comfort mainly includes temperature and relative humidity. In homes, the desired temperature of different rooms varies from 20 to 25 °C and relative humidity is around 50 %. Hence, automated measurement of these parameters to ensure comfort assumes significance. To achieve this, a fuzzy logic controller using Arduino was developed using MATLAB. In the present work, soft sensor was introduced in this system that can indirectly measure temperature and humidity and can be used for processing several measurements. The Sugeno method was used in the soft sensor in MATLAB and then interfaced to the Arduino, which is again interfaced to the temperature and humidity sensor DHT11. The comfort sensor developed was able to measure temperature and relative humidity correctly. Depending on the comfort percentage, the air conditioners and the coolers in the room were controlled. The main highlight of the project is its cost efficiency.

Keywords Arduino · DHT11 · Soft sensor · Sugeno

1 Introduction

Home automation is the process by which basic and simple home functions are controlled either automatically or remotely using single or multiple computers. These simple home functions could include yard watering, centralized control of lighting, control of home entertainment systems and pet feeding etc. The need for automation has gradually increased over the years with advancement in computer systems. The present day systems can process values acquired from the environ-

S. Sharanya · S. John (✉)
Department of Electronics and Instrumentation, SRM University, SRM Nagar,
Kattankulathur, Chennai 603203, Tamil Nadu, India
e-mail: samjohn.jk@gmail.com

© Springer Science+Business Media Singapore 2017
P. Deiva Sundari et al. (eds.), *Proceedings of 2nd International Conference on Intelligent Computing and Applications*, Advances in Intelligent Systems and Computing 467, DOI 10.1007/978-981-10-1645-5_13

155

ment around them and control the process depending on the set value. Home automation systems are also used in security applications, where an alarm is set along with the GSM communication system. If an intruder is detected then the alarm alerts the neighbors and also sends a message to the owners. One of the major functions to be controlled in a house is the atmospheric comfort level. The temperature varies from 0 to 25 °C and humidity is around 50 %. The atmospheric comfort is characterized by the temperature and the humidity present in the room. Depending on the values the system predicts the comfort value which is based on the Sugeno system. Unlike all home automation systems which make use of PLC and PIC controllers, this system uses an embedded microcontroller (Arduino) which is used to acquire the values through the temperature and humidity sensor and with help of MATLAB fuzzy computing system the comfort level is predicted.

2 Hardware Design

2.1 Arduino

It is an open source platform which is used as the microcontroller. It uses an AT mega 328 or 168 as its IC. It has analog inputs and digital I/O pins. In addition to this, it has on board voltage regulator and can be used to supply +5 V to any other external device. Arduino is portable in nature and highly efficient in operation. Arduino can be used as a standalone device or can be interfaced to sensors, actuators and other external peripheral devices. The Arduino takes input from these sensors, processes the data and can be used for controlling the motors or any other physical process. The Arduino can be programmed using Arduino IDE software which is basically consisting of C, C++ and java.

2.2 DHT11

The DHT11 is a sensor which gives a digital signal that can be calibrated, measuring both temperature and humidity. This sensor confirms long term stability as well as high reliability. The sensor comprises of a resistive element and with the help of a microcontroller senses the corresponding parameters. In addition to this, it has a rapid response, anti-interference ability and high cost performance advantages. The sensor also provides a set of features such as low power, small size and can transmit the signal up to 20 m.

2.3 Soft Sensor

This is a virtual sensor where several measurements can be processed together. They make use of the system knowledge and variable measurements in order to improve availability, reliability and accuracy of the measurements of the variables of interest. The software accesses and analyses measurements of variables correlated with the variables of interest. The soft sensor is designed using matlab and the inputs are fed to the sensor through the matlab with the help of a sensor interface. The soft sensor can be used for providing the estimate of process variables where the variables cannot be measured frequently. Soft sensors are especially used in data fusion, where measurements of different characteristics and dynamics are combined. Soft sensors are also used for fault detection as well as control applications. The implications of soft sensor use neural networks or fuzzy computing.

2.4 Proposed Model

The system consists of a soft sensor which has been designed in matlab using the fuzzy tool kit. The inputs to the soft sensor are obtained from the DHT11 through MATLAB and Arduino. In order to predict the comfort level, we have used the Sugeno system. We have used three membership functions for the inputs namely temperature and humidity. The system uses triangle shaped membership functions. The output set consists of three singleton membership functions: very uncomfortable ($u1 = 0$), uncomfortable ($u2 = 0.5$), comfortable ($u3 = 1$). There are nine set of rules set and developed for the system (Fig. 1).

The system has three stages of operation: fuzzification, prediction and defuzzification. Fuzzification is the process where the affiliation of the input variables to a membership function is calculated. The predefined set of rules set for the system has been used in the prediction stage to predict what the output can be. Defuzzification gives an output value between 0 and 1. The defuzzification is also known as the sharpening of the output value. The Arduino is programmed using the Arduino IDE software for acquiring the data form the DHT11 sensor. The Arduino together with the DHT11 forms the sensor interface part to the soft sensor. The atmospheric comfort level is predicted as a result of the two input variables namely temperature and humidity (Fig. 2).

The system displays the comfort level (Fig. 3) which in turn sends a signal to the cooling systems present in the room.

The cooling system adapts according to the comfort level present in the room. The developed system adapts itself to the surrounding environment and thereby predicts the comfort level.

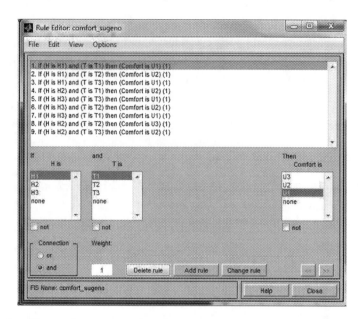

Fig. 1 Set of fuzzy rules

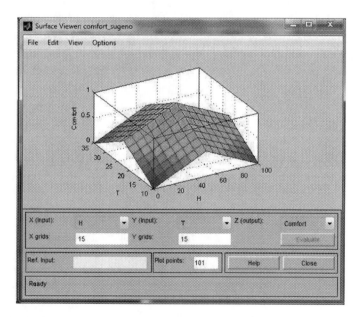

Fig. 2 Graph of comfort in reliance of temperature and relative humidity

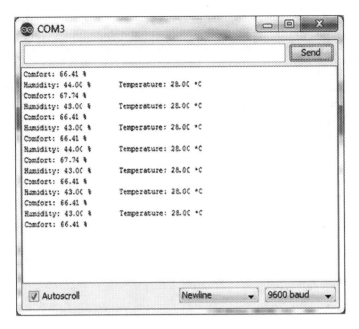

Fig. 3 The output values obtained from the Arduino serial monitor

3 Future Work

The system can used to infer the humidity-temperature value of the soil which can be used to determine the need for irrigation. Wireless transmission of data from the system to the external peripherals can be implemented.

4 Background

In estimating the comfort level of the room, very little work has been done in order to implement the system. Jassar et al. [1] proposed a system on the basis of ANFIS (adaptive neural fuzzy inference system) for evaluating the median air temperature in space heating systems. This method combines the benefit of expert knowledge of fuzzy inference systems (FISs) and the learning capability which is found in artificial neural networks (ANNs). It describes how an adaptive network based inferential sensor can be utilized to design closed-loop control for space heating systems. This main aim of this system is to progress the complete performance of heating systems, with respect to energy efficiency and thermal comfort.

This system is successful in calculating the thermal comfort by measuring only the temperature, but comfort level of a room depends on humidity and temperature.

Moreover since this is a neural approach which needs training and the modelling of the system is very complex compared to the proposed system.

Wang and Jin [2] developed a model which provided optimal control of VAV air-conditioning system using genetic algorithm. It used a control mechanism using a system approach based on estimating the responses of the overall system environment and energy performance to the changes of control settings of VAV air-conditioning systems.

The above planned system is able to control the air conditioning system of a room using genetic approach but the system is very complex as it involves the use of control valves and much more. Whereas the comfort sensor is implemented using Arduino which can be self-implemented by the user and can be programmed according to his purpose. Ming Tham [3], had stressed the use of a soft sensor for process approximation and inferential control.

The proposed comfort soft sensor has been developed to reduce the complexity of the system and also to make it very cost efficient.

5 Conclusion

This is an ongoing project. Our primary focus is to reduce the cost of the system and also make it very efficient in operation. Compared to all other system this system uses Arduino as its controller which can be easily programmed by the user according to his own specifications. Unlike all other home related activities like yard watering, controlling the lights, the atmospheric comfort is an important parameter which needs to be controlled. Compared to all other systems which are complex in nature, this system makes home automation more affordable and more cost efficient to the user.

References

1. Jassar, S., Liao, Z., Zhao, L.: Adaptive neuro-fuzzy based inferential sensor model for estimating the average air temperature in space heating systems. Build. Environ. **44**, 1609–1616 (2009)
2. Wang, S., Jin, X.: Model-based optimal control of VAV air-conditioning system using genetic algorithm. Build. Environ. **35**, 471–487 (2000)
3. Tham, M.T., Montague, G.A., Morris, A.J., Lant, P.A.: Soft sensors for process estimation and inferential control. J. Process Control **1**, 3–14 (1991)

A Novel Integrated Converter Based Hybrid System for Alternative Street Lighting

Aashish Nikhil Ghosh, A. Ajay Rangan, Nikhil Mathai Thomas
and V. Rajani

Abstract With the expansion of cities and conversion of rural to urban areas, one of the most fundamental requirements is the installation of street-lighting. However with increased expansion, this also creates an extra load requirement that is not easily met by existing grid systems. This puts an undue stress on power production. A solution to this issue is proposed by a hybrid self-sustained solution for street lighting. This makes use of wind and solar energy. The focus of this paper is the fused converter topology used in our solution.

Keywords Fused converter · Hybrid solution · Wind · Solar · Street lighting

1 Introduction

Most rural expansion projects will always include some form of street lighting in the project. And with the rapid growth of the Indian economy, expansion and urbanization is reaching a knee point on an exponential curve. But with this rapid rise in economy, there is a need for electric power to meet demands. This has led to the power crisis in India with current plants being unable to provide for the increasing power demand. And even if new plants were built in perfect synchronization with load rise, the conventional fuels used like coal, oil and gas are quickly

A.N. Ghosh (✉) · A. Ajay Rangan · N.M. Thomas · V. Rajani
Department of Electrical and Electronics Engineering, SSN College of Engineering,
Chennai, Tamil Nadu, India
e-mail: aashish.n.ghosh@gmail.com

A. Ajay Rangan
e-mail: ranganajay@gmail.com

N.M. Thomas
e-mail: nikhilmthomas@gmail.com

V. Rajani
e-mail: rajaniv@ssn.edu.in

© Springer Science+Business Media Singapore 2017   161
P. Deiva Sundari et al. (eds.), *Proceedings of 2nd International Conference
on Intelligent Computing and Applications*, Advances in Intelligent Systems
and Computing 467, DOI 10.1007/978-981-10-1645-5_14

being exhausted. Hence it has become essential now to make as much of the load as possible independent of the main grids. This lets the transition from conventional to alternative fuels become gradual. Because of this there has been a worldwide movement to shift towards safer and cleaner sources of energy. Solar, wind, geo-thermal as examples have all been considered. Another advantage of the system is scalability and independence. However when these sources are used independently they are very unreliable. They have a seasonal dependence with solar being the most productive in summer and wind the most productive in autumn and the monsoon months. This disadvantage can be mitigated by the use of complimentary sources, the most popular being wind and solar. The use of complimentary sources improves productivity because when one source is inactive or unavailable, the other compensates. If both are unavailable then the energy storage elements of the system would take over. One problem that is solved in this case is pollution, these sources are almost completely pollution free (except for the manufacturing process) as opposed to the conventional sources which always involve an element of pollution that is highly detrimental to the environment. The effects often have far reaching effects like climate change. Even Nuclear energy fails in this respect as the used fuel and waste is highly toxic to the environment. And possibly the greatest advantage of alternative energy systems is the fact that they are fully scalable and completely independent. Solar panels may be used to provide lighting and fan loads to households. Or may be used in large solar farms. Solar based cell phone chargers also exist. Wind energy is less scalable than solar and depends on the smallest size of a single wind turbine; as an example Wind turbines have been included on top of large housing apartments to provide energy for the common areas.

The hybrid generation system is a development based on the complimentary source concept. A Cuk-Buck fused system [1, 2, 3], has been previously proposed. And in our case, the fused topology [1] of a Cuk-SEPIC is a hybrid dc-dc converter fusing the conventional Cuk and SEPIC topologies and hence reducing the total number of components. The Cuk will be supplied by wind and the SEPIC by solar. The battery will be supplied by the converter and the battery in turn will supply the lighting solution during low light conditions. The fused topology has better efficiency compared to two parallel converters.

The LED street lighting solution is chosen because of its immense advantages as opposed to the conventional High Pressure Sodium Vapor Lamp (HPSV). It has a significantly longer lifespan. And better color rendering properties and luminous efficacy. All this allows us to use alternative energy in a small scale design to power each street light since the energy requirement is low.

2 System Description

The fused topology consists of a Cuk and a SEPIC topology (Figs. 1 and 2).

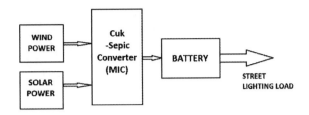

Fig. 1 The block-diagram representation of the proposed solution

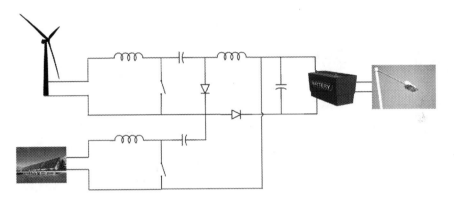

Fig. 2 A representation of the system with the converter visible

2.1 Cuk Converter Topology

The Cuk converter is used to convert rectified the electric power generated by the wind Turbine Generator to a controlled DC output for the load. The Cuk converter can act as both a step up as well as a step down converter. In the presence of wind, the Cuk converter will always operate in the step-down mode of operation. The Fig. 3 shows the basic circuit diagram of the Cuk converter. The voltage and current waveforms are depicted in Figs. 4, 5, 6 and 7.

Fig. 3 The Cuk converter

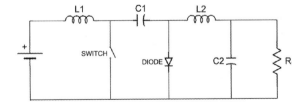

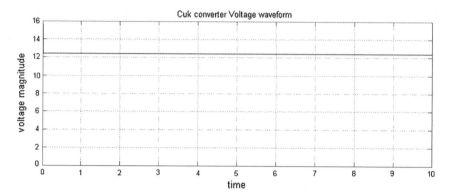

Fig. 4 Voltage waveform for the Cuk converter

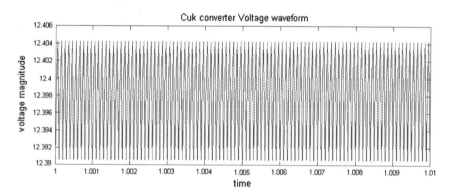

Fig. 5 Voltage waveform zoomed in

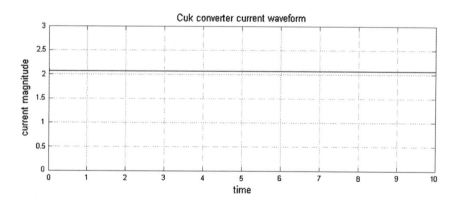

Fig. 6 Current waveform

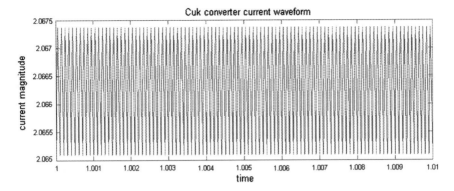

Fig. 7   Current waveform zoomed in

2.2   SEPIC Converter Topology

The SEPIC converter is used to convert the electric power from the solar panel to a controlled DC output for the load. Similar to the Cuk converter the SEPIC can also act in the step-up and step-down modes of operation. However in most cases when adequate sunlight is available it operates in the step-down mode. Figure 8 shows the basic circuit diagram of the Cuk converter. The voltage and current waveforms are depicted in Figs. 9, 10, 11 and 12.

These two converters were chosen because they offer the desired operation with small values of duty cycle and voltage ripple for their respective sources.

2.3   Fused Cuk-SEPIC Topology

The circuit diagram for the fused topology is shown in Fig. 13. The Cuk converter is the upper part of the topology while the SEPIC is the lower part. The fused topology has two distinct advantages over the two individual circuits connected in parallel. They are:

Fig. 8   The SEPIC converter

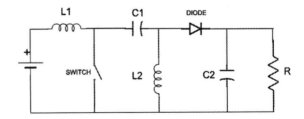

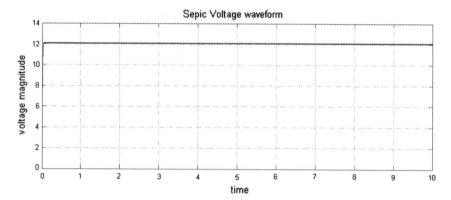

Fig. 9 The SEPIC output voltage waveform

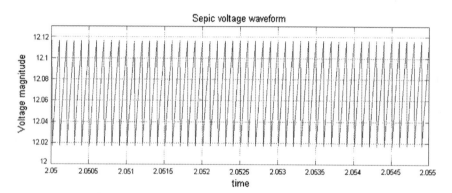

Fig. 10 The voltage waveform zoomed in

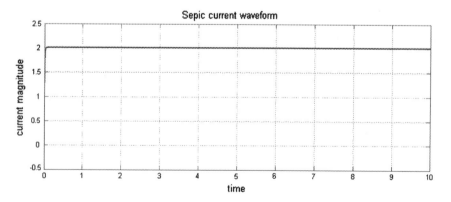

Fig. 11 The current waveform

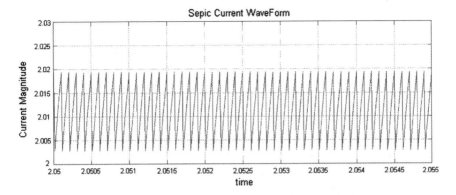

Fig. 12 The current waveform zoomed in

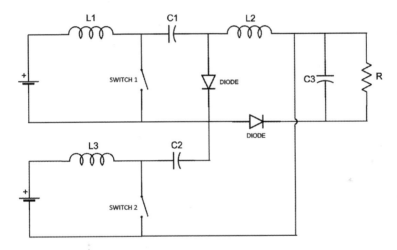

Fig. 13 The fused topology

1. Lower losses in the circuit.
2. Reduction in the number of components and hence also a reduction in size of the converter (significant when large power rating devices are used).

As can be seen from the circuit, the second inductor of the SEPIC converter has been removed in the fused topology. This highlights the reduction in the number of components that was stated above. The topology described is the topology suggested as the converter for the street lighting solution.

3 Parallel Power Transfer

The circuit performance may be improved by providing a parallel path to convert Solar panel and the load side ground connection. The current was seen to rise between 2 and 5 % for the same voltage levels (load side). These figures are based on experimentation and simulation in MATLAB/SIMULINK.

4 Working of the Circuit

The operation of the topology is given by two equations:

$$V_o = \left(d_{pv} / \left(1 - d_{pv} \right) \right) * V_{pv} = \left(d_w / \left(1 - d_w \right) \right) * V_w \tag{1}$$

$$I_{load} = I_{pv} + I_{wind} \tag{2}$$

where,

d_{pv} duty cycle of the solar power converter (SEPIC converter)
d_w duty cycle of the wind power converter (Cuk converter)
V_o output voltage of the converter
V_{pv} Solar panel voltage
V_w Wind generator voltage
I_{load} load current
I_{wind} wind source current contribution to I_{load}
I_{pv} solar cell contribution to I_{load}

   Note: the negative sign of Cuk output has been neglected due to the arrangement of the topology.

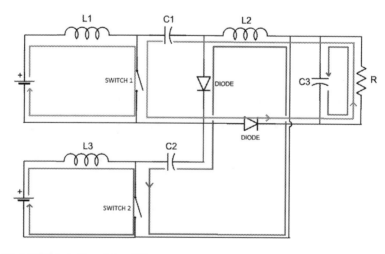

Fig. 14 MODE 1: both switches ON

The circuit has four modes of operation, based on the states of the two switches in the converter. Mode 1 with the switches ON and Mode 2 with the switches OFF. Mode 3 is when Switch 1 is ON and Switch 2 is OFF and Mode 4 when Switch 1 is OFF and Switch 2 is ON.

Figures 14, 15, 16 and 17 represent these states.

The converter has four different possible operating conditions depending on the availability of sources. Table 1 shows all possible conditions. 4 represents the condition (darkness without wind) when the battery is powering the load (LED lamp) and is not being charged by the converter.

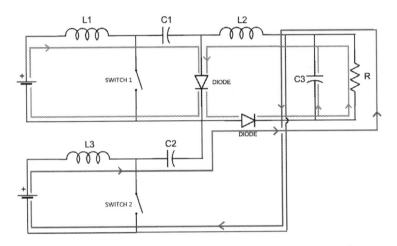

Fig. 15 MODE 2: both switches OFF

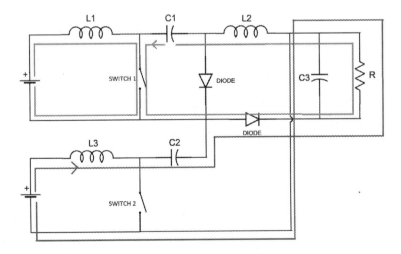

Fig. 16 MODE 3: S1 closed and S2 open

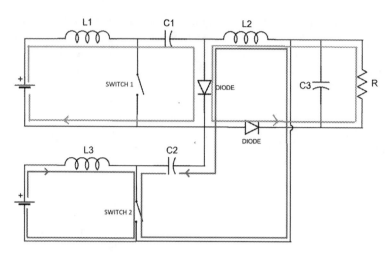

Fig. 17 MODE 4: S1 open and S2 closed

Table 1 Various source availability conditions

S. no.	Solar	Wind	Output voltage (V_0)	Load current (I_{LOAD})
1	Yes	Yes	V_0	$I_{WIND} + I_{PV}$
2	Yes	No	V_0	I_{PV}
3	No	Yes	V_0	I_{WIND}
4	No	No	0	0

5 Simulation Results

The above circuit was simulated in MATLAB simulator for different conditions of solar/wind availabilities.

In the first case both sources are active, while in the other two cases it is seen that one of the two sources is absent.

CASE 1: Solar + Wind sources both present. Figures 18, 19, 20 and 21 depicts the performance curves when both sources are available.
CASE 2: Solar as a Source alone.Figures 22, 23, 24 and 25 depicts the performance curves when only solar is available.
CASE 3: Wind as a Source alone. Figures 26, 27, 28 and 29 depicts the performance curves when only wind is available.

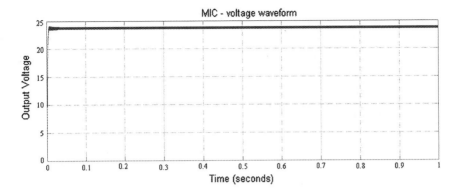

Fig. 18 Voltage waveform

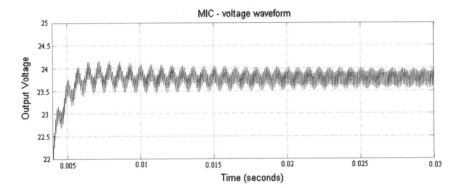

Fig. 19 Voltage waveform zoomed in

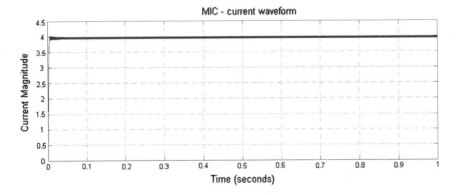

Fig. 20 Current waveform

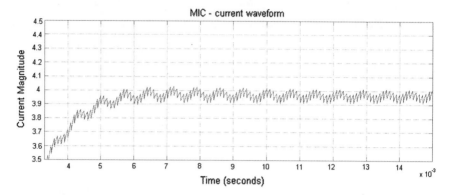

Fig. 21 Current waveform zoomed in

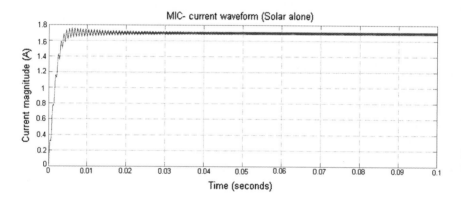

Fig. 22 Current waveform

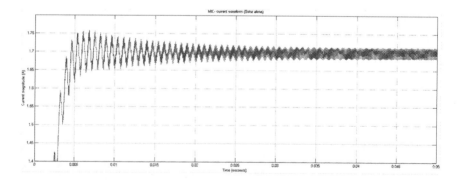

Fig. 23 Current waveform zoomed in

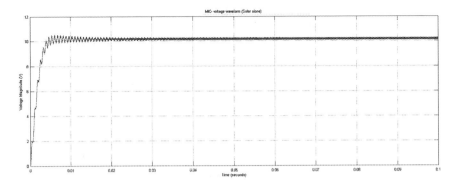

Fig. 24 Voltage waveform

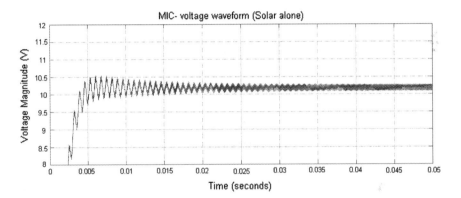

Fig. 25 Voltage waveform zoomed in

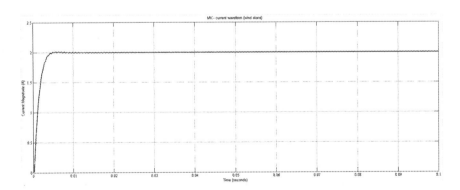

Fig. 26 Current waveform

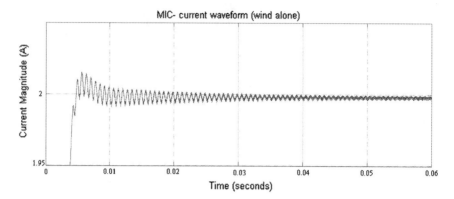

Fig. 27 Current waveform zoomed in

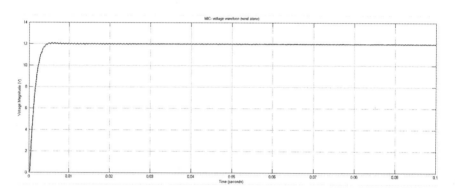

Fig. 28 Voltage waveform

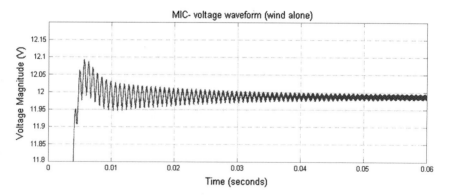

Fig. 29 Voltage waveform zoomed in

6   Conclusion

A hybrid system has been proposed using the fused converter topology based on the Cuk and SEPIC converters. This converter overcomes the drawbacks of previous designs. This topology allows the use of two different sources simultaneously or separately. This topology will find best application as proposed in the solution, for the lighting of remote regions, since the solution proposed is a grid-independent one.

References

1. Mangu, B., Kiran Kumar, K., Fernandes, B.G.: Efficiency improvement of solar-wind based dual-input converter for telecom power supply. In: IEEE, 978-1-4577-1829-8/12
2. Renugadevi, V., Margaret Amutha, W., Rajini, V.: A novel microgrid based DC-DC converter for rural telephony. Int. J. Res. Electr. Electron. Eng. 2(2), 25–32 (2014)
3. Amutha, W.M., Rajini, V., Renugadevi, V.: A novel fused converter based hybrid system with MPPT control for rural telephony. In: 2014 International Conference on Circuit, Power and Computing Technologies [ICCPCT]

Physiological Modeling of Retinal Layers for Detecting the Level of Perception of People with Strabismus

T. Rajalakshmi and Shanthi Prince

Abstract In the process of visual recognition, the human visual system (HVS) model provides a pathway on evaluating the image quality and there by detecting the performance of the human observer. Squint eye is the term associated with the misalignment in projection of the eye and hence provides binocular vision. Mathematical modeling of retina plays a vital role to produce better visual perception. The main aim of this work is to develop an HVS (human visual system) model for detecting the perceptual level of people with strabismus (squint). People with normal vision is capable of perceiving as one single image when the image falls on the retina. The situation is not the same in case of people with strabismus. One of the main symptoms of strabismus people is double vision, which is the simultaneous perception of two images of a single object. This paper focuses on the mathematical modeling of retinal layers namely photoreceptor, outer-plexiform and inner-plexiform layers and to provide an insight knowledge on the occurrence of double vision in the case of people with strabismus (medical term used for squint eye). It is proved through the proposed model that the energy level of normal vision person is higher than people with strabismus and the intensity level is approximately the same after processing through the photoreceptor layer.

Keywords Human visual system (HVS) · Strabismus · Binocular vision · Compression · Perception

T. Rajalakshmi (✉)
Department of Biomedical Engineering, SRM University,
Kattankulathur 603 203, India
e-mail: rajalakshmi.t@ktr.srmuniv.ac.in

S. Prince
Department of ECE, SRM University, Kattankulathur 603 203, India
e-mail: shanthi.p@ktr.srmuniv.ac.in

© Springer Science+Business Media Singapore 2017
P. Deiva Sundari et al. (eds.), *Proceedings of 2nd International Conference on Intelligent Computing and Applications*, Advances in Intelligent Systems and Computing 467, DOI 10.1007/978-981-10-1645-5_15

177

1 Introduction

Human retina is considered to be one of the most essential parts of the visual system. In view of retinal processing, optimal balance is required on data accuracy, processing on real time data, energy and structural complexity. The structure of retinal layer in HVS model is shown in Fig. 1. The first and foremost layer of retina is the photoreceptor layer which is responsible in acquiring information of the data visually and further compressing the luminance of the image using logarithmic compression technique. Photoreceptor layer includes two types of receptors namely the cones and the rods. Cone is responsible for photopic vision and is more sensitive to color. Rods are mainly achromatic. Central region of retina are occupied by cones. Cones are mainly used in day light because of their low light illumination. Visual response produced by rod at low level of illumination gives rise to scotopic vision. Photoreceptor layer accepts the incoming light information and enhances the contrast value of an image. This contrast enhanced image is then passed to the successive layers of retina.

Signals from the photoreceptor layer are transferred to the outer-plexiform layer which includes two types of cells namely, bipolar and horizontal cell layer. The main function of the outer-plexiform layer is to enhance the contour value of the incoming image. Bipolar cells are joined to the ganglion cells in the 'Inner-plexiform Layer' (IPL). Hence through inner plexiform layer finer details of the image are extracted. This work is an extension of the previous work on physiological modeling of retina for perceptual studies. The main aim of this study is to develop HVS model for detecting the level of perception of people with strabismus. This work is an extension of the previous work on physiological modeling of retina for perceptual studies. Strabismus is an abnormal condition which results in misalignment of the eye. Common name used for strabismus is crossed eye, lazy eye, double vision, squint and floating eye. Reasons for occurrence of squint includes, damages found in the muscles that controls the eyes movement, damages found in the nerves that is responsible for controlling the muscles, damage found in the brain due to non proper functioning of muscle control centers, Lack of vision in the eye which stops keeping the eyes together by the brain [1]. Strabismus affects about 2–4 % of the U.S. population approximately. In the U.S. alone 6–12 million people has lived with eyes apart. A world wide estimate shows that 130–260 million people are affected by strabismus [2]. Barry et al. proposed a photographic method for evaluating the angle of squint in both children and infants based on the reflection patterns using a reflex camera the angle of squint

Fig. 1 Retinal layer in HVS model

or angle kappa was calculated. Vertical angles of squint were calculated from the same photograph [3]. Thomas Lehmann et al. used hough transform and automatic strabtometry to detect whether the eye is normal or squint. Using cross covariance filtering method the center of iris was localized and reflex positions were located. The angle of squint was calculated by the MS-DOS and the calculation time was about two minutes. But the drawback that arises here is that this method cannot be applied to blurred images [4]. Seira Tak proposed artificial neural network approach using support vector machine for detecting squint eye detection problem which was based on training and testing features. The author carried out the work in two different phases namely training and testing set. In the training set the input eye image was compressed for the removal of distortion thereby extracting relevant geometrical features, these features were used in the testing phase for the detection of squint eye problem [5]. Lot of works in the literature was carried out to detect squint and to determine the angle of squint. This work was carried out as an extension of our previous work on contour contrast enhancement based on retinal layer processing. In the study spatial frequency was obtained by calculating DCT of the input image and temporal frequency was equated to zero [6]. No work was proposed to detect the level of perception of double vision or strabismus. This paper aims in detecting the level of perception of people with strabismus.

2 Mathematical Modeling of Photoreceptor Layer

Modified version of Michaelis–Menten relation [7, 8] which include a local adaption effect and to have a normalized value of luminance ranging from of [0, Vmax] is considered for the study:

$$A p = (p)p + (p).Vmax + Co(p) \tag{1}$$

$$Co p = So.Lp + Vmax\,1 - So \tag{2}$$

In Eq. 1, the adjusted luminance value of the original image Ap of the photoreceptor p depends mainly on the current luminance value of the image Cp and also on the compression parameter value Cop, which is further linearly linked to the luminance value Lp of the neighbouring photoreceptor p as shown in Eq. 2 [9]. In this work luminance value Lp is computed by calculating spatial and temporal frequencies as shown in Eq. (5)

$$Fh(fs, ft) = 1/[1 + \beta\,h + 2\alpha\,h(1 - \cos(2\pi f\,s)) + j2\,\pi\,\tau\,pf\,t] \tag{3}$$

$Fh(fs, ft)$ is the horizontal cell output which has only low spatial frequencies content of the image is used for evaluating the local luminance value $L\,p$ [10]. Spatial frequency (fs) is computed by applying Discrete Cosine transform (DCT) to the input image and temporal frequency (ft) is considered to be merely dc value since for this work a static image is considered. Equation 2 shows that the compression parameter value *Cop* depends mainly on the luminance value *Lp*. To enhance the flexibility and accuracy of the system, compression parameter *So* is considered whose value ranges from [0, 1]. *Vmax* Indicates the highest pixel value in an image and its value is 255, spatial cut off frequency αh is set at 7 and temporal constant τp is equated to 1 to minimize high frequency noise [11]. β h is the gain of Fh is set to 0.7 to provide result [12].

3 Mathematical Modeling of Outer-Plexiform Layer

Outer plexiform layer is modelled using spatial temporal filter whose transfer function is given by the expression.

$$F_{OPL}(fs) = F_{ph}(fs)[1 - F_h(f_s)] \tag{4}$$

where,

$$F_{ph}(f_s, f_t) = 1 / \left[1 + \beta_{ph} + 2\alpha_{ph}(1 - cos(2\pi fs)) + j2\,\pi\tau_{ph}f_t \right] \tag{5}$$

$$F_h(f_s, f_t) = 1 / [1 + \beta_h + 2\alpha_h(1 - cos(2\pi f_s)) + j2\,\pi\tau_h f_t] \tag{6}$$

As shown in [8] the spatial temporal filter is derived as a difference in the low pass filter which models the photoreceptor network and a low pass filter which model the horizontal cell network h of the retina. Response of the outer-plexiform is evaluated by calculating the difference between Fph and Fh and denoted as BON and BOFF. βph is the gain of photoreceptor and its value is equated to zero. βh is the gain of horizontal cell layer in order to extract contour information alone its value is taken as zero. fs & ft are spatial and temporal frequencies respectively. Spatial frequency is computed by applying DCT to the input image and since in this work only one single frame of image is considered the value for temporal frequency is equated to zero. τph & τh are temporal frequency constants. Some of the parameters has been taken from [13, 14].

4 Mathematical Modeling of Inner-Plexiform Layer

Contour contrast enhanced image from the outer-plexiform layer is subjected as an input image to the inner-plexiform layer. Information from the bipolar cell is subdivided mainly into two main channels namely ON and OFF, each of the ON and OFF channel is independently enhanced using logarithmic transformation as shown in Fig. 2.

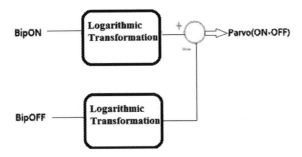

Fig. 2 Parvo channel modeling

Photoreceptor layer is modeled using Michaelis–Menten law in a similar manner modeling of retinal parvo channel was also carried out in the proposed study [15]. Logarithmic transformation is basically used to expand the value of dark pixel values in an image and to compress the dynamic range of image. This in turn leads to image contrast equalization. The difference of the two bipolar channels results in parvo (ON-OFF) output. Since the incoming information is about contours the parvo channel results in contour enhancement.

5 Methodology for Processing Srabismus Vision

In the proposed work physiological stages of retina were modeled through HVS. The flow diagram of the proposed retinal model to show the process involved in the visual perception of both the normal and strabismus person is shown in Figs. 3 and 4. For understanding the level of perception of a normal vision person the input image is passed to the retina of right and left eye, in the proposed methodology retinal layers like photoreceptor, outer plexiform and inner plexiform has been modeled mathematically. In the photoreceptor layer the image gets compressed and contrast value of the image is enhanced. The range of compression parameter varies from 0–1. In our study the compression parameter value is taken as 0.9 for which better compression and contrast enhancement occurs. The contrast enhanced image from both the eyes is processed by the outer plexiform layer for contour enhancement. Further the output from the outer plexiform layer is processed by the inner plexiform layer to extract the finer details of the image. The processed image from both the eyes combines together and hence the person is able to perceive as one single image. Figure 3 illustrates the level of perception of a normal person using the proposed methodology.

In the case of people with strabismus the situation is entirely different. People may have squint in the right or left eye. Hence in the proposed technique one such situation is considered, right eye is affected with strabismus and left eye is normal. In our study the mathematical model is developed such that the image passes in a normal fashion to the photoreceptor layer for a normal vision eye and the image is transversed, and then passed to the photoreceptor layer for studying the visual perception of a strabismus person.

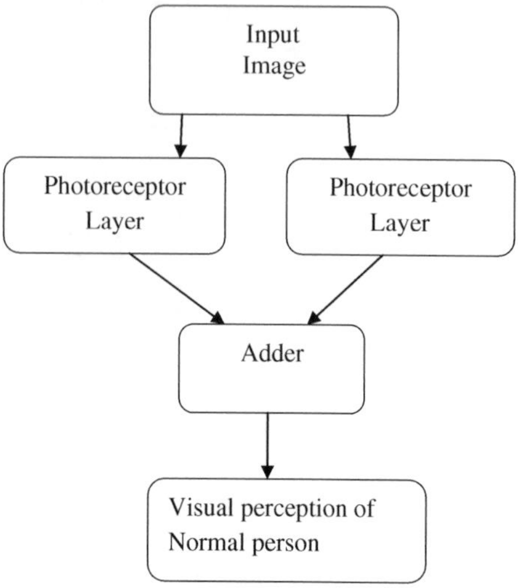

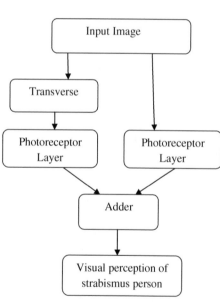

Fig. 3 Flow diagram that illustrates the visual perception of normal person

Fig. 4 Flow diagram that illustrates the visual perception of a strabismus person

In the case of people with strabismus double vision is one of the major symptoms. This paper aims in illustrating the reason of occurrence of double vision in a person. In the photoreceptor layer the image gets compressed and contrast value of the image is enhanced. The contrast enhanced image from both the eyes is processed by the outer-plexiform layer for contour enhancement. Further the output from the outer-plexiform layer is processed by the inner-plexiform layer to extract the finer details of the image. The processed image from both the eyes combines together and it is shown using the proposed technique since both the eye did not receive the input image in a similar fashion, this resulted in the occurrence of double image. Figure 4 illustrates the level of perception of a strabismus person using the proposed methodology.

6 Result

Natural image of size 256*256 is considered for the analysis to detect the level of perception of a normal person and people with strabismus. On an average of six to ten images were compiled for the study. Input images are fed to the retinal photoreceptor layer and processed separately by both right and left eye to enhance the contrast value of image for a compression parameter value of $V_0 = 0.6$. Simulation was carried out for both normal vision person and strabismus person. Simulation results are shown in Fig. 5. Figure 5a shows the original input image. Figure 5b

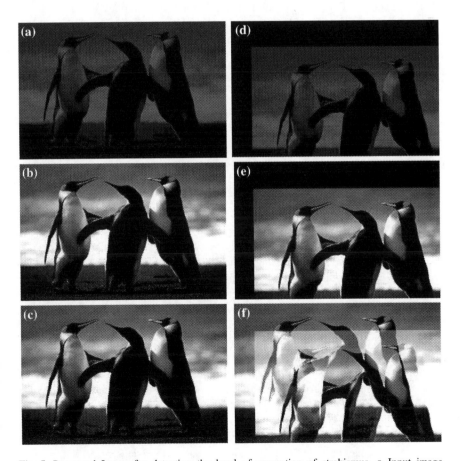

Fig. 5 Processed Image for detecting the level of perception of strabismus. **a** Input image. **b** Processed image by photoreceptor layer. **c** Perception of a normal vision person. **d** Translated image. **e** Processed image after translation. **f** Perception of a strabismus person

184 T. Rajalakshmi and S. Prince

Table 1 Feature values of the processed image

Image texture feature		Normal vision person	Strabismus person
Photoreceptor layer	Energy	150,070	137,359
	Intensity	0.7693	0.7041

shows the processed image by the photoreceptor layer for a compression parameter value of V_0 = 0.6. Figure 5c shows the perceptual level of an image for a normal vision person. Figure 5d shows the translated image. Figure 5e shows the processed image by the photoreceptor layer for a compression parameter value of V_0 = 0.6 after translation. Figure 5f shows the perceptual level of an image for a strabismus person.

Hence from the simulation results it is proved through the physiological modeling of retina, people with normal vision visualize a single image at its output, where as people with strabismus after processing through the photoreceptor layer is prone to double vision. Double vision is basically defined as the perception of two images of a single object. In the proposed methodology binocular double vision is focused on where both the eyes of the patient do not work together. From the processed image, features like energy and intensity are extracted and the numerical values of the same are shown in Table 1.

It is observed from the tabulated result the intensity level of the processed image for a normal vision person and strabismus person is approximately the same. The energy level of a normal vision person is comparatively higher than that of a strabismus person. Since the energy level of a strabismus person is low compared with a normal person they could not perceive image at a better rate.

7 Conclusion

The knowledge in the study of mathematical modeling of human visual system for perception of strabismus person has been limited. In the literature there are methods or technique which proves that severe or less vision impairment is the malfunctioning of the components of photo-transduction network. In this study human visual system model for detecting the level of perception of normal and people with strabismus is performed considering physiological properties of retina is performed. The proposed retinal layer model includes some parts of the retinal functionalities like luminance, compression properties and spatial and temporal frequencies for visual processing. The proposed method is applied to different variety of images and from the results it is proved that this model of photoreceptor layer, outer plexiform layer and the inner plexiform layer compresses the image, enhances contrast visibility in dark area and thus maintaining the same in its bright area and thereby enhancing the contour information in successive steps respectively. It is observed that the double vision is the perception of two images of a single object. It is proved that the energy level of a normal vision person is comparatively higher than that of people with strabismus. Intensity level of a normal vision person and strabismus person is approximately the same.

References

1. Potdar, R.M., Mishra, A., Yadav, S.: Real time squint eye detection. Int. J. Eng. Adv. Technol. (IJEAT)
2. www.eyesapart.com/2005/06/08/strabismus-statistics/. Accessed on 25th october 2015.
3. Barry, J.C., Effert, R., Kaupp, A.: Objective measurement of strabismus in infants and children through photographic reflection pattern evaluation. Ophthalmology 99(3), 320–328 (1992)
4. Lehmann, T., Kaupp, A., Efert, R., Meyer-Ebrecht, D.: Automatic strabtometry and hough-transformation and covariance filtering. IEEE Trans. 0-8186-6950-01 (1994)
5. Tak, S., Satao, K.J.: A widespread study, analysis, and management of a squint eye using support vector machine (SVM). Int. J. Emerging Technol. Adv. Eng. 3(8) (2013)
6. Rajalakshmi, T., Prince, S.: Contour-contrast enhancement based on retinal layer processing. In: Second International Conference on Devices, Circuits and Systems, pp. 312–316 (2014)
7. Beaudot, W.H.A.: Sensory coding in the vertebrate retina: towards an adaptive control of visual sensitivity. Netw. Comput. Neural Syst. 7(2), 313–317 (1996)
8. Benoit, A., Le Caplier, P., Durette, B., Herault, J.: Using human visual system modeling for bio-inspired low level image processing. Comput. Vis. Image Understand. 114, 758–773 (2010)
9. Beaudot, W.H.A., Palagi, P., Hérault, J.: Realistic simulation tool for early visual processing including space, time and colour data. International Workshop on Artificial Neural Networks, Barcelona, June (1993)
10. Benoit, A., Alleysson, D., Herault, J., Le Herault, P.: Spatio-temporal tone mapping operator based on a retina model. Lect. Notes Comput. Sci. 5646, 12–22 (2009)
11. Lens, A., Nemeth, S.C., Ledford, J.K.: Ocular Anatomy and Physiology, 2nd edn. SLACK Incorporated, New Jersey (2008)
12. Dan, Y., et al.: Efficient coding of natural scenes in the lateral geniculate nucleus: experimental test of a computational theory. J. Neurosci. 16, 3351–3362 (1996)
13. Beaudot, W.H.A.: The neural information processing in the vertebrate retina: a melting pot of ideas for artificial vision. Ph.D. thesis in computer science, INPG, France, December (1994)
14. Barlow, H.B.: Redundancy reduction revisited. Comput. Neural Syst. 12, 241–253 (2001)
15. Smirnakis, S.M., Berry, M.J., Warland, D.K., Bialek, W., Meister, M.: Adaptation of retinal processing to image contrast and spatial scale. Nature 386, 69–73 (1997)

Despeckling of Medium Resolution ScanSAR Data

Y. Muralimohanbabu, M.V. Subramanyam and M.N. Giriprasad

Abstract The speckle noise shows more impact on the performance of the radar image while deciding the objects. The objective of despeckling is to remove speckles from the SAR image, to represent a noise-free image and maintain all significant features like textures, region borders etc. The objective of the work is to investigate on the spatial domain and transform domain despeckling methods and how far the speckle can be removed and how far the texture details can be maintained. The proposed approach aims to despeckle the speckle noise to the possible extent while preserving the edge characteristics. The major concentration of the research work is on the Indian microwave imagery. RISAT-1 (RADAR Imaging Satellite) is the first and only Indian microwave active mode satellite that is capable of operating all the day and in all weather conditions even during cloudy times. It is a C-band radar mainly designed for monitoring and analyzing the agriculture.

Keywords RADAR · BM-3D · CS-3D · LEE filter · RISAT-1 · Speckle

1 Introduction

The observation of the earth from some safe ground has always been an important task from the earliest times in human history. The high ground has been extended from trees to early observation balloons, reconnaissance aircraft, and finally to cameras in space. This desire to view the earth has been extended to the observation of things in ways that could never be achieved with our own eyesight. Observation using radiation invisible to the eye is also becoming very important. It is the ability of radar to penetrate most atmospheric barriers to observation that makes it very desirable.

Y. Muralimohanbabu (✉) · M.N. Giriprasad
ECE Department, JNTUA, Anantapur, AP, India
e-mail: kisnamohanece@gmail.com

M.V. Subramanyam
ECE Department, SREC, Nandyal, Kurnool, AP, India

© Springer Science+Business Media Singapore 2017
P. Deiva Sundari et al. (eds.), *Proceedings of 2nd International Conference on Intelligent Computing and Applications*, Advances in Intelligent Systems and Computing 467, DOI 10.1007/978-981-10-1645-5_16

Since its origin in the 1950s, SAR has been developed into a mature technology and is now recognized as a highly successful imaging tool for weather monitoring, crop growing, forest deformation, mine detection, mapping and military systems that need imaging at high resolutions [1–3]. It is an active microwave sensor that transmits signals in microwave region [4]. SAR operation is entirely different from regular sensors like optical sensors.

Microwave images give high resolution and high contrast of topographical features. Interpretation and analysis of the data is difficult in microwave imagery. For optical remote sensing and other surveillance systems, the images are taken when the sun is in favorable position during the day. This is not necessary in the active remote sensing like radar imaging. The microwave remote sensing is having the capability of using its own radiation, called active radiation.

Many others have derived statistical models of SAR imagery and used these to develop methods of reducing the high noise levels caused by the properties of the illuminating radar beam and the objects being imaged. Speckle noise is most commonly modeled as a noise term multiplied by the input signal i.e. multiplicative noise, when the speckle is said to be fully developed. Methods have also been developed that attempt to take advantage of the statistics available from the image itself and approximate an image formation process that is used to determine how to remove the noise. The main goal of despeckling is to remove the speckle effects while maintaining the structural details of the scene. Speckle degrades [5–8] the quality of image and its reduction is a compulsory step before further processing of the image.

Different types of methods have been developed in two major areas like spatial domain and frequency domain for reduction of speckle noise. In the present work the main concentration is on some major areas of de-speckling methods. Mainly the despeckling methods are broadly given as LEE filter, Frost filter, Kaun filter, Sigma filter, Refined LEE filter, Gamma Map filter, Wavelet transform, Curvelet transform, Principal Component based transform, Non Local Mean method, Block Matching 3-Dimentional method (BM-3D), Compressive Sensing 3-Dimentional method (CS-3D), Probabilistic Patch Based method etc. The BM-3D method is generally considered as state of art method in denoising of regular images and despeckling of SAR images.

The efficiency of spatial filters mainly depends on the choice of the size and orientation of the local window. All standard speckle filters make use of neighboring pixels statistical characteristic within the local window to calculate the expected value needed to replace the filtered pixel. The size of the filter window will determine the amount of speckle reduced and the visual quality of the denoised image [9–13]. The filter will start its computation within the filter window from the top left corner of the padded image [14, 15]. If the selected filter is a median filter, the value of the first pixel will be replaced by the median value of its surrounding pixels within the filter window.

2 A Modified Block Matching 3D Algorithm

1. Apply the non sub sampled wavelet transform to MRS image to evaluate the coefficients.
2. Thresholding is performed on selective coefficients and calculate inverse non sub sampled wavelet transform.
3. Divide the image into standard size blocks and calculate distance between all blocks.
4. If difference $\leq$ thresholding, apply wavelet transform on that block.
5. Evaluate the wavelet coefficients and apply hard thresholding on selective bands.
6. Apply inverse wavelet transform to reconstruct the image.
7. Apply wavelet transform and evaluate the coefficients once again.
8. Apply wiener filter on selective bands and apply inverse wavelet transform to reconstruct the final image.

The modified BM-3D method along with existing methods like compressive sensing theory based 3D, block matching based 3D, principal component analysis, etc. are considered and tested for Medium Resolution ScanSAR (MRS)—HV mode image data of RISAT-1 satellite. The SAR image is having the standard size of 512 × 512 pixels. The noise variance of 0.5 is added to the original image and obtained image is processed for despeckling.

3 Results and Discussions

The quality parameters like equivalent number of looks (EQNL), speckle suppression index (SSI), correlation coefficient (CC), edge saving index (ESI or EPI), mean square error (MSE) and peak signal to noise ratio (PSNR) values are measured for the taken MRS image. The value 11.85, equivalent number of looks obtained in this present method is the best among other existing techniques, which describes the SAR image quality. Comparison of quality factors obtained for various methods is provided in Table 1 with corresponding images shown in Fig. 6.

Table 1 Quantitative comparison of despeckling techniques for MRS-HV Image (size = 512 × 512 and variance = 0.5)

	EQNL	SSI	EPI	CC	PSNR
LEE filter	1.79443	0.25991	0.24561	0.62142	11.85488
Wavelet	0.17824	0.68324	0.16868	0.68436	12.48479
Curvelet	0.14160	0.86171	0.39323	0.87608	18.44182
PCA	0.14406	0.80361	0.50108	0.96371	21.62268
BM-3D	8.55110	0.40614	0.61196	0.91883	17.81902
CS-3D	9.75436	0.06951	0.61720	0.91902	17.81960
Proposed	11.8540	0.05586	0.68354	0.97700	27.51640

The value 0.052, speckle suppression index obtained in this present method is the best among other existing techniques, which describes speckle content remaining in the image. This value should be minimum and ideally zero. The correlation coefficient value measures the similarity between the original and despeckled images. Its value generally lies between 0 and 1. Its value should be maximum as much as possible and ideally 1. The value 0.977, correlation coefficient obtained in this present method is the best among other existing techniques, which depicts similarity between original SAR and despeckled images.

The value 0.683, edge saving or preserving index obtained in this present method is the best among other existing techniques, which illustrates the preservation of edges of the image. The despeckled images should contain the edge features that exist the original image. The measuring parameter of the edge preserving is edge saving index. The value 27.51, peak signal to noise ratio obtained in this present method is the best among other existing techniques, which describes quality of the image. Comparison of quality factors have been given for MRS data from Figs. 1, 2, 3, 4 and 5. The quality factors have been evaluated for various modes and given in Table 2.

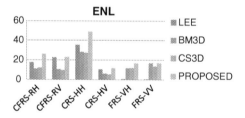

Fig. 1 Comparison of ENL values

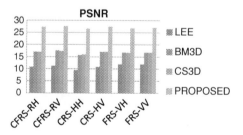

Fig. 2 Comparison of PSNR values

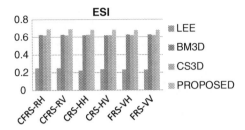

Fig. 3 Comparison of ESI values

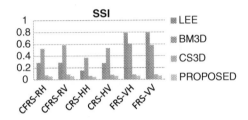

Fig. 4 Comparison of speckle suppression index values

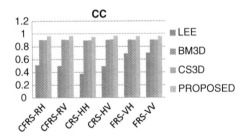

Fig. 5 Comparison of CC values

Table 2 Quantitative comparison of despeckling techniques for various RISAT-1 image modes (size = 512 × 512 and variance = 0.5)

	CFRS-RH	CFRS-RV	CRS-HH	CRS-HV	FRS-VH	FRS-VV	MRS-HH	MRS-HV	MRS-HH	MRS-HV
LEE	17.74	22.71	35.38	10.60	0.91	0.64	28.62	21.63	42.36	32.56
Wavelet	0.47	0.44	0.69	0.34	0.40	0.44	0.53	0.46	0.59	0.52
Curvelet	0.57	0.50	0.55	0.31	0.35	0.36	0.60	0.48	0.63	0.48
PCA-LPG	0.57	0.51	0.60	0.34	0.38	0.39	0.62	0.49	0.59	0.44
BM3D	11.34	10.57	28.17	6.37	11.82	16.72	20.05	13.29	31.10	23.32
CS3D	12.09	9.85	27.18	5.56	12.05	13.69	20.07	11.97	28.26	25.50
Proposed	25.97	22.98	48.88	11.73	16.59	16.80	28.76	21.98	43.78	33.54

4 Conclusion

The proposed method has been developed as an alternative to block matching with 3D transformation and compressive sensing based 3D transformation in which undecimation of image along with block matching method is used. Even though the LEE filter removes much speckle, it cannot preserve the edge details and it does not maintain the correlation with input image. It is a major drawback in statistical filters (Fig. 6).

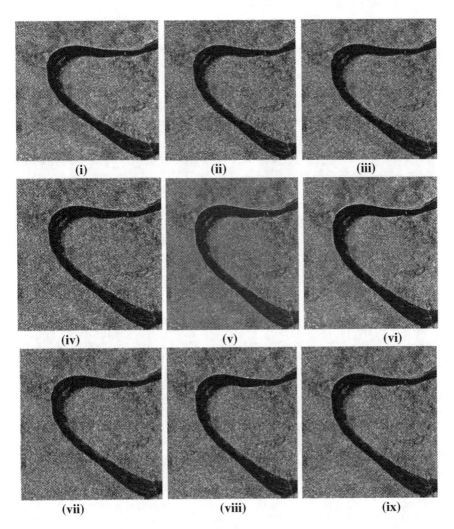

(i)   (ii)   (iii)

(iv)   (v)   (vi)

(vii)   (viii)   (ix)

Fig. 6 **i** Initial SAR image (MRS-HV), **ii** speckle noisy image, **iii** LEE despeckled image, **iv** wavelet despeckled image, **v** curvelet despeckled image, **vi** PCA despeckled image, **vii** BM-3D despeckled image, **viii** CS-3D despeckled image and **ix** proposed despeckled image

Whereas the latest block based techniques BM-3D and CS-3D are preserving the edge details and suppress the speckle better, the proposed method preserves edges and suppresses the speckle much better than BM-3D and CS-3D techniques. Un decimation of the data in the processing, increases the quality of the output image. Because of this reason, the edge details are persevered. It is observed that the proposed despeckling technique performs better than other latest methods in terms of EQNL, SSI, CC, ESI and PSNR.

Acknowledgments The authors would like to thank Indian Space Research Organization (ISRO) and German Aerospace Center for providing Synthetic Aperture Radar data.

References

1. Lee, J.S., Wen, J.H., Ainsworth, T.L., Chen, K.S., Chen, A.J.: Improved sigma filter for speckle filtering of SAR imagery. IEEE Trans. Geosci. Remote Sens. **47**(1), 202–213 (2009)
2. Gagnon, L., Jouan, A.: Speckle filtering of SAR images: a comparative study between complex-wavelet-based and standard filters. SPIE Proc. Wavelet Appl. Signal Image Process. **3**(169), 80–87 (1997)
3. MuraliMohanBabu, Y., Subramanyam, M.V., Giriprasad, M.N.: PCA based image denoising. Signal Image Process. Int J. (SIPIJ) **3**(2), 236–244 (2012)
4. Wang, Y., Yang, J., Yin, W., Zhang, Y.: A new alternating minimization algorithm for total variation image re-construction. SIAM J. Imag. Sci. **1**(3), 248–272 (2008)
5. Lee, J.S.: Refined filtering of image noise using local statistics. Comput. Graph. Image Process. **15**(4), 380–389 (1981)
6. Deledalle, C., Denis, L., Tupin, F.: Iterative weighted maximum likelihood denoising with probabilistic patch-based weights. IEEE Trans. Image Process. **18**(12), 2661–2672 (2009)
7. Parrilli, S., Poderico, M., Angelino, C., Verdoliva, L.: A nonlocal SAR image denoising algorithm based on LLMMSE wavelet shrinkage. IEEE Trans. Geosci. Remote Sens. **50**(2), 606–616 (2012)
8. MuraliMohanBabu, Y., Subramanyam, M.V., Giriprasad, M.N.: Fusion and texture based classification of indian microwave data—a comparative study. Int. J. Appl. Eng. Res. **10**(1), 1003–1009 (2015)
9. Koo, V.C., Chan, Y.K., Vetharatnam, G., Chua, M.Y., Lim, C.H., Lim, C.S., Thum, C.C., Lim, T.S., Bin Ahmad, Z., Mahmood, K.A., Bin Shahid, M.H., Ang, C.Y., Tan, W.Q., Tan, P. N., Yee, K.S., Cheaw, W.G., Boey, H.S., Choo, A.L., Sew, B.C.: A new unmanned aerial vehicle synthetic aperture radar for environmental monitoring. Prog. Electromagn. Res. **122**, 245–268 (2012)
10. Fan, C., Huang, X.T., Jin, T., Yang, J.G., An, D.X.: Novel pre-processing techniques for coherence improving in along-track dual-channel low frequency SAR. Prog. Electromagn. Res. **128**, 171–193 (2012)
11. Ren, S., Chang, W., Jin, T., Wang, Z.: Automated SAR reference image preparation for navigation. Prog. Electromagn. Res. **121**, 535–555 (2011)
12. Li, J., Zhang, S., Chang, J.: Applications of compressed sensing for multiple transmitters multiple azimuth beams SAR imaging. Prog. Electromagn. Res. **127**, 259–275 (2012)
13. Chen, J., Gao, J., Zhu, Y., Yang, W., Wang, P.: A novel image formation algorithm for high-resolution wide-swath spaceborne SAR using compressed sensing on azimuth displacement phase center antenna. Prog. Electromagn. Res. **125**, 527–543 (2012)

14. Xu, L., Lu, C., Xu, Y., Jia, J.: Image smoothing via l_0 gradient minimization. ACM Trans. Graph. **30**(6), 174–188 (2011)
15. MuraliMohanBabu, Y., Subramanyam, M.V., Giriprasad, M.N.: Bayesian denoising of SAR image. Int. J. Comput. Sci. Technol. **2**(1), 72–74 (2011)

Morphological Analysis for Breast Cancer Detection

Priya Darshini Velusamy, Porkumaran Karandharaj
and S. Prabakar

Abstract In this paper, the morphological analysis of breast cancer detection is been discussed. The various parameters like area, perimeter and diameter of the tumor are found using modified Adaptive Fuzzy C means and modified region growing algorithm. This method of analysis is used for effective identification of different parameters that helps in easy diagnosis of cancer at various stages also adapting suitable and earlier therapeutic techniques.

Keywords Morphological analysis · Adaptive fuzzy C means · Modified region growing algorithms

1 Introduction

With reference to the report published by National Cancer Registry Programme (NCRP) and Indian Council of Medical Research (ICMR), Bangalore-India, the breast cancer is expected to cross 1,00,000 in 2020 and the survey state that in 2010 the total cancer cases was around 9,79,786 and in 2020 it may increase to approximately 11,48,757 [1]. The research in medical imaging is very much important nowadays because of different food habits and stress due to multiple, reason, many people are getting different diseases, out of which cancer is one of the major disease. The different organs affected by of cancers are liver, brain, lung, breast etc. In total, numbers of breast cancer cases are getting gradually increased

P.D. Velusamy (✉)
Amrita Vishwa Vidyapeetham-Amrita School of Engineering,
Bangalore, Karnataka, India
e-mail: darshinime@gmail.com

P. Karandharaj · S. Prabakar
Dr. N.G.P. Institute of Technology, Coimbatore, Tamil Nadu, India
e-mail: porkumaran@ieee.org

S. Prabakar
e-mail: srisornaprabu@gmail.com

© Springer Science+Business Media Singapore 2017
P. Deiva Sundari et al. (eds.), *Proceedings of 2nd International Conference on Intelligent Computing and Applications*, Advances in Intelligent Systems and Computing 467, DOI 10.1007/978-981-10-1645-5_17

year after year. As a preventive for breast cancer early detection is essential. Medical image processing is one of the vital areas in science and technology and able to handle with set of techniques that produce images of the internal organs of the body without spreading. In this restricted sense, medical imaging is seen as the solution of mathematical inverse problems.

2   Existing Method

The existing method consists of thresholding technique along with region growing technique. In detection of tumor, a spatial characteristic plays a vital role [2]. In the existing method the spatial characteristics are difficult for observer to identify [3]. The segmentation using thresholding technique is having two possible combinations either black or white. But some of the images specifically the bit map image which has 0–255 gray scale values. Because of this most of the time the cancer cells or the pixel which represent the cancer may be ignored. For some cases, from the MR/CR image tumor area is visible but it is not clear enough for further treatment. For the better diagnoses and treatment more information needs to be extracted from the image. The thresholding method has two gray values which is one (1) for white and zero (0) for black. The value zero (0) is assigned to background image and the value one (1) is assigned to object. So the extraction of tumor from the image becomes more difficult. Due to this constraint we have proposed the image segmentation technique for the tumor detection.

3   Methods and Materials

3.1   Fuzzy C Means

The method of clustering used in medical image processing is to classify the same combination of information which is present in the large data set to give an accurate performance of an image analysis. Fuzzy C-means (FCM) is one of the methods in clustering technique, in a group of data set one data belongs to two or more clusters. This method is commonly used in pattern recognition [4, 5]. In this algorithm based on the distance between the each cluster and data information's the membership is assigned. If the data points are near to the center of the cluster that data points will have the high degree of membership to that specific cluster and the data points which are far away from the data points that will have the low degree of membership to that cluster. The addition of membership data points should be equated to one [6].

The following function is used for the algorithm

$$J_m = \sum_{i=1}^{N} \sum_{j=1}^{C} u_{ij}^m \|x_i - c_j\|^2, \quad 1 \le m < \infty \tag{1}$$

In Eq. 1 m (the Fuzziness Exponent) is the real number which should be greater than one, N represents Number of data, C represents Number of cluster, degree of membership of x_i in the cluster j is represented by u_{ij}, the ith of d-dimensional measured data is represented by x_i, the d-dimension center of the cluster represented by c_j, and $\|*\|$ represent the similarity between the center and the measured data [7]. With the help of iterative optimization of the above function the fuzzy partitioning is performed, along with the values of membership u_{ij} and the cluster centers c_j. Equation 2 is for the calculations of uij and c_j

$$u_{ij} = \frac{1}{\sum_{k=1}^{c} \left(\frac{\|x_i - c_j\|}{\|x_i - c_k\|}\right)^{\frac{2}{m-1}}} = \frac{1}{\left(\frac{\|x_i - c_j\|}{\|x_i - c_1\|}\right)^{\frac{2}{m-1}} + \left(\frac{\|x_i - c_j\|}{\|x_i - c_2\|}\right)^{\frac{2}{m-1}} + \cdots + \left(\frac{\|x_i - c_j\|}{\|x_i - c_k\|}\right)^{\frac{2}{m-1}}} \tag{2}$$

The Distance from point i to current cluster centre j is represented as $\|x_i - c_j\|$ and point i to next cluster centers k is $\|x_i - c_k\|$. Equation 3 is for the calculation of c_j.

$$c_j = \frac{\sum_{i=1}^{N} u_{ij}^m \cdot x_i}{\sum_{i=1}^{N} u_{ij}^m} \tag{3}$$

For Eq. 4 condition, the iteration will get stopped,

$$\max_{ij} \left\{ \left| u_{ij}^{(k+1)} - u_{ij}(k) \right| \right\} < \varepsilon \tag{4}$$

In Eq. 4 ε is between 0 and 1 which is the termination criterion, where the iteration steps are represented as k. This procedure converges to a local minimum or a saddle point J_m.

The algorithm consists of the following steps:

1. Select the cluster centre randomly
2. Calculate u_{ij} with the following formula and by initializing U = [u_{ij}] matrix, U(0)

$$u_{ij} = \frac{1}{\sum_{k=1}^{c} \left(\frac{\|x_i - c_j\|}{\|x_i - c_k\|}\right)^{\frac{2}{m-1}}}$$

3. Calculate the centre vectors C(k) = [c_j] with U(k) at k-step

$$c_j = \frac{\sum_{i=1}^{N} u_{ij}^m \cdot x_i}{\sum_{i=1}^{N} u_{ij}^m}$$

4. Update U(k) and U(k + 1)

$$u_{ij} = \frac{1}{\sum_{k=1}^{c} \left(\frac{\|x_i - c_j\|}{\|x_i - c_k\|}\right)^{\frac{2}{m-1}}}$$

5. If the J is minimum or if $\|U(k + 1) - U(k)\| < \varepsilon$, then STOP; otherwise return to step 2.

The parameters need to be specified before using the FCM algorithm are the number of clusters c, the fuzziness exponent m and the termination tolerance ε.

3.2 Adaptive Fuzzy C Means

The Adaptive fuzzy C Means algorithm(AFCM) is used for obtaining fuzzy segmentations of images with intensity inhomogeneities and an iterative algorithm for minimizing the brightness variations occurred due to inhomogeneities. For each iteration of the algorithm the multiplier field is getting added in order to avoid this multigrid methods are used [8].

The FCM algorithm for scalar data needs the centroids v_k and membership functions u_k, based on that the objective function is reduced. The function is shown in Eq. 5.

$$J_{FCM} = \sum_{i,j} \sum_{k=1}^{c} u_k(i,j)^q \|y(i,j) - v_k\|^2, \tag{5}$$

In Eq. 6 the membership value is represented as $u_k(i, j)$ at pixel location (i, j) for class k such that

$$\sum_{k=1}^{c} u_k(i,j) = 1, \tag{6}$$

The observed image intensity is y(i, j) at location (i, j) and centroid of class k is v_k. The value c is assumed to be known that is called the total number of classes. In each fuzzy membership a weighting exponent is represented as parameter q which determines the "fuzziness" amount of the resulting classification. the standard Euclidean distance is represented as norm operator $\|*\|$. For the minimization of

FCM function the pixels intensities which is close to the centroid is having high membership values, and the pixel information which is far away from the centroid is represented by low membership values. When the pixel is affected by the noise, then small amount of result change will occur in the segmentation part this is one of the merit of FCM method but for hard segmentation, the complete classification itself will get change. In the medical imaging segmentation technique, Fuzzy membership function [9] is used as a pointer for the partial averaging of the volume, which will be present in a single pixel which has multiple classes [7]. Taking the first derivatives of Eq. 5 with respect to v_k and $u_k(i, j)$ and equating to 0 gives necessary conditions for (5) to be reduced. By using Picard iteration through these two required conditions lead to an iterative method for decreasing the objective function [4, 5]. This is the standard FCM algorithm. The hard or crisp segmentation can be obtained from the result of fuzzy segmentation by assigning every pixel exclusively to the class that has the highest membership value for that pixel. This is known as maximum membership segmentation.

The objective function which conserve the advantages of FCM it will be suitable to images with intensity inhomogeneities. The brightness variation can be modeled by multiplying the centroids with unknown multiplier field m (i, j) which will vary smoothly and slowly with respect to i and j. The two-dimensional AFCM algorithm will minimize the following function with respect to u, v and m:

$$
\begin{aligned}
J_{AFCM} = &\sum_{i,j} \sum_{k=1}^{c} u_k(i,j)^2 \| y(i,j) - m(i,j) v_k \|^2 \\
&+ \lambda_1 \sum_{ij} ((D_i * m(i,j))^2 + (D_j * m(i,j))^2) \\
&+ \lambda_2 \sum_{ij} ((D_{ii} * m(i,j))^2 + 2(D_{ij} ** m(i,j))^2) + (D_{jj} * m(i,j)^2),
\end{aligned}
\tag{7}
$$

In Eq. 7 the standard forward finite difference operators is represented as D_i and D_j along with the rows and columns, and the second order finite differences are $D_{ii} = D_i * D_i$, $D_{ij} = D_i ** D_j$ and $D_{jj} = D_j * D_j$. The symbols * and ** represents the one and two dimensional discrete convolution operators, respectively. The parameter k_1 and k_2 will control the last two terms, k_1 and k_2 are first and second order regularization terms which is operating on the multiplier field. The multiplier fields that has a large amount of variation will be penalized by the first order regularization term. The multiplier fields which have discontinuities will be penalized by the second order term. When the membership functions centroids v_k and $u_k(i, j)$ are known values, then the multiplier field that reduces J_{AFCM} is the field which will makes the centroids near to the data, but is also gradually varying and smooth. According to the magnitude and the smoothness of the intensity inhomogeneity of the image the parameters k_1 and k_2 should be set. k_1 and k_2 value should be larger for an image with little or no inhomogeneities.

4 Proposed Method

The proposed system is divided into four sections preprocessing, Segmentation, Feature extraction and modified region growing. The block diagram is shown in Fig. 1. The Preprocessing is the process of eliminating the noise present in that image for that filtering technique is used [10]. Segmentation is used to locate the boundaries of the object in the image in this method advanced Adaptive Fuzzy C-means algorithms is used.

Feature extraction is done by thresholding and finally, region growing method is to recognize the tumor portion and edges of the tumor area in computed radiography (CR) image.

According to the literature survey there are many algorithms were developed for segmentation. But they are not suitable for all types of the image modality. The DIOCM breast cancer image was taken for analysis in that the advanced Adaptive Fuzzy C-means algorithm is applied in order to cluster the cancer portions. The same algorithm is also tested for non DIOCM images that are png images.

Fig. 1 Proposed method block diagram

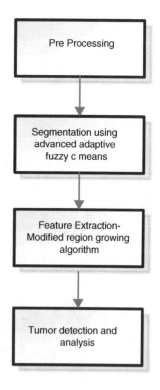

4.1   Algorithm Steps for Advanced Adaptive Fuzzy C-Means

The function J_{AFCM} will be reduced by taking the first derivatives of J_{AFCM} with respect to $u_k(i, j)$, $m(i, j)$ and v_k and equating to zero results in three necessary conditions for J_{AFCM} to be at a minimum. Using these conditions, the steps for the advanced AFCM algorithm can then be illustrated as follows:

1. Initial values for centroids is to be provide, v_k, $k = 1,\ldots, C$, and set the multiplier field $m(i, j)$ equal to one for all (i, j)
2. Calculate the memberships as follows:

$$u_k(i,j) = \frac{\|y(i,j) - m(i,j)v_k\|^{-2}}{\sum_{l=1}^{c}\|y(i,j) - m(i,j)v_l\|^{-2}} \tag{8}$$

   For all (i, j) and $k = 1,\ldots,c$. The k value set is 3 for this proposed method
3. Determine new centroids as follows:

$$v_k = \frac{\sum_{i,j} u_k(i,j)^2 m(i,j)y(i,j)}{\sum_{i,j} u_k(i,j)^2 m(i,j)^2}, \quad k = 1,\ldots,c. \tag{9}$$

4. Calculate new multiplier field by solving the following Space-varying difference equation for m (i, j)

$$\begin{aligned}
y(i,j)\sum_{k=1}^{c} u_k^2(i,j)v_k &= m(i,j)\sum_{k=1}^{c} u_k^2(i,j)v_k^2 \\
&+ \lambda_1(m(i,j) * * H_1(i,j)) \\
&+ \lambda_2(m(i,j) * * H_2(i,j)),
\end{aligned} \tag{10}$$

   where $H_1(i, j) = D_i *^{\wedge} D_j + ^{\wedge} D_j$ and $H_2(i, j) = D_{ii} *^{\wedge} D_{ii} + 2(D_{ij} * D_{ij}) + (D_{jj} * ^{\wedge} D_{jj})$. Here the notation$^{\wedge}$ f(i) = f(−i) is used. D_i, D_j are the standard forward finite difference operators like derivatives in the continous domain it is along with rows and columns. i, j are the pixel locations (rows and columns). The notation $^{\wedge}$ is a negative function.
5. If the algorithm satisfies the conditions then program stops at that step else proceed to Step 2. When the maximum change in the membership functions over all pixels between iterations is less than a threshold value then we define it as a convergence.

5 Results and Discussion

In this method, for the thresholding value of 0.551142 under cluster FCM0 the images where clustered properly but for the higher threshold levels the image is not clear. Otsu thresholding method is used to cluster the cancer image which is used to minimize the intra class variance and maximize the inter class variance. It is used for the reduction of gray level image into the binary image. The class used here is thresholding by 3 class advanced fuzzy c means. All the medical images will be having the speckle noises due to the acquisition and transmission of the images hence most of the image information's are difficult to analyze. For example the internal organ analysis and for the measurement of the defected portions the speckle noises need to be eliminated. In this paper we have done the most important preprocessing technique called speckle noise removal for the exact finding of the cancer portion and with respect to the edges. The result of the speckle noise removal is shown in Fig. 4. By using connected component technique the background and tumor portions are marked as black (background image) and white (tumor portion). Modified region growing-pixel based image segmentation technique is used for the identification of the tumor section. In this the neighboring pixels initial seed points are selected and it is compared with the other pixels and finally based on the seed points the pixels are grouped. The GUI for reading the input DICOM image, result of advanced adaptive fuzzy C means, the segmented and detected portions of the tumor, area [11], perimeter and diameter of the tumor are shown in Figs. 2–7 respectively.

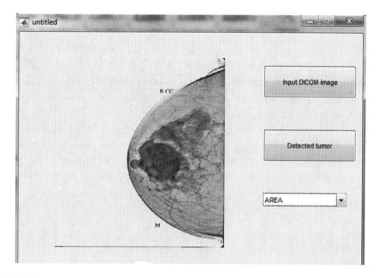

Fig. 2 GUI for reading the input DICOM image

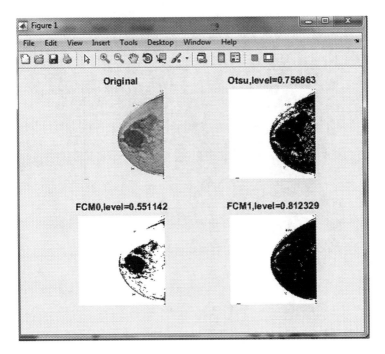

Fig. 3 Advanced adaptive fuzzy C means

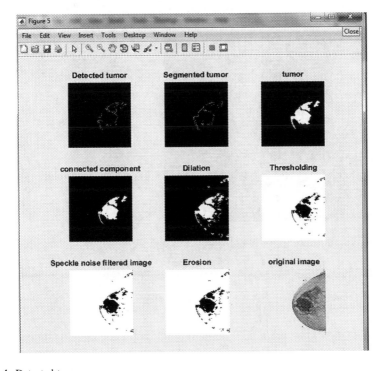

Fig. 4 Detected tumor

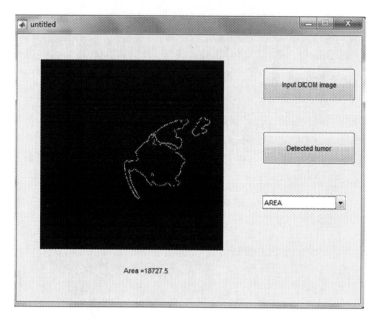

Fig. 5 GUI For the detection of tumor portion and displaying the area of the tumor

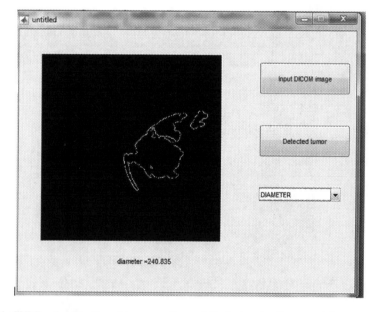

Fig. 6 GUI For the detection of tumor portion and displaying the diameter of the tumor

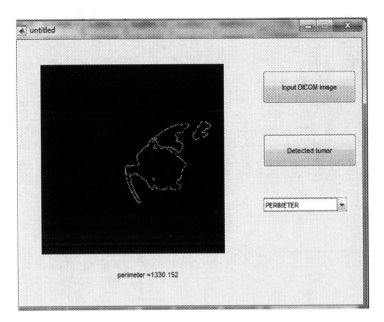

Fig. 7 GUI For the detection of tumor portion and the diameter of the tumor displaying the perimeter of the tumor

The detected tumor area has 18727.5 pixels, Perimeter value is 1330.152 pixels and the diameter value is 240.835 pixels. The same algorithm is applied for different data sets and the values of tumor area, perimeter and diameter are calculated.

In order to convert the tumor pixel values to the exact measurement size the following calculation is performed

Area = $\sqrt{p}$ * 0.264 mm^2
Perimeter = p * 0.264 mm
Diameter = p * 0.264 mm

According to the above formulas the area, perimeter and diameter is calculated for this proposed work.

Area = 36.13 mm^2
Perimeter = 351.16 mm
Diameter = 63.36 mm

According to the survey when the area is greater than 6 mm^2 then the cancer is in the critical stage [11].

6 Conclusion and Future Work

In this work, we presented the advanced adaptive fuzzy C means algorithm along with modified region growing algorithm which gives the best result for the images which has more information and comparatively better than k-means and normal Fuzzy C means algorithm because in k-means the data point particularly belong to one cluster center but in advanced AFCM the data point has assigned membership which is specified to each cluster center so that the data point will belongs to multiple cluster center. Also we have calculated the area, perimeter and diameter of the tumor in order to find the exact portion and the edges of the tumor. Based on these values the doctors can conclude the stage of the tumor and they can start the treatment.

Acknowledgment We would like to sincerely thank, Dr. K.S. Murugan, Chief Doctor, Clarity Imaging Centre, Coimbatore for providing the real time DICOM cancer images and our special thanks to the Chairman and Secretary of Dr. N.G.P Institute of Technology, Coimbatore, Amrita school of Engineering Bengaluru and Kovai Medical Center and Hospital for their eternal help and support.

References

1. http://www.ncbi.nlm.nih.gov
2. Arjun, P., Monisha, M.K., Mullaiyarasi, A., Kavitha, G.: Analysis of the liver in CT images using an improved region growing technique. In: 2015 International Conference on Industrial Instrumentation and Control (ICIC) College of Engineering Pune, India. 28–30 May 2015
3. Zulaikha Beevi, S., Sathik, M.: An effective approach for segmentation of MRI images: combining spatial information with fuzzy C-means clustering. Eur. J. Sci. Res. **41**(3), 437–451 (2010). ISSN I450-2I6X
4. Dunn, J.C.: A fuzzy relative of the ISODATA process and its use in detecting compact well-separated clusters. J. Cybern. **3**(32–57), 2002 (1973)
5. Bezdek, J.C.: A convergence theorem for the fuzzy ISODATA clustering algorithms. IEEE Trans. Pattern Anal. Mach. Intell. PAMI **2**, 1–8 (1980)
6. Chuang, K.-S., Tzeng, H.-L., Chen, S., Wu, J., Chen, T.-J.: Fuzzy C-means clustering with spatial information for image segmentation. Comput. Med. Imaging Graph. **30**, 9–15 (2006)
7. Brandt, M.E., Bohan, T.P., Kramer, L.A., Fletcher, J.M.: Estimation of CSF, white and gray matter volumes in hydrocephalic children using fuzzy clustering of MR images. Comput. Med. Imaging Graph. **20**, 25–34 (1994)
8. Pham, D.L., Prince, J.L.: An adaptive fuzzy C-means algorithm for image segmentation in the presence of intensity inhomogeneities. Pattern Recogn. Lett. **20**, 57–68 (1999)
9. Mendel, J.M., John, R.I.: Type-2 fuzzy sets made simple. IEEE Trans. Fuzzy Syst. **1**, 117–127 (2002)
10. Ramani, R., Vanitha, N.S., Valarmathy, S.: The pre-processing techniques for breast cancer detection in mammography images. Int. J. Image Graph. Signal Process. **5**, 47–54 (2013)
11. Selvakumar, J., Lakshmi, A., Arivoli, T.: Brain tumor segmentation and its area calculation in brain MR images using K-mean clustering and fuzzy C-mean algorithm. In: IEEE-International Conference on Advances in Engineering, Science and Management (ICAESM-2012), 30, 31 Mar 2012, pp. 186–190 (2012). ISBN: 978-81-909042-2-3

Recommendation Based P2P File Sharing on Disconnected MANET

V. Sesha Bhargavi and T. Spandana

Abstract In recent days, P2P networks have become the main focus of research in the field of communication. A P2P network is a dynamic open network that connects devices over the edge of internet and hence invites the operations of malicious nodes, which has lead the need to focus on security issues in these type of networks. Our ultimate aim is to improve the file sharing system with reduced file searching cost and delay. MANET's are dynamic networks in which the nodes constantly move from one location to another which leads to the formation of disconnected networks. This paper proposes a file transfer mechanism for such type of disconnected networks by using the information obtained from the nodes regarding file searching and also provides a better security for the data in the file being transferred among the peers.

Keywords P2P · Security · File sharing · Load balancing

1 Introduction

A Peer to Peer network is a kind of network in which all the nodes in the network shares the load and processes the requests. The availability of resources will be the same for all peers in this network and hence we can say that every peer is equally important as another. The nodes in this kind of network are usually heterogeneous devices such as laptops, PC's, mobile or smart phones etc. which can be easily interconnected through a wired or wireless physical medium. This can also be a type of distributed network or a centralized network in which each node shares its part of resources like disk storage for files, bandwidth, processing cycles, cache etc.

V. Sesha Bhargavi (✉) · T. Spandana
Department of IT, G. Narayanamma Institute of Technology and Sciences,
Hyderabad, India
e-mail: b.velagaleti@gmail.com

T. Spandana
e-mail: spandanareddyteegala@gmail.com

© Springer Science+Business Media Singapore 2017
P. Deiva Sundari et al. (eds.), *Proceedings of 2nd International Conference on Intelligent Computing and Applications*, Advances in Intelligent Systems and Computing 467, DOI 10.1007/978-981-10-1645-5_18

to other nodes in the network in case of any requirement for them. Nodes will have varying capabilities.

The nodes in this kind of network sometimes may act as client or server or both depending on the processes they have to fulfil. This is an improvement over many existing client–server architectures. Also many of these client–server architectures face the challenges such as scalability, having a single point of failure, requiring centralized coordination and management; resources not optimally utilized which can be addressed by P2P networks. Hence much research is focused on achieving better reliable P2P systems in modern days.

As every node can act as both client and server, Hybrid Peer-2-Peer systems permit such infrastructure nodes to exist often called super nodes. With ease of administration as nodes self-configure and better reliability, scalability and better usage of resources, Wired Peer to Peer file sharing systems have already become successful model for file sharing among millions of users. Due to P2P network's confidentiality and open characteristics, it's complicated to find an effective and

Fig. 1 Peer to Peer and
client/server model

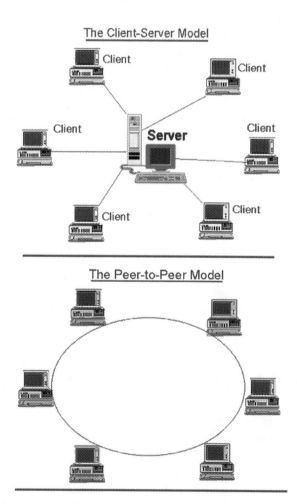

workable method to solve P2P network security problems. Also because of the nodes being dynamic and the network being autonomous, Conventional network security technologies are also difficult to apply to the P2P network. Hence our work mainly focuses on the recommendations done by the nodes regarding file sharing and searching. Figure 1 shows basic client server model as well as a P2P model.

2 Related Works

In [1], an approach that uses combined reputations of servants and resources, providing more informative polling's and overcoming the limitations of servant based only solutions was proposed. Servant reputations are associated with the servant identifier, which has to be tamper resistant.

A Bayesian network-based trust model and a method for building reputation based on recommendations in peer-2-peer networks has been illustrated in [2]. Since trust is multi-faceted, peers need to develop differentiated trust in different aspects of other peers' capability. The peer's needs are different in different situations. Depending on the situation, a peer may need to consider its trust in a specific aspect of another peer's capability or in multiple aspects.

A P2P file sharing system based on Swarm Intelligence for MANET [5], referred to as P2PSI, which uses a hybrid push-and-pull approach had been proposed. In the process of issuing advertisements, each file holder regularly broadcasts an advertisement message to inform surrounding nodes about what files are to be shared. Subsequent file searches can be easily processed with the help of discovery process which locates the desired file. The advantage of P2PSI is that it can easily shift to dynamically changing topology and discover nearby file holders timely. This paper considers the hybrid technique to reduce the overhead, but periodic advertisement and reactive searching both are increasing the overhead in dynamic nature.

In previous model, there is no trusted server to validate the peer. At same time trust mechanism is needed to punish peers that exhibit malicious behavior and furthermore, an access control mechanism is developed to secure the files sharing P2P network. In that model, each system stores the experience of file sharing in its own memory for future use. In this type, peer know about upload peer whether good or bad, which already downloaded file from that peer, another peers only considering the reputation in that uploading peer, it may chances to hack by malicious node. And there is no time specification, so it'll make some problem in transaction.

3 Proposed Work

In our proposed method, the file is grouped based on the frequent searching processes regarding the files. In our proposed technique, we are considering the disconnected MANET as group. In our proposed model, we consider different types of node mobility for file sharing. Group Local leader and Global leader nodes are proposed and defined in the view of a social network. In our proposed model we introduce the Trusting peer, Evaluate peer, Dictionary access control in P2P reference model. This may result in a more secure system than many existing models and it satisfies the requirements of access control for P2P file sharing system. In our work, peers send reputation queries to peers interacted in the past, which reduces network traffic comparing to flooding-based approaches. Furthermore, each peer expands its trust network with time and can obtain more credible recommendations from acquaintances.

3.1 Modules

We have divided our proposed technique into small modules, they are given below, Network design (Global leader Node, Local leader Node, Member node), Group Formation (File type, File searching), Own risk model, P2P rep model, Trusting peer, Evaluate peer, Volunteer recommendation.

3.1.1 Network Design

Each node can act with any one of the three different properties according to situation. (1) Global leader Node (The node which capable to collect the neighbor foreign group information. This node can connect the different groups to share the file). (2) Local leader Node (The node which is stable in the group, and contacting to the group node frequently. These nodes which are capable to collect the information of file availability in own group.) (3) Normal node (The node which maintaining only the own information).

3.1.2 Group Formation

In this module, we planned to group the nodes, based on file Information. The group formation depends on the file information; group of members contains the different type of files. So the group will be formed based on the file availability and searching process to improve the file searching system.

In this module, we divided the file searching scheme into two sub-modules, the file search will be done by the interest oriented file searching algorithm. In this module, the Local leader collects the information of file availability in the group. So if any member needs the file files then the nodes can ask to the Local leader is called intra community file searching and retrieval. If searching file information is not available in Local leader node, the file may available in other group. That file information will be collected by using Global leader node from other group (Fig. 2).

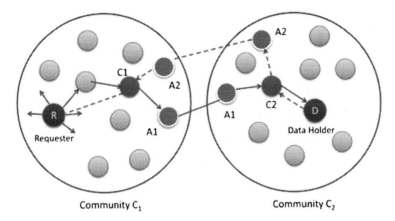

Fig. 2 Example model of different properties of network devices

3.1.3   Own Risk Model

All of the nodes in network not having any other node information at initial time. Therefore node can't believe wither node is good or bad. The requesting peer will select the uploading peer based on the downloading agreement (file size, packet size, bandwidth allocation, total duration to upload the file).

$$st_{lm} = \frac{sh_{lm}}{sh_{max}}(cb_{lm} - ib_{lm}/2) + \left(1 - \frac{sh_{lm}}{sh_{max}}\right)r_{lm} \tag{1}$$

3.1.4   P2P Rep Model

The reputation metric measures a stranger's trustworthiness based on recommendations. If node finds number of peer width indented file then its need to confirm wither peer is good or bad, so it will request to all peer about indented peer. By receiving recommendation from other peers, node can calculate the reputation value. The requesting peer will select the best peer based on higher reputation value (if own downloading history is low).

$$r_{lm} = \frac{[\mu_{sh}]}{sh_{max}}(ecb_{lm} - eib_{lm}/2) + \left(1 - \frac{[\mu_{sh}]}{sh_{max}}\right)er_{lm} \tag{2}$$

3.1.5   Trusting Peer

In this model, we are introducing the method to accept the trust based peer selection. If requesting peer already done more transaction then it can believe the peer with less number of recommendation. Based on own history value, the node will select the best peer to download the file

$$rt_{ln} = \frac{rh_{lp}}{rh_{max}}\left(rcb_{lp} - rib_{lp}/2\right) + \frac{rh_{max} - rh_{lp}}{rh_{max}} r_{lp} \tag{3}$$

3.1.6 Enhanced Volunteer Recommendation

In our base model, they have considered the limiting the number of downloader's to maintain the own trustworthiness in other peers. In base model, if limit is crossed then the uploading peer will ignore the req. To find the good peer, requester needs to spend most of the time on recommendation checking. We have enhanced base model to resolve the problem of delay. In this module, if limit is crossed then the node will check the highly trusted peer with requested file. If peer found then node will generates volunteer recommendation.

If any volunteer recommendation is received, then the node will check recommender is highly trusted node or not. If yes then the node won't make **Recom_req**, directly it will download the file from recommended node. In this module, the node evaluates the peer in two ways, (1) Service based, (2) Recommendation based. After each download, the peer will verify the agreement with final download level. Based on the performance, service info will be updated. After receiving recommendation from number of peer, the node will verify the peer's bad recommendation by comparing all recommendation. Then new value will be updated. By checking service and recommendation, the best peer will be considered for file download and recommendation.

Selection of best service provider may overload few peers while other peers having same resources are idle. A load balancing mechanism is implemented in this work, to utilize the resources of eligible good peers. In this method, each peer's simultaneous operations are limited to a maximum. If a peer reaches its maximum number of simultaneous operations, instead of simply rejecting the incoming requests, it suggests another good peer having the same resource to the service requester. Hence, this method avoids time required for the service requester to find another good service provider and also reduces the network traffic (Fig. 3).

Fig. 3 File sharing request

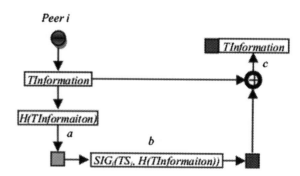

Algorithm:

U_c is Uploading count,
$N_{L_{Th}}$ is Max up limit threshold
W_p is working period
NW_p is non working period
B bandwidth, B_{U_R} bandwidth upload reserved,
B_{us} bandwith used

STEP 1

If (peer has to download file)
1. Generate file request and send
2. Wait for reply and good per selection for a time

STEP 2

If (any peer received the request)
1. Check for upload limit (ENHANCEMENT)

$$U_c = \begin{cases} 0, & At\ initial \\ \displaystyle\sum_{i=1}^{n} xl, x = 1, n = 1,2,3..n \end{cases}$$

If   (limit   is   not   crossed)
$U_c \le N_{L_{Th}}$
1. If (file found)
   Set the bandwidth possibilities
   $B = B_m - (B_{U_R} * Rand) - B_{us}$
   Give reply
2. else
   Ignore

If (limit is crossed)
$U_c > N_{L_{Th}}$
1. check for good peer

$$st_{lm} = \frac{sh_{lm}}{sh_{max}}\left(cb_{lm} - {ib_{lm}}/_2\right)$$

For each $j \epsilon Ai$
If $St_{lm} > S_{Th}$ found
Recom **m**
else
Ignore **m**

STEP 3

If (request received)
1. add peer in to lis

1. $P_i \cup P_{List}$
2. send recommendation request to all other peers

STEP 4

If (recommendation request received)
1. Checks the history
    1. For-each $H_i \in SH_{List_m}$
        a. if (info found)
           Send recommendation info
        b. else
           Ignore

STEP 5

If (recommendation received)
1. add the recommendation info in to a list
    $R_m \cup R_{List}$

STEP 6

If (best peer recommendation received)
1. send data request
2. collect the data
3. set the satisfaction
    $= W_p/(W_p + NW_p)$

STEP 7

If (data request)
1. check crossed the limit
    $U_c > N_{LTh}$
    I. If crossed limit
        a. send objection message

STEP 8

Time out for check best peer
1. filter the recommendation by SORT algorithm
2. Select best peer

4   Results

We have tested our proposed network with popular simulation tool called NS2. We have used the Single PC with configuration of 20 GB Hard disc space, 1 GB RAM, software's Linux OS (Ubuntu 10.04) and NS2.34. We have written the program by

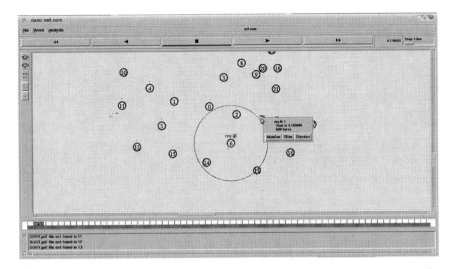

Fig. 4 File searching request

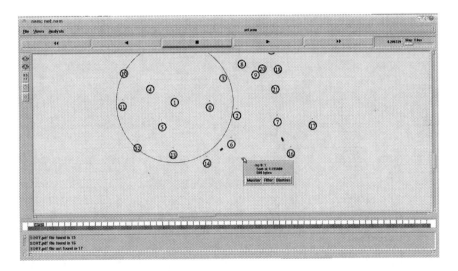

Fig. 5 File availability reply

TCL (**Front End language**). We simulated our proposed system with two types of results. One is Nam and Xgraph.

In this section, we present main result steps in Figs. 4, 5, 6, 7 and 8, which shows the different packets used in the trust management process.

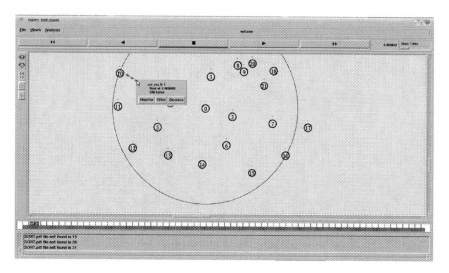

Fig. 6 Requesting for best peer recommendation

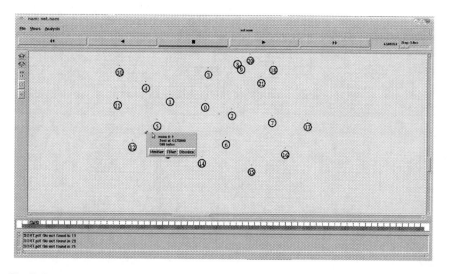

Fig. 7 Recommendation about best peer

The graphs in Figs. 9 and 10 show the packet delivery and overhead comparison between AODV and the proposed P2P model. It can be observed that the proposed P2P model works well in un-trusted environment, hence offering better reliability and security.

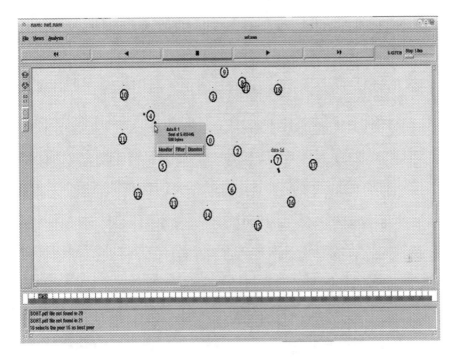

Fig. 8 Downloading the file from best peer

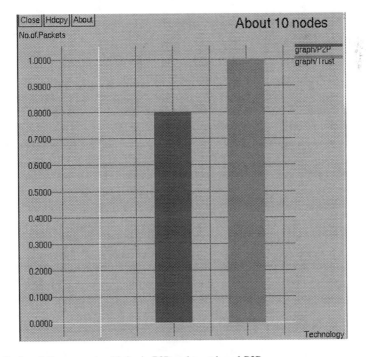

Fig. 9 Packet delivery graph with basic P2P and trust based P2P

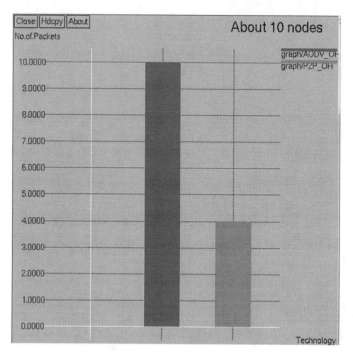

Fig. 10 Overhead comparison AODV and P2P method

5 Conclusion and Future Work

In our proposed method, the file is Information based on the frequent query processes regarding the files. In our proposed technique, we are considered the disconnected MANET as group groups. And we have implemented less overhead file searching system and we have tested successfully. The problem of identifying wrong recommendations is reduced in this work. It reduces the service based attacks and it also reduces the recommendation based attacks if there are not more than 50 % malicious nodes in the P2P network. It uses three types of metrics, service, and reputation and recommendation trust metrics to create a trust network in a peer's proximity. This work also implements the load balancing mechanism to utilize the network resources effectively. When the best service provider in the network reaches its maximum number of simultaneous operations, it suggests another good peer having the same service to the service requester. Hence, the time required for service requester to choose a different peer is reduced.

It helps reducing large amounts of attacks but, this work does not solve all the security issues of a P2P network. This issue should be focused in future work to use the trust model in various applications. Future work may consist of using different load balancing algorithms to reduce the delay of getting a resource.

References

1. Selcuk, A.A., Uzun, E., Pariente, M.R.: A reputation-based trust management system for P2P networks. In: Proceedings of IEEE/ACM Fourth International Symposium on Cluster Computing and the Grid (CCGRID) (2004)
2. Hoffman, K., Zage, D., Nita-Rotaru, C.: A survey of attack and defense techniques for reputation systems. ACM Comput. Surv. **42**(1), 11–131 (2009)
3. Zhou, R., Hwang, K.: Power trust: a robust and scalable reputation system for trusted peer-2-peer computing. IEEE Trans. Parallel Distrib. Syst. **18**(4), 460–473 (2007)
4. Can, A.B., Bhargava, B.: SORT: a self-organizing trust model for peer-2-peer systems. IEEE Trans. Dependable Secure Comput. **10**(1), 14–25 (2013)
5. Hoh, C.-C., Hwang, R.-H.: P2P file sharing system over MANET based on swarm intelligence: a cross-layer design (2007)
6. Marsh, S.: Formalising trust as a computational concept. Ph.D. thesis, Department of Mathematics and Computer Science, University of Stirling (1994)
7. Abdul-Rahman, A., Hailes, S.: Supporting trust in virtual communities. In: Proceedings of 33rd Hawaii International Conference System Sciences (HICSS) (2000)
8. Yu, B., Singh, M.: A social mechanism of reputation management in electronic communities. In: Proceedings of Cooperative Information Agents (CIA) (2000)
9. Despotovic, Z., Aberer, K.: Trust-aware delivery of composite goods. In: Proceedings First International Conference Agents and Peer-2-Peer Computing (2002)
10. Zhou, R., Hwang, K., Cai, M.: Gossiptrust for fast reputation aggregation in peer-2-peer networks. IEEE Transactions on Knowledge and Data Engineering, vol. 20, no. 9, pp. 1282–1295 (2008)
11. Ooi, B., Liau, C., Tan, K.: Managing trust in peer-2-peer systems using reputation-based techniques. In: Proceedings of Fourth International Conference on Web Age Information Management (2003)

PSO Application to Optimal Placement of UPFC for Loss Minimization in Power System

C. Subramani, A.A. Jimoh, Subhransu Sekhar Dash
and S. Harishkiran

Abstract This paper presents the efficient contribution of Particle Swarm Optimization (PSO) to select the optimal location and the optimal parameters setting of Unified Power Flow Controller (UPFC) which minimize the active power losses and improve the voltage stability in the power network. UPFC is the most promising FACTS devices for power flow control. The simulations are run on the IEEE 14 bus system to validate the proposed technique. The results shows that the optimal placement and parameter setting of FACTS device.

Keywords PSO · UPFC · FACTS · Real power loss · Voltage profile

1 Introduction

The Powerflow control and bus voltage regulations on the power system network have been effectively achieved by Flexible Alternating Current Transmission Systems (FACTS) devices. The influence of FACT devices, resulting in power transfer capability improvement, reducing the system loss and stability enhancement in the system. The performance of these FACTS devices highly depending on placement and parameter setting. In general, for power flow control in the network, the Unified Power Flow Controller (UPFC) and STATCOMS are preferable according to their operations.

C. Subramani (✉) · A.A. Jimoh
Department of Electrical Engineering, Tshwane University of Technology,
Pretoria Campus, Pretoria 0001, South Africa
e-mail: csmsrm@gmail.com

C. Subramani · S.S. Dash · S. Harishkiran
Department of Electrical and Electronics Engineering, SRM University,
Kattankulathur 603203, India

© Springer Science+Business Media Singapore 2017
P. Deiva Sundari et al. (eds.), *Proceedings of 2nd International Conference on Intelligent Computing and Applications*, Advances in Intelligent Systems and Computing 467, DOI 10.1007/978-981-10-1645-5_19

223

Shunt connected FACTS are playing an important role in reactive power flow control in the power system network to avoid the voltage fluctuation and stability improvement. Hence the reactive power compensation is a major issue in power system network [1]. Several blackouts in the power network due to the voltage collapse problems. The maximum loadability point is generally identified as voltage collapse point. If the load increase the beyond the maximum loadability in the power system network, the system experiencing the voltage collapse. The FACTS device or reactive compensators are helpful to control the power flow and steady state limit in the power system [2, 3]. Hence the contribution of these devices is very important in power system for avoiding the voltage collapse.

The location of FACTS devices is very important in the system. Different types of optimization techniques were used by the researchers to identify the optimal placement of FACTS devices in power systems [4]. UPFC is controlling the real power and reactive power autonomously or concurrently a versatile FACTS's device which can control the active power, the reactive power. Also, it controls the system voltage for avoiding the voltage fluctuations and improve the stability in the system. This device helps in the system on steady state and dynamic operations. It is important that to determine the location of these devices in the system, also the optimal parameter setting. The researchers developed and incorporated new algorithms for determining the suitable location of UPFC [4–8].

In this paper, the PSO algorithm is effectively implemented for determining the optimal location of FACTS devices. The optimal control parameters setting also successfully identified. The results show the voltage profile improvement and real power loss reduction with the help of UPFC with optimal location and accurate control parameters.

2 Optimization Techniques: Particle Swarm Optimization

2.1 Basic PSO

PSO technique is a search technique based on the emergent movement of a group of birds searching food. Swarm is constituted by the number of particles. The global minimum (or maximum) search looking by each particle. Multidimensional search space used by particles for flying in this optimization technique. During this search, each particles try to adjust their position according to their own experience. The best position caught by the particles with their own experience and its neighbors experience. The particles history experience will decide the swarm direction of the particles. P_{best} is recorded by the previous best particle position and G_{best} is noted by best particle among all the particles in the group.

2.2 PSO Algorithm

The steps involving in PSO technique.

- Set the system input parameters, variables with limits and constraint.
- Randomly initialize the particles of the population.
- Within the limit, generate the Initial searching points and velocities.
- Each initial searching point P_{best} is to be set. The best values from the P_{best} are set to G_{best}.
- Calculate the new velocities as follows

$$V_i^{K+1} = wV_i^k + c_1 \text{ rand}_1 \times \left(P_{best} - X_i^k\right) + c_2 \text{ rand}_2 \times \left(g_{best} - X_i^k\right)$$

- Update the value of P_{best} and G_{best} of each particle
  By using the following equation the P_{best} value of each particle updated in iteration $k + 1$

$$X_i^{k+1} = X_i^k + V_i^{k+1}.$$

- Stop the iteration in the constrains are satisfied, if not satisfied, again start from new velocity calculation.

3 UPFC Model and Problem Formulation

The main objective of optimization is to achieve the better utilization of the accessible power network. The FACTS devices are placed to improve the load-ability of the system with satisfying the voltage and thermal limit. This is meant that, the effect of FACTS device should be helpful for increasing as much as possible the power transmitted to the consumers by the power system network. The power system should be in a secure state in terms of improving loading level, reducing the total loss and maintain the load bus voltage within acceptable limits. The objective function is developed as optimize the maximum loadability and minimizing the voltage variations. In this study and analysis only the technological benefits of the UPFC are considered. The basic model of UPFC is shown in Fig. 1. The UPFC model having two VSCs (voltage-source converters). These two VCSs are connected back-back through a DC capacitor.

The UPFC is modeled by the simultaneous presence of series and shunt FACTS devices in the one transmission line. A UPFC is the combinations of TCSC and SVC, A TCSC in series in the line and SVC at a bus.

With the help decreasing or increasing the reactance of the branch, the operating mode of TCSC changed as inductive or capacitive. The value of reactance is the

226 C. Subramani et al.

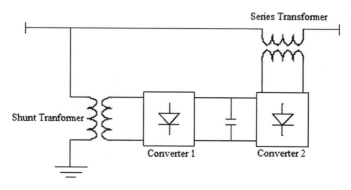

Series Transformer

Shunt Tranformer

Converter 1   Converter 2

Fig. 1 Basic structure of UPFC model

function of reactance of the line where the TCSC is placed. The impedance of the transmission line

$$Z_{ij} = Z_L + j X_{TCSC} \text{ and } X_{TCSC} = r_{TCSC} X_L$$

where
Z_L Transmission line Impedance
X_{TCSC} reactance of the line where TCSC is located
r_{TCSC} compensation degree of TCSC (Coefficient)

The FACTS device SVC is operated as both inductive and capacitive mode and control bus voltage by absorbing or injecting reactive power. A shunt variable susceptance added at both ends of the line for model the SVC.

The injected reactive power at bus i is

$$\Delta Q_{is} = Q_{svc}$$

Q_{SVC} = reactive power injected by SVC in MVAR and
$Q_{SVC} = Q_{min} \sim Q_{max}.$

The constrain limit of the UPFC is,

$$X_{TCSC} = -0.8X_L \text{ to } 0.2X_L \text{ and } Q_{SVC} = -200\,\text{MVAR to } 200\,\text{MVAR}$$

The placement of the UPFC in the power system network to be determined and the control parameters of UPFC is to be optimized. Locations of FACTS devices in the power system are obtained by the PSO technique performance. To reduce the stressed condition in the power system, the UPFCs devices placed in the weakest buses and heavy loaded areas.

Objective Function:

$$\text{Minimize} \quad F(x, u)$$

where F(x, u) is the real power loss in the system. X is the depended variable vector and u is the control variable vector.

Constraints:

Subjected to $g(x, u) = 0$, nonlinear equality constraints (power flow)
  $h(x, u) \leq 0$, nonlinear inequality constraints (bus voltage,
    transformer MVA limit, line MVA limit)

If, a series connected voltage source is placed between nodes i and j in a power system, the VSC can be modeled with an ideal series voltage V_s in series with a reactance Xs. Consider V_i^l represents a fictitious voltage behind the series reactance. Then, $V_i^l = V_s + V_i$. The series voltage source V_s is controllable in magnitude and phase. V_s is the function of control parameters r and. r values are between 0 and r_{max}, γ value is 0 to 2π. UPFC can control the power flow and reduce the real power loss in the system by selecting the control parameters r and γ in a proper manner.

4   Results and Discussion

The effectiveness and performance of the PSO algorithm is verified with IEEE 14 bus bus test power system. The single line diagram of the IEEE 14 bus is shown in Fig. 2. The algorithms are implemented in MATLAB for different loading conditions for finding the optimal location and settings of UPFC to achieve minimum real power loss. Optimal settings and total real and reactive power loss for IEEE 14 bus test system in PSO is represented in Table 1. For the comparative purpose, the weak bus identified with respect to the maximum loadability of load buses. The maximum loadability of all the load buses of the IEEE 14 bus system is shown in Table 1 for different three cases.

Case 1: Reactive power load change (the reactive power demand at selected load bus varies while the others remain fixed)
Case 2: Real power load change (the real power demand at selected load bus varies while the others remain fixed)
Case 3: Real and reactive power load change (both real and reactive power demand at selected load bus varies while the others remain fixed)

The location of UPFC is determined as line 5–6 at the base case loading and the loading is 200 % of base load. The line 3–4 is identified as the optimal location during the loading condition is 150 % of base load. The real power loss also reduced by the effect of UPFC with optimum control parameters. Then in the

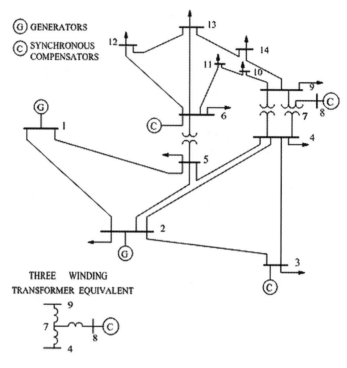

Fig. 2 Single line diagram of IEEE 14 bus system

Table 1 Maximum loadability estimation for IEEE 14 bus system

Bus no.	Case 1	Case 2	Case 3	
	Q_{max} (MVAR)	P_{max} (MW)	Q_{max} (MVAR)	P_{max} (MW)
4	387	532	101.9	487
5	375	462	86.2	409
7	180	294	94.4	188
9	176	272	104	184
10	144	223	93.6	145.2
11	119	223	74.3	144
12	100	181	39	148
13	175	271	175	217
14	86.6	146	42.7	113
Weak bus	Bus no. 14	Bus no. 14	Bus no. 12	
Critical line	13–14	13–14	6–12	
Critical voltage (p.u)	0.7304	0.6272	0.7033	

Table 2 Performance of PSO on IEEE 14 bus system with UPFC

Loading conditions	Without UPFC	With UPFC			
	Power loss (MW)	Power loss (MW)	Location	Control parameters	
				r	γ (°)
Base load	16.8320	15.5390	5–6	0.058	161.12
150 % of base load	65.6448	63.9871	3–4	0.042	211.22
200 % of base load	87.5264	86.1598	5–6	0.037	248.17
$Q_{14} = 86.6$ MVAR	30.2461	30.1006	13–14	0.041	228.36
$P_{14} = 146$ MW	35.3472	35.1932	13–14	0.058	209.91
$Q_{12} = 39$ MVAR, $P_{12} = 148$ MW	28.6144	27.9863	12–13	0.032	159.76

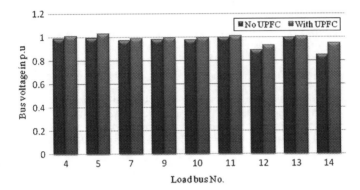

Fig. 3 Voltage profile of IEEE 14 bus system

special case, the load increase only in weak bus identified at different three loading cases and the location of UPFC has determined. The real power loss is effectively reduced by the suitable location of UPFC, which shown in Table 2. The voltage profile improvement with UPFC on IEEE 14 bus system with 150 % base load condition is shown in Fig. 3.

5   Conclusion

This paper attempted to determine the optimal location and optimized parameters for the UPFC device to minimize total real power loss using PSO techniques. The PSO algorithm is effectively implemented for optimal location and control parameter estimation of UPFC. The results show that the real power loss is minimized and voltage profile has improved as better level. With this technique, it is possible to locate the UPFC in the transmission line to proper planning and operation of the power system with minimum transmission losses.

References

1. Bose, D., Biswas, S., Vasilakos, A.V., Laha, S.: Optimal filter design using an improved artificial bee colony algorithm. Inf. Sci. **281**, 443–461 (2014)
2. Dixit, G.P., Dubey, H.M., Pandit, M., Panigrahi, B.K.: Artificial bee colony optimization for combined economic load and emission dispatch. In: IET Conference Publications, (583 CP), pp. 340–345 (2011)
3. Biswas, S., Kundu, S., Bose, D., Das, S., Suganthan, P.N., Panigrahi, B.K.: Migrating forager population in a multi-population artificial bee colony algorithm with modified perturbation schemes. In: IEEE Symposium on Swarm Intelligence (SIS), pp. 248–255 (2013)
4. Das, S., Biswas, S., Panigrahi, B.K., Kundu, S., Basu, D.: A spatially informative optic flow model of bee colony with saccadic flight strategy for global optimization. IEEE Trans. Cybern. **44**(10), 1884–1897 (2014)
5. Gao, W.-f., Liu, S.-Y., Huang, L.-L.: A novel artificial bee colony algorithm based on modified search equation and orthogonal learning. IEEE Trans. Cybern. **43**(3), 1011–1024 (2012)
6. Hingorani, N.G., Gyugyi, L.: Understanding FACTS: Concepts and Technology of Flexible AC Transmission Systems. IEEE Press, New York (2000)
7. Canizares, C.A.: Modeling and implementation of TCR and VSI based FACTS controllers. Internal report, ENEL and Politecnico di Milano, Milan, Italy (1999)
8. Mori, H., Goto, Y.: A parallel tabu search based method for determining optimal allocation of FACTS in power systems. In: Proceedings of the International Conference on Power System Technology (PowerCon 2000), vol. 2, pp. 1077–1082 (2000)

An Empirical Study on Fingerprint Image Enhancement Using Filtering Techniques and Minutiae Extraction

J. Shiny Priyadarshini and D. Gladis

Abstract The uniqueness of fingerprint provides a secured personal identification mark. Contact less acquisition of fingerprints is gaining popularity due to the ease in capturing the image. They can be easily blurred with noise and distortion. This paper defines filtering techniques in removal of noise and distortion from the blurred images. The paper highlights double enhancing method for image clarity and extraction of true minutiae points which are tabulated and analyzed.

Keywords Fingerprint · Lucy-Richardson · Wiener · Unsharp

1 Introduction

Fingerprints are made up of neat biometric pattern such as arches, whorls and valleys found on each human finger. They are so unique that not even two people in the globe can have the uniform pattern. Even the DNA found in identical twins can be same but cannot have the same fingerprints. Fingerprint identification is the most widely used secured, safe and economical biometric identity. They are formed in a baby's small developing fingers inside the womb thus making it an ease in sample collection for all human beings irrespective of adult or a small baby.

The impressions on the fingerprints are classified as ridges. The valleys are the gap found between two ridges. The pattern begins on one side of a finger, moves around or upward, and slides on the other side. The fingerprint patterns are analyzed which results in very minute details called minutiae that cannot be visualized by normal human eye. The unique feature of fingerprint is that the pattern remains unchanged throughout the lifetime of an individual unless destroyed by unexpected circumstances. A fingerprint sample image is shown in Fig. 1.

J.S. Priyadarshini (✉)
Madras Christian College, East Tambaram, Chennai 600059, India
e-mail: shinymcc02@gmail.com

D. Gladis
Presidency College, Chennai 600005, India

© Springer Science+Business Media Singapore 2017
P. Deiva Sundari et al. (eds.), *Proceedings of 2nd International Conference on Intelligent Computing and Applications*, Advances in Intelligent Systems and Computing 467, DOI 10.1007/978-981-10-1645-5_20

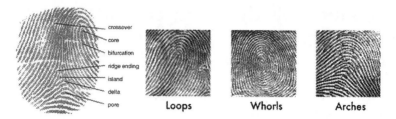

Fig. 1 A sample fingerprint image

2 Existing Work

Chauhan and Kumar [1] presented fingerprint enhancement techniques such as Histogram Equalization and Fast Fourier Transformation. Soni and Siddiqua [2] presented filtering techniques for blurred images. Patil and Bhatt [3] presented a survey on Latent fingerprint Matching techniques. The paper explains five steps such as Histogram, Segmentation, Thinning, Smoothing and Binarization. Garg and Bansal [4] presented minutiae extraction by combining image processing and fast fourier. Bana and Kaur [5] presented different techniques for pre and post processing in extracting the minutiae points. Rajkumar and Hemachandra [6] presented a primary and secondary enchancement technique for removing noise from the fingerprints. Patil and Chandel [7] presented a performance evaluation for Second derivative Gaussian for fingerprint enhancement. Chouthmal Bhosale and Kale [8] presented a matching score between fingerprints and also Ridge frequency estimation in fingerprint enhancement. Tarar and Kumar [9] presented a fingerprint image enhancement algorithm by using Iterative Fast Fourier Transform (IFFT). IFFT plays a key role in fingerprint enhancement systems.

3 Proposed Methodology

This paper mainly focuses on the pre-processing methods for image enhancement to obtain clarity of fingerprint image. This is done by fingerprint enhancement and Deblurring techniques. Enhancement is used for increasing the contrast of the images whereas Deblurring increases clarity in blurred images [2]. Figure 2 depicts the different stages in extracting the characteristics of fingerprints.

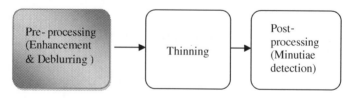

Fig. 2 Stages in fingerprint characteristics extraction

3.1 Fingerprint Enhancement

Image enhancement is carried out to improve the contrast of the image and to sharpen the values of the image. Image filtering is a technique applied to make images appear sharper and clear by smoothing the details of the image. To reduce the blurness in an image, the intensity values should be changed using filtering techniques. A study is carried out by applying different filtering techniques.

3.1.1 Lucy-Richardson

The Lucy-Richardson algorithm is used for deblurring images [2]. The algorithm uses a Point spread function (PSF) which is a 2D Gaussian filter with convolution. After convolution the image is restored using de-convolution (Fig. 3).

3.1.2 Wiener Filter

The second filter applied is Wiener Filter for deblurring images [2]. It can be used to filter additive noise from the image. During the process of filtering and noise smoothening the mean square error is minimised by the Wiener filter.

3.1.3 Unsharp

The third filter applied is Unsharp filter. Unsharp is an image sharpening 2D filtering technique. As it creates a mask of the original image for the blurred and unsharp images it has got the name "Unsharp". Unsharp results to an edge image m (x, y) from an input image n(x, y)

$$m(x, y) = n(x, y) - n_{smooth}(x, y)$$

where $n_{smooth}(x, y)$ is a smooth version of n(x, y).

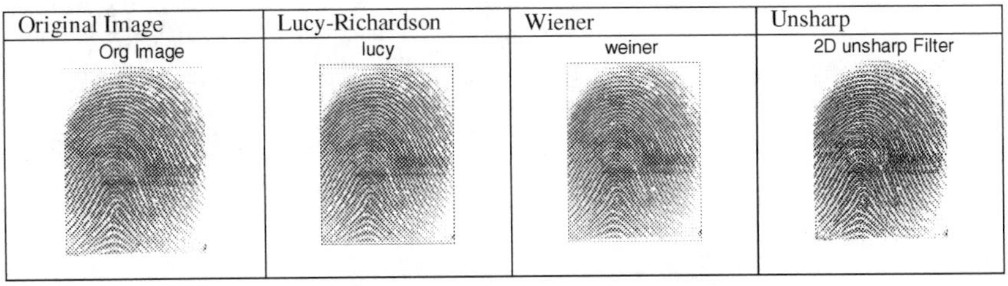

Original Image	Lucy-Richardson	Wiener	Unsharp
Org Image	lucy	weiner	2D unsharp Filter

Fig. 3 Original image, Lucy-Richardson, Wiener and Unsharp image

3.1.4 Fast Fourier Transformation

A Fourier transform converts an image to domain frequency from its original spatial domain. An Inverse Fourier transform converts the frequency domain components back into the original spatial domain. The fingerprint image is divided into 32×32 window. For each 32×32 pixels the Fourier transform is applied, for $u = 0, 1, 2, \ldots\ldots 31$ and $v = 0, 1, 2 \ldots\ldots 31$.

$$F(u, v) = \sum_{x=0} \sum_{y=0} f(x, y) \exp\{-j2\pi(ux/M + vy/N)\}$$

To enhance each block the magnitude of FFT is multiplied with FFT as

$$g(x, y) = F^{-1}\{F(u, v) * |F(u, v)|^k \text{where}, k = 0.4$$

The Inverse Fourier Transform regains the original image. The FFT enhances the image by connecting the false ridges and removing the spurious connections between the ridges [5–7].

3.2 Binarization

Binarization is done in the pre-processing stage of an image [4, 5]. Binary image is an image with only two possible values, zero's and one's and usually depicted as black and white image. The fingerprint image is binarized where the 8 bit gray image is transformed into a 1 bit image with 0 bit value for ridges and 1 bit value for valleys. The image is transformed into black and white colour, where the black colour represents ridges and white colour represents valleys. Figure 4 depicts an original image and its corresponding binary image.

Fig. 4 Original image and binary image

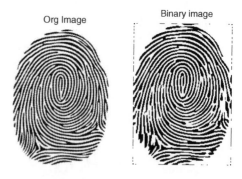

Org Image

Binary image

3.3 Thinning

Thinning results in skeletonization of an image. It is a morphological operation performed on the binary images to remove the selected foreground pixels. Ridge thinning is a process where the image is thinned by discarding the redundant pixels of ridges until it becomes one pixel wide. The thinned image is again filtered to remove H breaks, isolated points and spikes using the morphological functions [4, 5].

3.4 Minutiae Marking

The fingerprint verification uses two types of minutia called termination and bifurcation. Termination is the abrupt ending of a ridge whereas bifurcation refers to a ridge point from which two branches emerge [8]. A minutia marking is a process to mark Ridge bifurcations and Ridge Terminations. Using crossing number concept the Fingerprint image is divided into a 3×3 window. In this 3×3 matrix if the central pixel is equivalent to one, along with three adjacent one-valued neighbours then it represents a ridge branch. Whereas if the central pixel is one and has only one-valued adjacent pixel then it represents a ridge ending [4, 5].

3.5 Spurious Minutiae Removal

False Minutiae removal is important to maintain the accuracy of the image [4]. False minutiae removal is done by calculating the average inter ridge distance D. It is done by taking the width D which is the average distance between two neighbouring ridges for each row.

$$\text{Inter} - \text{ridge distance } (D) = \text{Sum of all pixels with value } 1 / \text{ row length}$$

If the distance between two bifurcations and two terminations is less than six then the spurious minutiae's are eliminated respectively. Similarly the spurious minutiae's are eliminated if the distance between terminations and bifurcations is less than six.

4 Simulation Results

The minutiae extraction is done by enhancing the fingerprint by using filtering techniques in the pre-processing stage. The image is further improved with Fast Fourier transformation as second enhancement technique. The filtering techniques used are Lucy-Richardson, Wiener and Unsharp. The techniques are applied on

Table 1 Minutiae count—spurious and non spurious ridge terminations and ridge bifurcations for FFT and Unsharp

S. no	FP names	FFT				Unsharp			
		Spur		No Spur		Spur		No Spur	
		T	B	T	B	T	B	T	B
	Noiseless images								
1	F6	64	42	*44*	*17*	180	256	42	26
2	F5	69	29	*56*	*9*	64	40	48	12
3	F4	46	87	*36*	*34*	47	114	34	37
4	F7	35	22	*27*	*8*	37	205	23	20
	Noisy images								
5	Fp2	158	60	*57*	*19*	162	86	56	23
6	F2	130	44	*48*	*11*	180	256	42	26
7	Noisy5	685	2121	*45*	*96*	1251	4621	28	50

randomly selected images from Fingerprint Verification Competition 2004, which include some noisy images too. In tables, Spur represents minutiae count with Spurious Terminations (T) and Bifurcations (B) and No-Spur represents the count for Terminations (T) and Bifurcations (B) with false minutiae removal.

Table 1 represent the minutiae count taken for the FFT and Unsharp filtering techniques. Figure 5a represents a sample pictorial representation for a noiseless and a noisy image. Table 2 represent the minutiae count taken for the Lucy-Richardson and Wiener filtering techniques with its pictorial representation in Fig. 5b. Table 3

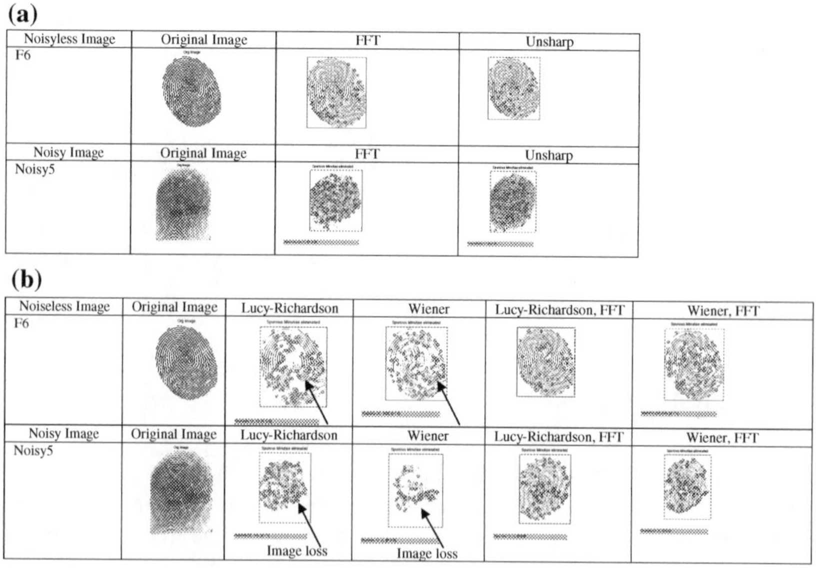

(a)

(b)

Fig. 5 a,b, c Double enhancement with spurious and nonspurious representation

(c)

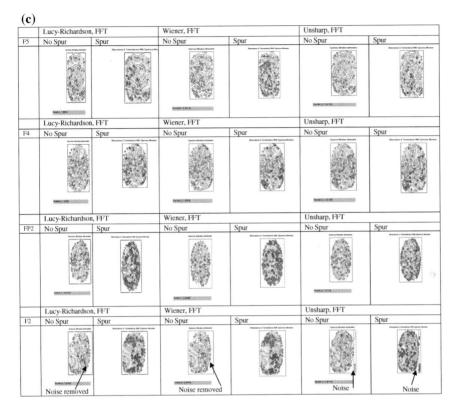

Fig. 5 (continued)

Table 2 Minutiae count—Lucy-Richardson and Wiener

S. no.	FP names	Lucy-Richardson				Wiener			
		Spur		No Spur		Spur		No Spur	
		T	B	T	B	T	B	T	B
	Noiseless images								
1	F6	311	4	*143*	*0*	464	22	*117*	*1*
2	F5	151	31	*111*	*5*	157	52	*102*	*10*
3	F4	195	23	*111*	*5*	98	51	*54*	*17*
4	F7	51	8	*40*	*4*	44	17	*36*	*7*
	Noisy images								
5	Fp2	413	29	*133*	*2*	362	54	*76*	*8*
6	F2	213	40	*77*	*7*	225	61	*64*	*12*
7	Noisy5	454	37	*143*	*2*	278	85	*66*	*6*

Table 3 Minutiae count—Lucy-Richardson and FFT, Wiener and FFT, Unsharp and FFT

S.no.	FP names	Lucy-Richardson, FFT				Wiener, FFT				Unsharp, FFT			
		Spur		No Spur		Spur		No Spur		Spur		No Spur	
		T	B	T	B	T	B	T	B	T	B	T	B
	Noiseless images												
1	F6	169	37	66	7	318	39	107	7	63	40	43	18
2	F5	96	65	77	15	156	50	96	10	64	48	45	11
3	F4	94	71	52	17	76	87	41	22	47	102	35	36
4	F7	36	32	26	9	38	29	32	10	38	188	25	18
	Noisy images												
5	Fp2	227	199	73	24	224	160	55	20	163	83	55	22
6	F2	150	220	57	18	177	168	57	15	181	224	43	25
7	Noisy5	194	390	75	43	288	326	98	39	681	5253	18	74

represents minutiae count taken with a combination of filters and FFT with its image representation in Fig. 5c. Table 1 clearly shows that the Unsharp filter works better when compared to FFT for normal images. But when noisy images are chosen there is a loss of image with less terminations and bifurcations. Similarly when Lucy-Richardson and Wiener are applied individually, there is a loss of image. By comparing Tables 1, 2 and 3 the results for noiseless images are found to be better in Table 1. But Table 3 a combination of filters and FFT shows better results for noisy images. Lucy-Richardson and Wiener along with FFT shows the best result. Though there is moderate increase in Terminations and Bifurcations count it has been noted that there is no loss of image.

5 Conclusion

The Unsharp filtering technique performs better for noiseless images but results in image loss for noisy images. Similarly by applying the filtering techniques individually, also results in image loss. Hence Lucy-Richardson and Wiener along with FFT were applied and the minutiae count which has been taken proves to be the best result for noisy images.

References

1. Chauhan, P., Kumar, A.: Survey on fingerprint enhancement technique. 4 (2014). ISSN: 227128X
2. Soni, N., Siddiqua, A.: Filtering techniques used for blurred images in fingerprint recognition. Int. J. Sci. Res. Publ. 3(5) (2013). ISSN: 2250-3153

3. Patil, S.G., Bhatt, M.: A survey on latent fingerprint matching techniques. **3** (2014). ISSN: 2319-5940
4. Garg, M., Bansal, H.: Fingerprint recognition system using minutiae estimation. **2** (2013). ISSN: 2319-4847
5. Bana, S., Kaur, D.: Fingerprint recognition system using image segmentation. Int. J. Adv. Eng. Sci. Technol. **5**(0) (2011). ISSN: 2230-7818
6. Rajkumar, R., Hemachandra, K.: A secondary fingerprint enhancement and minutiae extraction. Signal & Image Process.**3**(2) (2012)
7. Patil, S.S., Chandel, G.S., Gupta, R.: Fingerprint image enhancement techniques and performance evaluation of the SDG and FFT fingerprint enhancement techniques. **2**. ISSN: 2249-6343
8. Chouthmal Bhosale, P.P., Kale, K.V.: A novel approach for fingerprint recognition. **4** (2014). ISSN: 2277 128X
9. Tarar, S., Kumar, E.: Fingerprint image enhancement: iterative fast fourier transform algorithm and performance evaluation. Int. J. Hybrid. Inf. Technol. **6**(4) (2013)

Multi-application Antenna for Indoor Distribution Antenna System like Wi-Fi, Wi-max and Bluetooth

Hina D. Pal and Balamurugan Kavitha

Abstract The multi-application antenna is proposed for Bluetooth, Wi-Fi and Wi-max. The design is used for three different frequencies 2.45, 5 and 7 GHz. As a substrate material RT_Duroid 6006 is used. Antenna is designed by placing two slots in rectangular patch, so three dipoles are obtained. The simulation is done for this antenna by changing the length and width of three dipoles. The results shows that rectangular patch antenna with two slots offers return loss ≤10 dB with the input feed 50 Ω. The proposed antenna can be used in indoor distributed antenna systems, such as Bluetooth, Wi-Fi and Wi-max. This patch antenna was designed and simulated by ADS software. ADS supports of the design process—Engineer can characterize schematic capture, frequency and time-domain simulation, layout, and electromagnetic field simulation, allow engineer to characterize and RF design with common tool.

Keywords Wi-Fi · Wi-max and bluetooth · ADS · Indoor distribution system · Microstrip patch

1 Introduction

With the rapid development of modern wireless communications, such as Bluetooth, Wi-max and Wi-Fi, there has been on demand to cover different frequencies without the need for tuning any specific frequency because it is an omni directional antenna. The wideband omni directional antennas, which radiate equally in all directions over the whole operation bands, are the most popular antennas designed for in-building ceiling mount installations of indoor distributed antenna systems (DAS). In this paper good efficiency is offered by modified Microstrip

H.D. Pal (✉) · B. Kavitha
KCG College of Technology, Karapakkam, Chennai, India
e-mail: heenapal22@gmail.com

B. Kavitha
e-mail: kavitha@kcgcollege.com

© Springer Science+Business Media Singapore 2017
P. Deiva Sundari et al. (eds.), *Proceedings of 2nd International Conference on Intelligent Computing and Applications*, Advances in Intelligent Systems and Computing 467, DOI 10.1007/978-981-10-1645-5_21

241

Patch antenna which operates at frequency range 2–11 GHz has been discussed in detail. Microstrip patch antenna has becomes a main component in the development of Wi-Fi, Wi-max and Bluetooth, this is the reason for the attention of research form past two decades.

Compared to radiators this are preferred because of its light weight but its main disadvantage is it's low antenna efficiency and low gain. Overcoming this problems has become the motivation of research among researchers around the world. Here an attempt was made to improve the efficiency of the patch antenna. It has been noticed that there is some remarkable change in antenna efficiency measurement.

2   Antenna Design

The multi-application antenna is designed with the two slots in the rectangular patch. So, three dipoles can be achieved in the patch.

Here length of dipole is calculated using this formula:

$$\lambda g = \frac{C}{f\sqrt{\frac{\varepsilon r + 1}{2}}} \tag{1}$$

Here
λg is guide wave length,
εr is dielectric constant of the substrate material,
C   is velocity of light
f   is the frequency

In the design guided wave length is used as length of all three dipoles and the spaces between the all three dipoles are set as 5 mm. For the feed the central dipole was extended 5 mm. This 5 mm value chosen randomly. Cu-clad, RT-Duroid, FR-4 were the different dielectric substrate materials used for design. Here, first dipole designed with a length of $3\lambda g/2$ the second dipole with λg and third one with $\lambda g/2$. The feed used here is having $\lambda g/8$ length. Here the spacing between three dipoles was 5 mm. Cu-clad, RT-Duroid, FR-4 were the different dielectric substrate materials used for design. This design was used for Wi-Fi, Wi-max and Bluetooth. So, for the design every time the different frequency was used. So the resultant designed patch is shown in Fig. 1. Then the same design used for all three application simultaneously here the first dipole was used for Bluetooth, second for Wi-Fi and third for Wi-max. Here RT_Duroid 6006 was chosen as the dielectric substrate for the antenna. This material is lightweight, and possesses a low dielectric constant ($\varepsilon r = 6.6$).

Fig. 1 Designed patch antenna

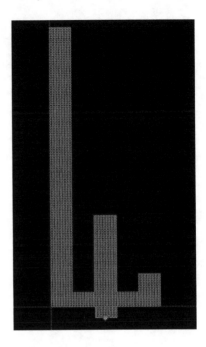

3 Fundamental Specifications of Patch Antennas

A micro strip or patch antenna is a low profile because of its light weight, low cost, and easy to integrity with related electronics, it is preferred compared to other antenna. The antenna can be 3D in structure, the elements are usually flat hence the other name is planar antennas.

3.1 Antenna Efficiency and Gain

Antenna gain is defined as antenna directivity times a factor representing the radiation efficiency. So here the antenna efficiency is 8.685 at 2.45 GHz (shown in Fig. 2), 21.518 at 5 GHz (shown in Fig. 3), and 34.385 at 7 GHz (shown in Fig. 4). The over all efficiency of antenna is 11.301 which is shown in Fig. 5. The gain for 2.4 GHz is 3.652 dB, for 5 GHz it is 5.102 dB and for 7 GHz it is 6.162 dB.

3.2 Radiation Pattern

The patch's radiation at the fringing fields results in a certain far field radiation pattern. This radiation pattern indicates that the antenna radiates how much power

244   H.D. Pal and B. Kavitha

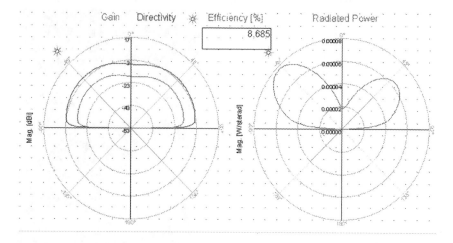

Fig. 2 Gain and efficiency for 2.45 GHz

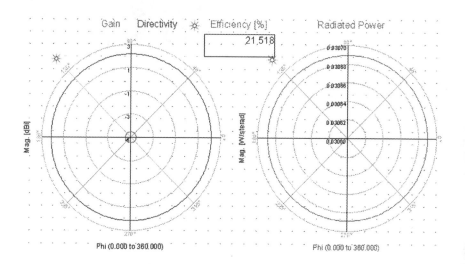

Fig. 3 Gain and efficiency for 5 GHz

in which direction. This is call antenna directivity and it is expressed in dB. The radiation pattern for 2.45 GHz is having two main lobes and two very small side lobes which is shown in Fig. 3. For 5 and 7 GHz radiation pattern is the wave type which is shown in Figs. 7 and 8. Multi-application patch antenna is having over all somehow omni directional radiation pattern (shown in Fig. 9) which is highly useful for indoor application Fig. 6.

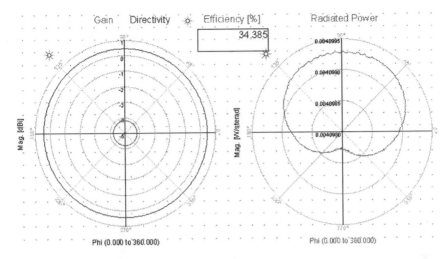

Fig. 4 Gain and efficiency for 7 GHz

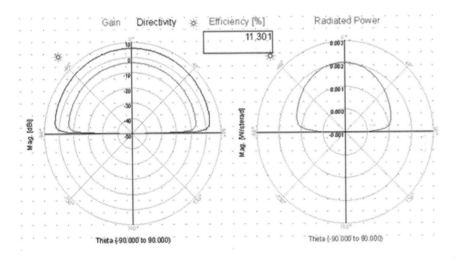

Fig. 5 Over all Gain and efficiency

3.3 Return Loss

In telecommunications, return loss is loss of power in the signal reflected back because of un uniform surface in a transmission line or optical fiber and it is expressed in decibels (dB). The results indicate that rectangular patch antenna with two slots offers impedance -10 dB for 2.4 GHz, -10 dB for 5.5 GHz and -22 dB (shown in Fig. 10) for 7 GHz with the input feed 50 Ω.

Fig. 6 3D radiation pattern
of patch antenna for
2.45 GHz

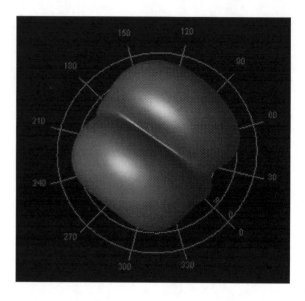

Fig. 7 3D radiation pattern
of patch antenna for 5 GHz

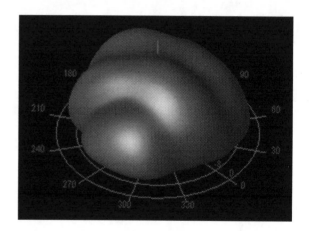

3.4 Antenna Parameters

It represents the values of the different parameters like radiated power, effective
angle, directivity, gain etc. Here Figs. 11, 12, 13 and 14 shows antenna parameters
for 2.45, 5, 7 GHz and for overall patch antenna respectively. Here accurate
numerical value is presented in the antenna parameter.

Fig. 8 3D radiation pattern of patch antenna for 7 GHz

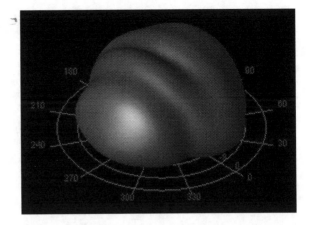

Fig. 9 Omni directional radiation pattern

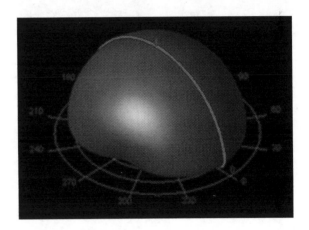

Fig. 10 Return loss of Multi-application patch antenna

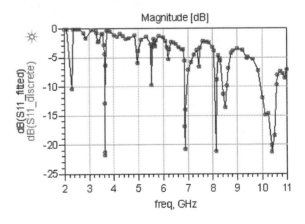

Fig. 11 Antenna parameters
for 2.45 GHz

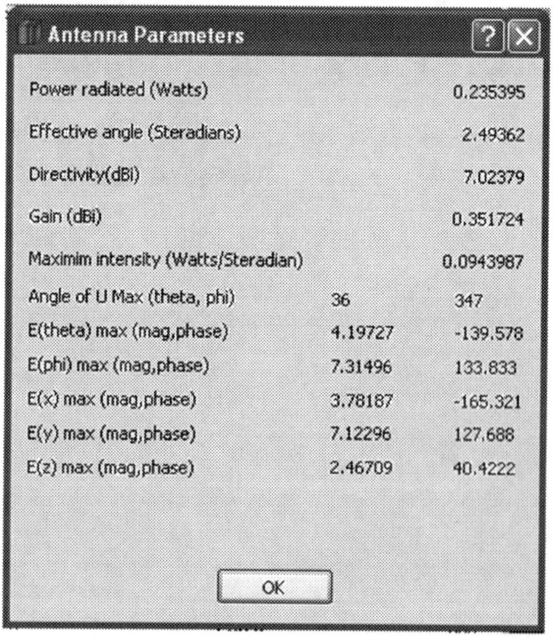

Fig. 12 Antenna parameters
for 5 GHz

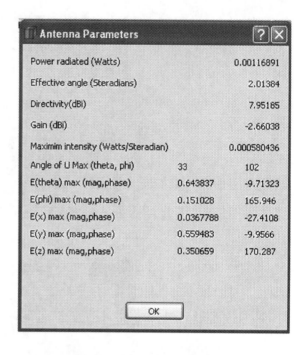

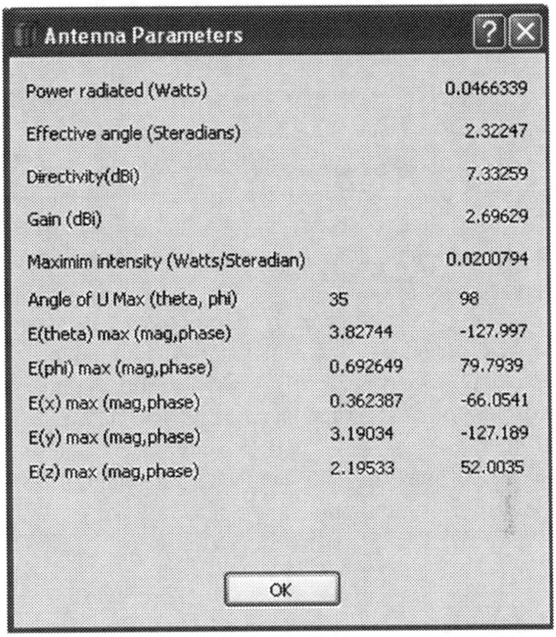

Fig. 13 Antenna parameters
for 7 GHz

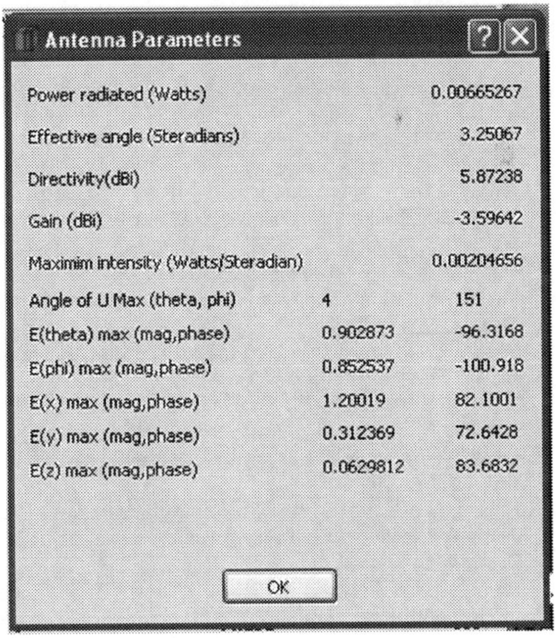

Fig. 14 Antenna parameters
for multi-application patch
antenna

250 H.D. Pal and B. Kavitha

4 Result and Conclusion

The research inspiration of this project is to design multi-application patch antenna for Wi-Fi, Wi-max and Bluetooth applications which operates at 2.45, 5 and 7 GHz. The rectangular patch antenna with two slots with 50Ω feed has been designed. It's operating range is 2–11 GHz and substrate material used here is RT_Duroid 6006. Here size of antenna is 102 mm × 25 mm and over all thickness of antenna is 2.375 mm only so it is light weight and flexible for indoor applications.

Over all antenna gives the return loss ≤−10 dB and Antenna efficiency is 11.30. Multi-application patch antenna is having over all somehow omni directional radiation pattern which is highly useful for indoor distribute antenna system. A design of slots on the patch achieving the multiband operation which is the part of this project is very good for future aspects. Even more number of applications can be included in multi-application antenna. ADS simulation software was used for design of multi-application antenna.

References

1. Balanis, C.A.: Antenna Theory-Analysis and Design, 2nd edn, pp. 454–456. Wiley, New York (1997)
2. Levine, E., Shtrikman, S., Treves, D.: Double-sided printed arrays with large bandwidth. In: IEEE Proceedings of Microwaves, Antennas and Propagation, vol. 135(1), pp. 54–59 (1988). doi:10.1049/ip-h-2.1988.0010
3. DeJean, G.R., Thai, T.T., Nikolaou, S., Tentzeris, M.M.: Design and analysis of microstrip Bi-Yagi and Quad-Yagi antenna arrays for WLAN applications. IEEE Antennas Wirel. Propag. Lett. 6, 244–248 (2007). doi:10.1109/LAWP.2007.893104
4. Zheng, G., Kishk, A.A., Yakovlev, A.B., Glisson, A.W.: Simplified feeding for a modified printed Yagi antenna. In: IEEE Antennas and Propagation Society, International Symposium Digest, vol. 3, pp. 934–937 (2003)
5. Floc'h, J.M., Rmili, H.: Design of multi-band printed dipole antennas using parasitic elements. Microw. Optical Technol. Lett. 48(8), 1639–1645 (2006). doi:10.1002/mop.21714
6. Wong, K.L.: Compact and Broadband Microstrip Antennas. Wiley, New York (2003)
7. Bailey, M.C.: Broad-band half-wave dipole. IEEE Trans. Antennas Propag. 32(4), 410–412 (1984). doi:10.1109/TAP.1984.1143318
8. Kaneda, N., Deal, W.R., Qian, Y.X., Waterhouse, R., Hoh, T.: A broad-band planar Quasi-Yagi antenna. IEEE Trans. Antennas Propag. 50(8), 1158–1160 (2002). doi:10.1109/TAP.2002.801299
9. Kramer, O., Djerafi, T., Wu, K.: Vertically multi-layer-stacked Yagi antenna with single and dual polarizations. IEEE Trans. Antenna Propag. 99, 1022–1030 (2010)
10. Deal, W.R., Kaneda, N., Sor, J., Qian, Y., Itoh, T.: A new Quasi-Yagi antenna for planar active antenna arrays. IEEE Trans. Microw. Theory Tech. 48(6), 910–918 (2000). doi:10.1109/22.846

11. Jagnid, S., Kumar, M.: Compact planar UWB patch antenna with integrated bandpass filter & band notched characteristics. In: International Conference on Communication Systems and Network Technologies (CSNT) (2012)
12. Zhou, L., Jiao, Y., Qi, Y., Weng, Z., Lu, L.: Wideband ceiling-mount omnidirectional antenna for indoor distributed antenna systems. IEEE Antennas Wirel. Propag. Lett. **13** (2014)

Application of Voltage Stability Index for Congestion Management

Madhvi Gupta, Vivek Kumar, N.K. Sharma and G.K. Banerjee

Abstract Voltage stability is a relevant problem in power system as far as congestion management is concerned. Ranking is calculated using the line voltage stability index. All the system buses have been loaded one by one up-to maximum load-ability limit and maximum load-ability limit is based on voltage stability index. The bus having least maximum load-ability has been considered as weakest bus of the system. This bus may be considered as the optimal location for the placement of a FACTS Device. The amount of reactive power to be injected will be considered in the further work. The proposed congestion management has been tested on standard IEEE 14-bus and 30-bus transmission test system and a real time 24-bus Indian transmission system.

Keywords Congestion management · Voltage stability index · Maximum load-ability · Voltage stability

1 Introduction

Traditionally, the problem of stability has been one of maintaining synchronous operation of 'Rotor Angle Stability'. In synchronous machines, whenever the angular position of rotor is effected with respect to synchronously rotating reference frame of the machine, there is a mismatch between the mechanical power input and electrical power output leading to instability of the system. Voltage stability is the ability of power system by virtue of which it maintains adequate magnitudes of voltage at various buses, so that when the nominal load of the system is increased, actual power transferred to the load also increases. However, if this does not happen, the system

M. Gupta (✉) · V. Kumar · G.K. Banerjee
Department of Electrical Engineering, IFTM University, Moradabad, India
e-mail: madhavigupta0510@gmail.com

N.K. Sharma
Department of Electrical Engineering, GL Bajaj Institute of Technology
and Management, Greater Noida, India

© Springer Science+Business Media Singapore 2017
P. Deiva Sundari et al. (eds.), *Proceedings of 2nd International Conference on Intelligent Computing and Applications*, Advances in Intelligent Systems and Computing 467, DOI 10.1007/978-981-10-1645-5_22

voltage will collapse which may lead to partial or full power interruption of the system. Several incidents of voltage instability and collapse have been observed throughout the world leading to major system break-down. It has been observed as a slow phenomenon in several cases and also fast phenomenon in few cases. The voltage collapse is a very complex phenomenon especially with the growing interconnected system and therefore in the present scenario it is of prime concern.

Depending upon the simulated time, voltage stability has been analyzed as (i) static voltage stability which involves the solutions of algebraic equations and (ii) Dynamic voltage stability which occurs when changes are very fast and require appropriate modeling of the system to provide the actual dynamics of voltage instability. In static voltage analysis, reactive power imbalance is mainly considered and therefore loading ability of the system bus depends upon the injection of reactive power locally. Since as the system approaches maximum load-ability limit or voltage collapse point, both the real and reactive power losses increase rapidly.

For the prediction of voltage collapse, various methods are used. The most basic and commonly used method is based on P–V curve and Q–V curve i.e. nose curves of the system under consideration along with its load characteristic. In second method, the index for a desired bus is calculated which indicates the impending voltage collapse at that bus. The distance to the point where the voltage collapse will take place can be calculated using nose curves. The distances in MW or MVAR to the critical point of the nose curves are used as indices of voltage stability and as the distances decrease, the chances of voltage collapse increases. In another method the power flow jacobian matrix is determined and index is calculated by calculating the minimum singular value of this matrix. Balamourgougan et al. [1] have proposed a voltage collapse prediction index (VCPI) for any bus in an interconnected system based on the measured bus voltage phasors and the network admittance matrix. They have shown the VCPI for kth bus will vary from 0 to 1 and if index is zero, the bus voltage is stable and if index is 1, the voltage at the bus has collapsed.

$$\text{VCPI} = \left| 1 - \frac{\sum_{\substack{m=1 \\ m \neq k}}^{n} v'_m}{V_k} \right|$$

where

$$v'_m = \frac{Y_{km}}{Y_{kk}} V_m$$

Musirin et al. [2] have proposed a Fast Voltage Stability Index method to predict the occurrence of voltage collapse and contingency analysis caused by the line voltage in power system. They have referred the voltage stability index to a time and formed the power or voltage quadratic equations. The discriminant of the roots of voltage or power quadratic equation is set to be greater than zero. If the discriminant is less than zero, it causes the roots of the quadratic equations to be imaginary which in turn cause the value of the FVSI index such that the system voltage may collapse. The line index evaluated close to 1.00 will indicate the limit of voltage instability.

2 Line Voltage Stability Index

In this work the Fast voltage stability index (FVSI) as derived by Musirin et al. [2] is applied to a real time Indian transmission system. Since due to economic and environmental pressures the systems are being operated under stressed conditions, a better congestion management can be provided by analyzing the power system under these conditions. A well analyzed system can be considered as more secured which in turn will be easy to manage. The Fast voltage stability index [2] is used to analyze the transmission system under consideration. The voltage stability index is given by

$$FVSI = \frac{4Z^2 Q_j}{V_s^2 X}$$

where Z = Line impedance, X = Line reactance, Q_j = Reactive power at the receiving end, V_s = Sending end voltage. If the value of FVSI is '0' it means system is stable and if '1' the system is unstable. The value of index is closer to 1 means; system is in the verge of instability. Hence it should be always less than one.

3 Test Systems

For the purpose of analysis, three transmission test systems have been considered. One is modified IEEE 14-bus system, second is IEEE-30 bus test system and third is a real time 24-bus Indian transmission system. By using MATLAB program, the critical bus having least value among maximum load-ability limit list has been determined. This least load-ability limit of the bus will decide the weakest bus in whole transmission system and it requires the proper placement of a FACTS Device.

The modified IEEE 14-bus test system is shown in Fig. 1. It consists of five generators on buses 1, 2, 3, 6 & 8 and 11 loads consuming total active and reactive powers of 259 MW and 77 MVAR respectively through 14 transmission lines. It is also having a synchronous generator at bus 8. Bus 1 is considered as slack bus.

The 24-bus Indian transmission system is shown in Fig. 2. It consist of 4 generators, 16 transmission lines, 4 shunt compensators, 17 reactors, 2620 MW Peak load and 980 MVAR reactive load. With the help of computation, we have

Fig. 1 Single line diagram of
IEEE 14-bus test system

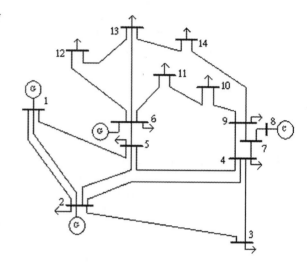

Fig. 2 Single line diagram of
24-bus transmission system

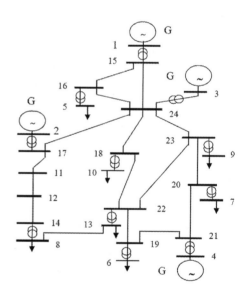

tabulated the given data of standard 24-bus Indian transmission system and calculated the maximum load-ability limit. Bus 1 is considered as a slack bus. Bus 2, 3,
4, 5 is considered as generator bus.

In the next step we have considered IEEE-30 bus transmission system. The IEEE
30-bus transmission system is a considered here; consist of 6 generators, 41
transmission lines and 20 loads. With the help of computation, we have tabulated
the given data of IEEE 30 bus transmission system and calculated maximum
load-ability limit. Bus 1 is considered as a slack bus. Bus 1, 2, 5, 8, 11, 13 is
considered as generator bus as shown in Fig. 3.

Fig. 3 Single line diagram of
standard IEEE 30 bus test
system

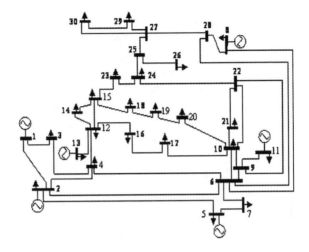

In the present work, all the buses have been analyzed separately by increasing the loads in steps to provide better understanding of the system Fast Voltage Stability Index of the line in real time. In the proposed method of analysis ranking of buses have been based on their stability. It has been shown that the last numbered bus in the system is not always the weakest bus. In IEEE 30 bus transmission system 26th bus is weakest, which is not the last bus, and in 24-bus Indian transmission system 10th is the weakest bus.

4 Simulation Results

Maximum load-abilities were calculated for each load bus and the bus having least values were considered as weakest bus. The 10th bus has least maximum load-ability, hence it is the weakest bus in the 24-bus transmission system. The 24-bus system also has shown the semi transmission line property for few buses. The load flow did not converge up to 100 iterations for some buses. However, the load flow converged within 500 iterations for the same buses. But this happened for only some buses so those buses are considered as inadmissible. The weakest bus found this way may be considered as the optimum location for the placement of a FACTS device.

For all the systems, voltage stability index have been increased until the index reached 0.99 limit. The value 1 indicated the voltage collapse in the system. The corresponding voltage magnitudes represents the system behavior for congestion management (Table 1 and Fig. 4).

In the Table 2, given below, the maximum load-ability limit for all the load buses is calculated for 24-bus system. Figure 5 suggest that buses-10 and 14 are the places where special care should be taken as far as stability is concerned.

Table 1 Maximum load-ability limit calculation for 14-bus test system

Bus number	Maximum load-ability limit	Voltage index	Voltage magnitude
1	Generator bus	–	–
2	Generator bus	–	–
3	Generator bus	–	–
4	387	0.9995	0.7580
5	377.8	0.9999	0.7719
6	Generator bus	–	–
7	180.8	0.9999	0.8320
8	Generator bus	–	–
9	176.39	0.9999	0.7540
10	144.44	0.9999	0.6914
11	120.25	0.9999	0.7975
12	100.37	0.9999	0.8266
13	175.25	0.9999	0.8016
14	86.81	0.9998	0.7301

Fig. 4 Bus number versus maximum load-ability plot of IEEE 14 bus test system

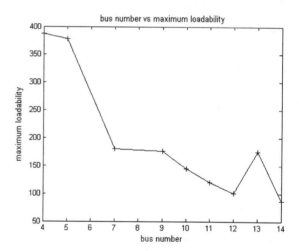

Table 2 Maximum load-ability limit calculation for 24-bus test system

Bus number	Maximum load-ability limit	Voltage index	Voltage magnitude
1	Slack bus	–	–
2	Generator bus	–	–
3	Generator bus	–	–
4	Generator bus	–	–
5	720	0.9869	0.7103
6	700.1	0.9574	0.6699
7	567.3	0.6717	0.7908

(continued)

Table 2 (continued)

Bus number	Maximum load-ability limit	Voltage index	Voltage magnitude
8	610.9	0.9119	0.5928
9	491.4	0.9995	0.8065
10	450	0.9494	0.7593
11	717.5	0.9984	0.7716
12	582.6	0.9994	0.7756
13	795.6	0.6615	0.8325
14	505	0.5687	0.7867
15	3739	0.9999	0.8489
16	1047.9	0.9998	0.7652
17	998.6	0.9537	0.9066
18	691.5	0.9382	0.7524
19	844.5	0.8696	0.9477
20	699	0.9265	0.7133
21	1188	0.9986	0.7756
22	787.4	0.9995	0.7902
23	1009	0.9890	0.6913
24	2231.3	0.9999	0.7722

Fig. 5 Bus number versus maximum load-ability plot of 24-bus test system

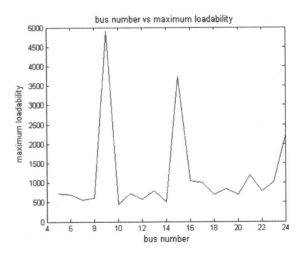

Table 3 and Fig. 6 shows the similar comparative results for a 30-bus Indian transmission system where bus no. 26 and 30 were found to be the weakest.

Table 3 Maximum load-ability limit calculation for IEEE-30 bus test system

Bus number	Maximum load-ability limit	Voltage index	Voltage magnitude
1	Generator bus	–	–
2	Generator bus	–	–
3	286.4	0.9997	0.7755
4	478.7	0.9997	0.7325
5	Generator bus	–	–
6	687.2	0.9998	0.7838
7	282	0.9999	0.7552
8	Generator bus	–	–
9	176.5	0.9922	0.8269
10	173.9	0.9999	0.7462
11	Generator bus	–	–
12	207.4	0.9993	0.8169
13	Generator bus	–	–
14	82.3	0.9998	0.7784
15	149.7	0.9998	0.6747
16	112.9	0.9999	0.7180
17	154.8	0.9987	0.5798
18	87.3	0.9975	0.6625
19	92.5	0.9995	0.6135
20	88.7	0.9986	0.6890
21	159.4	0.9985	0.6220
22	148.8	0.9995	0.6126
23	88.4	0.9996	0.6637
24	105.3	0.9990	0.6551
25	60.9	0.9976	0.6710
26	28.5	0.9936	0.7079
27	57.2	0.9997	0.7480
28	293.4	0.9995	0.7352
29	34.6	0.9990	0.6492
30	31	0.9894	0.6617

Fig. 6 Plot of bus number versus maximum load-ability limit of IEEE-30 bus test system

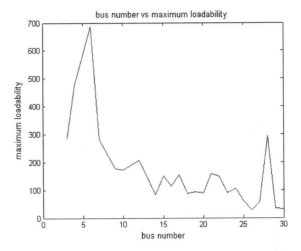

5 Conclusion

In the present work a thorough study on the voltage stability based congestion management has been presented. The voltage stability analysis has been carried out using Fast Voltage Stability Index and the weakest bus has been determined so that FACTS device may be placed there.

In 14 bus transmission systems, it has been observed that the 12th and 14th are the weakest bus and the maximum load-ability limit is 100.37 and 86.81 MVAR respectively, therefore these are the optimum locations for the placement of FACTS device.

In 24-bus test system, it has been observed that 10th and 9th buses are the weakest buses having maximum load-ability limit of 450 and 491.4 MVAR respectively. Therefore these are the optimum locations of the placement of FACTS device.

Similarly, in IEEE 30 bus test system, it has been observed that 26th and 30th are the weakest buses having maximum load-ability limit of 28.5 and 31 MVAR respectively. Therefore these are the optimum locations of the placement of FACTS device.

Appendix

The system data for transmission systems considered are given as follows. The bus code-1, 2 and 0 indicates slack bus, generator bus and load bus respectively. The tap setting ratios were considered in between the desired limits (Tables 4, 5).

Table 4 Bus data for 24-bus system

Bus no	Bus code	Voltage magnitude	Angle degree	Load		Generator		Injected		
				Mw	Mvar	Mw	Mvar	Q_{min}	Q_{max}	Mvar
1	1	1.06	0	0	0	1820	0	450	−150	100
2	2	1.04	0	0	0	160	0	320	−50	150
3	2	1.04	0	0	0	350	0	400	−100	50
4	2	1.05	0	0	0	520	0	400	−90	75
5	0	1	0	430	170	0	0	0	0	0
6	0	1	0	280	90	0	0	0	0	0
7	0	1	0	320	110	0	0	0	0	0
8	0	1	0	180	70	0	0	0	0	0
9	0	1	0	120	40	0	0	0	0	0
10	0	1	0	60	20	0	0	0	0	0
11	0	1	0	0	0	0	0	0	0	0
12	0	1	0	0	0	0	0	0	0	0
13	0	1	0	450	180	0	0	0	0	0
14	0	1	0	0	0	0	0	0	0	0
15	0	1	0	780	300	0	0	0	0	0
16	0	1	0	0	0	0	0	0	0	0
17	0	1	0	0	0	0	0	0	0	0
18	0	1	0	0	0	0	0	0	0	0
19	0	1	0	0	0	0	0	0	0	0
20	0	1	0	0	0	0	0	0	0	0
21	0	1	0	0	0	0	0	0	0	0
22	0	1	0	0	0	0	0	0	0	0
23	0	1	0	0	0	0	0	0	0	0
24	0	1	0	0	0	0	0	0	0	0

Table 5 Line data for 24 bus system

Bus nl	Bus nr	R pu	X pu	1/2 B pu	1 for Line code or tap setting value
1	15	0.00033	0.00670	0.0	0.97
15	16	0.00372	0.03931	0.53139	1
16	5	0.00099	0.01984	0.0	0.97
24	16	0.00245	0.02587	0.34966	1
15	24	0.00261	0.02780	1.48500	1
24	3	0.00099	0.01984	0.0	0.97
17	24	0.00477	0.05103	0.72673	1
24	18	0.00569	0.06008	0.79414	1
24	23	0.00272	0.02872	1.51830	1
21	4	0.00099	0.01984	0.0	0.97
23	9	0.00198	0.03968	0.0	0.97
22	23	0.00430	0.04770	0.63700	1
18	10	0.00198	0.03968	0.0	0.97
22	18	0.00589	0.05995	0.78410	1
11	17	0.00280	0.02998	0.42699	1
11	12	0.00198	0.02471	0.32304	1
12	14	0.00546	0.06794	0.88836	1
14	8	0.00125	0.02500	0.0	0.97
13	8	0.00315	0.01569	0.05274	1
22	13	0.00063	0.01250	0.0	0.97
23	20	0.00388	0.04834	0.65470	1
21	20	0.00297	0.03706	0.47543	1
20	7	0.00099	0.01984	0.0	0.97
19	6	0.00099	0.01984	0.0	0.97
21	19	0.00145	0.01802	0.93968	1
22	19	0.00289	0.03603	0.46222	1
17	2	0.00198	0.03960	0.0	0.97

References

1. Balamourougan, V., Sidhu, T.S., Sachdev, M.S.: Techniques for online prediction of voltage collapse. IEEE Proc. Gener. Transm. Distrib. **151**(4), 453–460 (2004)
2. Musirin, I., Rahman, T.K.A.: Novel fast voltage stability index (FVSI) for voltage stability analysis in power transmission system. In: 2002 Student Conference on Research and Development Proceedings, Shah Alam, Malasia, July (2002)
3. Kumar, V., Subramani, C.: Successive zooming optimization and its application to voltage stability problems. Int. Rev. Model. Simul. **6**(2), 426–430 (2013)
4. Mehdi, O.H.: Fast prediction of voltage stability index based on radial basis function neural network: Iraqi super grid network, 400-kV. Modern Appl. Sci. **5**(4), 190–199 (2011)
5. Niazi, K.R.: A comparison of voltage stability indices for placing shunt FACTS controllers. In: 2008 First International Conference on Emerging Trends in Engineering and Technology (2008)

6. Jyothsna, T.R.: Effects of strong resonance in multimachine power systems with SSSC supplementary modulationcontroller: strong resonance inpower systems with SSSC. Eur. Trans. Electr. Power (2011)
7. Abidin, I.Z.: Adaptive protection for voltage instability mitigation scenario. In: 2009 IEEE Student Conference on Research and Development (SCOReD) (2009)
8. Ibrahim, A.A., Mohamed, A., Shareef, H., Ghoshal, S.P.: Optimal power quality monitor placement in power systems based on particle swarm optimization and artificial immune system. 2011 3rd Conference on Data Mining and Optimization (DMO) (2011)
9. Singh, G.: Genetic algorithm-based artificial neural network for voltage stability assessment. Adv. Artif. Neural Syst. (532785), 9 (2011). doi:10.1155/2011/532785
10. Ghardash Khani, N., Abedi, M., Gharehpetian, G.B., Riahy, G.H.: Analyzing the effect of wind farm to improve transmission line stability in contingencies. Indian J. Sci. Technol. 8(11) (2015)
11. Wartana, I.M., Singh, J.G., Ongsakul, W., Sreedharan, S.: Optimal placement of FACTS controllers for maximising system loadability by PSO. Int. J. Power Energy Convers. 4(1), 9–33 (2013)
12. Babu, N., Venkata, A., Sivanagaraju, S.: Optimal power flow with FACTS device using two step initialization based algorithm for security enhancement considering credible contingencies. In: 2012 International Conference on Advances in Power Conversion and Energy Technologies (APCET) (2012)
13. Reis, C.: Line indices for voltage stability assessment. 2009 IEEE Bucharest PowerTech, (2009)
14. Shashank, T.R., Rajesh, N.B.: Analysis of fast voltage stability index on long transmission line using power world simulator. Int. Rev. Model. Simul. 6(3) (2013)
15. Sanz, F.A., Ramírez, J.M., Correa, R.E.: Experimental design for a large power system vulnerability estimation. Electr. Power Syst. Res. 20–27 (2015)
16. Swarup, K.S.: Artificial neural network using pattern recognition for security assessment and analysis. Neurocomputing 71(4–6), 983–998 (2008)
17. Musirin, I.: Estimation of maximum loadability in power systems by using fast voltage stability index. Int. J. Power Energy Syst. 118, 1127–1136 (2005)
18. Ampofo, D.O., Ai-Hinai, A., El Moursi, M.S.: Utilization of reactive power resources of distributed generation for voltage collapse prevention in optimal power flow. In: 2015 International Conference on Solar Energy and Building (ICSoEB), 1–5 (2015)
19. Sakthivel, S., Mary, D.: Voltage stability limit improvement by static VAR compensator (SVC) under line outage contingencies through particle swarm optimization algorithm. Int. Rev. Model. Simul. 20(1), 0975–8887 (2011)
20. Subramani, C., Dash, S.S.: Line stability index for steady state stability enhancement using FACTS device. Int. Rev. Model. Simul. 8(4), 585–590 (2012)

Fuzzy Logic-Based Control in Wireless Sensor Network for Cultivation

V. Sittakul, S. Chunwiphat and P. Tiawongsombat

Abstract This paper presents a cultivation control system on wireless sensor network using fuzzy logic approach. The system is controlled with the use of a sprinkler for moisture and lamp for sunlight and fan for temperature. The wireless sensor network consists of temperature, sunlight and moisture sensors directly connected to a microcontroller (Arduino). The signals obtained from these sensors are converted as digital data by the microcontroller before transmitted on air via both X-bee radio modules (Compatible with Zigbee and IEEE 802.15.4) to a tiny computer (Raspberry Pi2). Finally, the digital data is analyzed based on Fuzzy logic on the Linux Operation System (OS) of the Raspberry Pi2 and the generated output signals are used to control the fan, sprinkler and lamp to decrease or increase the levels of temperature, moisture and sunlight respectively.

Keywords Wireless sensor network · Fuzzy logic · IEEE 802.15.4 · Zigbee

1 Introduction

Wireless Sensor Network (WSN) plays an important role in recent years for monitoring and tracking data. Most of wireless sensors are small, inexpensive and low power consumption, trading off with limited processing and computer resource. Also, each sensor node has ability to detect, monitor and gather data from the

V. Sittakul (✉) · S. Chunwiphat · P. Tiawongsombat
Electronics Engineering Technology, College of Industrial Technology,
King Mongkut's University of Technology North Bangkok,
1518 Pracharat 1 Road, Wongsawang, Bangsue, Bangkok 10800, Thailand
e-mail: Vitawat.sittakul@gmail.com

S. Chunwiphat
e-mail: suphot.chunwiphat@gmail.com

P. Tiawongsombat
e-mail: tiawongsombat@gmail.com

© Springer Science+Business Media Singapore 2017
P. Deiva Sundari et al. (eds.), *Proceedings of 2nd International Conference on Intelligent Computing and Applications*, Advances in Intelligent Systems and Computing 467, DOI 10.1007/978-981-10-1645-5_23

environment and transmit them back to the user [1–3]. Most of mechanical, thermal, chemical, optical and magnetic sensors may be attached to a wireless sensor node to send the monitored data among the sensor node and finally to the base station such as a computer or server to store the data as a logfile due to the limited memory of the sensor node [4]. Further more, each sensor node consumes very low energy and this enables the portable battery to be used as the main power supply. This allows each sensor node can be independently placed close to the monitored point. As for agriculture countries, farmers are very concern about the cultivation and farming and consequently many wireless sensor network researches on agriculture technologies have been carried out. For example, in Egypt, the wireless sensor network has been used to save the resources for cultivation such as fertilizer and irrigation water [5]. In Ethiopia, it was used to monitor and control the farms via mobile phones [4]. In India, it was used to control the soil moisture via sprinkles [6]. However, all previous researches have to be manually accomplished and their systems cannot be self-controlled. Therefore, in this paper, the concept of Artificial Intelligence (AI) based on fuzzy logic is applied to the wireless sensor network. The system is controlled with the use of a sprinkler for moisture and lamp for sunlight and fan for temperature. The wireless sensor network consists of temperature, sunlight and moisture sensors directly connected to a microcontroller (Arduino). The signals obtained from these sensors are converted as digital data by the microcontroller before transmitted on air via both X-bee radio modules (Compatible with Zigbee and IEEE 802.15.4) to a tiny computer (Raspberry Pi2). Finally, the digital data is analyzed based on Fuzzy logic on the Linux Operation System (OS) of the Raspberry Pi2 and the generated output signals are used to control the fan, sprinkler and lamp to decrease or increase the levels of temperature, moisture and sunlight respectively. Here the fuzzy logic-based algorithm can be varied by changing the fuzzy rules regarding the types of crops. The fuzzy logic parameters are collected from the experience in farming and cultivation. The paper can be organized as follows. Section 2 shows the device characterization and Sect. 3 shows the fuzzy logic theory. Section 4 demonstrates the experimental setup and Sect. 5 shows the experimental results. Finally, Sect. 6 summarizes all results.

2 Device Characterization

2.1 Microcontroller

The microcontroller (Arduino Mega 2560) as shown in Fig. 1. functions as a receiver and converter to receive the data from the sensors and transmit them to the computer or server.

Fig. 1 Microcontroller

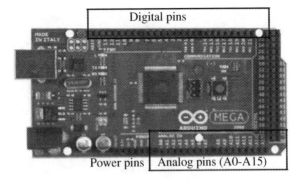

Digital pins

Power pins | Analog pins (A0-A15)

2.2 Moisture Sensor

The moisture sensor manufactured from Seeedstudio Company with model of sku: SEN92355P is used to measure the soil moisture through its conductance. The moisture sensor transmits the measured data as 10-bit output to the pin#10 of microcontroller as shown in Fig. 2.

2.3 Temperature Sensor

The temperature sensor manufactured from D-Robotics UK Company with model of DHT11 is used to measure the temperature by means of thermister principle as shown in Fig. 3. Its accuracy is ±2 °C with measured data of 8-bit output.

Fig. 2 Moisture sensor module

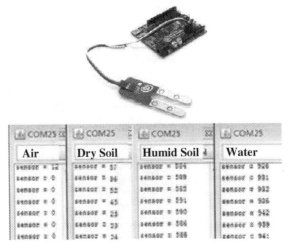

Air	Dry Soil	Humid Soil	Water

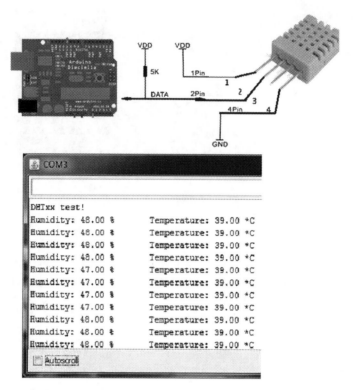

Fig. 3 Temperature sensor module

2.4 Light Sensor

The light sensor manufactured from szhaiwang company is used to measure the sunlight by means of light independent resistor as shown in Fig. 4. The light independent resistor is made by CdS placed on the ceramic base.

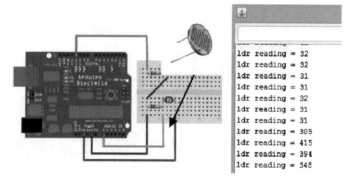

Fig. 4 Light sensor module

3 Fuzzy Logic Theory

Fuzzy logic theory was proposed by Zadeh [7] and Bellman and Zadeh [8] based on the fuzzy probability. The membership of an element is not only strictly false or true {1,0} as same as Boolean logic, but rather gradual as shown in Fig. 5. The degree of its membership in fuzzy logic can be any real number in the interval [0,1] whereas those in Boolean logic can be only 0 or 1. This is to deal with the vagueness and imprecision of many reality and it can be used to simulate the human ability of making decision based on not so precise information.

A formal definition of fuzzy sets (A) in universe (U) can be written as:

$$A \equiv \{ <x, \mu A(x) > |x \in U\} \tag{1}$$

where $\mu_A\ (x)$ is called the membership function for the set of all objects x in U.

There are four common member functions as shown in Fig. 6.

The fuzzy set operations of the union, intersection and complement can be expressed in terms of logical operations; disjunction, conjunction and complement as follows [9] (Figs. 7, 8, and 9):

$$Disjunction\ (OR) : \mu_A(x) V \mu_B(x) = \max\{\mu_A(x), \mu_B(x)\}$$

$$Conjunction\ (AND) : \mu_A(x) \wedge \mu_B(x) = \min\{\mu_A(x), \mu_B(x)\}$$

$$Complement\ (NOT) : \mu_{\neg A}(x) = 1 - \mu_A(x)$$

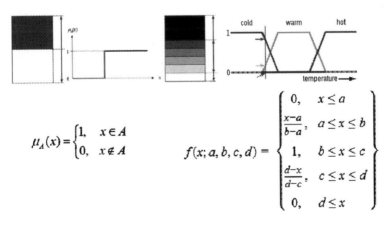

$$\mu_A(x) = \begin{cases} 1, & x \in A \\ 0, & x \notin A \end{cases} \qquad f(x; a, b, c, d) = \begin{cases} 0, & x \leq a \\ \dfrac{x-a}{b-a}, & a \leq x \leq b \\ 1, & b \leq x \leq c \\ \dfrac{d-x}{d-c}, & c \leq x \leq d \\ 0, & d \leq x \end{cases}$$

Fig. 5 Boolean logic (*left*) and Fuzzy logic (*Right*)

Fig. 6 Four common
membership functions

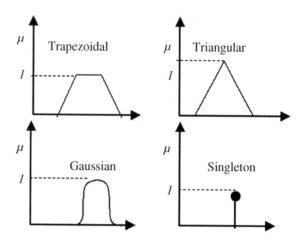

Fig. 7 *Disjunction (OR)*
property of $\mu_A(x)$ and $\mu_B(x)$

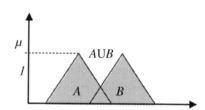

Fig. 8 *Conjunction (AND)*
property of $\mu_A(x)$ and $\mu_B(x)$

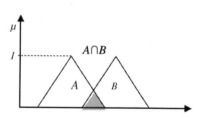

Fig. 9 *Complement (NOT)*
property of $\mu_A(x)$

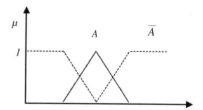

4 Experimental Setup

In this section, all components as shown in Sect. 2 are now combined as a system as can be seen in Fig. 10. All sensors (Temperature, Light and Moisture) are connected to the input ports of the ardunio. The ardunio now converts a parallel-data inputs into a serial-data output and it is then transmitted to a tiny computer (Rasberrry Pi2) via X-bee radio modules on air. It can be noted that the rasberry Pi2 has been used to reduce the size and the cost of the remote unit. In reality, this unit has to be place close to the controlled devices (Lamp, Sprinkler and Fan) where the size and cost are very critical. The received data from the X-bee (Rx) is now computed on the Java software (based on fuzzy logic) to generate control output signals. Finally, the control signals are used to control the lamp, sprinkler and fan to adjust the light intensity, moisture and temperature for cultivation.

4.1 Design of Fuzzy control

In order to achieve the fuzzy control, it is important that the inputs and outputs of the system have to be assigned. The inputs of the system are temperature, moisture and light intensity obtained from the sensors. The level of temperature, moisture and light intensity in this work are divided into three triangular memberships as shown in Fig. 11 which are Negative (N) from −50 to 50 %, No Change (NC) from

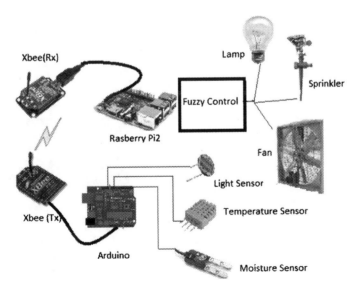

Fig. 10 Experimental setup

Fig. 11 Membership functions of inputs

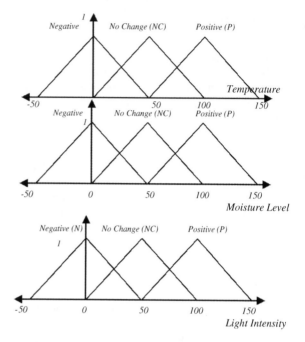

0 to 100 % and Positive (*P*) from 50 to 150 % respectively. The triangular memberships are chosen here to reduce the complexity of the calculation. Since the level of the parameters is defined within the range of 0–100 %, the maximum and minimum values of the measured data have to be normalized [9].

Fig. 12 Membership functions of outputs

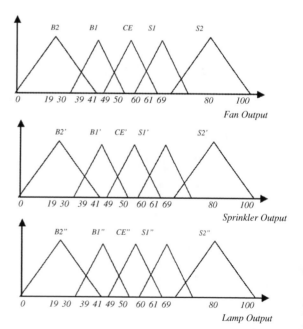

The outputs of the system are used to control the fan, the pump of sprinkler and lamp respectively. The output levels of fan, pump of sprinkler and lamp are divided into five triangular memberships as shown in Fig. 12 which are B2 (0–39 %), B1 (30–49 %), CE (41–60 %), S1 (50–69 %), S2 (61–100 %) respectively as shown in Fig. 8.

To allow the system to make a decision correctly, the fuzzy rules have to be created as shown on Tables 1, 2 and 3, based on computer programming like if-then statements [9, 10]. The rules as in this work have been generated from the survey of

Table 1 Rules of fuzzy logic for Fan control

Rule #	IF	Temp	Operator	Moisture	Operator	Light	Operator	Output1
1	IF	P	AND	P	AND	P	THEN	S2
2	IF	P	AND	P	AND	NC	THEN	S2
3	IF	P	AND	P	AND	N	THEN	S1
4	IF	P	AND	NC	AND	P	THEN	S2
5	IF	P	AND	NC	AND	NC	THEN	S1
6	IF	P	AND	NC	AND	N	THEN	S1
7	IF	P	AND	N	AND	P	THEN	S1
8	IF	P	AND	N	AND	NC	THEN	S1
9	IF	P	AND	N	AND	N	THEN	CE
10	IF	NC	AND	P	AND	P	THEN	S1
11	IF	NC	AND	P	AND	NC	THEN	CE
12	IF	NC	AND	P	AND	N	THEN	CE
13	IF	NC	AND	NC	AND	P	THEN	CE
14	IF	NC	AND	NC	AND	NC	THEN	CE
15	IF	NC	AND	NC	AND	N	THEN	B1
16	IF	NC	AND	N	AND	P	THEN	CE
17	IF	NC	AND	N	AND	NC	THEN	B1
18	IF	NC	AND	N	AND	N	THEN	B1
19	IF	N	AND	P	AND	P	THEN	B1
20	IF	N	AND	P	AND	NC	THEN	B1
21	IF	N	AND	P	AND	N	THEN	B2
22	IF	N	AND	NC	AND	P	THEN	B1
23	IF	N	AND	NC	AND	NC	THEN	B2
24	IF	N	AND	NC	AND	N	THEN	B2
25	IF	N	AND	N	AND	P	THEN	B1
26	IF	N	AND	N	AND	NC	THEN	B2
27	IF	N	AND	N	AND	N	THEN	B2

Fan scale	Fan control signal
S2 = 61–100 %	7.32–12.00 V
S1 = 50–69 %	6.00–8.28 V
NC = 40–60 %	4.80–7.20 V
B1 = 30–49 %	3.60–5.88 V
B2 = 0–39 %	0–4.68 V

Table 2 Rules of fuzzy logic for pump control (Sprinkler)

Rule #	IF	Temp	Operator	Moisture	Operator	Light	Operator	Output2
1	IF	P	AND	P	AND	P	THEN	CE'
2	IF	P	AND	P	AND	NC	THEN	CE'
3	IF	P	AND	P	AND	N	THEN	B2'
4	IF	P	AND	NC	AND	P	THEN	B1'
5	IF	P	AND	NC	AND	NC	THEN	B2'
6	IF	P	AND	NC	AND	N	THEN	CE'
7	IF	P	AND	N	AND	P	THEN	S2'
8	IF	P	AND	N	AND	NC	THEN	S2'
9	IF	P	AND	N	AND	N	THEN	S1'
10	IF	NC	AND	P	AND	P	THEN	CE'
11	IF	NC	AND	P	AND	NC	THEN	CE'
12	IF	NC	AND	P	AND	N	THEN	CE'
13	IF	NC	AND	NC	AND	P	THEN	B2'
14	IF	NC	AND	NC	AND	NC	THEN	CE'
15	IF	NC	AND	NC	AND	N	THEN	CE'
16	IF	NC	AND	N	AND	P	THEN	S1'
17	IF	NC	AND	N	AND	NC	THEN	S1'
18	IF	NC	AND	N	AND	N	THEN	S1'
19	IF	N	AND	P	AND	P	THEN	CE'
20	IF	N	AND	P	AND	NC	THEN	CE'
21	IF	N	AND	P	AND	N	THEN	CE'
22	IF	N	AND	NC	AND	P	THEN	CE'
23	IF	N	AND	NC	AND	NC	THEN	CE'
24	IF	N	AND	NC	AND	N	THEN	CE'
25	IF	N	AND	N	AND	P	THEN	CE'
26	IF	N	AND	N	AND	NC	THEN	CE'
27	IF	N	AND	N	AND	N	THEN	S1'

Sprinklerscale %	Sprinkler control signal
S2' = 61–100 %	0 V
S1' = 50–69 %	0 V
NC' = 40–60 %	0 V
B2' = 30–49 %	1–9.5 V
B1' = 0–39 %	7.3–12 V

farming of lettuces in the central region of Thailand and from the experts and they may be varied depending on the types of crops, regions and countries. However, these rules can be easily changed inside the rule tables on the software and this allows the possibility to change the types of crops regarding the seasons.

Table 1 displays the rules of fuzzy logic for fan control. It can be seen that there are 27 rules being used here to create the fan control output (Output1). For instance,

Table 3 Rules of fuzzy logic for Lamp control

Rule #	IF	Temp	Operator	Moisture	Operator	Light	Operator	Output3
1	IF	P	AND	P	AND	P	THEN	$B2''$
2	IF	P	AND	P	AND	NC	THEN	$B1''$
3	IF	P	AND	P	AND	N	THEN	N/C''
4	IF	P	AND	NC	AND	P	THEN	$B1''$
5	IF	P	AND	NC	AND	NC	THEN	CE''
6	IF	P	AND	NC	AND	N	THEN	CE''
7	IF	P	AND	N	AND	P	THEN	$B2''$
8	IF	P	AND	N	AND	NC	THEN	$B1''$
9	IF	P	AND	N	AND	N	THEN	$B1''$
10	IF	NC	AND	P	AND	P	THEN	$B1''$
11	IF	NC	AND	P	AND	NC	THEN	CE''
12	IF	NC	AND	P	AND	N	THEN	CE''
13	IF	NC	AND	NC	AND	P	THEN	$B1''$
14	IF	NC	AND	NC	AND	NC	THEN	CE''
15	IF	NC	AND	NC	AND	N	THEN	CE''
16	IF	NC	AND	N	AND	P	THEN	CE''
17	IF	NC	AND	N	AND	NC	THEN	CE''
18	IF	NC	AND	N	AND	N	THEN	CE''
19	IF	N	AND	P	AND	P	THEN	$B1''$
20	IF	N	AND	P	AND	NC	THEN	N/C''
21	IF	N	AND	P	AND	N	THEN	$S1''$
22	IF	N	AND	NC	AND	P	THEN	$B1''$
23	IF	N	AND	NC	AND	NC	THEN	N/C''
24	IF	N	AND	NC	AND	N	THEN	$B1''$
25	IF	N	AND	N	AND	P	THEN	$B1''$
26	IF	N	AND	N	AND	NC	THEN	$S1''$
27	IF	N	AND	N	AND	N	THEN	$S2''$

Lamp scale	Lamp control signal
$S2'' = 61$–100 %	0–4.68 V
$S1'' = 50$–69 %	3.6–5.88 V
$NC'' = 40$–60 %	4.8–7.2 V
$B1'' = 30$–49 %	6–8.28 V
$B2'' = 0$–39 %	7.32–12 V

if the inputs from temperature sensor (Temp), moisture sensor (Moisture) and light sensor (Light) are positive (50–150 %), the output1 is set to $S2$. The $S2$ now is referred to the output power of fan 61–100 %.

Table 2 displays the rules of fuzzy logic for pump control (sprinkler). Similarly, it can be seen that there are 27 rules being used to create the pump control output (Output2). For instance, if the inputs from temperature sensor (Temp), moisture sensor (Moisture) and light sensor (Light) are positive (50–150 %), the output2 is set to *N/C'*. The *N/C'* is referred to the output power of the pump to sprinkler 40–60 %.

Table 3 displays the rules of fuzzy logic for lamp control. It can be seen that there are 27 rules being used here to create the lamp control output (Output3). For instance, if the inputs from temperature sensor (Temp), moisture sensor (Moisture) and light sensor (Light) are positive (50–150 %), the output3 is set to *B2″*. The *B2″* is referred to the output power of lamp 0–39 %.

5 Experimental Results

In order to configure the parameters of the fuzzy logic on the raspberry pi2, a java software is written and run on the Linux OS as shown in Fig. 13. It is necessary here to validate if the software can function correctly. Therefore, the simple test inputs can be used here to check the results.

As shown in Fig. 13, the received data of temperature, moisture and light are set to 25 °C (25 %), 0 RH (0 %) and 0 Lux (0 %) respectively. This can be graphically explained by Fig. 14. At the temperature of 25 %, the levels of membership function can be divided into 0.5 of triangular memberships *N* and *NC* respectively. Since the moisture and light has been set to 0 %, their memberships are 1.0 of triangular membership *NC*. To calculate the fan output (output1), the rules #14 and 23 in Table 1 are applied. This results in the memships output *B1* and *B2*. To calculate the fan output level, the areas of *B1* and *B2* are now averaged and the average point of fan output of 25.5 % has successfully been found.

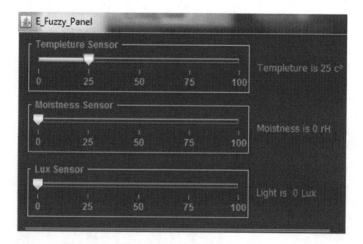

Fig. 13 User Interface of Java software on Rasberry Pi2

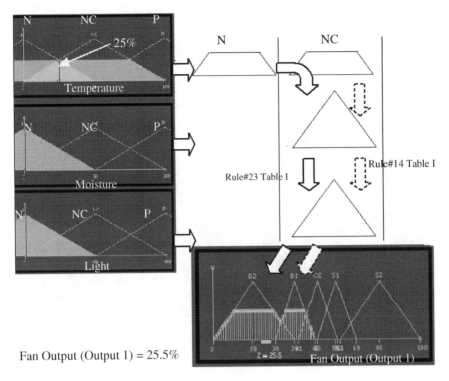

Fan Output (Output 1) = 25.5%

Fig. 14 Concept of fuzzy logic control

Finally, the test inputs are now replaced by the real sensors inputs from the system in Fig. 10. Here, the temperature of 24 °C (24 %), moisture of 46RH (46 %) and light of 600 Lux (13.3 %) are measured as shown in Fig. 15. By using the fuzzy logic control theory, the outputs of fan, pump of sprinkler and lamp of

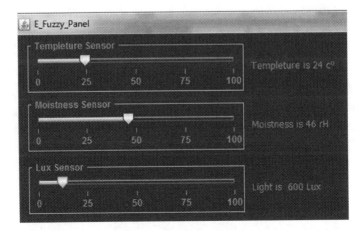

Fig. 15 Measured sensor data of the Java software on Rasberry Pi2

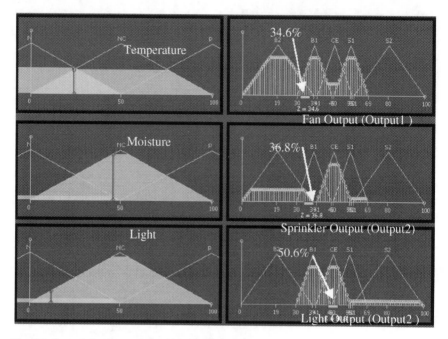

Fig. 16 Control signal outputs of the system

34.6, 36.8 and 50.6 % can be calculated (Fig. 16). The output signal in Fig. 16 can now be converted to the controlled signals in voltage. Compared with the data in Tables 1, 2 and 3, the output signals in voltage are in the range of (3.6–5.88 V) for fan output, (1–9.5 V) for pump output and (4.8–7.2 V) for lamp output respectively.

6 Conclusion

The fuzzy logic-based control in wireless sensor network for cultivation has been successfully demonstrated. Temperature, moisture and light sensors can be used to monitor and control the growth of the crops such as lettuce. The concept of fuzzy logic can be suitably applied to the wireless sensor network to enhance the ability to control the crop environment. This system is also achieved a low cost design since all inexpensive devices can be used. Moreover, it is very flexible for changing fuzzy logic rules for different types of crops in future.

References

1. Akyldiz, F., Su, W., Sankarasubramanian, Y., Cayirci, E.: A survey of sensor networks. In: IEEE Communications Magazine, pp. 102–114, August 2002
2. Culler, D., Srivastava, D.E.M.: Overview of sensor network. IEEE Comput Mag **37**(8), 41–49 (2004)
3. Siva Ram Murthy, C., Manoj, B.: Ad Hoc Wireless Networks: Architectures and Protocols. Prentice Hall, Englewood Cliffs (2004)
4. Satyanarayana, G.V.: Wireless sensor based remote monitoring system for agriculture using ZigBee and GPS. In: Conference on Advances in Communication and Control Systems 2013 (CAC2S 2013), pp. 110–114
5. Sherine, M., Abd El-kader, Basma, M., El-Basioni, Mohammad: Precision farming solution in Egypt using the wireless sensor network technology. Egypt. Inform. J. **14**(3), 221–233 (2013)
6. Hussain, R., Sahgal, J.L., Gangwar, A., Riyaj, Md.: Control of irrigation automatically by using wireless sensor network. Int. J. Soft Comput. Eng. (IJSCE) **3**(1) (2013)
7. Zadeh, L.A.: Fuzzy sets. Inf. Control **8**(3), 338–353 (1965)
8. Bellman, R.E., Zadeh, L.A.: Decision making in a fuzzy environment. Manag. Sci. **17**(4), B141–B164 (1970)
9. Iancu, I.: A Mamdani Type Fuzzy Logic Controller, Fuzzy Logic—Controls, Concepts, Theories and Applications. In: Prof. Dadios, E. (Ed.), ISBN: 978-953-51-0396-7, Chapter 16 (2012)
10. Bai, Y., Wang, D.: Fundamentals of Fuzzy Logic Control–Fuzzy Sets, Fuzzy Rules and Defuzzifications, a text book on fuzzy logic applied to engineering Roger C. Dugan, Mark F. McGranagham, Surya Santoso, H.Wayne Beaty, " Electrical Power Systems Quality", a textbook copyrighted from Mc Graw–Hill

Simulation of a Seven Level Inverter and Its Comparison with a Conventional Inverter

Revanth Mallavarapu, Meenakshi Jayaraman and V.T. Sreedevi

Abstract Reducing the number of switches in multilevel inverter topologies is drawing incredible interest in renewable energy, motor drive and reactive power compensation applications. This paper focuses on the simulation of a seven level inverter with reduced number of switches. The seven level inverter is based on transistor clamped topology consisting of a single H-bridge inverter with two IGBT switches between two diode bridges. Triple reference single carrier modulation technique is adopted to generate gating pulses for the seven level multilevel inverter with reduced switches. Using this modulation technique, output voltage, output current and voltage stress across the switches are obtained for a modulation index of 0.9 and 1.25. The total harmonic distortion obtained for the various values of modulation index is presented. In addition, a comparison is established with a seven level cascaded H-Bridge inverter with respect to complexity of the circuit topology and total harmonic distortion. Results are obtained and observed using simulations done through MATLAB Simulink simulation tool.

Keywords Total harmonic distortion (THD) · Cascaded H-bridge (CHB) · Triple reference single carrier modulation · Transistor clamped H-bridge (TCHB)

1 Introduction

Multilevel inverters are preferred in medium and high voltage applications since they offer advantages like improved output waveforms, lower switching loss, low voltage stress, smaller filter size and less Total Harmonic Distortion (THD) [1–3]. They can realize high power output through low voltage switches without using

R. Mallavarapu (✉) · M. Jayaraman · V.T. Sreedevi
School of Electrical Engineering, VIT University, Chennai, India
e-mail: revanth518@gmail.com

© Springer Science+Business Media Singapore 2017
P. Deiva Sundari et al. (eds.), *Proceedings of 2nd International Conference on Intelligent Computing and Applications*, Advances in Intelligent Systems and Computing 467, DOI 10.1007/978-981-10-1645-5_24

281

transformers and with increase in the number of output levels. The basic multilevel inverter topologies include the diode clamped multilevel inverter [3–5], capacitor clamped multilevel inverter [6, 7] and cascaded multilevel inverter [8–10]. To increase the number of output levels, the diode clamped inverter topology requires more number of clamping diodes. This makes the circuit complex, costly and increases the size of the inverter. Further, voltage unbalance of DC link capacitors occurs inherently in a diode clamped topology. Capacitors are involved in a capacitor clamped multilevel inverter topology to increase the number of voltage levels. However, these inverters require excessive number of storage capacitors for high voltage steps. For an m-level capacitor clamped multilevel inverter, (m − 1) DC-link capacitor are required. This topology also suffers from voltage unbalance issues. A Cascaded H-Bridge (CHB) multilevel inverter topology consists of many single phase inverters connected in series [11–13]. The number of inverters to be connected in series depends on the output voltage level. An 'n' level CHB multilevel inverter requires (n − 1) single phase inverters in series. It has a modular structure and it is more reliable than other topologies. However each H-Bridge is to be supplied with an isolated DC power supply in a cascaded inverter.

The firing instants for these multilevel inverters are generated by using a suitable modulation technique. Different modulation techniques have been proposed to get a better output voltage waveform with lesser harmonic content. Any multilevel inverter's efficiency, switching losses and harmonic reduction depends on the modulation technique applied. Modulation strategies applied to multilevel inverters include Sinusoidal Pulse Width Modulation (SPWM) [14, 15], space vector modulation [16], staircase or fundamental frequency modulation and selective harmonic elimination technique [17–19]. Currently, more focus is towards development of multilevel inverter topologies with reduced number of switches without compromising on the levels of output. On the other front, modulation strategies are being explored to reduce the harmonic content and to obtain lower switching losses. As the number of levels increase in a multilevel inverter, the harmonic content on the output waveforms decreases. A seven level inverter produces lower harmonic distortion than three level and five level inverters.

This paper presents the simulation and analysis of a seven level inverter with reduced number of switches. The seven level inverter consists of a single H-bridge inverter with two diode embedded IGBT switches [20–22]. The conventional seven level H-cascaded inverter requires an individual DC voltage for each H-bridge and twelve switching devices. The seven level transistor clamped structure requires only one DC source. The switching pulses for the inverter is generated using triple reference single carrier modulation technique. The output voltage, load current and voltage stress across the switches of the multilevel inverter topology with reduced switches are obtained for a modulation index of 0.9 and 1.25. Further, the transistor clamped inverter is compared with a conventional CHB inverter with regard to the number of switches, voltage stress across the switches and THD.

The organization of the paper is as follows: Sect. 2 provides the topology of a seven level cascaded H-Bridge multilevel inverter. Section 3 discuses on the transistor clamped topology with triple reference single carrier modulation technique. Section 4 shows the simulation results obtained with both the topologies followed by conclusion.

2 Seven Level CHB Inverter

Figure 1 shows a conventional seven level CHB inverter. It consists of three series H-Bridge inverters with three DC sources. The switching pulses are generated by using a single reference signal and three carrier signals. The six carrier signals are

Fig. 1 Seven level CHB multilevel inverter

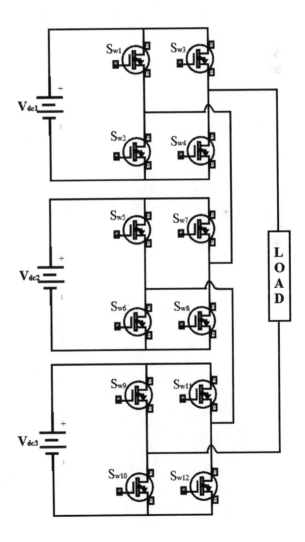

Table 1 Modes of operation of seven level CHB inverter

V_0	S_{w1}	S_{w2}	S_{w3}	S_{w4}	S_{w5}	S_{w6}	S_{w7}	S_{w8}	S_{w9}	S_{w10}	S_{w11}	S_{w12}
V_{dc}	1	0	0	1	1	0	0	1	1	0	0	1
$\frac{2V_{dc}}{3}$	1	0	0	1	1	0	0	1	0	1	0	1
$\frac{V_{dc}}{3}$	1	0	0	1	0	1	0	1	0	1	0	1
0	1	0	1	0	1	0	1	0	1	0	1	0
$\frac{-V_{dc}}{3}$	1	0	1	0	1	0	1	0	0	1	0	1
$\frac{-2V_{dc}}{3}$	1	0	1	0	0	1	1	0	0	1	0	1
$-V_{dc}$	0	1	1	0	0	1	1	0	0	1	0	1

level shifted to obtain output voltages of V_{dc}, $2V_{dc}/3$, $V_{dc}/3$, 0, $-V_{dc}/3$, $-2V_{dc}/3$, $-V_{dc}$. Table 1 shows the modes of operation of a seven level CHB inverter.

'1' indicates the ON condition of the switch, '0' indicates the OFF condition of the switch in Table 1.

The CHB multilevel inverter is operated in one of the seven modes according to their output voltage level. In mode I the switches S_{w1}, S_{w4}, S_{w5}, S_{w8}, S_{w9} and S_{w12} are ON and the remaining switches are OFF. In mode II the switches S_{w1}, S_{w4}, S_{w5}, S_{w8}, S_{w10} and S_{w12} are ON and remaining switches are OFF. In mode III the switches S_{w1}, S_{w4}, S_{w6}, S_{w8}, S_{w10} and S_{w12} are ON and remaining switches are OFF. In mode IV the switches S_{w1}, S_{w3}, S_{w5}, S_{w7}, S_{w9} and S_{w11} are ON and remaining switches are OFF. In mode V the switches S_{w1}, S_{w3}, S_{w5}, S_{w7}, S_{w10} and S_{w12} are ON and remaining switches are OFF. In mode VI the switches S_{w1}, S_{w3}, S_{w6}, S_{w7}, S_{w10} and S_{w12} are ON and remaining switches are OFF. In mode VII the switches S_{w2}, S_{w3}, S_{w6}, S_{w7}, S_{w10} and S_{w12} are ON and remaining switches are OFF.

Figure 2 shows the comparison of a single reference signal with six level shifted carrier signals [23]. The six triangular carrier signals are of equal amplitude and are

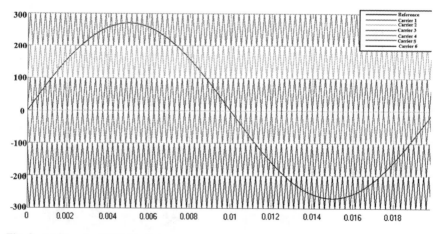

Fig. 2 Multicarrier SPWM technique

arranged one over another and compared with the reference sinusoidal signal. The intersection points are used to generate the switching instants for the cascaded inverter based on Table 1.

The voltage output of a seven level cascaded H-Bridge inverter is the sum of three voltages given to each of the H-bridge inverters. The output equation of the seven level cascaded H-bridge inverter is given in Eq. (1).

$$V_o = V_{dc1} + V_{dc2} + V_{dc3} \tag{1}$$

The actual number of levels on the output voltage is given using Eq. (2).

$$M_{steps} = 2m + 1 \tag{2}$$

where 'm' is the number of H-Bridge inverters.

3 Seven Level Inverter with Reduced Number of Switches: Transistor Clamped Topology

Figure 3 shows a seven level TCHB inverter consisting of a normal H-Bridge inverter and an auxiliary circuit [20–22]. The auxiliary circuit contains two diode embedded IGBT switches. The seven level TCHB inverter produces seven output levels with reduced number of switches.

The transistor clamped topology is operated in as per the modes presented in Table 2 to obtain seven levels on the output waveforms. In mode I $(0 < \omega t < \phi_1$ and $\phi_4 < \omega t < \pi)$, the switches S_{w1} and S_{w4} are ON and remaining switches are OFF. In mode II $(\phi_1 < \omega t < \phi_2$ and $\phi_3 < \omega t < \phi_4)$, the switches S_{w4} and S_{w5} are ON and remaining switches are OFF. In mode III $(\phi_2 < \omega t < \phi_3)$, the switches S_{w4} and S_{w6} are ON and remaining switches are OFF. In mode IV, the

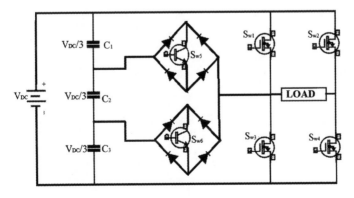

Fig. 3 Seven level inverter with transistor clamping

Table 2 Modes of operation of transistor clamped inverter

V_0	S_{w1}	S_{w2}	S_{w3}	S_{w4}	S_{w5}	S_{w6}
V_{DC}	On	Off	Off	On	Off	Off
$2V_{DC}/3$	Off	Off	Off	On	On	Off
$V_{DC}/3$	Off	Off	Off	On	Off	On
0	Off	Off	On	On	Off	Off
0^*	On	On	Off	Off	Off	Off
$-V_{DC}/3$	Off	On	Off	Off	On	Off
$2V_{DC}/3$	Off	On	Off	Off	Off	On
V_{DC}	Off	On	On	Off	Off	Off

* represents another possible mode to obtain zero state in the inverter

switches (S_{w3} and S_{w4}) or (S_{w1} and S_{w2}) are ON and remaining switches are OFF. In mode V ($\pi < \omega t < \phi_5$ and $\phi_8 < \omega t < 2\pi$), the switches S_{w2} and S_{w5} are ON and remaining switches are OFF. In mode VI ($\phi_5 < \omega t < \phi_6$ and $\phi_7 < \omega t < \phi_8$), the switches S_{w2} and S_{w6} are ON and remaining switches are OFF. In mode VII ($\phi_6 < \omega t < \phi_7$), the switches S_{w2} and S_{w3} are ON and remaining switches are OFF. $\phi_1, \phi_2, \phi_3, \phi_4, \phi_5, \phi_6, \phi_7$ and ϕ_8 are the switch conduction angles.

The gating pulses for the transistor clamped topology are generated using triple reference single carrier modulation technique [20]. The technique involves comparison of a single triangular carrier signal with three sinusoidal reference signals. The modulation index is given in Eq. (3).

$$modulation\ index = \frac{D_m}{3D_c} \tag{3}$$

where D_m is the magnitude of the reference sinusoidal signal wave, D_c is the amplitude of the triangular carrier signal. Figures 4 and 6 shows the triple reference

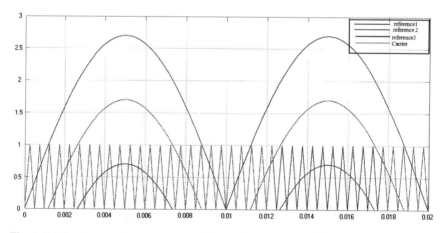

Fig. 4 Triple reference single carrier modulation technique for m = 0.9

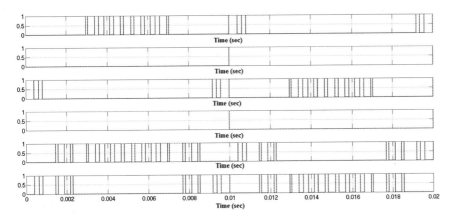

Fig. 5 Gate signals for switches S_{w1}–S_{w6}

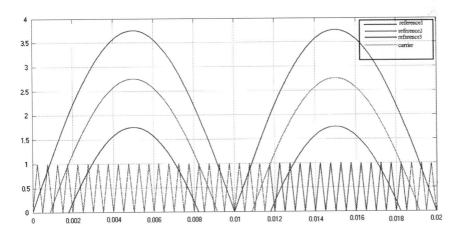

Fig. 6 Triple reference single carrier modulation for m = 1.25

single carrier modulation technique for modulation index < 1 and modulation index > 1 respectively. In this modulation scheme, V_{ref1}, V_{ref2} and V_{ref3} are assumed to have the same magnitude. They are shifted by an offset value of (0, −1, −2) which is equal to the amplitude of the triangular carrier signal. With this modulation technique, the inverter is capable of generating seven levels of output levels: $\left(V_{DC}, \frac{2V_{DC}}{3}, \frac{V_{DC}}{3}, 0, -V_{DC}, \frac{2V_{DC}}{3}, \frac{V_{DC}}{3}\right)$. The gating pulses generated for the seven level transistor clamped inverter for modulation index of 0.9 and 1.25 is shown in Figs. 5 and 7 respectively.

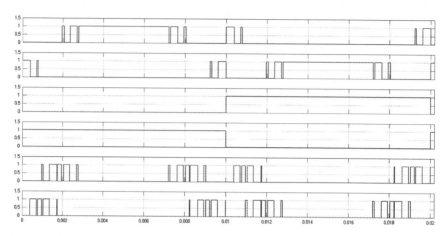

Fig. 7 Switching pulses from S_{w1}–S_{w6}

The output voltage of the seven level inverter is expressed using Fourier series as given in Eq. (4).

$$V_0 = A_m\cos{(m\omega t)} + B_m\sin{(m\omega t)} \tag{4}$$

where A_m, B_m and ωt are,

$$A_m = \frac{4}{\pi}\int_0^{\frac{\pi}{2}} \cos(m\theta)d\theta \text{ for even values of `m'} \tag{5}$$

$$B_m = \frac{4}{\pi}\int_0^{\frac{\pi}{2}} \sin(m\theta)d\theta \text{ for odd values of `m'} \tag{6}$$

and $\omega t = \theta$.

4 Simulation Results

The simulation of the multilevel inverter topologies is done using MATLAB-Simulink. The simulations are carried out for modulation index of 0.9 and 1.25 for a switching frequency of 2 kHz. The obtained results with the transistor clamped topology is compared with a conventional seven level CHB inverter for the same values of modulation index and load specifications. The simulations parameters are shown in Table 3.

Table 3 Simulation parameters

DC voltage $(V_{DC1} + V_{DC2} + V_{DC3})$	300 V
Modulation index (m)	0.9 < m < 1.25
Switching frequency	2 kHz
RL load	50 Ω, 10 mH

4.1 Results Obtained with Transistor Clamped Topology for Modulation Index of 1.25

Figure 9 shows the output voltage and current waveforms for a modulation index of 1.25. It is evident that seven levels of voltage output is synthesized. The current waveforms are more towards sinusoidal due to inductive load. Figure 10 displays the harmonic spectrum of the output voltage and current waveforms. The voltage THD is 16.40 % and current THD is 9.60 % with a transistor clamped topology. Figure 8 shows the voltage stress across the switches of the multilevel inverter for a modulation index of 1.25. It is observed that the applied total DC voltage of 300 V appears as stress across the switches S_{w1}, S_{w2}, S_{w3} and S_{w4} and for the remaining switches S_{w5} and S_{w6}, the voltage stress is 200 V.

4.2 Results Obtained with Transistor Clamped Topology for Modulation Index of 0.9

Figure 12 shows the output voltage and current waveforms for a modulation index of 0.9. It is evident that seven levels of voltage output are synthesized. The current waveforms are more towards sinusoidal due to inductive load. Figure 13 displays the harmonic spectrum of the output voltage and current waveforms. The voltage THD is 22.76 % and current THD is 7.27 % with a transistor clamped topology. Figure 11 shows the voltage stress across the switches of the multilevel inverter for a modulation index of 0.9. It is observed that the applied total DC voltage of 300 V appears as stress across the switches S_{w1}, S_{w2}, S_{w3} and S_{w4} and for the remaining switches S_{w5} and S_{w6} voltage stress is 200 V.

4.3 Results Obtained with CHB Multilevel Inverter Topology for m = 0.9

Figure 15 shows the output voltage and current waveforms for a modulation index of 0.9. It is evident that seven levels of voltage output are synthesized. The current waveforms are more towards sinusoidal due to inductive load. Figure 16 displays the harmonic spectrum of the output voltage and current waveforms. The voltage

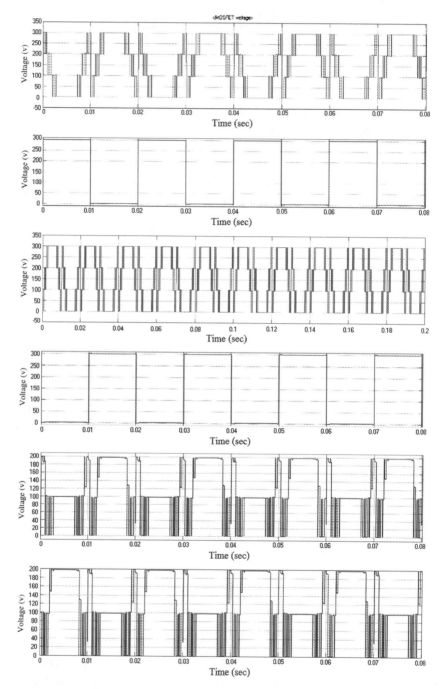

Fig. 8 Voltage stress across switches S_{w1}–S_{w6} of transistor clamped topology for m = 1.25

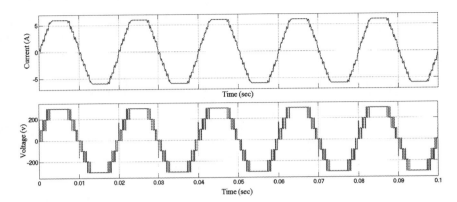

Fig. 9 Output voltage and current of transistor clamped topology for m = 1.25

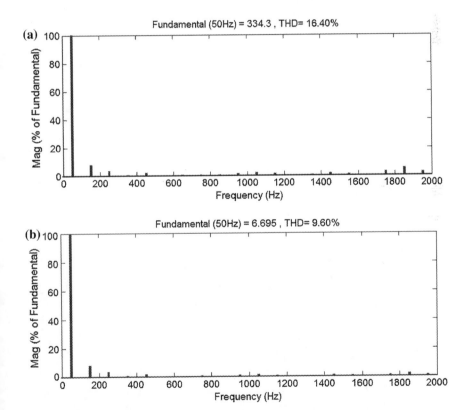

Fig. 10 Harmonic spectrum obtained with transistor clamped topology for m = 1.25. **a** Output voltage, **b** output current

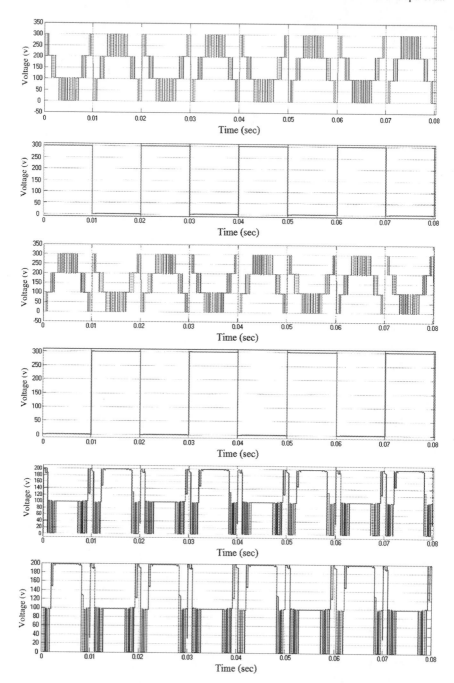

Fig. 11 Voltage stress across switches $S_{w1}-S_{w6}$ of transistor clamped topology for m = 0.9

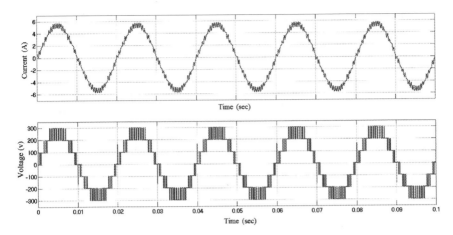

Fig. 12 Output voltage and current waveforms of multilevel inverter with transistor clamped topology for m = 0.9

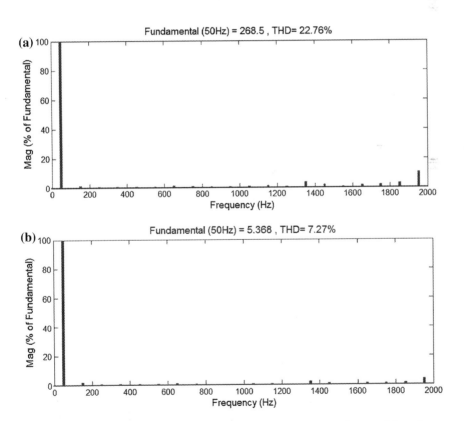

Fig. 13 Harmonic spectrum obtained with transistor clamped topology for m = 0.9. **a** Output voltage, **b** output current

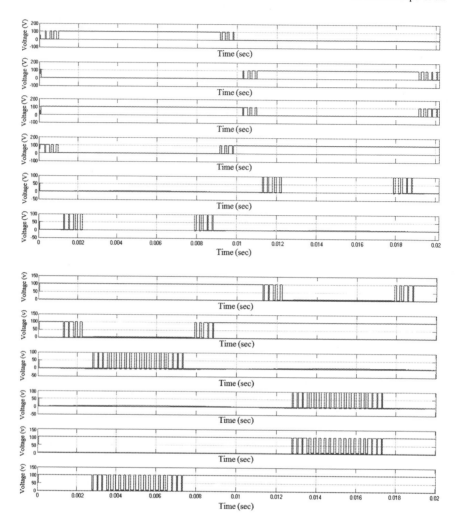

Fig. 14 Voltage stress across the switches of CHB multilevel inverter topology for m = 0.9

THD is 23.32 % and current THD is 4.95 % with a CHB multilevel inverter. Figure 14 shows the voltage stress across the switches of the CHB multilevel inverter for a modulation index of 0.9. It is observed that the applied DC voltage of 100 V appears as stress across the switches S_{w1}, S_{w2}, S_{w3}, S_{w4}, S_{w5}, S_{w6}, S_{w7}, S_{w8}, S_{w9}, S_{w10}, S_{w11} and S_{w12}.

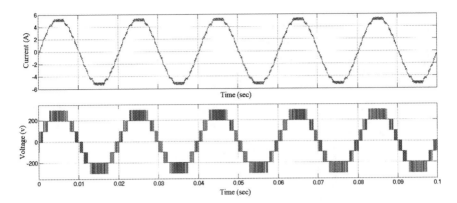

Fig. 15 Output voltage and current of CHB multilevel inverter topology for m = 0.9

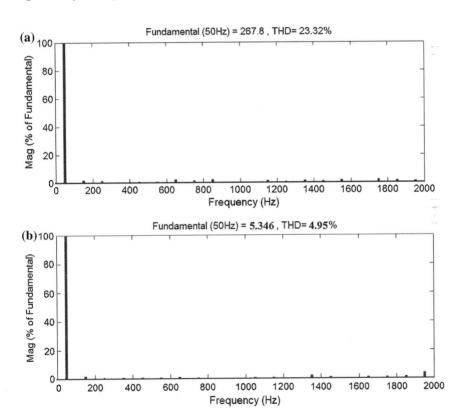

Fig. 16 Harmonic spectrum of CHB multilevel inverter topology for m = 0.9. **a** Output voltage, **b** output current

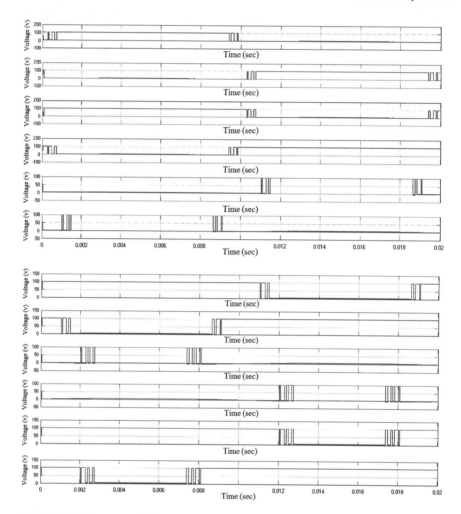

Fig. 17 Switching stress of CHB multilevel inverter topology for m = 1.25

4.4 Results Obtained with CHB Multilevel Inverter Topology for m = 1.25

Figure 18 shows the output voltage and current waveforms for a modulation index of 1.25. It is evident that seven levels of voltage output are synthesized. The current waveforms are more towards sinusoidal due to inductive load. Figure 19 displays the harmonic spectrum of the output voltage and current waveforms. The voltage THD is 15.95 % and current THD is 9.09 % with a CHB multilevel inverter. Figure 17 shows the voltage stress across the switches of the CHB multilevel inverter for a modulation index of 1.25. It is observed that the applied DC voltage

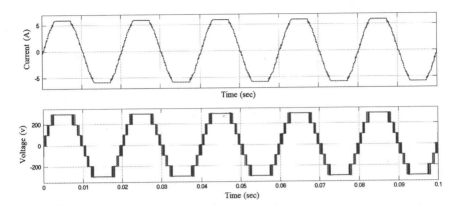

Fig. 18 Output voltage and current of CHB multilevel inverter topology for m = 1.25

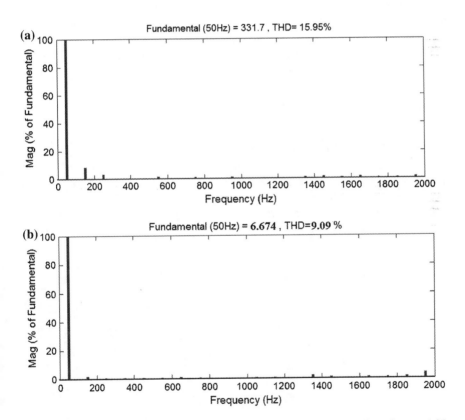

Fig. 19 Harmonic spectrum analysis of CHB multilevel inverter topology voltage for m = 1.25.
a Output voltage, **b** output current

Table 4 Comparison of THD with CHB multilevel inverter and transistor clamped multilevel inverter topology

Modulation index (m)	Topology	THD% (output voltage) (%)	THD% (output current) (%)	Output current (A)
0.9	CHB	23.32	4.95	5.346
	TCHB	22.76	7.27	5.359
1.25	CHB	15.95	9.09	6.674
	TCHB	16.64	9.60	6.62

of 100 V appears as stress across the switches S_{w1}, S_{w2}, S_{w3}, S_{w4}, S_{w5}, S_{w6}, S_{w7}, S_{w8}, S_{w9}, S_{w10}, S_{w11} and S_{w12}.

Table 4 shows the THD obtained with the conventional seven level inverter and transistor clamped inverter for modulation index of 0.9 and 1.25. It is clear that, when the modulation index is greater than 1 (m > 1), the THD decreases and when the modulation index is lesser than 1, the THD increases. In case of transistor clamped seven level inverter topology, the output THD is 16.40 % for m = 1.25 and the THD is 22.76 % for m = 0.9. In case of CHB seven level inverter topology, when the modulation index more than 1, the voltage THD 15.95 % and when the modulation index is less than 1 the THD is 23.32 %.

Table 5 shows the voltage stress across the switches of the conventional and transistor clamped seven level inverter topologies for modulation index of 0.9 and 1.25. From the table, it is clear that the voltage stress across the switch doesn't depend on the modulation index. It is observed that a constant voltage stress of 100 V is obtained across the switches of the CHB multilevel inverter. In case of transistor topology, the voltage stress across the switches is 300 V (except the switches S_{w1} and S_{w6} due to transistor clamped, they have a voltage stress of 200 V) which is thrice than that of a CHB multilevel inverter. Though the switch stress levels of transistor clamped topology is higher when compared to that of CHB

Table 5 Comparison of voltage stress across the switches of CHB and transistor clamped multilevel inverter topology

Topology	CHB	TCHB	CHB	TCHB
Modulation index	0.9		1.25	
S_{w1} (V)	100	300	100	300
S_{w2} (V)	100	300	100	300
S_{w3} (V)	100	300	100	300
S_{w4} (V)	100	300	100	300
S_{w5} (V)	100	200	100	200
S_{w6} (V)	100	200	100	200
S_{w7} (V)	100	–	100	–
S_{w8} (V)	100	–	100	–
S_{w9} (V)	100	–	100	–
S_{w10} (V)	100	–	100	–
S_{w11} (V)	100	–	100	–
S_{w12} (V)	100	–	100	–

multilevel inverter, however the number of switches required for a transistor clamped multilevel inverter topology is very less. The transistor clamped topology requires only six switches, whereas conventional CHB multilevel inverter topology needs twelve power switches to produce a seven level output.

5 Conclusion

In this paper, the analysis of a seven level inverter with reduced switches is carried out. Triple reference single carrier pulse width modulation technique is employed to generate gating pulses for the inverter. The output voltage, current and voltage stress across the switches of the multilevel inverter are obtained for a modulation index of 0.9 and 1.25. It is found that the seven level inverter based on transistor clamped topology synthesizes same levels of output as a conventional seven level inverter with reduced number of switches. Both the topologies produce a lesser THD with increase in the modulation index.

References

1. Chinnaiyan, V.K., Jovitha, J., Karpagam, J.: An experimental investigation on a multilevel inerter for solar applications. Int. J. Power Electron. 47, 157–164 (2013)
2. Daher, S., Schmid, J., Antunes, F.L.M.: Multilevel inverter topologies for stand-alone PV systems. IEEE Trans. Ind. Electron. 55(7), 2703–2712 (2008)
3. Cheng, Y., Qian, C., Crow, M.L., Pekarek, S., Atcitty, S.: A comparison of diode-clamped and cascaded multilevel converters for a STATCOM with energy storage. IEEE Trans. Ind. Electron. 53(5), 1512–1521 (2006)
4. Holmes, D.G., Lipo, T.A.: Pulse Width Modulation for Power Converters Principles and Practice, vol. 18. Wiley, New York (2003)
5. Gayathri Devi, K.S., Arun, S., Sreeja, C.: Comparative study on different five level inverter topologies. Int. J. Electr. Power Energy Syst. 63, 363–372 (2014)
6. Thielemans, S., Ruderman, A., Reznikov, B., Melkebeek, J.: Simple time domain analysis of a 4-level H-bridge flying capacitor converter voltage balancing. In: Proceedings of the IEEE International Conference on Industrial Technology (ICIT), pp. 818–823 (2010)
7. Ruderman, A., Reznikov, B., Thielemans, S.: Four-level H-bridge flying capacitor converter voltage balance dynamics analysis. In: Proceedings of the IEEE 35th Annual Conference on Industrial Electronics (ICIE), pp. 498–503 (2009)
8. Ding, K., Cheng, K.W.E., Zou, Y.P.: Analysis of an asymmetric modulation method for cascaded multilevel inverters. IET Power Electron. 5(1), 74–85 (2012)
9. Govindaraju, C., Baskaran, K.: Efficient sequential switching hybrid–modulation techniques for cascade multilevel inverters. IEEE Trans. Power Electron. 26(6), 1639–1648 (2011)
10. Gupta, V.K., Mahanty, R.: Optimised switching scheme of Cascaded H-bridge multilevel inverter using PSO. Int. J. Power Electron. Energy Syst. 64, 699–707 (2015)
11. Villanueva, E., Correa, P., Rodriguez, J., Pacas, M.: Control of single-phase cascaded H-bridge multilevel inverter for grid connected photovoltaic system. IEEE Trans. Ind. Electron. 56(11), 4399–4406 (2009)

300   R. Mallavarapu et al.

12. Du, Z., Tolbert, L.M., Opineci, B., Chaisson, J.N.: Fundamental frequency switching strategies of a seven-level hybrid cascaded H-bridge multilevel inverter. IEEE Trans. Power Electron. **24**(1), 25–33 (2009)
13. Shalini, B.A.P., Sethuraman, S.S.: Cascaded multilevel inverter for industrial applications. Commun. Comput. Inf. Sci. **296**, 339–344 (2013)
14. Rahim, N.A., Selvaraj, J., Krismadinata, C.: Five-level inverter with dual reference modulation technique for grid-connected PV system. Renew. Energy **35**(3), 712–720 (2010)
15. Palanivel, P., Dash, S.S.: Analysis of THD and output voltage performance for cascaded multilevel inverter using carrier pulse width modulation techniques. IET Power Electron. **4**(8), 951–958 (2011)
16. Hasan, M., Mekhilef, S., Ahmed, M.: Three-phase hybrid multilevel inverter with less power electronic components using space vector modulation. IET Power Electron. **7**(5), 1256–1265 (2014)
17. Maia, H.Z., Mateus, T.H., Ozpineci, B., Tolbert, L.M., Pinto, J.O.: Adaptive selective harmonic minimization based on ANNs for cascaded multilevel inverters with varying DC sources. IEEE Trans. Ind. Electron. **60**(5), 1955–1962 (2013)
18. Wells, J.R., Nee, B.M., Chapman, P.L., Krein, P.T.: Selective harmonic control: a general problem formulation and selected solutions. IEEE Trans. Power Electron. **20**(6), 1337–1345 (2005)
19. Sultana, W.R., Sahoo, S.K., Karthikeyan, S.P., Reddy, P.H.V., Reddy, G.T.R.: Elimination of harmonic in seven-level cascaded multilevel inverter using particle swarm optimization technique. Adv. Intell. Syst. Comput. **324**, 265–274 (2014)
20. Rahim, N.A., Chaniago, K., Selvaraj, J.: Single-phase seven-level grid-connected inverter for photovoltaic system. IEEE Trans. Ind. Electron. **58**(6), 2435–2443 (2011)
21. Rahim, N.A., Elias, M.F.M., Hew, W.P.: Transistor clamped H-bridge cascaded multilevel inverter with new method of capacitor voltage balancing. IEEE Trans. Ind. Electron. **60**(8), 2943–2956 (2013)
22. Babei, E., Hosseini, S.H.: New cascaded multilevel inverter topology with minimum number of switches. Energy Convers. Manag. **50**(11), 2761–2767 (2009)
23. McGrath, B.P., Holmes, D.G.: Multicarrier PWM strategies for multilevel inverters. IEEE Trans. Ind. Electron. **49**(4), 858–867 (2002)

Distinct Exploration on Two-Level and Hybrid Multilevel Converter for Standalone Solar PV Systems

R. Uthirasamy, P.S. Mayurappriyan and C. Krishnakumar

Abstract This paper mainly focuses on the comparative analysis of Power Spectral Density (PSD) and harmonics between Boost Two-Level Inverter (BTLI) and Multi String Hybrid Capacitor Clamped Multilevel Converter (MSHCCMLC) configurations. Compared with conventional BTLI, the proposed system employs the reduction of harmonics and PSD. Maximum utilization of solar power is achieved through Maximum Power Point Tracking (MPPT) control mechanism. Modified Sinusoidal Pulse Width Modulation (MSPWM) scheme is adopted to attain better quality of output power. The proposed MSHCCMLC requires small LC filter component at the inverter end which makes the system more effective. Simulation and hardware models of 5-level Hybrid Multilevel Converter (HMLC) are developed and the system performances are validated.

Keywords Multi-string solar PV · Maximum power point tracking · Capacitor clamped inverter · Sinusoidal pulse width modulation · Power spectral density and harmonics

1 Introduction

Nowadays, researchers are mainly paying attention to propose new energy sources because of increase in cost of fossil fuels and Non-renewable energy resources. On other hand, they have an enormous negative impact on the environment and increased power generation rate. In this perspective, an alternate energy sources are

R. Uthirasamy (✉) · C. Krishnakumar
Department of EEE, KPR Institute of Engineering and Technology,
Coimbatore, India
e-mail: rusamy83@gmail.com

C. Krishnakumar
e-mail: ckk1973@gmail.com

P.S. Mayurappriyan
Department of EEE, KCG College of Technology, Chennai, India
e-mail: mayurappriyan.eee@kcgcollege.com

© Springer Science+Business Media Singapore 2017
P. Deiva Sundari et al. (eds.), *Proceedings of 2nd International Conference on Intelligent Computing and Applications*, Advances in Intelligent Systems and Computing 467, DOI 10.1007/978-981-10-1645-5_25

essential to meet the demand with cost effective. It is anticipated that the electrical energy generation from renewable sources will enlarge from 19 %, in 2010, to 32 %, in 2030, with the reduction of CO_2 emission. Among the energy resource, there has been a noticeable increase in use of PV based systems for power generation, given its renewable nature. Development of power converters have been one of the most active areas in research and development of power electronics. In recent years, multilevel and hybrid converters have become more attractive for researchers and manufacturers due to their advantages over conventional inverters [1]. Multilevel converter offers lower EMI and lower Total Harmonic Distortion (THD).

PV array/module characterizes the primary power renovation unit of a PV generator unit. Solar irradiance, fill factor and cell/panel temperature of solar modules decide the I–V characteristics of a PV module. The output voltage of a PV panel/module is degraded due to increase in panel temperature. The magnitude of panel current is varied with respect to solar irradiance. So, PV module has nonlinear I–V characteristics, and it is essential to make linear characteristics through Maximum Power Point Tracking (MPPT) based controllers [2–5]. Mathematical modeling of PV module is being continuously updated to enable researcher to have a better understanding of its working. MSHCCMLC can operate at both fundamental switching frequency and high switching frequency. MSHCCMLC is proposed for utilizing solar power with reduced PSD and harmonics. This paper is organized as five sections are as follows: Introduction of the solar PV is reviewed in Sect. 1. Mathematical modelling of a PV system is addressed in Sect. 2. Structure of MSHCCMLC system is reviewed in Sect. 3. Performance of BTLI and MSHCCMLC configurations are addressed in Sect. 4. Section 5 concludes the paper.

2 Mathematical Modeling of a Photovoltaic Module

Solar energy is received on earth in cyclic manner and intermittent with low power density from 0 to 1 kW/m^2 [6–8]. Received energy is affected by atmospheric precision, quantity of latitude, etc. Bandwidth of solar energy is ranging from infra-red to ultraviolet frequencies. A solar cell can generate a voltage and power of 0.5 V and 0.3 W respectively. The electron hole pair generation is proportional to the solar irradiance incident on the solar cell. The equivalent circuit of interconnected solar PV cell is as shown in Fig. 1.

The current source I_{ph} stands for the photo-current. R_{sh} and R_s are the inherent shunt and series resistance of the solar cell respectively. Typically, the value of R_{sh} is very large compared to that of R_s, hence R_s may be ignored to simplify the investigation. PV cells are grouped in larger units called PV modules which are further interconnected in a parallel-series configuration to form PV arrays. The V–I characteristics of PV array is as shown in Fig. 2. The mathematical modeling of PV cell is given by the following equations:

Fig. 1 Equivalent circuit of a PV cell

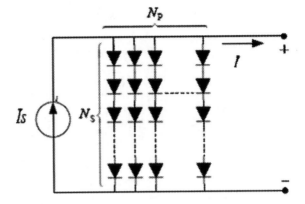

Fig. 2 V–I characteristics of PV array

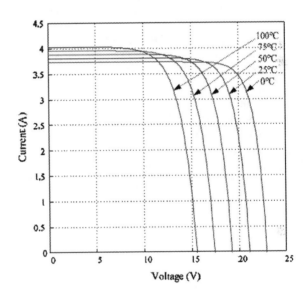

Module photo current (I_{ph}) is expressed using Eq. (1)

$$I_{ph} = [I_{scr} + K_i(T - 298)] * \lambda/1000 \qquad (1)$$

Module reverse saturation current (I_{rs}) is expressed using Eq. (2)

$$I_{rs} = I_{scr}/[\exp(qV_{oc}/N_s kAT) - 1] \qquad (2)$$

Module saturation current (I_o) varies with the cell temperature is given by Eq. (3)

$$I_o = I_{rs}[T_r/T]^3 \exp\left[\frac{qE_{go}}{gk}\left\{\frac{1}{T_r} - \frac{1}{T}\right\}\right] \qquad (3)$$

Output current of PV module is expressed using Eq. (4)

$$I_{pv} = N_P * I_{ph} - N_p * I_o \left[\exp\left\{ \frac{q * (V_{pv} + I_{pv}R_s)}{N_s kAT} \right\} - 1 \right] \tag{4}$$

where 'T' is the p-n junction, 'q' is the electron charge [1.60217646 * 10^{-19} C], 'k' is the Boltzmann constant [1.3806503 * 10^{-23} J/K], 'A' is the diode ideality constant. If the array is composed of N_p parallel connections of cells the photovoltaic and saturation currents may be expressed as:

$$I_{pvNp} = (I_{pv,n} + K_i \Delta T) \frac{G}{G_n} \tag{5}$$

where G (W/m^2) is the irradiation on the device surface and G_n is the nominal irradiation.

Power generated by the solar PV arrays is getting varied with respect to the intensity of solar radiation. So, an efficient controller is required to utilize of the solar irradiance properly [9–15]. Currently, a number of MPPT algorithm based controllers are designed and implemented to harvest maximum solar PV power. The main theme of MPPT controller is to generate firing pulses to DC–DC controllers. In case of battery backup provided PV applications, MPPT controllers are designed to perform as charge controllers. In this paper, the Perturb & Observe (P&O) algorithm is used to extort maximum power from the PV arrays and deliver it to the inverter. The feedback controller used for the inverter is based on Proportional Integral (PI) algorithm. Grid current I_g, is sensed and fed back to a comparator, which compares it with reference current I_{ref}. I_{ref} is obtained by sensing utility voltage V_u. The sensed V_u signal is converted into a reference signal before it is multiplied with variable m.

Therefore the reference current is expressed as

$$I_{ref} = V_u * m \tag{6}$$

3 Multi String Hybrid Capacitor Clamped Multilevel Converter

(a) Structure of MSHCCMLC

The generalized structure of MSHCCMLC system is shown in Fig. 3. The equivalent structure consists of DC–DC converter and HMLC units. In all DC–DC stages, each DC–DC converter is used to connect a PV array to the HMLC.

HMLC is integrated with passive component like capacitor and switching devices. To obtain 5-level AC output from MSHCCMLC, two capacitors are required along with switching devices. The equivalent circuit of the single-phase

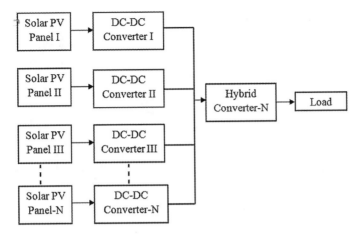

Fig. 3 Generalized structure of MSHCCMLC system

5-level MSHCCMLC topology is shown in Fig. 4. The inverter approves a full-bridge arrangement with supplementary circuit. DC–DC converters boost up the solar PV voltages and fed its output to the capacitor clamping circuit.

Clamping and supplementary circuits' synthesis stepped DC voltage from the capacitor voltages. The switching of supplementary circuit is achieved through PI controller. PWM switching pulses of H-bridge inverter are obtained by comparing a high frequency carrier signal with a fundamental frequency sinusoidal signal. Combinations of PV strings are used as the input voltage sources. The voltage across the PV strings is considered as V_{pv1}, V_{pv2} and V_{pv3}.

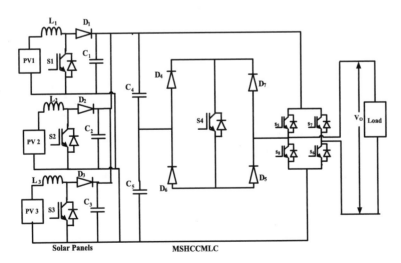

Fig. 4 Equivalent circuit of 5-level MSHCCMLC

(b) **Analysis of DC–DC converters**

At $t = T_1$, the DC–DC converter I switch S_1 is turned ON and the inductor L_1 current raises linearly from I_1 to I_2. The voltage across the inductor L_1 is expressed as

$$V_{L1} = L_1 \frac{I_2 - I_1}{T_1} \qquad (7)$$

At $t = T_2$, the DC–DC converter I switch S_1 is turned OFF and the inductor L_1 current falls linearly from I_2 to I_1. The average output voltage of DC–DC converter I can be expressed using Eq. (8)

$$V_1 = V_{pv1} + V_{L1} \qquad (8)$$

Considering the system to be ideal, the average output voltage of DC–DC converter I is obtained using Eq. (9)

$$V_1 = \frac{V_{pv1}}{1 - D} \qquad (9)$$

where, V_{pv1} is the DC source voltage I, D is the duty cycle of DC–DC converter I, V_1 is the output voltage DC–DC converter I.

(c) **Sinusoidal Pulse Width Modulation**

Pulse Width Modulation (PWM) technique is the commonly adopted switching scheme for power converters. Many soft computing and optimization techniques like fuzzy, particle swarm are developed to attain the converter triggering pulses [16]. The switching period T and the frequency modulation ratio p is given by Eqs. (10) and (11)

$$T = \frac{2\pi}{M_f} \qquad (10)$$

$$M_f = \frac{f_s}{f_1} \qquad (11)$$

where f_s and f_1 are switching and fundamental frequency respectively. The quarter period of pulse δ_0 is given as

$$\delta_0 = \frac{T}{4} \qquad (12)$$

4 Performance Evaluation

(a) Single Phase Boost Two-Level Inverter System

The simulation model of BTLI fed AC load system is shown in Fig. 5. The entire circuit is simulated using SIMULINK tools and its performance is investigated. BTLI circuit is integrated with boost chopper and H-bridge inverter. SPWM schemes are adopted for generating triggering pulses to H-bridge inverter switches. The output voltage of BTLI fed resistive load system is shown in Fig. 6. From the result, it is found that the voltage magnitude and frequency of the inverter output voltage are 230 V and 50 Hz respectively. The power spectral density for the output voltage of BTLI system is shown in Fig. 7. The switching frequency the inverter is considered to be 2 kHz. At 2 kHz, the power spectral density for the output voltage of BTLI fed resistive load is 28 dB/Hz. From the PSD analysis, it is

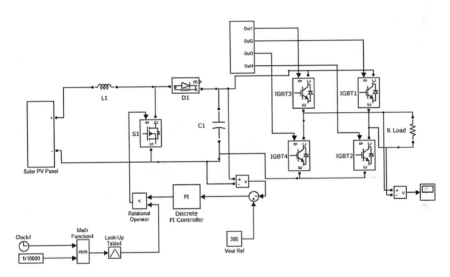

Fig. 5 Simulation model of BTLI system

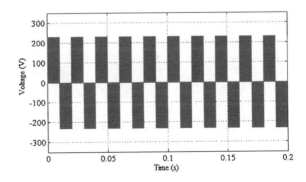

Fig. 6 Inverter output voltage of BTLI

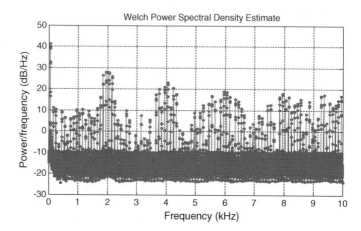

Fig. 7 Output voltage power spectral density of BTLI

Fig. 8 Experimental inverter
output voltage of BTLI

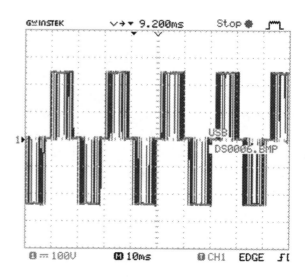

understood that the density of power is maximum in the multiples of switching
frequency which leads to switching loss. Also, the analysis clearly elucidate that the
BTLI system has high switching stress and switching loss in the multiples of
switching frequency. The experimental output voltage of BTLI is shown in Fig. 8.
From the result it is observed that the voltage magnitude and frequency of single
phase BTLI is 230 V, 50 Hz respectively. The experimental voltage FFT spectrum
of BTLI is shown in Fig. 9.

(b) **Single Phase MSHCCMLC System**

The simulation model of 5-level MSHCCMLC system is shown in Fig. 10. The
entire circuit is simulated using SIMULINK tools and their performance is

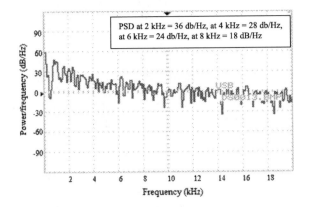

Fig. 9 Experimental output voltage FFT spectrum of BTLI

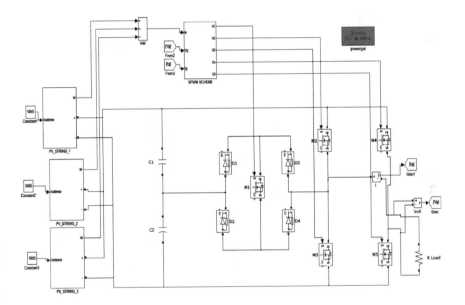

Fig. 10 Simulation model of MSHCCMLC system

evaluated. Three PV panels are connected with boost chopper units. The outputs of boost chopper units act as input for the clamping and supplementary circuit. The sub-system model of P&O based MPPT algorithm is shown in Fig. 11. Panel voltage V_{PV} and current I_{pv} is sensed periodically and fed back to the MPPT controller for developing the firing pulses to the DC–DC converters with the duty cycle D. Separate MPPT controller are implemented for each PV module. Subsystem Model of Solar Panel I is shown in Fig. 12. The output voltage of

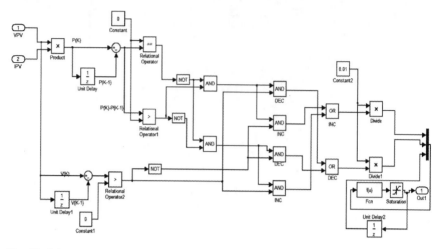

Fig. 11 Subsystem model of MPPT P&O algorithm

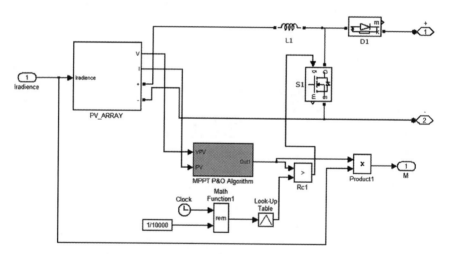

Fig. 12 Subsystem model of solar panel I

MSHCCMLC is shown in Fig. 13. From the waveform it is found that the magnitude of inverter output voltage is 230 V.

The power spectral density for output voltage of MSHCCMLC system is shown in Fig. 14. The switching frequency of inverter switches is 20 kHz. At 20 kHz, the power spectral density for the output voltage of MSHCCMLC is 8 dB/Hz. From the PSD analysis, it is understood that the density of power in the multiples of switching frequency are 20 dB/Hz reduced in MSHCCMLC compared to BTLI system.

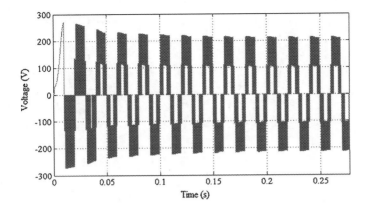

Fig. 13 Inverter output voltage of MSHCCMLC system

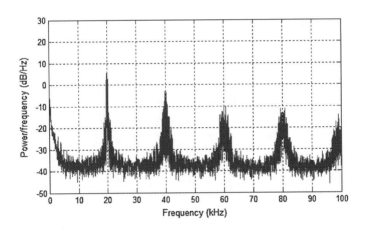

Fig. 14 Output voltage power spectral density of MSHCCMLC

(c) Experimental Analysis

The gating pulses are generated by Peripheral Integral Controller 16F877 (PIC) and the magnitude of the pulses are amplified by the driver unit. The gate pulses to the inverter switches S_5 and S_7 are shown in Figs. 15 and 16 respectively.

Output voltage of the 5-level MSHCCMLC system is shown in the Fig. 17. From the obtained result, it is inferred that the output voltage and the frequency of the MSHCCMLC is 200 V and 50.266 Hz.

The experimental voltage FFT spectrum of BTLI is shown in Fig. 18. From the spectrum it is inferred that the 5-level MSHCCMLC system has reduced voltage stress in the multiples of switching frequency. The experimental setup of

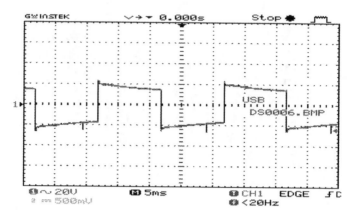

Fig. 15 Gating signals to inverter switch S_5

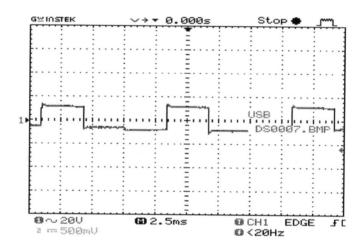

Fig. 16 Gating signals to inverter switch S_7

MSHCCMLC system is shown in Fig. 19. It comprises of power supply, regulator, controller, supplementary converter and H-bridge inverter units.

(d) Comparative Analysis

Comparative analysis between conventional BTLI and 5-level MSHCCMLC is carried out to highlight the merits of the proposed system. From the Tables 1 and 2, it is clear that MSHCCMLC systems have reduced PSD in the multiples of switching frequency. Also, it is found that the experimental FFT spectrum almost close to the simulated PSD spectrum. Thus the proposed MSHCCMLC system is more suitable for solar power applications.

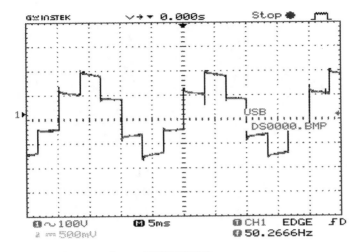

Fig. 17 Experimental output voltage of MSHCCMLC

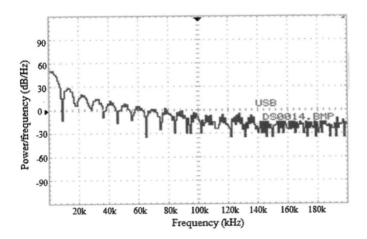

Fig. 18 Experimental output voltage FFT spectrum of MSHCCMLC

The output voltage THD spectrum of the BTLI and MSHCCMLC system is shown in Figs. 20 and 21 respectively. From the analysis it is observed that the conventional BTLI generates the voltage THD of 34.28 % but the 5-level MSHCCMLC generates the voltage THD of 4.64 %. Thus the proposed MSHCCMLC reduces the voltage THD of 29.64 %. Also, THD spectrum of MSHCCMLC satisfies the IEEE 519:1992 standard.

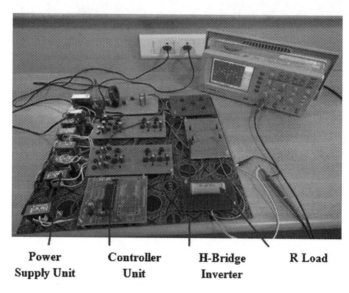

| Power | Controller | H-Bridge | R Load |
| Supply Unit | Unit | Inverter | |

Fig. 19 Experimental setup of MSHCCMLC

Table 1 PSD analysis for BTLI

Switching frequency (kHZ)	Power spectral density dB/Hz	
	BTLI	
	Simulated	Experimental
2	28	36
4	24	28
6	20	24
8	18	18

Table 2 PSD analysis for MSHCCMLC

Switching frequency (kHZ)	Power spectral density dB/Hz	
	MSHCCMLC	
	Simulated	Experimental
20	8	26
40	−4	8
60	−10	0
80	−12	−10

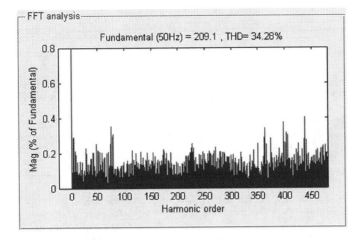

Fig. 20 Output voltage THD spectrum of BTLI

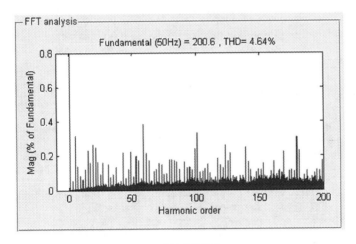

Fig. 21 Output voltage THD spectrum of MSHCCMLC

5   Conclusion

Through this work, comparative analysis between the configurations of single-phase BTLI and 5-level MSHCCMLC are carried out for standalone solar powered utilities. MSHCCMLC systems have reduced PSD in the multiples of switching frequency. Proposed converter reduces the voltage THD of 29.64 %. Better utilization of the solar PV is achieved through P&O MPPT algorithm technique. The proposed MSHCCMLC eliminates the LC filter at the inverter end. Thus the power quality issues like voltage swell and sag is mitigated in MSHCCMLC.

References

1. Walker, G.R., Sernia, P.C.: Cascaded DC–DC converter connection of photovoltaic modules. IEEE Trans. Power Electron. 19(4), 1130–1139 (2004)
2. Selvaraj, J., Rahim, N.A.: Senior member, IEEE "multilevel inverter for grid-connected PV system employing digital PI controller". IEEE Trans. Ind. Electron. 56(1), 149–158 (2009)
3. Park, S.J., Kang, F.S., Lee, M.H., Kim, C.U.: A new single-phase five-level PWM inverter employing a deadbeat control scheme. IEEE Trans. Power Electron. 18(3), 831–843 (2003)
4. Sadigh, A.K., Gharehpetian, G.B., Hosseini, S.H.: New method for estimating flying capacitor voltages in stacked multicell and flying capacitor multicell converters. J. Zhejiang Univ. Sci. C. 11(8), 654–662 (2010). doi:10.1631/jzus.C0910559
5. Çelebi, Me., Alan, İ.: A novel approach for a sinusoidal output inverter. J. Electr. Eng. 92, 239–244 (2010)
6. Rahim, N.A., Chaniago, K., Selvaraj, J.: Single-phase seven-level grid connected inverter for photovoltaic system. IEEE Trans. Ind. Electron. 58(6), 2435–2443 (2011)
7. Uthirasamy, R., Ragupathy, U.S., Mithra, R.: Design and investigation of solar powered soft switched Z-source inverter. Lect. Notes Electr. Eng. 326 (2015). doi:10.1007/978-81-322-2119-7_13
8. Agelidis, V.G., Baker, D.M., Lawrance, W.B., Nayar, C.V.: A multilevel inverter topology for photovoltaic applications. In: IEEE International Symposium on Industrial Electronics, ISIE '97, vol. 2, pp. 589–594, July (1997)
9. Esram, T.: Student Member, IEEE, and Patrick L. Chapman, Senior Member, IEEE comparison of photovoltaic array maximum power point tracking techniques. IEEE Trans Energy Convers. 22(2) (2007)
10. Sera, D., Teodorescu, R., Rodriguez, P.: PV panel model based on datasheet values. In: IEEE International Symposium on Industrial Electron (2007)
11. Jensen, M., Louie, R., Etezadi-Amoli, M., Sami Fadali, M.: Model and simulation of a 75KW PV solar array. In: IEEE Conference on Power Electron (2010)
12. Esram, T., Chapman, P.L.: Comparison of photovoltaic array maximum power point tracking techniques. IEEE Trans. Energy Convers. 22(2) (2007)
13. Calais, M., Agelidis, V.G., Dymond, M.S.: A cascaded inverter for transformer less single phase grid-connected photovoltaic systems. In: Proceedings of 31st Annual IEEE PESC, vol. 3, pp. 1173–1178 (2001)
14. Veerachary, M.: Power tracking for nonlinear PV sources with coupled inductor SEPIC converter. IEEE Trans. Aerosp. Electron. Syst. 41(3), 1019–1029 (2005)
15. Altas, H., Sharaf, A.M.: A photovoltaic array simulation model for matlab-simulink GUI environment. In: IEEE, Clean Electrical Power, International Conference on Clean Electrical Power (ICCEP '07). Ischia, Italy 14–16 June 2007
16. Barkat, S., Berkouk, E.M., Boucherit, M.S.: Particle swarm optimization for harmonic elimination in multilevel inverters. J. Electr. Eng. 91, 221–228 (2009)

Smart Plug for Household Appliances

N. Prayongpun and V. Sittakul

Abstract A common home consists of some electrical loads such as lighting, laundry appliances, kitchen appliances, climate control equipment, entertainment devices, etc. All devices consume lots of electric power in spite of being on their standby mode. In this chapter, programmable smart plugs with bundled ZigBee (IEEE 802.15.4) are introduced without the need of replacing the present appliances. The implemented software allows programming, controlling and real-time monitoring over TCP/IP platform. Furthermore, an energy management system with smart plugs is also proposed in the system framework to efficiently utilize the energy consumption.

Keywords Smart plug · IEEE 802.15.4 · TCP/IP · Programming · Controlling · Real-time monitoring · ZigBee

1 Introduction

Resident and commercial buildings are the main users where consume the electrical energy over 20 % of the overall energy used. To monitor and control the energy consumption in each area, an intelligent technology with efficient planning of

N. Prayongpun (✉) · V. Sittakul
Electronics Engineering Technology, College of Industrial Technology,
King Mongkut's University of Technology North Bangkok, 1518 Pracharat 1 Road,
Wongsawang, Bangsue, Bangkok 10800, Thailand
e-mail: nuttapol.p@cit.kmutnb.ac.th

V. Sittakul
e-mail: vitawat.s@cit.kmutnb.ac.th

© Springer Science+Business Media Singapore 2017
P. Deiva Sundari et al. (eds.), *Proceedings of 2nd International Conference on Intelligent Computing and Applications*, Advances in Intelligent Systems and Computing 467, DOI 10.1007/978-981-10-1645-5_26

317

energy utilization is indispensable. Therefore, many researches focus on smart grid, energy management, zero waste technologies. Smart and green technologies are one of the most important research issues in order to save energy and decrease pollution emission [1, 2].

In the last decade, many smart electrical grid technologies with smart meters are introduced in many countries. Its objective is to reduce operation cost and to improve the quality of electricity network [3]. However, conventional smart meters are not sufficiently developed nowadays because they can only measure the electrical energy consumption for whole buildings. Thank to the high speed development in electronic and communication technologies, each appliance in buildings can be investigated for the overall energy consumption [4]. A typical home consists of some electrical loads such as lighting, laundry appliances, kitchen appliances, climate control equipment, entertainment devices, etc. A smart device attached to each appliance in this work is called "smart plug". All smart plugs will intelligently be linked inside buildings for energy monitoring, safety and automation management.

In this chapter, the framework of smart plug is proposed in Sect. 3 and implemented with the aim of measuring the energy consumption with high accuracy and low cost in Sect. 4. In Sect. 5, the conclusion will be presented.

2 Related Works

Data acquisition and control for smart plug has been demonstrated in 2013 [1]. In [1], a smart plug with a real-time energy monitoring system has been developed using Arduino microcontroller board. In 2014, the data acquisition and control for smart plug has also been developed so that the global users can check the status via the GSM technology. As for the local users, the energy meter around the area can be controlled for the load usage using Bluetooth technology to reduce the cost of GSM rental fees [2].

However, as shown in the previous works, all smart plugs were directly wired-connected to the Ethernet controllers to begin the data acquisition via web browsers. In [5], as the electrical energy sources become prominent, the framework home controller system to monitor and control efficiently home energy consumption has been presented. The proposed technique can gain overall between 10 and 22% in the energy consumption load management.

In practice, there are many smart plugs have to be installed in a house and they shall be wired-connected. Therefore, in this work, the novel concept of data acquisition and control for smart plug has been designed based on ZigBee connection among the smart plugs and embedded web server unit. The X-bee radio modules (Compliance with ZigBee) are chosen due to their low cost and make it possible to install in the house, based on IEEE 802.15.4. As shown in the next section, the framework system is introduced.

3   Proposed Framework

The proposed hardware system comprises two device categories: one is Smart Plug (*SP*) and other one is Embedded Web Server (*EWS*). Both categories can communicate with the other on the IEEE 802.15.4 wireless sensor network as shown in Figs. 1 and 2. A smart plug can be employed into two modes: manual and automatic modes. In manual mode, a smart plug is similar to the conventional plug. But in automatic mode, the smart plugs can be controlled via a web browser.

Figure 1 demonstrates the block diagram of the proposed smart plug composed of a current-voltage sensing circuit (based on ACS712), microcontroller board (based on ATmega328), LCD display and wireless sensor module (based on 2mW X-bee). In Fig. 2, the embedded web server system consists of the microcontroller (ATmega 2460), wireless sensor module, DP83848V Ethernet module and display module.

Figure 3 shows the sequent flow of the communication between a smart plug and an embedded web server. When an appliance is connected to the smart plug (*SP*), the *SP* will send the registration message to the embedded web server (*EWS*) via a ZigBee network. The *EWS* will collect the data and reply with the time slot and the time reference. After finishing the initialization, the *SP* can regularly transmit the energy consumption data at the given time slot until the appliance is

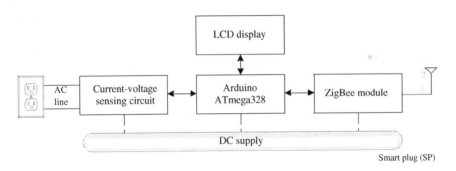

Fig. 1 Block diagram of proposed smart plug

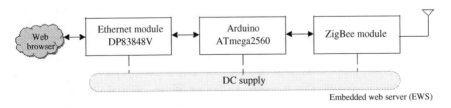

Fig. 2 Block diagram of embedded web server

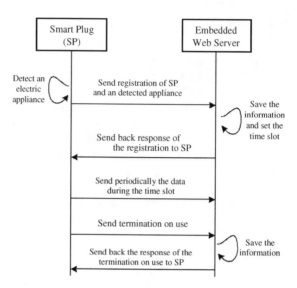

Fig. 3 Communication sequence between *SP* and *EWS*

unused. All mobile devices with web browser, such as laptop, tablet and mobile-phone, can monitor the energy consumption in real-time and also control the electric transmission by turning ON/OFF.

Moreover, the electronic smart plugs are designed in order to guarantee the system failure. While the appliance connected to smart plug consume unusually the energy, the designed system can switch off automatically the corresponding plug. Furthermore, in the case of the malfunction smart plug system, the smart plug system can be disabled and it will function as the traditional electronic plug.

4 System Implement and Results

Figure 4 shows photos of the prototype of smart plug and embedded web server. The software used for web server development is based on *HTML5* and *PHP* languages as can be seen in Fig. 5. The electrical data are collected by smart plug for every two seconds and are preprocessed for calculating the min-max energy consumption before the smart plug will send the data back to the embedded web server.

The next step of implementation of smart plug is the calibration to obtain the highest accuracy of the power consumption measurement as much as possible. As the microcontroller is used to calculate the power consumption of appliance using the current-voltage sensing circuit, the calibration process would be done properly. The power analyzer (HIOKI 3380) with a variable dummy load is employed for the calibration process. In order to reduce the tolerance less than 2 %, the polynomial estimation technique has to be used.

(a)

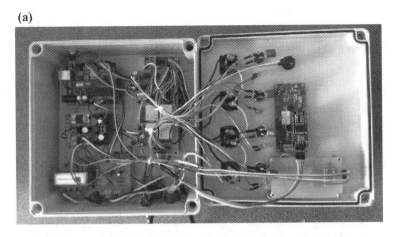

(b)

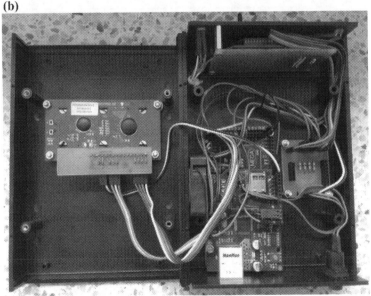

Fig. 4 **a** Prototype of the smart plug and **b** prototype of the embedded web server

Table 1 shows the comparisons of the electric load, current and voltage from power analyzer HIOKI 3380, and current and voltage form *SP* sensors. In Fig. 6, the voltage calibration equation is given by

$$y^3 = -333{,}488x^3 + 1{,}081{,}650x^2 - 1{,}168{,}877x + 421{,}064 \qquad (1)$$

where x is the voltage from *SP* and y denotes the calibrated voltage and also, in Fig. 7, the current calibration equation is given by

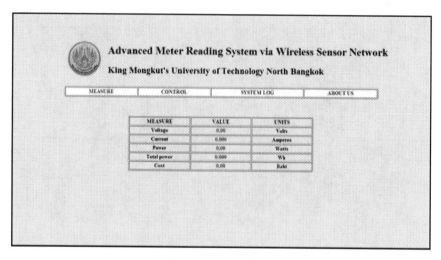

Fig. 5 Web page for controlling via smart plug

Table 1 Comparison of the electric load, current and voltage from power analyzer HIOKI 3380, and current and voltage form proposed smart plug

Load Lamp (W)	Current power from analyzer, I_a (Amp)	Current meter from SP, I_m (Amp)	Voltage power from analyzer, V_a (Volt)	Voltage meter from SP, V_m (Volt)
0	0.000	0.030	-	-
25	0.108	0.041	223.20	1.0979
50	0.213	0.056	223.10	1.0978
75	0.321	0.074	222.56	1.0968
100	0.437	0.095	222.21	1.0951
125	0.544	0.114	222.06	1.0945
150	0.650	0.133	221.75	1.0931
175	0.756	0.153	221.81	1.0941
200	0.863	0.172	221.41	1.0926
300	1.293	0.252	220.34	1.0909
400	1.721	0.332	220.01	1.0899
500	2.178	0.417	219.44	1.0891
600	2.599	0.495	218.57	1.0863
700	3.032	0.577	218.04	1.0854
800	3.474	0.658	217.21	1.0849

$$y = 56.513x^5 - 104.4x^4 + 71.463x^3 - 22.296x^2 + 8.4897x - 0.2181 \quad (2)$$

where x is the current from *SP* and y denotes the calibrated current. After applied the polynomial estimation technique, the tolerance is less than 2 %.

Fig. 6 Comparison between reference voltage and various orders of the calibration equations

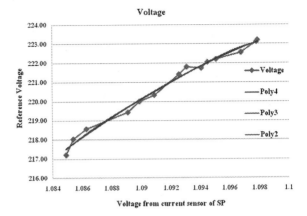

Fig. 7 Comparison between reference current and various orders of the calibration equations

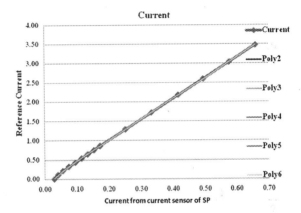

Table 2 shows the testing comparison between the power analyzer and smart plug for different loads. Consequently, the average tolerance of current, voltage and power are 1.143, 0.142 and 1.174 %, respectively. The measured power values from smart plug are compared to that from HIOKI 3380 as shown in Fig. 8.

Table 2 Tolerance of current, voltage and power

Load (W)	HIOKI 3380			Smart plug			Error		
	Voltage (V)	Current (A)	Power (W)	Voltage (V)	Current (A)	Power (W)	Voltage (%)	Current (%)	Power (%)
25	223.35	0.110	0.025	223.02	0.102	0.023	0.148	7.273	7.198
50	222.90	0.216	0.048	222.26	0.213	0.047	0.287	1.389	1.871
75	222.51	0.323	0.072	221.94	0.327	0.072	0.256	1.238	0.836
100	222.08	0.441	0.098	222.35	0.449	0.100	0.122	1.814	1.839
125	222.05	0.548	0.122	221.41	0.555	0.123	0.288	1.277	1.235
150	221.88	0.654	0.145	221.32	0.660	0.146	0.252	0.917	0.482

(continued)

Table 2 (continued)

Load (W)	HIOKI 3380			Smart plug			Error		
	Voltage (V)	Current (A)	Power (W)	Voltage (V)	Current (A)	Power (W)	Voltage (%)	Current (%)	Power (%)
175	222.39	0.761	0.169	222.34	0.764	0.170	0.022	0.394	0.354
200	221.96	0.886	0.197	221.63	0.884	0.196	0.149	0.226	0.560
300	221.36	1.315	0.292	221.59	1.311	0.290	0.104	0.304	0.412
400	220.50	1.752	0.386	220.41	1.757	0.387	0.041	0.285	0.285
500	220.00	2.184	0.481	220.13	2.190	0.482	0.059	0.275	0.229
600	219.26	2.615	0.573	219.38	2.613	0.573	0.055	0.076	0.035
700	218.67	3.045	0.667	218.63	3.049	0.669	0.018	0.131	0.330
800	218.72	3.485	0.762	219.12	3.499	0.768	0.183	0.402	0.774

Fig. 8 Comparison of measured power values between *SP* and HIOKI3380

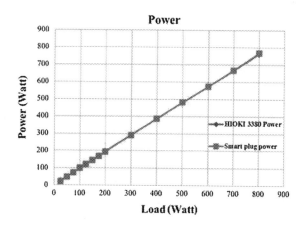

5 Conclusion

This chapter presents a low cost smart plug with a tolerance under 2 %. The simple algorithm is implemented for monitoring and controlling the energy consumption. In this work, it is possible to use the ZigBee concept to implement the wireless communication between the smart plugs and the embedded base server for monitoring and controlling the electrical power energy usage.

Acknowledgment The authors would like to thank the departments of electronic and electric engineering technology for their collaboration, financial support and equipment support.

References

1. Shajahan, A.H., Anand, A.: Data acquisition and control using Arduino-Android platform: Smart plug. In: International Conference on Energy Efficient Technologies for Sustainability (ICEETS), pp. 241–244, Nagercoil (2013)

2. Bhagyalakshmi, D., Sarathchandra, S.: Data acquisition and loads controlling using android mobile and GSM technology. Int. J. Sci. Eng. Technol. Res. **3**(35), 6979–6984 (2014)
3. Brenkus, J., Stopjakova, V., Zalusky, R., Mihalov, J., Majer, L.: Power-efficient smart metering plug for intelligent households. In: 25th International Conference on Radioelektronika (RADIOELEKTRONIKA), pp. 110–113, Pardubice (2015)
4. Fan, Z., Kulkarni, P., Gormus, S., Efthymiou, C.: Smart grid communications: overview of research challenges, solutions, and standardization activities. Commun. Surv. Tutor. IEEE **15**(1), 21–38 (2012)
5. Morsali, H., Shekarabi, S.M., Ardekani, K., Khayam, H., Fereidunian, A., Ghassemian, M.: Smart plugs for building Energy Management Systems. In: 2nd Iranian Conference on Smart Grids, IEEE pp. 1–5, (2012)

Modeling and CFD Analysis of Gerotor Pump

G.S. Kumarasamy, V.P.M. Baskaralal and S. Arunkumar

Abstract A Gerotor pump is a positive displacement pump which will deliver a predetermined quantity of fluid in proportion to speed. The main advantages of this pump are: (i) having only two moving parts, (ii) high speed operation, (iii) constant flow rate. The outer and inner rotor profiles are formed by epicycloids and hypocycloids and by using the inter relationship between the outer rotor and inner rotor. The span angle design of tooth profile affect the motion and pumping performance of the Gerotor pump. The rotor profiles are built using the mathematical models and CFD Analysis was conducted for various span angle designs using ANSYS CFX. The overall fluid analysis gives the fluid behavior inside the pump. In this paper, the various flow characteristics are analyzed for different span angle designs of Gerotor pump.

Keywords Gerotor pump · CFD analysis · ANSYS CFX

1 Introduction

The Gerotor pump operates much similar way of internal gear pump. The Gerotor pump consists of: inner rotor as a driver, outer rotor as the follower and housing. The outer rotor has one tooth more than the inner rotor. Both the rotors rotate in the same direction with different centers of rotation. The power driven inner rotor draws the outer rotor around when they mesh together. Each tooth of the inner rotor is always in sliding contact with the surface of the outer. This sliding contact provides the sealing effect.

G.S. Kumarasamy (✉) · V.P.M. Baskaralal · S. Arunkumar
KCG College of Technology, Karapakkam, Chennai 600097, India
e-mail: hoded@kcgcollege.com

V.P.M. Baskaralal
e-mail: baskaralal@kcgcollege.com

S. Arunkumar
e-mail: arunkumar@kcgcollege.com

© Springer Science+Business Media Singapore 2017
P. Deiva Sundari et al. (eds.), *Proceedings of 2nd International Conference on Intelligent Computing and Applications*, Advances in Intelligent Systems and Computing 467, DOI 10.1007/978-981-10-1645-5_27

327

As the teeth disengage, the space between them increases. This creates a partial vacuum and hence the oil is sucked into the chamber through the suction port. When the space or chamber reaches its maximum volume, the suction will stop. As the space reduces with the meshing of the teeth, it forces the oil to discharge from the pump through the outlet port.

Earlier work on Gerotor pumps studied the various aspects of generation of geometrical profiles of rotors with emphasis on theory of envelope for a family of parametric curves and analysis of profile meshing [1, 2].

The theory of gearing is applied to generate the profile of the rotors by developing the inner rotor relationship equation [3]. Deviation function method is developed and applied to design the outer rotor and inner rotor [4].

2 Gerotor Modeling Phase

A cycloidal profile is considered for the rotors as in Fig. 1. Tooth profile σ_1 of outer rotor is a circular arc and the neighboring teeth of outer rotor are connected with a curve that forms the dedendum fillet of outer rotor. Profile σ_2 of inner rotor is conjugated to circular arc profile σ_1 of outer rotor. Introchoid ratio ($\lambda_1 = \lambda_2 = 1$) [2] of the epicycloid and hypocycloid form respectively, so that the epicycloid $\sigma_1^{(1)}$ and hypocycloid $\sigma_1^{(2)}$ form the profile of the outer rotor as in Fig. 2.

Each rotor has a common half tangent O_1E. The Gerotor is analysed for different span angle parameters to assess the effect of span angle design variations on gerotor motion and pumping characteristics.

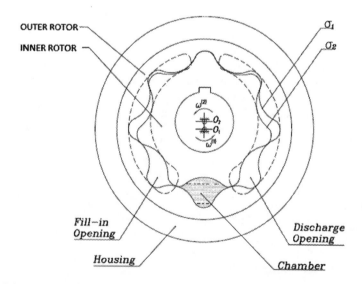

Fig. 1 Schematic of the cycloidal profiles

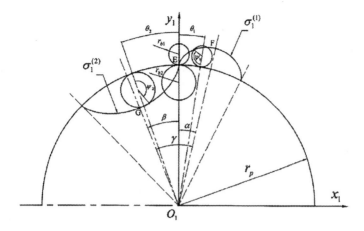

Fig. 2 Outer rotor profile

The tooth numbers N_1 and N_2 of pump rotor are chosen as $N_2 = N_1 - 1$. All rotor teeth are in mesh simultaneously and form closed chambers as the space between rotor profiles. Two opening in the housing are foreseen for the fill-in (inlet) and discharge (outlet) of chambers.

To generate the inner rotor, coordinate systems are to be created, as shown in Fig. 3. The following coordinate systems are considered: (i) movable coordinate systems $S_1(x_1, y_1)$ and $S_2(x_2, y_2)$, rigidly connected to outer rotor and inner rotor and (ii) fixed coordinate system $S_f(x_f, y_f)$ rigidly connected to the housing of the

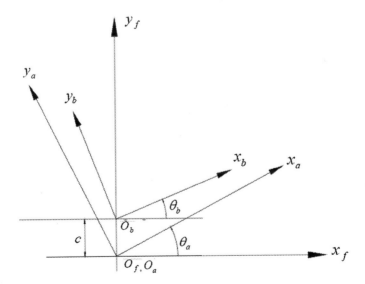

Fig. 3 Coordinate systems for gerotor

pump. Rotors of pump perform rotation with constant angular velocity in the same direction. The centrode in planar motion represent two circles of radii r_p (outer rotor) being in internal tangency.

Point E of tangency of centrode is the instantaneous center of rotation. The ratio of centrode radii or angular velocity ratio can be determined by:

$$m_{21} = \frac{\emptyset_2}{\emptyset_1} = \frac{N_1}{N_2} = \frac{\omega_1}{\omega_2} \tag{1}$$

where m_{21} is the angular velocity ratio and N_1 and N_2 are the tooth numbers of the outer and inner rotors. $\emptyset_1$ and $\emptyset_2$ represent the rotation angle of the outer and inner rotors respectively. ω_1 and ω_2 represent the angular velocity of the outer and inner rotors.

The position vectors of the addendum $\sigma_1^{(1)}$ and dedendum $\sigma_1^{(2)}$ are represented in coordinate S_1 as follows [3]:

$$r_1^{(1)} = \begin{bmatrix} r_{1x}^{(1)} \\ r_{1y}^{(1)} \\ 1 \end{bmatrix} = \begin{bmatrix} -r_{b1}\sin(\theta_1 + \psi_1) + (r_{b1} + r_p)\sin\theta_1 \\ -r_{b1}\cos(\theta_1 + \psi_1) + (r_{b1} + r_p)\cos\theta_1 \\ 1 \end{bmatrix} \tag{2}$$

$$r_1^{(2)} = \begin{bmatrix} r_{1x}^{(2)} \\ r_{1y}^{(2)} \\ 1 \end{bmatrix} = \begin{bmatrix} -r_{b2}\sin(\theta_2 - \psi_2) + (r_{b2} - r_p)\sin\theta_2 \\ -r_{b2}\cos(\theta_2 - \psi_2) + (-r_{b2} + r_p)\cos\theta_2 \\ 1 \end{bmatrix} \tag{3}$$

where r_{b1} and r_{b2} are the radii of the rolling circle, θ_1 and θ_2 are the angles of the base circle, ψ_1 and ψ_2 are the angles of the rolling circle for curves $\sigma_1^{(1)}$ and $\sigma_1^{(2)}$, respectively. r_p is the pitch radius of the outer rotor. r_{e1} and r_{e2} are changed because in this case ($\lambda_1 = \lambda_2 = 1$), so

$$r_{e1} = \lambda_1 r_{b1} = r_{b1}, \quad r_{e2} = \lambda_2 r_{b2} = r_{b2}$$

The following some of the parameters which are used to calculate the basic values:

$$\psi_1 = \frac{r_p}{r_{b1}}\theta_1 \tag{4}$$

$$\psi_2 = \frac{r_p}{r_{b2}}\theta_2 \tag{5}$$

$$C = r_{b1} + r_{b2} \tag{6}$$

$$r_p = cN_1 \tag{7}$$

$$\alpha = \frac{\pi r_{b1}}{r_p} \tag{8}$$

$$\beta = \frac{\pi r_{b2}}{r_p} \tag{9}$$

$$\gamma = \alpha + \beta = \frac{\pi}{N_1} \tag{10}$$

The coordinate transformation yields the equation of the inner rotor:

$$r_2^{(i)} = M_{21}(\emptyset_1) r_1^{(i)}, \quad i = 1, 2 \tag{11}$$

$\emptyset_1$ is the generalized parameter of motion and M_{21} is the coordinate transformation from $S_1(x_1, y_1)$ and $S_2(x_2, y_2)$.

$$M_{21}(\emptyset_1) = \begin{bmatrix} \cos(\emptyset_1 - \emptyset_2) & -\sin(\emptyset_1 - \emptyset_2) & -c\sin\emptyset_2 \\ \sin(\emptyset_1 - \emptyset_2) & \cos(\emptyset_1 - \emptyset_2) & -c\cos\emptyset_2 \\ 0 & 0 & 1 \end{bmatrix} \tag{12}$$

After transformation of outer rotor equation,

$$r_2^{(1)} = \begin{bmatrix} r_{2x}^{(1)} \\ r_{2y}^{(1)} \\ 1 \end{bmatrix}$$
$$= \begin{bmatrix} (r_{b1} + r_p)\sin(\theta_1 - \phi_1 + \phi_2) - r_{b1}\sin(\theta_1 - \phi_1 + \phi_2 + \psi_1) - c\sin\phi_2 \\ (r_{b1} + r_p)\cos(\theta_1 - \phi_1 + \phi_2) - r_{b1}\cos(\theta_1 - \phi_1 + \phi_2 + \psi_1) - c\cos\phi_2 \\ 1 \end{bmatrix} \tag{13}$$

$$r_2^{(2)} = \begin{bmatrix} r_{2x}^{(1)} \\ r_{2y}^{(1)} \\ 1 \end{bmatrix}$$
$$= \begin{bmatrix} (r_{b2} - r_p)\sin(\theta_2 + \phi_1 - \phi_2) - r_{b2}\sin(\theta_2 + \phi_1 - \phi_2 - \psi_2) - c\sin\phi_2 \\ (-r_{b2} + r_p)\cos(\theta_2 + \phi_1 - \phi_2) - r_{b2}\cos(\theta_2 + \phi_1 - \phi_2 + \psi_1) - c\cos\phi_2 \\ 1 \end{bmatrix} \tag{14}$$

The inner rotor position vectors of the addendum $\sigma_2^{(1)}$ and dedendum $\sigma_2^{(2)}$ are represented in coordinate S_2. From the theory of gearing of meshing [2]

$$f^{(i)}(\theta_i, \emptyset_1) = N_1 \times v_1^{(12)} \tag{15}$$

where N_1 is the normal to the surface of rotor and $v_1^{(12)}$ relative velocity (velocity of sliding). However, solving the conjugate curve requires consideration of the equation of meshing, which can be represented as follows:

$$f^{(i)}(\theta_i, \emptyset_1) = \left(\frac{\partial r_2^{(i)}}{\partial \theta_i} \times k \right) \frac{\partial r_2^{(i)}}{\partial \emptyset_i} = 0, \quad i = 1, 2 \tag{16}$$

k is the unit vector in the z direction.

$$f^{(2)} = cm_{21}[r_{b2} \sin(\theta_2 + \emptyset_1) - r_{b2} \sin(\theta_2 + \emptyset_1 - \psi_2)] - r_{b2}r_p(m_{21} - 1) \sin \psi_2$$
$$f^{(1)} = cm_{21}[r_{b1} \sin(\theta_1 - \emptyset_1) - r_{b1} \sin(\theta_1 - \emptyset_1 + \psi_1)] + r_{b1}r_p(m_{21} - 1) \sin \psi_1$$

2.1 Equations of Curvature

The sealing performance of the Gerotor is assessed using curvature analysis. The equation of curvature of the outer rotor profile is as follows [3]:

$$k_1^{(i)} = \frac{r_1^{(i)} n_1^i}{|r_1^{(i)}|}, \quad i = 1, 2 \tag{17}$$

where

$$r_1^{(i)} = \frac{\partial^2 r_1^{(i)}}{\partial \theta_i^2}, \quad n_1^{(i)} = \frac{r_1^{(i)} \times k}{|r_1^{(i)}|}, \quad i = 1, 2$$

$$r_1^{(1)} = \frac{\partial^2 r_1^{(1)}}{\partial \theta_i^2} = \begin{bmatrix} -(r_{b1} + r_p) \sin \theta_1 + (r_{b1} + r_p)^2 \sin(\theta_1 + \psi_1) \\ -(r_{b1} + r_p) \cos \theta_1 + (r_{b1} + r_p)^2 \cos(\theta_1 + \psi_1) \\ 0 \end{bmatrix} \tag{18}$$

$$r_1^{(2)} = \frac{\partial^2 r_1^{(2)}}{\partial \theta_i^2} = \begin{bmatrix} -(r_{b2} - r_p) \sin \theta_2 + (r_{b2} - r_p)^2 \sin(\theta_2 + \psi_2) \\ -(r_{b2} - r_p) \cos \theta_2 + (r_{b1} - r_p)^2 \cos(\theta_2 + \psi_2) \\ 0 \end{bmatrix} \tag{19}$$

$$k_1^{(1)} = \frac{r_{b1}^3 + r_{b1}^2(r_{b1} + r_p) - r_{b1}^2(2r_{b1} + r_p) \cos \psi_1}{(r_{b1} + r_p)(r_{b1}^2 - 2r_{b1}^2 \cos \psi_1 + r_{b1}^2)^{1.5}} \tag{20}$$

$$k_1^{(2)} = \frac{-r_{b2}^3 - r_{b2}^2(r_{b2} - r_p) + r_{b2}^2(2r_{b2} + r_p)\cos\psi_2}{(r_{b2} - r_p)(r_{b2}^2 - 2r_{b2}^2\cos\psi_2 + r_{b2}^2)^{1.5}} \quad (21)$$

Similarly the curvature of the inner rotor profile is computed.

3 Modeling of Gerotor

Based on the mathematical equations, the tooth profiles are formed by an epicy-cloids and hypocycloid. For comparative purposes, the rotors are assumed to be equivalent in thickness 10 mm and the outer diameter is taken as 70 mm. The analysis uses three distinct rotor span angles ($2\beta = 20°$, $30°$ and $40°$) that is, the angle between the two inflection points of a concave curve on the outer rotor are varied (Table 1) [5, 6].

As an example, Case 2 is taken. By substituting the different values in the equations, the following parameters are calculated.

$$c = 4.125 + 4.125 = 8.25 \text{ mm}$$

$$r_p = 8.25 \times 6 = 49.5 \text{mm}$$

Next, By varying θ_1 and θ_2 values based on α ($0°–15°$ in case 2) and β ($0°–15°$ in case 2) the various coordinates are obtained for outer rotor as shown in Table 2.

Table 1 Input data of gerotor design

Case 1	Case 2	Case 3
$2\beta = 20°$	$2\beta = 30°$	$2\beta = 40°$
$N_1 = 6$ and $N_2 = 5$	$N_1 = 6$ and $N_2 = 5$	$N_1 = 6$ and $N_2 = 5$
$r_{b1} = 5.087$ mm	$r_{b1} = 4.125$ mm	$r_{b1} = 3$ mm
$r_{b2} = 2.543$ mm	$r_{b2} = 4.125$ mm	$r_{b2} = 6$ mm

Table 2 Outer rotor coordinate points

N1 = 6			N2 = 5	Span angle = 30°	
$\alpha = 15°$			$\beta = 15°$		
Outer rotor	Addendum		Dedendum		
θ_1	$r_{1x}^{(1)}$	$r_{1y}^{(1)}$	θ_2	$r_{1x}^{(2)}$	$r_{1y}^{(2)}$
0	0	49.5	0	0	49.5
3	0.21	50.34	3	−0.13	48.77
6	1.57	52.47	6	−0.97	46.81
7.5	2.91	53.7	7.5	−1.83	45.53
9	4.71	54.83	9	−3.02	44.18
12	9.46	56.22	12	−6.36	41.63
15	14.93	55.79	15	−10.66	39.85

Table 3 Inner Rotor coordinate points

N1 = 6			N2 = 5	Span angle = 30°	
α = 15°				β = 15°	
Inner rotor	Addendum		Dedendum		
θ_1	$r_{2x}^{(1)}$	$r_{2y}^{(1)}$	θ_2	$r_{2x}^{(2)}$	$r_{2y}^{(2)}$
0	0	41.25	0	0	41.25
3	0.21	42.09	3	−0.13	40.52
6	1.57	44.22	6	−0.97	38.56
7.5	2.91	45.45	7.5	−1.83	37.28
9	4.71	46.58	9	−3.02	35.93
12	9.46	47.97	12	−6.36	33.38
15	14.93	47.54	15	−10.66	31.6

Then varying the θ_1 and θ_2 value based on α (0°–15°) and β (0°–15° in case 2), and by substituting these values in the equations, the various coordinates are obtained for inner rotor as shown in Table 3. The solid model of the Gerotor including outer and inner rotor is developed using SolidWorks (Fig. 4).

Finally same procedures are followed for other two cases. The design results for the three cases are shown in Fig. 5. For the case 1 $r_{b1} > r_{b2}$ and a 20° span angle; case 2 $r_{b1} = r_{b2}$ and a 30° span angle; and case 3 $r_{b1} < r_{b2}$, and a 40° span angle and area efficiency also calculated. The area efficiency is defined by a simple equation for evaluating pumping capacity.

$$A_\eta = \frac{A_{Out} - A_{in}}{A_{Out}} \times 100\%$$

Fig. 4 Rotor assembly

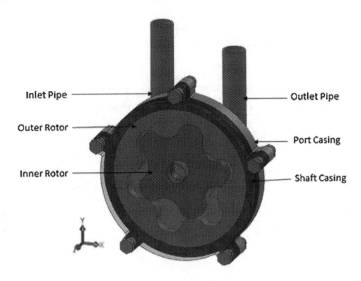

Fig. 5 Gerotor assembly

Table 4 Design calculation results

Parameter	Case 1	Case 2	Case 3
N_1	6	6	6
N_2	5	5	5
$\theta_{Span}^{Outer}(2\beta)$	20	30	40
r_{b1}	5.087	4.125	3
r_{b2}	2.543	4.125	6
A_{out}	7804.73 mm²	7804.82 mm²	7804.93 mm²
A_{in}	5609.21 mm²	5452.40 mm²	5258.05 mm²
A_η	28.13 %	30.14 %	32.63 %

where A_{Out} represents the cross-section area of the outer rotor and A_{in} represents the cross-section area of the inner rotor.

As Table 4. clearly shows, the area efficiency of case 3 is higher than that of the other cases, meaning that it produces greater fluid discharge capacity. Further CFD analysis and comparisons for three cases like outlet pressure, outlet flow rate and outlet flow velocity are carried out.

4 CFD Analysis

When the rotors rotate clockwise, the fluid passes through the inlet oil pipe into the suction oil area and the rotor volume gradually increases. In this suction stage, the Gerotor pump sucks the oil into fill in the rotor volume. Then, as the two rotors rotate continuously, the Gerotor pump moves the oil to the outlet area and the volume gradually decreases, creating a discharge stage in which the oil is moved to the outlet oil pipe [7].

The fluid dynamic inside the pump is illustrated using a design different span angle on 20°, 30° and 40° and the contour pressure diagram on span angle 20°, span angle 30° and span angle 40° respectively. The pressure levels are within the specific limits for all the span angle conditions but the pressure in the rotor with the span angle 40° is higher than other two span angle pressures [8, 9].

From the vector velocity diagram on span angle 20°, span angle 30° and span angle 40°, the velocity in the span angle 20° is more in the specific area compared to the other two span angle conditions and this will make a damage on the rotor. From the comparative results, velocity in the rotor with the span angle 40° is higher than other two span angle velocities and at same time the inner area velocity also within the specific limit (Figs. 6 and 7).

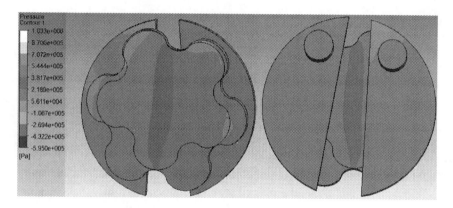

Fig. 6 Pressure control plot for span angle 20°

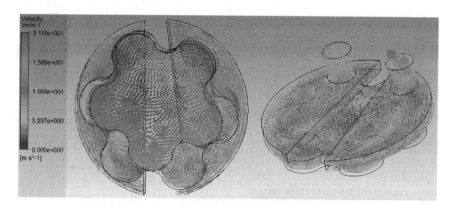

Fig. 7 Velocity contour plot for span angle 20°

Pressure in the discharge area may increase gradually at the beginning of the operation because the suction and discharge areas become unstable before the inner rotor finishes the two cycles. Hence, after two full cycles of rotations, the pressure may become periodic and stable on span angle 30° and 40° but span angle 20° takes nearly three full rotations after that only it became periodic and stable [10].

For the 30° span angle, the pressure of each point may be stable in the third cycle, at a pressure of nearly 0.533 MPa (Fig. 8).

Similarly whereas the simulations of the 20° and 40° design spans pressure 0.525 and 0.539 MPa respectively (Fig. 9).

The analysis of this flow velocity at outlet indicates that flow velocity may increase gradually at the beginning because of the unsteady condition from the suction to the discharge stage. After the Gerotor rotates one cycle, however, the

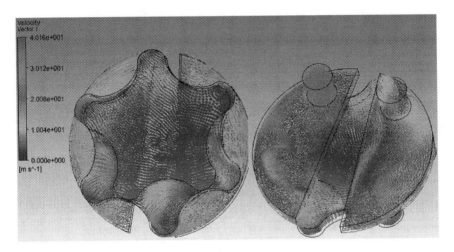

Fig. 8 Velocity contour plot for span angle 40°

Fig. 9 Comparison of outlet pressure

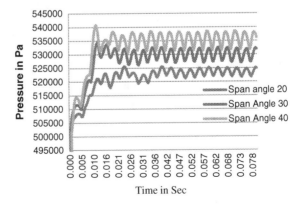

Fig. 10 Comparison of
outlet velocity

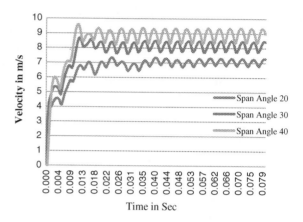

Fig. 11 Comparison of mass
flow rate

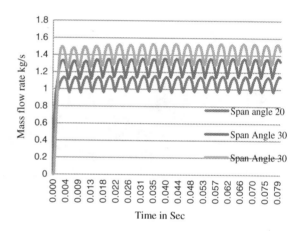

velocity oscillation becomes periodic because of the periodic variation in volume but span angle 20° takes two cycles (Fig. 10).

However, the output velocity in span angle 20°, with a highest value of approximately 7.32 m/s, is lower than that of span angle 30° and span angle 40°, whose corresponding highest values are 8.5 and 9.3 m/s (Fig. 11).

The equivalent relation holds true for the mass flow rate, which becomes periodic after one cycle of rotation the flow rate in span angle 20°, with a highest value of 1.15 kg/s, is lower than in span angle 30° and span angle 40°. The corresponding mass flow rates are 1.36 and 1.53 kg/s.

The area efficiency is increasing gradually for span angle 20°, 30° and 40°. Hence, the fluid analysis specifies that the higher span angle, the higher the area efficiency and the higher the output flow velocity, pressure and mass flow rate compared to other two span angle 20° and 30°.

Finally comparing all the span angle conditions, the highest values of output flow velocity, pressure and mass flow rate are obtained for span angle 40°. So the span angle 40° is a better design to compare to other two designs. But the increase in span angle further will not satisfy the geometrical relations for the Gerotor profile.

5 Conclusion

In this paper, three distinct rotor span angles for Gerotor are considered. Each of the rotor profile is formed by epicycloids and a hypocycloids ($\lambda_1 = \lambda_2 = 1$). The CFD analysis of the Gerotor pump has been carried out using ANSYS-CFX for three different span angles. The pump area efficiency can be increased if its span angle increases and the radii of the rolling circle can be changed accordingly. The pump outlet pressure, outlet velocity and mass flow rate are gradually increased for the higher value of the span angles.

References

1. Vecchiato, D., Demenego, A., Argyris, J., Litvin, F.L.: Geometry of a cycloidal pump. Comput. Methods Appl. Mech. Eng. **190**, 2309–2330 (2001)
2. Litvin, F.L., Demenego, A., Vecchiato, D.: Formation by branches of envelope to parametric families of surfaces and curves. Comput. Methods Appl. Mech. Eng. **190**, 4587–4608 (2001)
3. Hsieh, C.F., Hwang, Y.W.: Geometric design for a gerotor pump with high area efficiency. Trans. ASME J. Mech. Des. **129**, 1269–1277 (2007)
4. Hsieh, C.F.: Influence of gerotor performance in varied geometrical design parameters. Trans. ASME J. Mech. Des. **131**(12), 121008 (2009)
5. Gamez-Montero, P.J., Castilla, R., Khamashta, M., Codina, E.: Contact problems of a trochoidal-gear pump. Int. J. Mech. Sci. **48**(12), 1471–1480 (2006)
6. Suresh Kumar, M., Manonmani, K.: Computational fluid dynamics integrated development of gerotor pump inlet components for engine lubrication. J. Automob. Eng. **224**, 1555 (2010)
7. Tong, S.H., Yan, J., Yang, D.C.H.: Design of deviation-function based gerotors. Mech. Mach. Theory **44**(8), 1595–1606 (2009)
8. Suresh Kumar, M., Manonmani, K.: Numerical and experimental investigation of lubricating oil flow in a Gerotor pump. Int. J. Automotive Technol. **12**(6), 903 (2011)
9. Karthikeyan, N., Suresh Kumar, J., Ganesan, V.: Development of a gerotor oil pump. In: Proceedings of the 37th National & 4th International Conference on Fluid Mechanics and Fluid Power, IIT Madras, Chennai, India. FMFP10-TP-22 (2010)
10. Lorentz Fjellanger B.: CFD Analysis of a Pelton Turbine, Norwegian University of Science and Technology—Department of Energy and Process Engineering thesis EPT-M-2012-28 (2010)

Design of Area-Delay Efficient Parallel Adder

K.V. Ganesh and V. Malleswara Rao

Abstract This describes a parallel adder. It is related to a repetitive formulation for doing multi bit binary summation. This is operation is easier for the bits which does not require carry generation. Thus, the design achieves drastic speed over alternate operand conditions without any special speedup block or look-ahead schema. A complete execution is generated along with a detection unit. This accomplishes common and does not have any limitations of high fan outs. A high fan-in gate is required but this is not necessary for asynchronous logic and is connected by the transistors in parallel. Simulations have been performed using a DSCH and Micro Wind tool that validates the practicality and superiority of the proposed design over existing asynchronous adders.

Keywords Parallel adder (PADD) · Look-ahead schema · Fan out · DSCH · Micro wind

1 Introduction

Processor important operation is Binary Summation. Adder circuits have been designed for synchronous blocks even though there is strength of interest in clock less/asynchronous blocks [1]. Asynchronous designs do not acquire any of time. Therefore, they hold powerful and potential for logic design as they are risk free from different problems of clocked blocks. The logic flow in asynchronous circuits [2] is restrained by a request/ack, handshaking protocol to institute a pipeline in the absence of clocks. Observable handshaking designs for tiny elements, such as single bit adders, are expensive. Therefore, it is inherantly and efficiently managed

K.V. Ganesh (✉) · V. Malleswara Rao
Department of ECE, GITAM University,
Visakhapatnam, Andhra Pradesh, India
e-mail: kona.venkat13@gmail.com

V. Malleswara Rao
e-mail: mraoveera@yahoo.com

© Springer Science+Business Media Singapore 2017
P. Deiva Sundari et al. (eds.), *Proceedings of 2nd International Conference on Intelligent Computing and Applications*, Advances in Intelligent Systems and Computing 467, DOI 10.1007/978-981-10-1645-5_28

using double-rail carry propagation in adders. A valid double-rail carry output also generates acknowledgment from an adder block. Thus, asynchronous adders are either based on full double-rail encoding of all signals or pipelined operation using single-rail data encoding and double-rail carry representation for acknowledgments. While these constants add strength to circuit blocks, they introduce speed benefits of asynchronous adders. Therefore, a more healthy alternative resemble is good consideration that can address these problems. This presents an asynchronous parallel adder using the algorithm. The design of Parallel adder is simple and uses half-adders (HAs) along with mux's requiring minimum interconnections. Thus, it is suitable for Very Large Scale Integration execution. This design works with independent Carry chains. The execution in this paper is moderate it has a feedback from xor gate to generate a cyclic asynchronous adder. Cyclic circuits are more efficient than there acyclic blocks. Inputs are applied before the output get strengthen is known as wave pipelining. It manages automatic pipelining of the generated carry inputs separated by propagation and inertial delays of the gates in the circuit path. Single-rail pipelined blocks are different from double-rail blocks.

2 Background

There are numerous blocks of binary adders and we concentrate on asynchronous adders. The Self-Timed designs are nothing but industry standard designs. This type of adders are runs faster for data provided dynamically and early sensing can avoid worst case delay in synchronous circuits [3]. They are classified as.

2.1 Pipelined Adders Using Single-Rail Data Encoding

The asynchronous Request/Acknowledge handshake can be used to start the adder block as well as to establish the flow of carry generation signals. In most of the cases, a double-rail carry convention is used for internal bitwise flow of carry outputs. These double-rail signals can represent more than two logic values (invalid, 0, 1), and therefore can be used to propagate bit-level acknowledgment when a single-bit operation is completed. When all ack bits are high complete detection unit will sense. The carry-completion sensing adder blocks is an example of pipeline adders, which uses full adder (FA) functional designs, adapted for double-rail carry. A non-financially completion adder, It uses so-called different logic and early completion to select the number of delay lines for proper completion of response. However, the differed logic implementation is expensive due to high fan-in requirements.

2.2 Delay Insensitive Adders Using Dual-Rail Encoding

Delay insensitive (DI) adders are asynchronous adders that assign bundled constraints or DI operations [4]. They can operate correctly in presence of bounded but unknown gate and net delays. There are many variants of DI adders, such as ripple carry adder and carry look-ahead adder. These adders use double-rail encoding and are assumed to increase area. Though double-rail encoding doubles the net complexity, they can still be used to generate designs nearly as efficient as that of the single-rail variants using dynamic logic or nMOS designs. DIRCA uses 40 transistors whereas RCA uses only 28 transistors. Similar to CLA, the DICLA defines carry propagation and kill equations in terms of double-rail encoding. They do not connect carry signals as chain but in a hierarchical manner. Thus, they can strongly perform more when there is long carry chain in a tree. A further minimization is provided from the observation that double rail encoding logic can benefit from settling of either the 0 or 1 path. Double-rail logic does not wait for the both the paths to be executed. So it should speed up the CLA to send carry kill signals to any stage in the tree. This is developed and referred as DICLA with speedup circuitry (DICLASP).

3 Design of Pasta

In this section, theory and architecture of Parallel adder is presented. The adder first accepts two inputs and performs two half additions. Subsequently, it starts iterates using previously generated carry and sums to perform half-additions recursively until all carry bits are consumed and settled at zero value.

3.1 Architecture of PASTA

The design implementation of the adder is shown in Fig. 1. The sel input of muxs are used as initially it selects the operands and when sel = 1 used for carry paths. The feedback path from the HAs enables the multiple iterations to proceed till completion when all carry signals will get the values of zero [4].

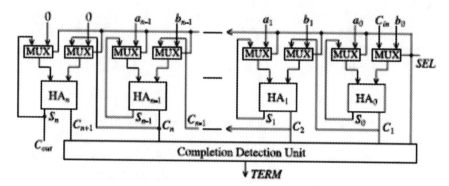

Fig. 1 General block diagram of PASTA

3.2 State Diagrams

In Fig. 2, 2 state diagrams are given for the initial phase and the iterative phase of the proposed design. Each state is represented by (C_{i+1}, S_i) pair where C_{i+1}, S_i is carryout and sum values, respectively, from the ith bit adder block. During the initial phase, the circuit works as a combinational half adder operating in normal mode. Due to the usage of half adders instead of full adders, state (11) cannot appear.

During the iterative phase (SEL = 1), the feedback path through mux block is activated. Till Completion of the recursion the carry transitions (C_i) are allowed. From the definition of normal mode designs, the present design cannot be considered as a normal mode circuit as the input–outputs will go through several transitions before producing the final output. Several transitions will take place, as shown in the state diagram. This is difference to cyclic sequential circuits where individual gate timings are used to differentiate single states.

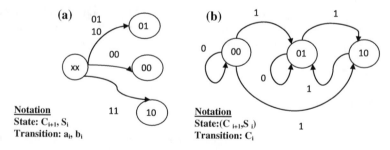

Fig. 2 State diagrams for PASTA. **a** Initial phase. **b** Iterative phase

3.3   Recursive Formula for Binary Addition

Let S_i^j and $C_{i+1}^{\ j}$ denote the sum and carry, respectively, for ith bit at the jth iteration. The initial condition ($j = 0$) is given as

$$S_i^0 = a_i^\wedge b_i.$$
$$C_{i+1}^0 = a_i \cdot b_i. \tag{1}$$

The jth iteration for the recursive addition is formulated by

$$S_i^j = S_i^j - 1^\wedge C_i^j - 1, \quad 0 \le I < n. \tag{2}$$

$$C_i + 1^j = S_i^j - 1^\wedge C_i^j - 1, \quad 0 \le I \le n. \tag{3}$$

When the below condition matches the recursion will get terminated at kth state:

$$C_n^k + C_{n-1}^k + \cdots + C_1^k = 0, \quad 0 \le k \le n. \tag{4}$$

4   Implementation

A CMOS implementation for the recursive circuit is shown in Fig. 3. For multiplexers and AND gates we have used DSCH implementations while for the XOR gate we have used the faster ten transistor implementation based on logic gate xor to match with the timing of AND gate [5]. The completion detection following (4) is neglected to get a enable TERM. This requires a large fan-in n-input NOR gate. Therefore, another nmos design is used. The resulting design is shown in Fig. 3d. Using the pseudo-nMOS design, all connections are parallel so high fan-in problems are avoided in complete detection unit. It acts as a load register when pmos design is connected to VDD, when Nmos are simultaneous it results the current drain in the circuit. In addition to the C_i s, the negative of SEL signal is also included for the TERM signal to guarantee that the completion cannot be turned on during the initial selection phase of the actual inputs. Hence, static current will flow for actual computation. VLSI layout has also been performed (Fig. 3e) for a standard cell environment using two metal layers. The layout occupies $270\ \lambda \times 130$

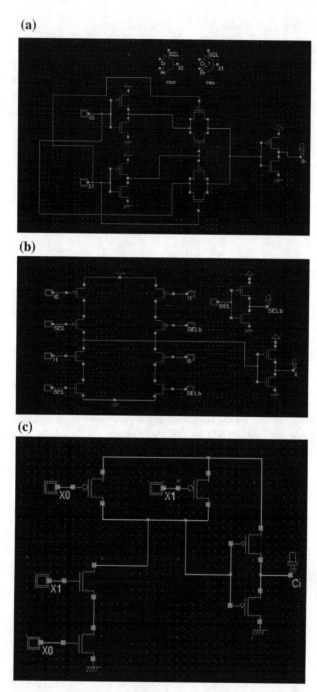

Fig. 3 a Single bit Sum module. **b** 2 × 1 Mux for the 1-bit adder. **c** Single bit Carry module. **d** Completion Signal Detection Circuit. **e** Layout

(d)

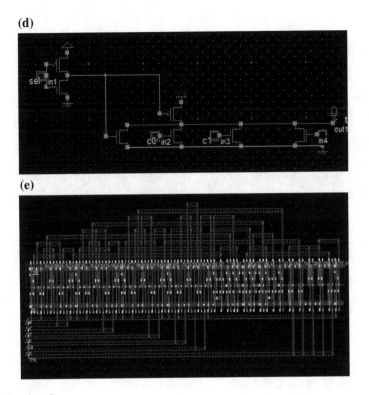

(e)

Fig. 3 (continued)

λ for 2-bit resulting in 1.123 Mλ2 area for 16-bit. The pull down transistors of the completion detection logic are included in the single-bit layout (the T terminal) while the pull-up transistor is additionally placed for the full 16-bit adder. It is nearly double the area required for RCA and is a little less than the most of the area efficient prefix tree adder.

5 Simulation Results

Initially, we show how the present design of PASTA can effectively perform binary addition for different voltage, time and frequencies as shown in Fig. 4.

(a)

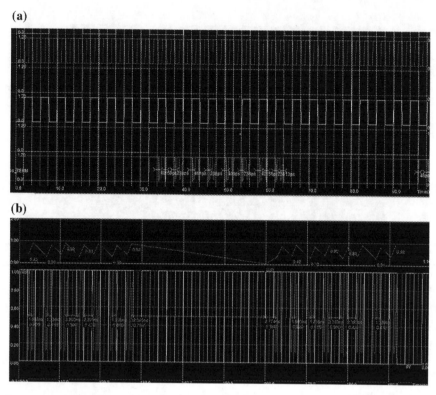

(b)

Fig. 4 **a** Voltage versus Time. **b** Frequency versus Time

6   Conclusion

This paper presents an expense execution of PASTA. The design acquires a very simple *16*-bit adder that is area efficient and fewer interconnections like Ripple Carry Adder. The circuit works in a way such that it achieves logarithmic average time speed over randomly generated values. The Detection unit of the Parallel adder is also consumes less area and delay as validated. Simulation results are also verified using DSCH and Micro Wind tools.

References

1. Geer, D.: Is it time for clockless chips? [Asynchronous processor chips]. IEEE Comput. **38**(3), 18–19 (2005)
2. Sparsø, J., Furber, S.: Principles of Asynchronous Circuit Design. Kluwer Academic, Boston (2001)

3. Choudhury, P., Sahoo, S., Chakraborty, M.: Implementation of basicarithmetic operations using cellular automaton. In: Proceedings of ICIT, pp. 79–80 (2008)
4. Rahman, M.Z., Kleeman, L.: A delay matched approach for the design of asynchronous sequential circuits. Department of Computer System and Technology, University of Malaya, Kuala Lumpur, Malaysia, Technical Report 05042013 (2013)
5. Riedel, M.D.: Cyclic combinational circuits. Ph.D. dissertation, Department of Computer Science, California Institute of Technology, Pasadena, CA, USA, May 2004

Author Biographies

K.V. Ganesh Received his B.E. degree in Electronics and Communication Engineering from Andhra university in the year 2009 and received M.Tech. degree in the year 2011 from JNT University, Kakinada. He is a Ph.D. scholar in GITAM Institute of Technology, GITAM University, Visakhapatnam, India. His research activities are related to Low Power VLSI Design.

DR. V. Malleswara Rao Received his B.E. degree in Electronics and Communication Engineering from Andhra University in the year 1985 and received M.E. degree in the year 1989 from Andhra University and completed his Ph.D. from J.N.T.U Kakinada, India and working in GITAM Institute of Technology, GITAM University, Visakhapatnam as Professor and H.O.D. He is a life member of AMIE. His research activities are related to Low Power VLSI Design, Microwave, Bio-Signal Processing.

Autonomous Quantum Reinforcement Learning for Robot Navigation

Arjun Mohan, Sudharsan Jayabalan and Archana Mohan

Abstract In order to achieve a safe and traffic free transportation from place to place it is mandatory that a vehicle should communicate with other vehicles autonomously to control their speed and movement. Fuzzy logic control architecture which is placed in the vehicle has been designed to take appropriate decision based on the various parameters inside the vehicle. Number of input and output parameters and rules of operation can be varied easily in the fuzzy logic toolbox in MATLAB. The output will be sent to the micro-controller using RS232 serial communication. Now the micro-controller will take the appropriate action to execute the command received. So in order to communicate with other vehicle we have used GPS receiver which will transmit the current position of the vehicle to nearby vehicle ranging within 30 or 100 m using Zigbee communication in a common transmission frequency. If there is any vehicles present in this range it will accept and communicate back by sending the corresponding GPS location of that vehicle and the path of its course. If it is found to be moving on the same direction or path a communication with the vehicle will be established and data will be transmitted among them to achieve the safe and smooth traffic travel to increase the vehicle efficiency. The same rules will apply if there is more than one vehicle is present in the transmitting range. All these transmitting data will be privacy protected. We have successfully controlled prototype model and found that architecture is working autonomously up to the expectation and provides efficient travel for the vehicle.

Keywords Matlab · Fuzzy logic · Arduino · Robot · GPS · Zigbee

A. Mohan (✉) · S. Jayabalan · A. Mohan
Department of Mechanical Engineering, CEG Campus, Anna University, Chennai, India
e-mail: arjunragavendh@gmail.com

S. Jayabalan
e-mail: sudharsan.jayabalan@gmail.com

A. Mohan
e-mail: archanamohan96@gmail.com

© Springer Science+Business Media Singapore 2017
P. Deiva Sundari et al. (eds.), *Proceedings of 2nd International Conference on Intelligent Computing and Applications*, Advances in Intelligent Systems and Computing 467, DOI 10.1007/978-981-10-1645-5_29

1 Introduction

Manual vehicle control system leads to increase in the traffic flow and at some times cause's accident due to unexpected circumstance and lack of proper response to control the vehicle. So in order to achieve safe and smooth traffic, autonomous vehicle control system should be incorporated in a vehicular system. Previously analog and mechanical transmission systems where used to control the vehicle. The advancement in digital technologies made changes in some of the measurement techniques like speedometer, fuel indicator. In high end vehicles we have digitally advanced facilities like anti-lock braking system (ABS), airbags, cruise control, climate control, automatic wiper control, light beam control, parking sensors and many more. All these control techniques provides comfort ability to driver and the travelers and save them when they are met with accidents. With the help of GPS system [1] can easily communicate to the control room where the accident has occurred to send a rescue team. But there is no efficient accident avoidance and traffic regulatory system in the vehicle which will provide free flow of traffic.

In this research work a collision avoidance system is designed using a fuzzy logic controller toolbox [2], which will provide appropriate decision based on the inputs and fuzzy rules that have been designed based on the expert system. Since the number of sensors and its purpose in a vehicle varies from model to model and the manufacturer. Moreover the system has to respond to the nature of the passenger this will be decided based on the interiors that vary from product to product. The expert system will be designed by the expert in the field and he will also suggest corresponding rule to be followed to take a decision. This is the major reason for choosing the fuzzy logic controller as it input condition and rules may vary.

In this proposed prototype the vehicle speed is controlled based on the passenger's condition. System will monitor the passenger's heart beat and temperature and will control the speed of the vehicle based on the output from fuzzy logic controller. Since it follows the lane system when it needs to speed up the vehicle it will move to the fast speed lane which will be in the right hand side of the driver. When it needs to decrease the speed of the vehicle for the passengers comfort the vehicle will move from the top right lane to the left lane which is comparably slower than the right side lane. The main reason for consider the passengers safety is that, as the vehicle is going to be driven automatically it should know what is the health condition of the passengers, for example a pregnant lady, old aged persons, and heart patient needs to be driven slowly whereas for the youngster and in case of emergency situation the speed may be increased.

In order to achieve proper lane switchover and to avoid accidents we have designed autonomous path detection system [3–9]. In this a Zigbee based wireless communication system to communicate with the adjacent vehicles. This is achieved with the help of GPS and autonomous vehicle control system which will be explained in detail later.

2   Design of Hardware Architecture

In this section various hardware modules and their architecture and interconnections that has been used for the prototype has been explained in short. In this prototype two vehicular robot where designed. Each robot consists of the following modules. The robot chase consists of a metal plane/sheet which act as the housing for the entire circuit. The chase is driven with the help of 2-DC motor which is connected at the back wheel to act as back wheel drive vehicle. The front 2 wheel acts as a support to move the vehicle. These DC motors are controlled using PWM to achieve different speed variations. L293D based H-bridge control circuit helps to achieve the forward and reverse directions.

A laptop/pc has been used exclusively for the purpose of running Matlab application software. Software architecture will be explained in detail in next section. Atmega-328 micro-controller which runs on the Arduino platform is used for the controlling of entire modules and it is interfaced with PC using USB cable to communicate and transfer results from and to the Matlab application. Zigbee transceiver which works on 2.4 Ghz is connected to the Arduino which is used to communicate between vehicles as what are all the controls and how each vehicle should move (Fig. 1).

Prototype also has GPS modem which gives us the information where the vehicle is traveling. This GPS modem is used for testing when it was used outside the room and the distance between both the vehicles are assumed to be about 4-m. The accuracy of the GPS can be still increased once if we put a high end GPS module.

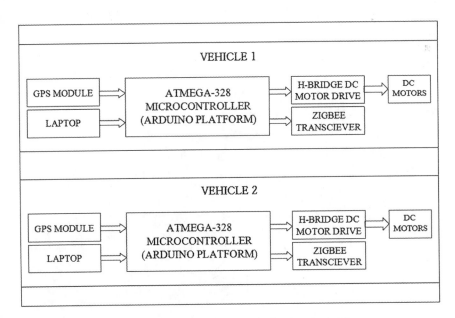

Fig. 1 Block diagram for the hardware architecture of robot 1 and robot 2

When the prototype is working in indoor for testing purpose we have assumed to represent the lanes as columns which consist of number of small segments which is of 1 sq.ft. area in size. The robot identifies in which segment and in which lane it is in by calculating the total distance moved, i.e. there will be three columns which are named as L1, L2, and L3 and the segments in each lane are identified by L11, L12 … L1n. etc. similarly for the lane 2/column 2 it is identified by L21, L22…L2n. etc. So using this assumption of lanes we have used to test the working in a room. You can also use this assumption of prototype to loco mote inside a factory, in mining areas, even inside a small lab etc.

3 Software Architecture and Experimental Flow

Using Matlab Software GUI toolbox, an GUI interface is created which will pass the necessary sensor values to Fuzzy logic toolbox which acts as a major decision taking tool and the output from the fuzzy inference system is passed to Arduino for which a serial RS232 object have been initialized based on the comport in which it is connected. In Arduino software embedded-C programming is coded and embedded in ATMEGA-328 (Arduino) micro-controller. The Arduino is designed to work in such a way that it read the command from the matlab and identifies what instruction it is to be carried out next (Fig. 2).

Fig. 2 Software architecture block diagram

(a)   **(b)**

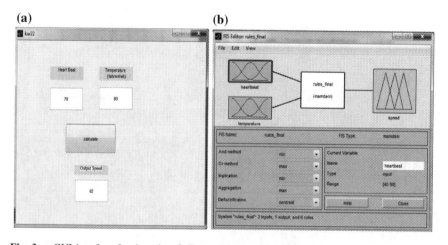

Fig. 3 **a** GUI interface for the robot. **b** Fuzzy inference system for the robot

Fig. 4 **a** Robot 1 should move in the path where it finds obstacle robot 2 through which it communicates using Zigbee and asks the robot 2 to slow down. **b** Robot 1 takes left turn after initiating proper instructions to robot 2 using Zigbee. **c** Robot 1 reached *lane 2* from *lane 1* to overtake robot 2. At this point robot 2 slows down and takes the minimum speed of lane 1. **d** Robot 1 overtakes the robot 2 and reaches its destination lane

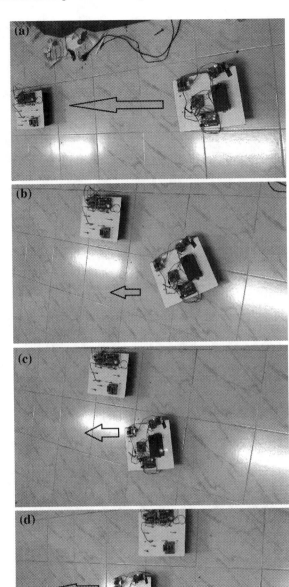

Suppose if the vehicle should slow down first it will check in which lane it is with the help of GPS or based on the distance and direction it has moved. Identify what is the maximum and minimum speed in that lane. If the recommended speed is well in that lane the vehicle will run in that speed in the same lane. Suppose if it needs to have a lane change it will first identify what are all the vehicles in that lane by establishing wireless communication with the help of Zigbee. Then it will identify the exact location of the vehicle and insist to slow down till the vehicle as changed the lane. Once the lane has been changed the vehicle will move in the prescribed speed as recommended in that new lane (Figs. 3 and 4).

All these working have been learned and all the conditions and decisions have been stored in a log file in the PC. So the next time when any similar situation arises the output can be directly invoked from Matlab without the need for any fuzzy inference system.

4 Conclusion

Prototype model has worked successfully by communicating with the other vehicle and taking necessary conditions based on both the vehicles. The log file gave us a lot of useful information which helps us to identify the way in which the system had operated. For the testing purpose the prototype had worked as expected and control decisions were taken as programmed.

5 Future Enhancements

With the help of the data from the log file the necessary rules of the fuzzy inference system can be varied. Currently next prototype of this research work is going on to make an android application which will eradicate the need for Zigbee and GPS modem by Smart phones or android based tablets. This will communicate much faster using 3G data communication.

References

1. Shair, S., Chandler, J.H., Gonzalez-Villela, V.J., et al.: The use of aerial images and GPS for mobile robot waypoint navigation. IEEE ASME Trans. Mechatron. **13**(6), 692–699 (2008)
2. Vengerov, D., Bambos, N., Berenji, H.: A fuzzy reinforcement learning approach to power control in wireless transmitters. IEEE Trans. Syst. Man Cybern. B Cybern. **35**(4), 768–778 (2005)
3. Sutton, R., Barto, A.G.: Reinforcement Learning: An Introduction. MIT Press, Cambridge (1998)

4. Kaelbling, L.P., Littman, M.L., Moore, A.W.: Reinforcement learning: a survey. J. Artif. Intell. Res. **4**, 237–287 (1996)
5. Watkins, C., Dayan, P.: Q-learning. Mach. Learn. **8**(3/4), 279–292 (1992)
6. Dong, D., Chen, C., Chen, Z.: Quantum reinforcement learning. In: Proceedings of First International Conference on Natural Computation, Lecture Notes in Computer Science, vol. 3611, pp. 686–689 (2005)
7. Dong, D., Chen, C., Li, H., Tarn, T.J.: Quantum reinforcement learning. IEEE Trans. Syst. Man Cybern. B Cybern. **38**(5), 1207–1220 (2008)
8. Chen, C., Dong, D., Chen, Z.: Quantum computation for action selection using reinforcement learning. Int. J. Quantum Inf. **4**(6), 1071–1083 (2006)
9. Dong, D., Chen, C.: Quantum-inspired reinforcement learning for decision-making of Markovian state transition. Presented at the 2010 International Conference on Intelligent Systems and Knowledge Engineering, Hangzhou, China, 15–16 Nov 2010

A Novel Approach to Detect XSS Attacks in Real Time Online Social Networking

K.G. Maheswari and R. Anita

Abstract In a real time network scenario, online social networks (OSN) play a significant role in connecting and growing business and technology. This technology gathers much information and share secret data among network. This attitude gives the intruders to exploit the original information. This paper contributes for major widely spread and critical OSN vulnerability. XSS, popularly noted as a one-click attack or session riding attack which is the most common malicious attack that exploits the trust that a site has in a user's browser. Proposed method is a XSS attack detection mechanism for the client side. It focuses on the matching of parameters and values present in a suspected request with a form's input fields and values that are being displayed on a webpage. Next to address concerns of offensive content over Internet. The proposed method analyzes the social network features integrating with textual features improving the accuracy of automatic detection of XSS.

Keywords Online social network · OSN · XSS · Intruder · SQL injection

1 Introduction

Today many latest technologies growing very fast. Online social networks (OSN) are one of the connecting technologies to serve for people working in web. Considerable big amount of secret data of social network users are shared in common without getting prior permission from original authors which carries

K.G. Maheswari (✉)
Department of MCA, Institute of Road and Transport Technology,
Anna University, Erode, Tamil Nadu, India
e-mail: kgmaheswari@gmail.com

R. Anita
Department of EEE, Institute of Road and Transport Technology,
Anna University, Erode, Tamil Nadu, India
e-mail: anita_irtt@yahoo.co.in

© Springer Science+Business Media Singapore 2017
P. Deiva Sundari et al. (eds.), *Proceedings of 2nd International Conference on Intelligent Computing and Applications*, Advances in Intelligent Systems and Computing 467, DOI 10.1007/978-981-10-1645-5_30

personal data opinions, views, audio etc. Hence, these online social networks have personal related big data of single or group of people. Because of the large data usage, there is large amount of threads also available in web. Therefore, there is a heavy need for protecting the data in online networking to implement certain authenticate methods. This paper discusses web intrusion detection in OSN which involves SQL injection attack, XSS and cyber-bullying attack. The proposed method analyzes the social network features integrating with textual features improving the accuracy of automatic detection of cyber-bullying. The recent generation have many social connections through networking account and each day working towards it.

The intruders easily corrupt the original data by taking correct information from your social networking profile [1] and posts, then plan their attacks based on the interests and likes of your profile. This personal data attack can be considered as "social engineering" and it creates individual data loss, more difficult to recognize. A user often has trusted "friends" with whom she is willing to share her private information and whom she trusts to perform computations accurately and protect the privacy of her information. The intruders may sometimes hack their friend's account and send several spam messages.

The purpose of social network is user should share their personal information to their friends circle. When the information available free in the social network, can be steal by other unauthorized person for a misusage. Once the information is downloaded, that can be misused widely such as creation of fake profile and the information can be sold to other socially disturbed websites. This kind of data staling activities in online social networks even leads the life to death [2, 3]. With these social network characteristics and the more assertiveness of attackers, privacy and security issues in social networks has become a critical issue in the cyber world. The predominant attacks in online social networks are as follows.

1.1 Cross-Site Scripting Attack

One of the worst thread run in web browser is Cross-site scripting (XSS). It is the premier web intrusion detection among various vulnerabilities of web. In this method victim's browser is feeds by JavaScript The intruder will write dynamic HTML code to make the web browser send victims cookies to attacker's server.

If the user access malicious websites, then XSS Worm is spread itself automatically among users. It uses the technique of spreading the malware to other users and steals the users personal information [4]. Since many users are connected through the platform like social networks, the XSS worm will spread very easily. To spread the worm, the intruders will select source node, that will start the scattering of the malware. Once the source node record into the social networking website, malware will take control of the browser and command it to perform some tasks.

1.2 SQL Attack

SQL Injections are one of the most common and easiest techniques adopted by the attackers, to attack the web server, data server and sometimes the network [5]. This category of attack is conducted by spammers for unauthorized web application access, breaking the role based accessibility and violating the integrity of the data storage. SQL attack poses a serious threat to the security of web applications.

2 Related Works

There are many work related to cross site scripting (XSS) in online social networking Fonseca et al. [6] proposed a methodology and a prototype tool to evaluate web application security mechanisms. The methodology is based on the idea that injecting realistic vulnerabilities in a web application and attacking them automatically can be used to support the assessment of existing security mechanisms and tools in custom setup scenarios. To provide true to life results, the proposed vulnerability and attack injection methodology relies on the study of a large number of vulnerabilities in real web applications.

Another work is proposed by Atoum et al. [7] based on hybrid technique for SQL injection attacks detection and prevention. In this paper is a reliable and accurate hybrid technique that secure systems from being exploited by SQL injection attacks is created. SQL injection is a type of attacks used to gain, manipulate, or delete information in any data-driven system whether this system is online or offline and whether this system is a web or non-web-based. It is distinguished by the multiplicity of its performing methods, so defence techniques could not detect or prevent such attacks.

This work talking about the tools proposed by Kumar and Indu [8]. Here Detection and Prevention of SQL Injection attack is discussed in a elaborate manner. SQL injection is a technique where the attacker injects an input in the query in order to change the structure of the query intended by the programmer and gaining the access of the database which results modification or deletion of the user's data. In this work vulnerable attacks such as SQL injection, XSS attack has discussed and the techniques to avoid those attacks are also been discussed.

3 Online Social Networks

Social network is an act of communication between users in many-to-many cardinality relationship for the purposes of sharing information with the network or subsets thereof [9]. A user often has trusted "friends" with whom she is willing to share her private information and whom she trusts to perform computations accurately and protect the privacy of her information.

Large amount of their private information is shared in their social network space. For example, the photos are freely available in social network, the unauthorized users can easily access the photos of others and download it. Once the photo is downloaded, that image can be misused widely such as creation of fake profile and the photo can be sold to other nuisance websites [10].

4 Proposed System

In order to preserve the privacy OSN users and to protect their data the proposed work contributes for the detection and prevention of two major widely spread and critical OSN vulnerabilities: SQL injection attack, XSS attack and cyber-bullying attack. Proposed method is concentrating on a new method a for classifying the web intrusion attack in which the query based request and detects SQL injection attack. The reverse proxy technique is used to sanitize the user's inputs that may transform into a database attack. This technique employs the use of a filter program, which redirects the user's input to the proxy server for sanitization, before it is sent to the application server. Next to address concerns of offensive content over Internet, automatic monitoring of cyber-bullying is proposed. The proposed method analyzes the social network features integrating with textual features improving the accuracy of automatic detection of cyber-bullying. The bag of words algorithm is used for the purpose of text-analysis. Social network features are analyzed with the computation of ego networks (Fig. 1).

The decision manager monitors the overall process of this proposed system. It acts as an intermediate between modules. One of the major responsibilities of the decision manager is its functionality in providing an expert decision support by interacting with rule manager module. By receiving a response from the feature selection module, it decides whether the request has been sent to the inference system or not. It takes the decisions based upon the information provided from the rule manager. It instructs the new facts and rules into the rule manager at the time of expert decision making process.

The precision of SQL Queries are checked by a Query Detector. It is a simple tool to detect the malicious request from user at the web server. The requests of users are validated before forwarding it to the web server for further execution and processing. The inputs from users are collected and stores in the repository, the Data collection subsystem does the above operation.

The malicious script embedded in the web application is detected by the Rule Manager. Before executing on the web server the HTML input are preprocessed or cleaned by the rule manager. All the invalid and unwanted tags from the user input are cleaned by this sanitization process. After the encoding process starts. Then encodes the remaining input into simple text thus preventing the execution of any malicious script. The rules for making decisions on offensive contents and injection

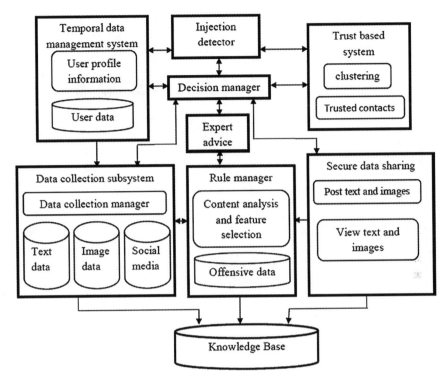

Fig. 1 System architecture diagram

attacks are farmed in Rulebase. This rule base is created by storing rules obtained from domain experts and also using the rules obtained from the training performed on the data set.

This module entirely focuses on the prevention of cyber-bullying. The users shall post inoffensive text and images. They are prohibited to share illegal contents. Analysis of the social network features integrating with textual features improving the accuracy of automatic detection of cyber-bullying. The bag of words algorithm is used for the purpose of text-analysis. Social network features are analyzed with the computation of ego networks.

It consists of user information. The temporal data manager is responsible for handling user profile and the user database respectively.

5 Result and Discussion

The proposed approach gives comparatively high percentage of detection rate for SQL and XSS attacks in the online social networks. The usage of unauthorized information by the trusted person is analyzed and experimented by the

Fig. 2 Comparison chart representing percentage of security impact over various social networking features

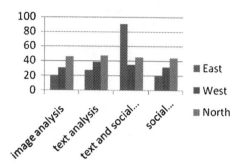

proposed algorithms which mainly focus on the detection of vulnerability in online attacks.

The comparative analyses of various social networks are given in Fig. 2, which represents the percentage of security impact of various social networking features. The chart depicts the image, text analysis in the information sharing between users by different constraints in social networking analysis.

6 Conclusion

In this work, an intelligent system for detection and prevention of most critical vulnerabilities in social network namely, cross site scripting attack been developed with the application of Reverse Proxy. In this technique a filter program is used to redirect the user's input to the proxy server before it is sent to the application server. The proposed system also enhances the automatic monitoring of XSS by incorporating social relationship features along with text analysis. It also provides effective follow-up strategy by detecting threats and signalling alarm to the trusted contacts.

References

1. Van Royen, K., Poels, K., Daelemans, W., Vandebosch, H.: Automatic monitoring of cyberbullying on social networking sites: from technological feasibility to desirability. Telemat. Inform. (2014)
2. Al Mazari, A.: Cyber-bullying taxonomies: definition, forms, consequences and mitigation strategies. In: IEEE International Conference on Computer Science and Information Technology (2013)
3. Rybnicek, M., Poisel, R., Tjoa, S.: Facebook watchdog: a research agenda for detecting online grooming and bullying activities. In: IEEE International Conference on Systems, Man, and Cybernetics (2013)
4. Shar, L.K., Tan, H.B.K.: Predicting SQL injection and cross site scripting vulnerabilities through mining input sanitization patterns. Inf. Softw. Technol. **55**(10), 1767–1780 (2013)

5. Kopecký, K.: Cyberbullying and other risks of internet communication focused on university students. In: Elsevier International Conference (2013)
6. Fonseca, J., Vieira, M., Madeira, H.: Evaluation of web security mechanisms using vulnerability and attack injection. IEEE Trans. Dependable Secure Comput. **11**(5), 440–453 (2014)
7. Atoum, J.O., Qaralleh, A.J.: A hybrid technique for SQL injection attacks detection and prevention. Int. J. Database Manag. Syst. (IJDMS) **6**(1), 21–28 (2014)
8. Kumar, M., Indu, L.: Detection and prevention of SQL injection attack. Int. J. Comput. Sci. Inf. Technol. (IJCSIT) **5**, 374–377 (2014)
9. Lee, I., Jeong, S., Yeo, S., Moon, J.: A novel method for SQL injection attack detection based on removing SQL query attribute values. J. Math. Comput. Modell. **55**(1–2), 56–56 (2011)
10. Zhou, C.V., Leckie, C., Karunasekera, S.: A survey of coordinated attacks and collaborative intrusion detection. Comput. Secur. **29**(1), 124–140 (2010)

Enhancement of Power Quality Problem in Grid Using Custom Power Devices

P. Sivaperumal and Subhransu Sekhar Dash

Abstract Power quality is of vital concern in the consumer and utility side problems like voltage sag, voltage swell and harmonics can cause malfunctioning of the end user equipment. To mitigating these problems a working model of an Interline Dynamic Voltage Restorer (IDVR), which maintains voltage of two or more feeders simultaneously is presented here. In this research instead of using conventional topology for the inverter, a Voltage Fed-Switched Coupled Inductor Inverter (VF-SCII) is proposed. A controller of Artificial Neural Network (ANN) strategy with PI controller is implemented for generating PWM signals. The model is simulated using MATLAB-SIMULINK and the results are discussed in detail.

Keywords Interline dynamic voltage restorer · Voltage fed-switched coupled inductor inverter · Artificial neural networks

1 Introduction

The electronics devices is most sensitive to the disturbance and are less tolerant to common power quality problems like voltage sag, voltage swell and harmonics. Custom power device like Dynamic Voltage Restorer (DVR), D-STATCOM and UPQC provides a solution to these problems. The DVR is most effective device to mitigating the voltage sag and swell on the lines by injecting voltage using series transformer. It can compensate amplitude, frequency and phase angle in the distribution side [1].

The Power Electronics is based on equipment of IDVR is to provide technical solution for these new operating challenges being presented today. The new

P. Sivaperumal · S.S. Dash (✉)
Department of EEE, SRM University, Kattankulathur, Tamil Nadu, India
e-mail: munu_dash_2k@yahoo.com

P. Sivaperumal
e-mail: Sivaperumal.me@gmail.com

© Springer Science+Business Media Singapore 2017   367
P. Deiva Sundari et al. (eds.), *Proceedings of 2nd International Conference on Intelligent Computing and Applications*, Advances in Intelligent Systems and Computing 467, DOI 10.1007/978-981-10-1645-5_31

technologies allow for improved transmission system operation with minimal infrastructure investment and environment impact. A better solution at the utility end user for mitigating these problems in two or more feeder is to use custom power device like Interline Dynamic Voltage Restorer (IDVR). The IDVR of Voltage Source Inverter (VSI) and Current Source Inverter (CSI) with series transformer is investigated and the results are compared. IDVR is used to provide compensation for voltage sag and swell [2, 3].

An IDVR consists of two or more DVRs are connected in a Grid system, which can prevent the sensitive load voltage disturbances in the feeders. When one of DVRs of IDVR system acts as power flow controller, while another one of DVRs acts for compensating voltage sag, voltage swell and harmonics in distribution side. It is replenishment of energy take place from one feeder to another one [4].

In this paper, IDVR using Voltage Fed-Switched Coupled Inductor Inverter (VF-SCII) with series injecting transformer, a combined control technique with Artificial Neural Networks (ANN) & PI controller is proposed to overcome PQ problems at distribution side [5]. It is provide a sinusoidal and regulate to load voltage in either linear or non-linear load. The VF-SCII has been improving boost function and its implementation gives an extra shoot through state on the inverter bridge. VFSCII have same principle of Z-Source Inverter (ZSI), but has higher boost ratio and less active switch voltage stress at same voltage gain, so far sustain minimum voltage and current stress at certain operation point [6, 7].

If utilities an inverter has wide range of buck/boost function, to regulate voltage across sensitive load for variation in the supply voltage due to any faults occurs on the system. The simulation for VF-SCII based IDVR with combined control strategies of an ANN and PI controller is done with MATLAB-SIMULINK. The performance of IDVR system is to maintain the voltage and its fast dynamic response.

2 Interline Dynamic Voltage Restorer

The schematic diagram of Interline Dynamic Voltage Restorer as shown in Fig. 1 is proposed to overcome PQ issues. The IDVR, which is connected between the two feeders are feeder1 and feeder 2. The IDVR consists of two or more DVRs, when anyone of the feeder is subjected to disturbances either due to fault occurring on a line or due to sudden changes in load, PQ problems like voltage sag, voltage swell with injection of harmonics on that feeder. In this case, the DVR of corresponding feeder acts in compensating mode while other DVRs of the healthy feeder act in power flow control mode.

A module of IDVR, which allows real and reactive power to be either supplied or absorbed its operating normal condition. If faults occur on the system, the DVR's compensate them during the depth of anticipated sag. The IDVR has capability of real power flow transfer between feeder lines in addition to execution of independent over reactive power compensation in each feeder. This makes it possible to

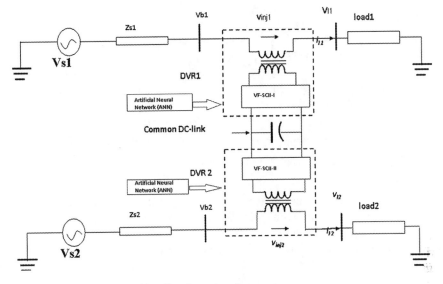

Fig. 1 Schematic diagram of interline dynamic voltage restorer

equalize the real and reactive power flow in each feeder, it can also transfer real power demand from loaded and unloaded to compensate against resistive line voltage drops by using voltage fed switched couple inductor inverter and ANN controller, effectiveness of the compensation is increased along with the response to faults.

The injecting voltage with series transformer is given by,

$$V_{in} = \left[\frac{N_P}{N_S}\right] V_C \tag{1}$$

Equation (1) is used to injecting the voltage in line through linear transformer and provides the sinusoidal voltage to the load.

3 Voltage Fed Switched Coupled Inductor Inverter

Voltage Fed-Switched Coupled Inductor Inverter is used to carry out the voltage Buck-Boost function can be carried out. It can also extend the duty cycle range of conventional inverter and can lower active device voltage stress a without affecting voltage gain. The VF-SCII is capable of extra shoot through state and has half number of passive components is compared to impedance source inverter. The model of VFSCII as shown in Fig. 2.

It has two operating states, namely shoot through and Non-shoot through state. In the case of conventional inverters, because of the switching delays of switches,

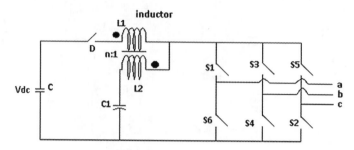

Fig. 2 Voltage fed-switched coupled inductor inverter

the upper and lower switches on the same leg S_1S_6, S_3S_4 and S_2S_5 may be ON at same time which can damage inverter. This happen during zero crossing of voltage or current of a phase When inverter bridge shoot through happens, front diode is automatically turn off and inductor 2, L_2 gets charged. In Non-shoot through state, the diode 'D' is on, and inductor 1, L_1 gets discharged. In these whole switching period, while energy is a coupled inductor get balanced, total turn current product $(n * 1)$ keep constant. During the switching period, each inductor current is discontinuous. The total flux is constant for the inductor and so continuous unit inductor current, due to jump switching period [6].

$$i_L = \frac{n i_{L1} + i_{L2}}{n+1} \tag{2}$$

The output voltage, V_L of non-shoot through state is given by

$$V_L = \left(\frac{1}{1 - nD_0}\right) V_{abc} \tag{3}$$

where, D_0-Shoot through duty cycle.

The boost ratio β of the inverter network is given by,

$$\beta = \left(\frac{1}{1 - nD_0}\right). \tag{4}$$

4 Control Technique

The VF-SCII based IDVR using combined control strategies of ANN with PI controller is shown in Fig. 3. The input signal phase voltage, V_{abc} transformation to PQ theory as direct-quadrature axis, V_{dqo} is aided with reference signal and this is given to ANN controller.

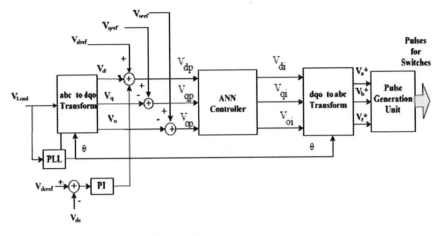

Fig. 3 Controller of ANN with PI controller

This signal trained by weight, W_{ij} and Biasis, Bij from each input signal as like V_{dref}, V_{qref}, V_{oref}, with comparing PI controller.

4.1 Artificial Neural Networks

An Artificial Neural Networks control strategy used consists of three neuron layers are input layer, hidden layer and output layer. The input layer is taken from load signal in power line as direct-quadrature axis voltage, V_{dqo} transmits the signal to hidden layer. An output layer is continuously learning process, is to be providing output signal. The output signal as like V_{di}, V_{qi}, and V_{0i} are transformation to phase voltage V^*_{abc} through generate the PWM signal to gate pulse as VF-SCII.

5 Simulation and Result

The schematic diagram of IDVR system is simulated using MATLAB-SIMULINK as shown in Fig. 4. The IDVR module is consists of two DVRs; each DVR is connected between the linear transformer and common DC link (Capacitor). The capacitance has higher range of value carried out, it is provides DC voltage to inverter. However, the capability of energy storage that usually consists of capacitor in DVR is limited and both DVRs operate simultaneously. The feeder 1 and feeder 2, input sources as 13 * $\sqrt{2}$ kV and using transformer is delta-star in transmission line. Hence, three phase fault is given to the system is generated by is it in form of voltage sag and harmonics. They also added external sources across the line is generated by swell for any one of the feeder, consider by feeder 1.

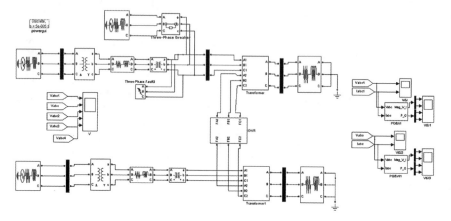

Fig. 4 Simulink diagram of IDVR systems

The inverter voltage V_{abc} is the injecting through linear transformer in distribution line and it's mitigating the voltage sag, voltage swell and eliminated by harmonic content. The desired output voltage is computing the simulation across sensitive load. The output signal which is used a control of ANN with PI controller as shown in Fig. 5. The Artificial Neural Network is comparing PI signal, V_{dc} and V_L, load voltage or actual voltage with set value or reference voltage.to computes a desired signal of the system. The error between the load voltage and set value is given to input signal for generating PWM signal which is generates the pulse for the inverter gate pulse.

DVR is regulating sinusoidal voltage across linear or non-linear load for the operating normal condition and its fast dynamic performance of these systems as shown Fig. 8. The VFSCII diode, mutual inductance (inductor) and capacitance is connected across the bridge is shown in Fig. 5. It can operate both function of Buck/Boost operation and low harmonics. When front diode, D is on condition, it inductor 1 gets discharged for the non-shoot through the switching period, each inductor current is discontinuous in nature.

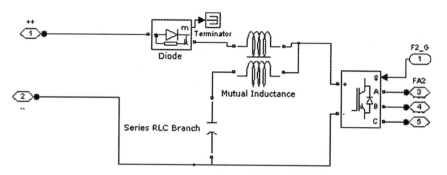

Fig. 5 VFSCII simulink diagram

Figures 6 and 7 shows that input source voltage of Feeder 1 and Feeder 2, normally, feeder 2 is not affected because its healthy one and Feeder 1 is affected, due to three phase faults (phase to ground), resulting in sag formation between range of 0.45–0.7 s and similarly additionally adding the external source, voltage swell is created range between 0.7 and 0.9 s in distribution line. At these instant, both DVRs are operated simultaneously, to mitigating voltage sag and voltage swell along with eliminated by harmonics.

Figure 8 shows that injection output voltage has compensation of sag/swell is using method of IDVR based VF-SCII is combined ANN and PI control technique is used and the results are verified. A feeder 2 output voltages, is not affected, it's due to disturbance in feeder 1 and system can be observed in Fig. 9.

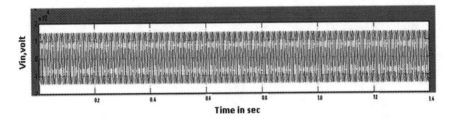

Fig. 6 Feeder 2 source voltage

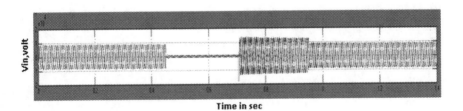

Fig. 7 Feeder 1 without injection voltage

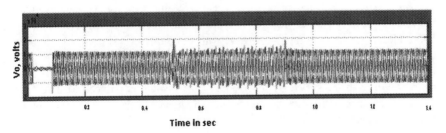

Fig. 8 Feeder 1 with injection output voltage

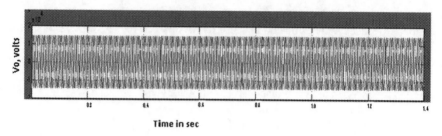

Fig. 9 Feeder 2 output voltage

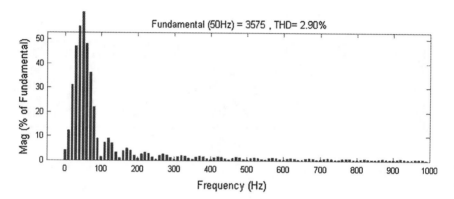

Fig. 10 FFT analysis of harmonic content

From Fig. 10 show that harmonics content, its reduced THD value of the IDVR based VF-SCII has 2.90 % in order to 3.57 kHz. It has finding results using FFT analysis.

6 Conclusion

The proposed VF-SCII based Interline Dynamic Voltage Restorer is simulated and the results show that it can effectively compensate voltage sag, voltage swell and harmonics. The control strategy of ANN with PI controller has been used in these systems to improve performance, its fast dynamic response and to provide low harmonic of the voltage restorer. VFSCII has higher boost ratio and lesser active device is used in the IDVR system. Hence, THD value is reduced to 2.90 % of harmonic content in order to that of 3.57 kHz frequency. This model can be applied real time implementation in three phase system to solve Power Quality issues in distribution networks for improving PQ problems.

References

1. Mahinda Vilathgamuwa, D.: A novel technique to compensate voltage sags in multiline distribution system—the interline dynamic voltage restorer. IEEE Trans. Ind. Electron. **53**(5), 1603–1611 (2006)
2. Usha Rani, P.: Voltage swell compensation in an interline dynamic voltage restorer. JSIR **73**, 29–32 (2014)
3. Nittala, R., Parimi, A.M., Rao, K.U.: Comparing the performance of IDVR for mitigating voltage sag and harmonics with VSI and CSI as its building blocks. In: IEEE International Conference, SPICES (2015)
4. Kanitpanyacharoean, W., Premrudeepreechacharn, S.: Power quality problem classification using wavelet transformation and artificial neural networks. In: IEEE PES Power System Conference and Exposition, vol. 2, pp. 1496–1501, 10–13 Oct 2004
5. Kinhal, V.G., Agarwal, P., Gupta, H.O.: Performance investigation of neural-network-based unified power-quality conditioner. IEEE Trans. Power Deliv. **26**(1), 431–437 (2011)
6. Lei, Q., Fang, Z., Peng, M.S.: Switched-coupled-inductor inverter. IEEE Trans. Power Deliv. **13**, 5280–5287 (2013)
7. Vilathgamuwa, M., Perera, A.A.D.R., Choi, S.S.: Performance improvement of the dynamic voltage restorer with closed-loop load voltage and current-mode control. IEEE Trans. Power Electron. **17**(5), 824–834 (2002)

Genetic Algorithm Based Placement and Sizing of Distributed Generators

R. Gayathri and P.S. Mayurappriyan

Abstract This paper reveals the impact of Distribution Generator (DG) present in electrical force circulation systems by taking IEEE 14 bus system as proposed frameworks. The investigation carried out is to inspect the impact on the general framework misfortunes and voltage profile. The point behind this study is to acquire the optimal area and entrance level of the new DG unit with a specific end goal to diminish the misfortunes and to upgrade the voltage profile. Genetic Algorithm is one the imperative stream of science and currently it is considered as most sweltering examination zone. Proficient techniques and advances are profoundly required to lessen the estimations and perform the operations in exact way. The force framework is extremely a recondite subject so that there is a need of ideal arrangements with which the framework gets to be upgraded and be prudent by taking care of complex issue. There are numerous advantages to introduce a DG in the framework to do complex count in arriving the size and the arrangement of DG. The estimation of ideal size and area of DG for a disseminated framework is the essential motivation behind this paper.

Keywords Distributed generation · Genetic algorithm · Voltage profile

1 Introduction

There is a basic progressive expansion of circled period resources which is upheld through change in power time headways and new environment regulation. Most of the offices and office structures have stand-by diesel generators as an emergency power hotspot for utilization amid power outages. Consequently an appropriated time structure has been used as a standby power hotspot for fundamental business. Then again, the diesel generators are beneficial. On the other hand, scattered time systems, similar to vitality segments, scaled down scale turbines, biomass, wind

R. Gayathri (✉) · P.S. Mayurappriyan
Department of EEE, KCG College of Technology, Chennai, India
e-mail: gaya3sbrcdm@gmail.com

© Springer Science+Business Media Singapore 2017
P. Deiva Sundari et al. (eds.), *Proceedings of 2nd International Conference on Intelligent Computing and Applications*, Advances in Intelligent Systems and Computing 467, DOI 10.1007/978-981-10-1645-5_32

energy, hydel power or solar power are ecological benevolent are thought to be the piece of answer for dealing with the growing interest of electricity and characteristic regulations on account of green-house gas release [1–3]. Right away 500 kW level circled DG systems are exceedingly used, due to development upgrades in little generators, power devices and stockpiling devices. Beneficial clean fossil force developments, renewable essentialness headways are growingly used for new passed on period structures. These DGs are joined with a standalone interconnected system, and have an extensive measure of focal points, as actually very much arranged electric time, extended reliability/security, high power quality, load organization, fuel versatility, uninterruptible organization, expense assets, and expandability.

1.1 Generalized Reduced Gradient

This improvement technique is used as a piece of the non-straight constraint issues, as it changes the obliged issue to an unconstrained one using direct substitution. It fits in with a gathering of various technique called the decreased slant systems. The idea behind this method is to fabricate backup structure that contains the auxiliaries of the limit concerning every variable. Moreover it utilizes sham variable to address the goals thusly and the count has novel variables (key) and fake variables (unnecessary). At the end of first cycle, figuring changes the estimations of the starting guess, as the improvement system goes on the variable qualities are overhauled after each accentuation, until the count accomplishes a tasteful worth or the bumble accomplishes its pre-described breaking point.

A drawback of this framework is the need of limit subordinate and its estimation which may be, in a couple issues, difficult to be found or registered. In like manner the issue may have far reaching number of variables which adds to the issue inconvenience in light of the considerable estimation of the backup structure [4].

1.2 Genetic Algorithm

This improvement technique was persuaded from the headway theory proposed by Charles Darwin (1809–1882) which communicates the eminent thought "survival of the fittest". The count codes the possible responses for the issue in sort of equal length bit strings, each piece string known as chromosomes or individuals and the bits confining the strings known as qualities. By then, the count structures mating pool which consists of each string applies three chairmen remembering the final objective to redesign the limit. These directors are decision, half breed and change [5–7].

1.2.1 Selection

This executive is used by the count to pick the best individual or chromosome in the mating pool in perspective of wellbeing qualities to outline the initial people. The framework used is a stochastic uniform decision due to its ability to give zero pre-mien and minimum spread in the determination of the general population encircling the mating pool [5–7].

1.2.2 Elitism

Elitism is considered as one of the determination techniques and its used to ensure the survival of the "super individuals" (individual with the most huge health regard). The picked individuals are called tip top youths; the point from the method is to keep the loss of potential courses of action, as the decision schedules may arrange this game plan from the mating pool. Additionally, the general population go to the front line direct.

1.2.3 Crossover

This overseer is used by the count to enhance the health estimation of the general population. Here, two self-assertively picked individual (people) exchanges some of their qualities together to get two new two individual known as successors. The framework used is cross breed probability as the head will be associated on 80 % of the general population in the present people. This strategy was picked because of its commonness among the other half and half technique as it promises the transportation of the best attributes in the present period to next one.

1.2.4 Mutation

This manager keeps the computation from being gotten in close-by minima as it associates in the recovering of the lost trait as a result of the half and half director. It moreover aides in the examination of the interest space more varying qualities to the masses. The system used as a piece of this investigation was the change probability in which the figuring flips particular number of bits of the picked individual to be changed in the purpose of growing the individual wellbeing. The probability worth used was 0.2 (i.e. 20 % of the general population will be changed). The value was low also with half and half extent in order to keep the misuse/examination equality [8, 9].

The GA won't run endless as it has various stopping criteria among which, it should convey particular number of periods (peoples), or to continue running for a couple of preset time, or the qualification in the health quality does not change for a couple of preset number of times.

2 Evaluation of Benefits

The course of action of records cleared up the particular favourable circumstances with respect to voltage and structure hardships [6, 7]. The voltage profile improvement index (VPII) and line mishap decline document (LLRI) detailed in the accompanying areas.

2.1 Voltage Profile Improvement Index (VPII)

The extent of the sum of all the voltage at all the vehicles when DG is joined with the system to the aggregate of all the voltage at every vehicle when structure is running without DG is portrayed as VPII. The measuring segment can in like manner be exhibited if we pick the largeness of the vehicles for a structure as from comparison (1).

$$\text{VPII} = \frac{\text{Voltage profile with DG}}{\text{Voltage profile without DG}} \tag{1}$$

where, the voltage profile of the structure with DG and without DG are with the same weights at the different weight transports. The general expression for VP is given by (2) and (3) as takes after:

$$\text{VP} = \sum_{i=1}^{M} V_i L_i K_i \tag{2}$$

with

$$\sum_{i=1}^{N} K_i = 1 \tag{3}$$

From comparisons (2) and (3), the voltage degree, load and measuring variables are VP, L i and k, N is the total number of weight transports in the allotment system. As described, the expression for VP allows to assess and aggregates the voltage levels at which loads are being supplied at the distinctive weight transports.

2.2 Line Loss Reduction Index (LLRI)

In the midst of the foundation of DG, it is vital to recollect the line disaster diminishment list LLRI in light of the way that it is the principle thought of DG. It identifies with the line adversities and subsequently for better result, it is should be minimum however much as could sensibly be normal. For the most part line setbacks are decreased when DG are joined with a course system. Then again, for dependent upon the examinations and regions of DG units, it is possible to have an addition in adversity at high passage levels. The proposed LLRI is portrayed as the extent of total line mishaps in the system with DG to the total line hardships in the structure without DG and is imparted as,

$$\text{LLRI} = \frac{\text{Line loss reduction index with DG}}{\text{Line loss reduction index without DG}} \tag{4}$$

where $LL_{w/DG}$ is the aggregate line misfortunes in the framework with the business of DG and is given as,

$$\text{LL}_{\text{W/DG}} = \sum_{i=1}^{M} R_i I_a i D_i \tag{5}$$

where I_a, i is the per-unit line current, R_i is the line resistance for line i (pu/km), Di is the scattering line length (km), and M is the amount of lines in the allocation structure. In this way, $LL_{wo/DG}$ is given as,

$$\text{LL}_{\text{WO/DG}} = \sum_{i=1}^{M} R_i I_a i D_i \tag{6}$$

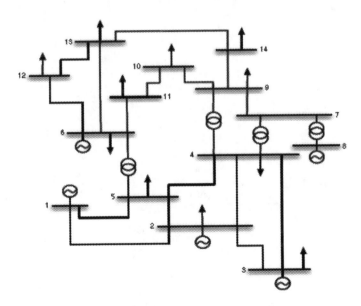

Fig. 1 Structure of IEEE 14 bus system

where I_a is the per-unit line current in appropriation line i without DG. The heaps at the diverse burden transports are assumed to be the same for both the cases i.e. with and without DG (Fig. 1).

3 Case Study

IEEE 14 transport test structure as represented in Fig. 2 is decided for examination. Tables 1 and 2 compress the line data and the vehicle data for given test system. The data in the tables are used as a piece of the MATLAB weight stream program. The contextual analysis fuses the going with steps:

Step 1: Read the line and transport data.
Step 2: Run the stack stream and store the results.
Step 3: Run the Genetic computation program in MATLAB apparatus compartment.

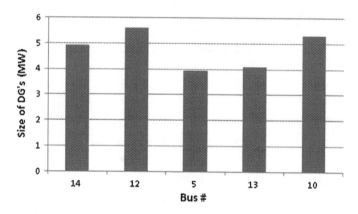

Fig. 2 Size of DGs on optimal placement

Table 1 Investigation results

Optimal placement	Size (MW)	Real power loss without DG (MW)	Real power loss with DG (MW)	Loss reduction (%)
14	4.92	16.80	15.78	6.071428571
12	5.61		14.83	11.72619048
5	3.96		14.08	16.19047619
13	4.10		13.82	17.73809524
10	5.33		12.56	25.23809524

Table 2 Summary of voltage profile

Load bus #	Without DG's	With DG's
4	0.9821	0.999
5	0.9711	1.012
7	0.9601	0.9981
9	0.9671	0.9871
10	0.9439	0.998
11	0.9567	1.032
12	0.9761	0.9987
13	0.9971	1.001
14	0.9881	1.031

Step 4: Connect the DG to distinctive transports and redesign the vehicle parameters.
Step 5: Run the stack stream and register the estimation of target limit.
Step 6: Repeat the strides 4 and 5 until the estimation of target gathering is opened up.

4 Result

The accompanying charts demonstrated, the examples acquired from the outcomes because of the examination that was connected on both frameworks under study utilizing the hereditary calculation. The ideal areas of burden transports are 14, 12, 5, 13 and 10 individually. The measure of the DGs and misfortunes are appeared in Tables 1 and 2. Table 1 demonstrates the genuine force misfortune diminishment with ideal area and size of DGs. Figure 2 speaks to the span of DGs on ideal arrangement.

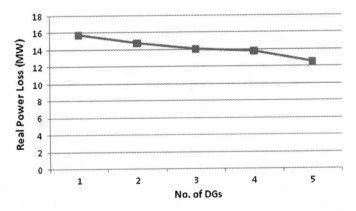

Fig. 3 Reduction of real power loss (MW) with number of DGs

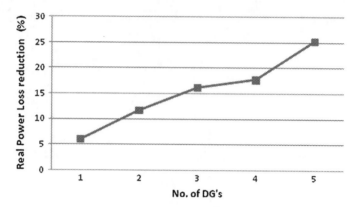

Fig. 4 Impact on real power loss reduction (%) with respect to number of DGs

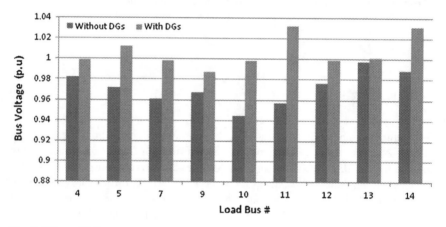

Fig. 5 Effect of DGs on voltage profile

As outlined in Fig. 3, decrease in force misfortune with 5 quantities of the DGs is from 16 MW to 12 MW. Figure 4 clarifies the change in genuine force misfortune lessening concerning the quantity of DGs, up to 25 % for 5 number of DGs. Voltage profile of all the heap transports with and without DGs are compressed in Table 2 and the voltage profile has been enhanced around 1.05 %. Figure 5 demonstrates the voltage level in all the heap transports with and without DGs. The ideal areas and size of the DGs derived from the definite examination are given beneath for the accepted IEEE 14 Bus System:

- The ideal areas for the heap transports are 14, 12, 5, 13 and 10 individually.
- The size of the DGs for the heap transports 14, 12, 5, 13 and 10 are 4.92, 5.6, 3.96, 4.10, 5.3 MW individually.
- The genuine force misfortune for the heap transports with DGs is 16.80 MW.
- The genuine force misfortune for the heap transports 14, 12, 5, 13 and 10 loss without DGs are 15.78, 14.83, 14.08, 13. 82, 12.56 MW.

5 Conclusion

In this paper, a diagram and key issues of diverse exploration ponders for ideal position of DG is displayed. DG ought to be put in the ideal area keeping in mind the end goal to give most extreme conservative, specialized and natural advantages. The ideal position of DG issue is entirely unpredictable. The investigation is finished by utilizing computerized reasoning one which is GA. In this investigation, the GA has been turned out to be more successful than ordinary improvement strategies. The examination has demonstrated that there are three arrangements of transports which are:

- Set one: Implementing of the DG unit to those transports will prompt consistent diminishing in the general framework misfortunes.
- Set two: executing of the DG unit to those transports will prompt ceaseless increment in the general framework misfortunes.
- Set three: actualizing of DG unit will prompt a "fluctuating" conduct which implies that the general misfortunes worth will rely on upon the entrance level (DG unit size).

References

1. Milligan, M.R., Graham, M.S.: An enumerated probabilistic simulation technique and case study: integrating wind power into utility production cost models. In: National Renewable Energy Laboratory for Wind Energy Program (1996)
2. Hoff, T., Shugar, D.S.: The value of grid support photovoltaic's in reducing distribution system losses. IEEE Trans. Energy Convers. **10**, 569–576 (1995)
3. Caire, R., Retiere, N., Martino, N., Andrieu, N., Hadjsaid, N.: Impact assessment of LV distributed generation on MV distribution network. In: IEEE Power Engineering Society Summer Meeting, Paper no. 02SM152 (2002)
4. Pike, R.W.: Optimization for Engineering Systems. Van No strand Reinhold Company (1986). ISBN 978 0442275815
5. Sivanandam, S.N., Deepa, S.N.: Introduction to Genetic Algorithms. Springer, Berlin (2008). ISBN 978-3-540-73189-4
6. Haupt, R.L., Haupt, S.E.: Practical Genetic Algorithms, 2nd edn. Wiley Interscience, Hoboken (2004). ISBN 0-471-45565-2
7. Melanie, M.: An Introduction to Genetic Algorithms. A Bradford Book, 5 edn. The MIT Press, Cambridge (1999). ISBN 0-262-63185-7
8. Acharya, N., Mahat, P., Mithulananthan, N.: An analytical approach for DG allocation in primary distribution network. Int. J. Electr. Power Energy Syst. **28**(10), 669–746 (2006)
9. Zhu, D., Broadwater, R.P., Tam, K., Seguin, R., Asgeirsson, H.: Impact of DG placement on reliability and efficiency with time-varying loads. IEEE Trans. Power Syst. **21**(1), 419–427 (2006)

Power Wastage Audit and Recommendation of Conservation Measures at University Library

N. Neelakandan, K. Sujan, Priyanka Kumari, Pooja Kumari,
Alok Kumar Mishra and L. Ramesh

Abstract Energy plays a important role in all day to day activities, especially those that are energy intensive. An Energy audit plays a major role in reduction of new power generation. It can help us to determine the energy wasting deficiencies in homes, firms, factories, industries and can show exactly how to address these problems. A detailed study to establish and investigate optimal utilization of lighting for specific department has been carried out in this work. The energy audit of University library has been executed with formulated procedure and proposed recommendation. The suggested implementation can improve the energy efficiency of library and thereby reducing the energy wastage.

Keywords Consumption · Energy audit · Illumination · Lighting · Lux

1 Introduction

Energy is critical, directly or indirectly, in the entire process of evolution, growth and survival of all living beings. Power availability plays a major role in economic development of the country. Energy is high priced in the today's world. India ranks third in the world total energy consumption [1]. All India installed capacity of electric power generating power stations is 278,734 MW till 30th October 2015 [2]. The detail break up share of different type of generating stations is follow: Hydro power plants—15.2 %, thermal power plants—69.7 %, nuclear power plants—2.1 %, renewable energy source—13.1 %. India's electricity generation touched the

N. Neelakandan (✉) · K. Sujan · P. Kumari · P. Kumari · L. Ramesh
EEE Department, Dr. M.G.R Educational and Research Institute (University),
Chennai, India
e-mail: mgrvision10mw@drmgrdu.ac.in

L. Ramesh
e-mail: rameshlekshmana@gmail.com

A.K. Mishra
Power Research and Development Consultants Pvt. Ltd, Bangalore, India

© Springer Science+Business Media Singapore 2017
P. Deiva Sundari et al. (eds.), *Proceedings of 2nd International Conference
on Intelligent Computing and Applications*, Advances in Intelligent Systems
and Computing 467, DOI 10.1007/978-981-10-1645-5_33

1 trillion units mark during 2014–15 this is for the first time in the history, showing a growth of 8.4 % over the previous year. Since 1991–95, the compounded annual growth rate of electricity generation has been around 5–6.6 %. The contribution of thermal sector was significant i.e. 20,830 MW (92 % of the total) but still 11 laces thousand people have no electricity. Where in an indication of growing appetite for electricity in a country like India with huge population, there has always been an appetite for electricity. According to Central Electricity Authority, the country's power usage were pitched up to 1010 kilowatt-hour (kWh) in 2014-15, compared with 957 kWh in 2013–14 and 914.41 kWh in 2012–13, according to the Central Electricity Authority (CEA), India's apex power sector planning body. Conservation of energy means reduction in energy consumption with less usage of power resources.

There are different ways of meeting generation and demand Generation of Power, Saving of Energy. We are going with Energy Audit for saving energy rather than generating power. We spend most of our time in buildings—homes, schools, offices, stores and libraries. But most people hardly notice details about the buildings, such as how they are designed, how many lights and fans are there in the building, whether they are used efficiently or not, how they are maintained. The details have shown a strong effect on how comfortable a building is and how much it costs to operate. Energy Audit provides the platform to analyze the building, determine wastage and to provide suitable recommendation. An Energy audit is used to determine pattern of energy use; and to enhance energy efficiency in the system. A study say that energy auditing and conservation in India can minimize the operation and production cost of Rs. 1750 Crore per year as a continuation of the statistics of this data the industrial sector and save installation equivalent to 5200 MV.

2 Audit Review

Different authors presented review papers related to library which are coming in the next section. This paper [3] shows the prefatory study of energy audit that has been done in UMP library. Energy audit has been done within 7 days time frame. In this study, energy consumption data has been recorded over a period of time by installing a data logger at the library main switch board. Programmed capacitor bank is installed at main switch board to improve the power factor, reduce the current consumption, voltage drop and electrical energy losses for actual implementation of energy efficiency. From the analysis, it was found that the level of energy efficiency in building is inversely proportionate to the energy losses that occur; the higher the loss, the lower the efficiency. The main objective of this audit is to propose energy efficiency method to reduce energy consumption with techniques and calculations.

The paper [4] presents the Energy Audit that has been done in the State Library of Tasmania, on 15 May 2008. The building energy consumption and GHG inventory for the period May 2007 to April 2008 was calculated from the energy usage and information provided by Department of Education Monthly Electricity Consumption of the library along with electricity index is studied. The HVAC system is controlled by a microprocessor based BMS system. High efficiency hot water storage heaters has been installed which replaced old normal hot water storage heaters. Occupancy sensors are installed in the toilets and Multi stack heat pumps replaced old boiler system. This paper [5] presents on a model and formulation for library load management on electricity consumption. Lighting is an essential service in all the libraries. Innovation and continuous improvement in the field of lighting has to be carried out on basic necessity basis. Energy plays a vital role in the socio-economic development and human welfare of a country but in present scenario most of the common people are waiting energy by different methods. To create energy awareness to the general public, Dr. M.G.R Educational and Research Institute, University Chennai has taken initiative called 'MGR Vision 10 MW' under leadership of Dr. L. Ramesh to save 10 MW in 10 years. In this pilot audit study-1 was conducted by the team of members in the year 2015 at various residential houses and industries. The outcomes of the studies are published in indexed conferences and Journals [6–11]. This work is the pilot study-2 of Vision 10 MW. This study covers the waste audit analysis and recommendation for the University library in the first stage. In the second stage, detailed power audit is conducted in the university library and the suitable recommendation for savings of energy is recommended for implementation.

3 Data Observation

Data observation and calculations are discussed in this section. This work executed with our own procedure to start and end the audit process. The basic structure of the data analysis and observation were taken on the basis of the reference taken from this paper

- Layout sketch for existing system
- Theoretical lux level is calculated by using the formula
- Practical lux level is calculated by using lux meter depending on the room index value
- Live practical lux value is calculated
- The constraints for designing optimal lighting is identified
- Present issues related to lighting system is found
- Design of proposed optimal design of lighting, fan, flux etc. is made
- Proposed layout lux level is studied
- Investment cost analysis
- Saving benefit analysis (Fig. 1)

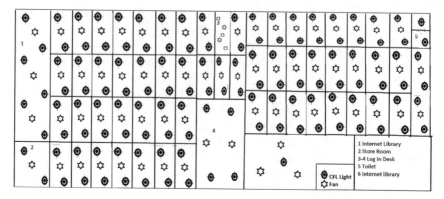

Fig. 1 Layout sketch of the library

Fig. 2 Daily unit
consumption on the university
library

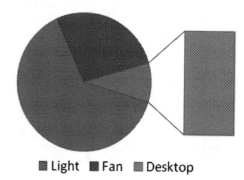

■ Light ■ Fan ▨ Desktop

Power Utilization Analysis: The usage of power across the library is presented in
the pi-chart. It shows that the lighting occupies 75 % of the library (Fig. 2). The
representation of total number of equipments used is shown in Fig. 3 .

Fig. 3 Number of equipment
fitted

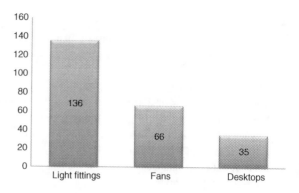

3.1　Index Level Calculation

Theoretical lux level and Room Index calculation [12–14] for the given area is calculated using the specified formula given by Installed lux = total no. of fitting * no. of lamps per fittings * L.D.L output of each lamp

Lux Level − LHS side of the library = 1074.34 lx/m^2
Lux Level − RHS side of the library = 1075.26 lx/m^2
Lux level − Central portion of the library = 1279.56 lx/m^2

Room Index: Room Index is given by Room Index = length * width/Mounting height * (length + width).

But here for this library we need to find the room index of both LHS and RHS side as width of both side is different.

Room index of LHS side = 4
Room index of RHS side = 4.9
Room index of central room = 1.17
Room index of main central = 2.37 (Table 1)

ILER (Installed Load Efficiency Ratio) Calculation: The procedure to calculate ILER is presented below in steps

Step 1:- Measure the floor area of the interior
Step 2:- Calculate the room index
Step 3:- Determine total circuits watts of installation
Step 4:- Calculate watt per m^2
Step 5:- Ascertain the average maintained illuminate
Step 6:- Divide Step 5 by 4 to calculate actual lux/watt/m^2
Step 7:- Obtain target lux/w/m^2
Step 8:- Calculate ILER (Divide step 6 by 7)

The calculated value of ILER in all the area are presented below

Lux level required = 53
ILER = 0.28 (LHS)
ILER = 0.28 (RHS)
ILER = 0.37 (Counter room)
ILER = 0.5 (Central room) (Table 2)

Table 1 Room index calculation	Room index value	No. of measurements
	Below one	9
	Between (1–2)	16
	Between (2–3)	25
	Above 3	36

Table 2 ILER assessment

ILER	Assessment
0.75 or over	Satisfactory or good
0.51–0.74	Review suggested
0.5 or less	Urgent action required

Table 3 Power wastage sample data

Timing	Day 1	Day 2	Day 3	Day 4
10–12 A.M.	10 fittings, 3 fans	10 fittings, 4 fans	14 fittings, 4 fans	8 fittings
12–2 P.M.	25 fittings, 14 fans, 2 desktop (sleep mode)	17 fittings, 20 fans, 3 desktop (sleep mode)	15 fittings, 18 fans (3 desktop sleep mode)	20 fittings, 15 fans
2–4 P.M.	16 fittings, 14 fans, 2 desktop (sleep mode)	15 fittings, 10 fans, 2 desktop (sleep mode)	14 fittings, 12 fans (2 desktop sleep mode)	10 fittings, 6 fans (4 desktops sleep mode)
4–6 P.M.	30 fittings, 14 fans, 3 Desktop (sleep mode)	28 fittings, 5 fans	30 fittings, 8 fans	20 fittings, 5 fans

With reference to the ILER ratio of all the rooms, it indicates that urgent action is needed for LHS, RHS and counter room. There is a review suggested for central room. This will help us to identify guarantee for strong recommendation.

3.2 Wastage Audit

The power wastage in the library is audited for the period of 1 month. The average analysis for 4 days is presented in the Table 3 (Fig. 4).

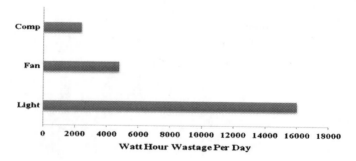

Fig. 4 Watt hour wastage

4 Recommendation

The recommendations which are give after a audit can be take as a best suggestion for the betterment of energy saving, especially the best suggestion you choose the better result you get. Mostly recommendations are based on the average of both particle and theoretical value. Auditing provide clear and reliable information on potential investment and saving the electricity in long term benefits, by calculating net present values cash flow and the resulting discounted saving over time. This enhances considerably the quality and value of the recommendations.

4.1 Recommendation Without Investment

According to the layout of the library, we have recommended some of the best saving tips by which electrical energy can be saved and tariff without an investment by proper utilization and also reduce tariff in their bills. These are the important tips to save energy in library.

- Unplug and switch off the entire electric device of appliance that is not in used to reduce no—load losses.
- Clean the fittings regularly at least once in a week as a heavy cost of dust can block 50 % of light output.
- Remove the cut covering used in the fitting by plain glass. It also reduce the amount of light output
- Clean the fan blades regularly as heavy coat of dust in fan blades reduces motor efficiency and output. The light control may consist of a row of switches at the main circulation desk provided that single switch is
- Connected to every single fitting. Adjustable window covering can be provided so that direct sunlight does not reach the stack or other sensitive materials.

With reference to the Fig. 5 indicated that the library can save above 300 units per month, if they maintain proper switching procedure.

4.2 Recommendation with Rearrangement

Calculations of No. of fittings required

$$N = E^* A/F^* UF^* LLF$$

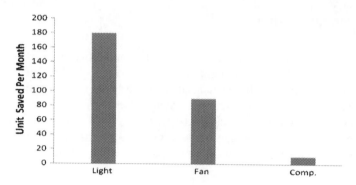

Fig. 5 Wastage audit saving graph

where,

E = lux level required on working plane
A = Area of the room
F = Total flux (lumens) from all the lamps in one fittings
UF = utilization factor
N = 500 * 1040.48/2800 * 3 * 0.75 * 0.63 = 131 fittings

So total 131 fittings are required in the library

Total 136 fittings are available in the library; Hence 5 fittings are extra in the library.
Unit used per day = 127.3

4.3 Recommendation with Investments

The lighting design is reworked for fixing of LED lighting and the proposed layout is shown in Fig. 6. It represented by 120 number of light fitting, which are 16 light set reductions when compared with the existing system (Fig. 7).

No. of LED light required (Type 15 W square LED) = 120 Fitting * 2 set light = 240
Unit consumed by 240 LEDs = 3600 watts = 28.8 units/day (Average 8 h/day)
Unit consumed/month = 720 unit/month (Figs. 8 and 9)

Money invested in buying total LED = Rs 80,000
Money Saved/month by using LED in the place of CFL = Rs 12,652
Money will be repaid in approx. 2 years

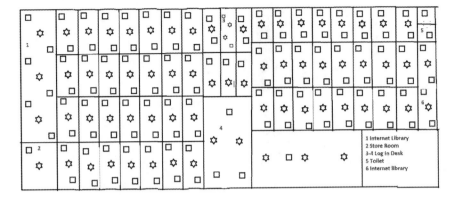

Fig. 6 Proposed layout with rearrangement

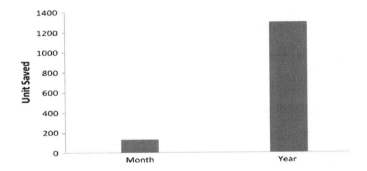

Fig. 7 Saving graph on the monthly and yearly basis

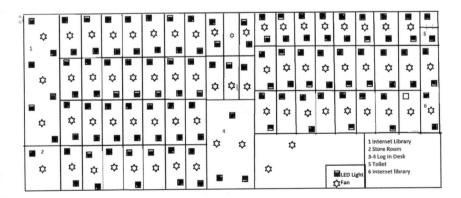

Fig. 8 Proposed lighting recommendation layout for the university library

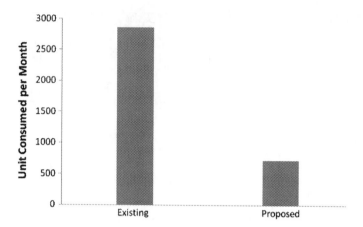

Fig. 9 Comparison of unit saved after proposed system

Recommendation with On-Grid Solar The study on the library states that the daily usage of library from 9 a.m. to 8 p.m. The study also reveals that the average number of light used per day in the library is 90–100. It is recommended to implement on-grid solar connected system to glow all the 100 LED light from 10 a.m. to 5 p.m.

Total watts required = 100 * 15 = 1500 watts
The number of solar panel required = 12
Total cost of solar panel and control equipment = 12 * Rs 14,700 = Rs 176,400
The unit saved per day by solar implementation = 20 unit (Fig. 10)

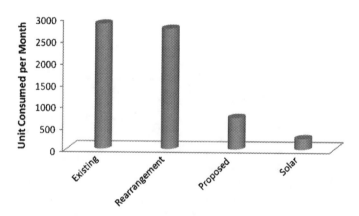

Fig. 10 Comparison of unit saved after proposed solar system

5 Conclusion

The Energy audit wastage analysis and implementation of conservation measures at University Library has been successfully completed and suggested the recommendation for implementation. The wastage audit recommendation was implemented immediately in the library. The proposed layout for lighting is agreed for implementation from June 2016. The estimated savings in the library is 55 %.

Acknowledgments The authors' expressed their valuable gratitude to Er. A.C.S. Arunkumar, President of Dr.M.G.R Educational and Research Institute, who provide constant supported to the MGR Vision 10 MW initiative. We convey our special thanks to the Rector, Registrar and Dean E&T for their valuable suggestions in the present work.

References

1. Bobdubley, S.: BP Statistical Review Report of World Energy Statistical Yearbook, 64th edn (2015)
2. Article on 'India's per capita electricity consumption touches 1010 kWh' published on Mon, July 2015
3. Ahmad Syakir, B.M.: Energy Audit for the UMP Library and Energy Efficiency. University Malaysia Pahang (2008)
4. Brinckerhoff, P.: Australia Pty Limited ABN 80 078 004 798
5. Singh, Malkiat, Singh, Gurpreet, Singh, Harmandeep: Energy audit: a case study to reduce lighting cost. Asian J. Comput. Sci. Inf. Technol. 2(5), 119–122 (2012)
6. Keerthi Jain, N., Kumar, Kishore, Ramesh, L., Raju, Madhusudhana: Comparatative analysis of residential houses for effective reduction in power demand. Int. J. Appl. Eng. Res. 10(6), 5489–5494 (2015)
7. Kumar, A., Raj, A., yadav, A., Ramesh, L.: Energy audit for a residential house with considerable recommendation for implementation. Int. J. Appl. Eng. Res. 10(20), 15537–15541 (2015)
8. Raju, M., Ramesh, L., Balamurugan: Residential house energy conservation analysis through proposed package. Int. J. Appl. Eng. Res. 10(6), 5526–5531 (2015)
9. Keerthi jain, K., Kishore Kumar, N., Ramesh, L., Madhusudhana Raju, M.: An analysis to save electrical energy in a residential house. Int. J. Eng. Sci. 6(2), 59–66 (2014)
10. Kumar, A., Thanigivelu, M., Yogaraj, R., Ramesh, L.: The impact of ETAP in residential house electrical energy audit. In: Proceedings of International Conference on Smart Grid Technologies (2015)
11. Keerthi Jain, K., Kishore Kumar, N., Muralikrishanan, S., Ramesh, L.: An analysis to save electrical energy in a residential house using ETAP. In: Proceedings of International Conference on Communication, Computing and Power Technologies (ICCCPT-2015)—April 2015—Chennai (2015)
12. Washington State University Energy Program Energy Audit Workbook
13. Code of practice of interior illumination—IS 3646-1 (1992)
14. Lynne, C.: Writing User-Friendly Energy Audit Reports. University of Florida Energy Analysis and Diagnostic Center, Department of Industrial and Systems Engineering, University of Florida (2009)

Illumination Level Study and Energy Assessment Analysis at University Office

Regu Narayanan, Ashok Kumar, Chandru Mahto, Omshivam and L. Ramesh

Abstract The power consumers of the modern world start to think about the concept of "Right appliance to the Right usage". Nationally now there is an unbridgeable gap between electricity generation capacity and the ever raising demand. The electrical energy audit is paving the way to conserve electrical energy by the way of analyzing and adopting the standards without any major investments. Energy saving of up to 80 % can be achieved by modernizing our lighting system by installing an intelligent light management system with day light triggered dimming function and occupancy sensors. By professionally restructuring the process flow cycle, lighting system, etc. more beneficial results can be achieved in industrial establishments too. Indeed electric power is precious energy which involves the generation, transmission and distribution of energy to the consumer via cumbersome process. So the question of monitoring its usage is affirmative on every single unit. Nowadays an audit in electric sector has become a mandatory requirement especially for large consumers. In the proposed project the audit was conducted in university office in prime two stages. The first stage dealt with wastage audit and second with detailed electrical energy audit. The necessary recommendation is suggested for optimal savings of energy in university office.

Keywords Consumption · Energy audit · Illumination · Lighting · Lux

R. Narayanan (✉) · A. Kumar · C. Mahto · L. Ramesh
EEE Department, Dr. M.G.R Educational and Research Institute (University), Chennai, India
e-mail: mgrvision10mw@drmgrdu.ac.in

L. Ramesh
e-mail: rameshlekshmana@gmail.com

Omshivam
EE Focus Instruments Pvt. Ltd, Chennai, India

© Springer Science+Business Media Singapore 2017
P. Deiva Sundari et al. (eds.), *Proceedings of 2nd International Conference on Intelligent Computing and Applications*, Advances in Intelligent Systems and Computing 467, DOI 10.1007/978-981-10-1645-5_34

1 Introduction

The development of economy of any country is fundamentally based on the capacity of generation of electricity via non conventional and conventional methods. The need of electrical energy audit and conservation of energy can be understood if one can just imagine the consequences of the whole world slipped to black out or even brown out of energy flow. No one could enlist the losses faced under these circumstances. One can appreciate the job of optimizing the energy consumption but that will be a challenging work in the endeavor as the cost of energy is in ascending trend although inevitable addition of new equipment will increase the total energy consumption as the consumer's demand is generally in ascending trend [1]. Optimization, beneficial utilization and improving consumption pattern by interfacing with electronic sub systems are the areas where exists a huge possibility of revamping the present status also use of some energy efficient appliances can be recommended against the existing system. In the present system huge possibility is there for saving energy through new generation and energy efficient equipments. This paper progressed through the analysis and review of existing audit research papers, collection of real data for the present audit through proposed procedure, suitable recommendations and conclusion of the present work. In connection to the energy conservation, audit and management, the views of the selected researchers are presented in the next section.

2 Audit Review

This section featured with selected work on audit done by existing researchers. Singh [2] outlined 'electrical energy audit' outcome of an industrial class load. According to his work the new generation innovation and tremendous improvement in the field of lighting has given rise to number of energy saving opportunities. According to him lighting is an area which has major scope of saving energy while conducting audit. According to his recommendation, electronics chokes can be used in place of electromagnetic chokes. They can be replaced one by one, when they became defective. metal halide lamp can be used in place of halogen and mercury lamps. The indicating lamps were recommended to be replaced in a phased manner to LED when existing lamps became defective. The tubes are not required during day and should be switched off and better arrangement for the use of natural daylight should be availed. The Right way to achieve energy efficiency is to start planning at the design stage, use of modern efficient lamp, luminaries and gears are also important apart from good practice. An industrial unit has been undertaken as a case study, as they are the major consumers of the power. After case study the author has provided data in the paper which shows the different ways of saving energy by incorporating certain changes and installation in the present structure can make the present system more energy efficient. The author has also out lined that an energy auditor should see all possibilities available in and around the proposed area. Energy conservation and exploring new methods to reduce the demand and to save more energy can fulfill the growing industrial demand in future. The author has also advocated that the implementation of suggestion of energy audit can improve efficiency and thus reduces the wastage.

Pramanik [3] working in the electrical engineering department, Kalyani Government Engineering College also conducted energy audit with similar audit recommendations. In his work he presented very simple ideas on energy conservation. In order to verify the ideas described in his paper, a room size (25' × 30') belongs to the faculty members of the Electrical Engineering has been considered as a case study. He envisaged that, the modern society is strongly based on the energy for their economic development. Production and supply of goods and energy consumption, exercising strong effect environmentally in local and global level which requires equitable balance between the energy usage for the development of social welfare and the environmental preservation. The misuse of energy and lavish handling may lead to negative environmental impacts. The author has also stressed the need of the energy management which is indeed the need of the hour. The conventional resources beyond our limits, might have been exhausted within some decades. The paper has explored solutions for the energy reduction. The author has recommended using energy efficient appliances and implementation of microcontroller based system, along with power electronics which can reduce the energy consumption. With these microcontroller based system included in the control system of air conditioner, the percentage saving in energy could have proven better.

Ahuja [4] along with his team conducted electrical energy audit in the IIT Roorkee Campus and data collected during May–June'09. They have conducted audit to find the new opportunities to improve the energy efficiency of the campus. The audit was not only done to identify the energy consumption pattern but also to find most energy efficient appliances. Moreover, some daily practices relating to common appliances have been provided which help them in reducing the energy consumption. The report gives a detailed information regarding the energy consumption pattern of the academic area, central facilities and bhawans, based on actual survey and detailed analysis during the audit. The work comprises the area wise consumption traced, using suitable equipments. The ELEKTRA software was used for their audit purpose. The report compiles a list of possible actions to conserve and efficiently access the available scarce resources and identifying to save the potentials. The author has looked forward towards optimization for adoption of set of mission for the authorities, students and staff should follow the recommendation in the best possible way. The report is based on certain generalizations and approximations wherever they found it necessary.

To create energy awareness to the general public, Dr. M.G.R Educational and Research Institute, University Chennai has taken initiative called 'MGR Vision 10 MW' under leadership of Dr. L. Ramesh to save 10 MW in 10 years. In this pilot audit study-1 was conducted by the team of members in the year 2015 at various residential houses and industries. The outcomes of the studies are published in indexed conferences and Journals [5–10].

This work is the pilot study-1 of Vision 10 MW. This study covers the waste audit analysis and recommendation for the University office in the first stage. In the second stage, detailed power audit is conducted in the university office and the suitable recommendation for saving of energy is recommended for implementation.

3 Data Observation

The first phase of an energy audit started with site inspection work. The measurements in all aspects have been taken for reckoning actual value of prevailing luminous intensity level of the office. In this paper the details of possible technical viability are analyzed and scope of saving both energy and cost has been done through auditing in the university office. The theoretical level of illumination required as per the standards which studied against the actual level. Every appliance has been subjected to audit and aggregate load details were prepared. The room index for every partition was calculated and LUX levels were recorded. The analysis of energy wastage in the university office was also done on the merits of data collected by visiting different timing over the span of a week. The data collected revealed the possibility of energy saving in the office envelope of total area of approximately 225 m^2. This paper projects with energy audit recommendations in the later part.

The steps involved [11] in the execution of electrical energy audit are as below.

- The preliminary study in the prospective area
- Wastage audit
- Theoretical lux level calculation for the proposed area
- Practical lux level calculation
- Constraints for optimal lighting
- Addressing present issues in lighting
- Proposed layout
- Recommendation part on cost analysis and saving benefits.

3.1 Preliminary Study About the Prospective Area

The following layout shows the university office and the way of spread of lighting system. The load details with respect to room tabulated. The office was subjected to electrical energy audit and solutions devised. The overall area was divided into various segments named from A to K (Fig. 1).

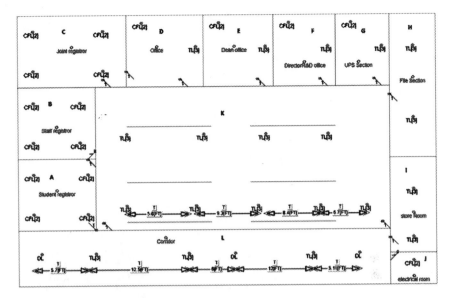

Fig. 1 Layout of the existing area

Here mainly seating arrangement for most of the staff is in 'K' segment. The prospective auditable area was calculated both in sq. feet and square meters. The total area of the office comprises of 2470 sq. feets with a staff strength of 40. Out of that the major area is covered by the area "K" which is around 903 sq. feets. "A" is of 121 sq. feets, "B" is of 110 sq. feets, "C" is of 130 sq. feets, "D" is of 100 sq. feets, "E" is of 110 sq. feets, "F" is of 120 sq. feets, "G" is of 80 sq. feets, "H" is of 264 sq. feets, "I" is of 121 sq. feets, "J" is of 33 sq. feets, "L" is of 378 sq. feets.

3.2 Wastage Audit

This step is to ascertain the scope of the possible saving of energy during the working hours of office. As to acquire reliable data, periodical visits have been paid in different time slots and observed the nature of load running waste and accounted its ratings. The computed consumption also worked out. An independent enquiry also conducted without revealing the purpose among the staff to ascertain their style of functioning with energy equipments and also collected the factors influencing the wastage of energy particularly in respect of the operation of Photostat, personal computers etc. The average pattern of energy running waste is tabulated below (Table 1 and Fig. 2)

Table 1 Wastage assessment with time

Time	Fans	Lights	Rating in Watts
9:15–10:15 A.M.	3	10	1260
10:30–11:15 A.M.	2	12	1416
11:15–12:15 P.M.	2	10	1200
12:15–1:15 P.M.	3	12	1476
01:30–2:30 P.M.	2	11	1308
02:30–4:00 P.M.	1	10	1140
04:00–05:00 P.M.	2	10	1200
05:00–06:00 P.M.	2	8	984

Fig. 2 Energy wastage percentage assessment

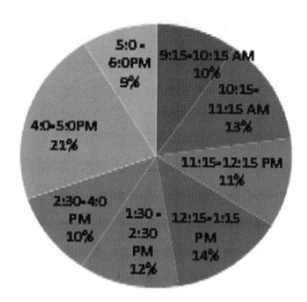

The graphical representation of the wastage audit with the data collected from the university office has shown. From the table it can be vividly seen that the maximum wastage occurs during 9.15–10.15 h, in the morning which is being because of the college work starts from 9:30 but the equipments are switched on by 9:00 h and there is only one person who is switching on and off all the lights and fans regularly. Considerable wastage occurs during 12:30 to 1:30 which is lunch time, while most of the equipments left in switched on state but less staff were found in the office. So here it can be seen that just due to unawareness and ignorance precious energy is being wasted which can be minimized by taking proper care. It is estimated that, by the way of effective utilization around 10 kWh per day can be saved which roughly works out to the figure of 3000 kWh units per annum. If the cost per unit is Rs 7 then estimated saving of Rs 21000 per annum which reminds the proverb 'Little drop makes an ocean'. Already the office enclave is air-conditioned fully; hence under normal conditions fans are not necessary.

Table 2 Aggregate connected load

S. no.	Description	Total nos.	Wastage	Total load
1	CFL fitting	16	36	576
2	Sq. fitting	19	108	2052
3	Doom fitting	3	120	360
4	Photostat machine	1	300	300
5	Ceiling fan	20	60	1200
6	Split A/c machine	3	2500	7500
7	Desk top computers	8	250	2000
8	Water purifier	1	300	300
9	UPS system 1 7.5 KVA	1	7.5 KVA	–
	Total load			14.288 KW

But considering extra ordinary occurrence in the event of air-conditioning failure, the usage of fans can be opted. However such a view point was not taken into account in this audit study for calculation of saving aspect. The team suggesting to keep the fans in off, when office enclave in air conditioned mode. Then the estimated saving will be higher.

Connected Load

The table given below shows the total connected load of the office. From the graph shown it can be easily depicts that the max energy consumption is due to the UPS, The lightning on a whole adds a total load of 3 KW in addition with fans, computers the minimum consumption by the printers (Table 2).

3.3 Lux Level Calculation

In fact it was conducted by adopting two methodologies.

1. Measurement of actual level of lux at the center of the every working surface and tabulated. The lux level is measured at every working table. Among these measurements it was noticed that the lux measurement level was below the required value in a8 and b5 working Table.
2. Measurement carried out in each cabin and hall, corridor etc. as per the calculated room Index and analyzed.

The available luminous intensity level of the corridor is more than the prescribed level. The illumination level of the office is non–uniform which can be improved by the way of installing distributed illumination using, single fixture or using louvers.

Room index According to the bureau of energy efficiency, 'room index' is the number that describes the ratio factor of the room length, width and height.

$$\text{Room index} = (L \times W)/(Hm(L \times W))$$

L = length of the room, W = width of room, Hm = mounting height.

It doesn't matters whether the dimensions are in meters or not, but the unit should be same for all. The minimum number of measurement point can be ascertained from the table shown below (Table 3).

   As per the Code of practice [12] of interior illumination – IS 3646-1 (1992) room index is required to get the numbers of reading required to measure the theoretical lux level of the proposed area. Using the above facts the required numbers of reading are taken and tabulated as shown above. Measurement carried out in each cabin and hall, corridor etc. as per the calculated room index and analyzed. Lux level measured at every working table. In such measurement it was noticed that the lux measurement level was below the stipulated value in a8 and b5 table. Hence both the seat may be rearranged accordingly.

   The available luminous intensity level of the corridor is more than the prescribed level. The illumination level of the office is not uniform which can be eliminated by intelligent lighting system.

Constraints for Optimal Lighting

The wall with brown wooden coverage even though appears good is not supporting for effective lighting spread due to more rate of diffusion and not supporting gross illumination. The lights are provided above the two fans in 'K' segment may be

Table 3 Room index calculation

Segment	Description	Prescribed level of lux. (advocated value)	Room index	Meas. Reqd.
A	Room	100–200	0.50	9
B	Room	100–200	0.48	9
C	Room	100–200	0.51	9
D	Room	100–200	0.45	9
E	Room	100–200	0.48	9
F	Room	100–200	0.50	9
G	Room	100–200	0.40	9
H	Room	100–200	0.69	9
I	Room	100–200	0.50	9
J	Room	40–60	0.21	9
K	Office	100–200	1.28	16
L	Corridor	50–75	0.56	9

Table 4 Practical lux level

Segment	Area	Lux level measured at working surface						
A	Room	104						
B	Room	101						
C	Room	132						
D	Room	250						
E	Room	233						
F	Room	205						
G	Room	214						
H	Room	140						
I	Room	198						
J	Room	255						
K	Office hall	a1	124	b1	125	c1	176	
		a2	122	b2	141	c2	160	
		a3	142	b3	127	c3	160	
		a4	129	b4	106	c4	138	
		a5	115	b5	95	c5	150	
		a6	135	b6	110	c6	155	
		a7	110	b7	127	c7	164	
		a8	92	b8	150	c8	155	
L	Corridor	No working table						

rearranged to improve shadowing effect which will be quiet annoying. In spite of its technological merits in all aspects, it seems impractical to suggest all lighting fittings by light emitting diode lamps owing to the initial high cost (Table 4).

4 Recommendation

Recommendation is an act of suggestion or proposal as to the best of course of action, especially one put forward by an authority's body [13]. Basically there is two modes of recommendations namely without investment and with investment.

Proposed Layout

Instead of recommending all lamps to be replaced by the present modern technology energy star rated LED lamps which possess long span of life up to 50,000 h and environmental friendly, but due to the factor of cost, it is suggested the first phase of conversion for k segment, the main staff working area and for the corridor as shown replacing 36×3 (108 W) fluorescent fitting by 24×3 (72 W). By these replacement, the luminous level will be more over same with lesser involvement of cost. The system of LED can also provided with presently available electronic control gear. The use of modern concept of lightning including the use of daylight offers up to 75 % potential to save energy (Fig. 3).

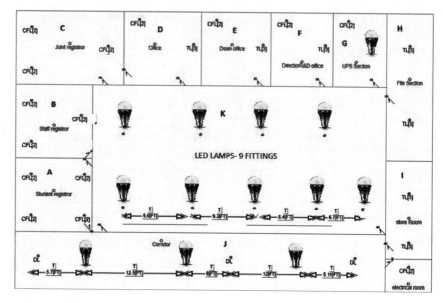

Fig. 3 Proposed lighting design

4.1 Recommendation Without Investment

1. **Wastage Recommendation**: The wastage audit reveals the fact that the practice should be imparted that when leaving working table, everyone should be bound to switch off the lights and fans. There is considerable saving in long run. In this audit, there will be estimated saving of 3000 kWh per annum which is roughly saving of Rs 21,000 per annum. Briefly to say that

   - All the equipment should be switched off when not in use.
   - Fans should be switched off when the AC is ON. It is advocated to keep a slogan display on energy conservation in every one's view with a advice of keeping fans off when air condition system is operative.
   - Usage of minimum lights in the segment C, D, E, F, G during day time because of sunlight availability (Fig. 4).

2. **Rearrangement Recommendation**: It is seen that two lights are connected above the fans in segment 'K' and should be rearranged to improve illumination and avoiding possible rotational shading while the fan is rotating.

   (a) Matching the proper lamp type to the respective work task, consistent with color, brightness control and other requirements.
   (b) Establishing adequate light level without compromising objective and safety.
   (c) The decorating lighting fitting (TL-3nos.) of 120 W in the corridor normally recommended to be kept off, except on special occasions.

Fig. 4 Wastage audit saving analysis

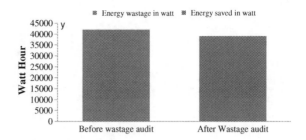

3. **Recommendation with PC and Photostat Machine**: According to the survey of the wastage audit conducted by the energy audit team, it is noticed that, most of the system and the Photostat machines are left in the sleep mode. Hence, energy is wastage in the form of no load loss. So the additional energy saved is used to shut down the unit when there will be no work with it for longer period. The computer systems, printers and Photostat machine are used effectively for 6 h in a day in the university office and left 2 h in sleep mode as observed in audit study then the energy loss will be more than 80 units per month.

Audit Observation: The energy saving per annum by avoiding sleep mode in PC and Photostat machine is 960 units.

4.2 Recommendation with Investment

1. **Recommendation with LED light**: In any energy audit report, if it fails to envisage to adopt modern efficient system at least at its preliminary level in the area of the audit, then the suggestions and directives are not up to the present technological yard stick. Hence it is recommended to replace the existing 9 florescent lamp fitting of 108 W into 72 W in 'K' segment and 1 number in 'G' segment. These lamps are environment friendly as it does not possess mercury and it will be energy efficient. The life of the star rated LED lamps will be more than 50,000 h. By this replacement using LED lamps, estimated energy saving will be 1290 kWh per year. Considering the cost implication for conversion to LED lamps apparently thrice of other fittings, first phase of conversion is suggested only for main staff working segment (Fig. 5).

The given chart stresses that if we properly switch on and off the lights and avoid using lights during day time will save the load of 1124 W without any investment.

Audit Observation: Cost equivalent of 1290 unit per annum can be saved by the LED lamp replacement.

2. **Recommendation with Fan**: The energy audit team found that the fan installed in the in office is not star rated. Available fan is consuming 60 W. It is recommended for replacement with energy efficient star rated fan which will be

Fig. 5 Saving with lights

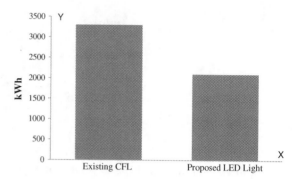

available for 50 W per fan. If the fans are replaced by star rated fans in a pace manner the savings in that aspect will be:

$$Difference\ in\ watt\ =\ 10\ watt$$
$$Saving\ =\ 700\ kwh\ per\ annum$$

Audit Observation: From the above data analysis, the possible saving of 700 units per annum is achieved by replacing old fans by new energy efficient fans.

3. **Recommendation with PC and Xerox machine**: The graph shown below gives the quantum of saving if the staff not preferred the sleep mode in personal computers and Photostat machine (Fig. 6).

Audit Observation: From the above data analysis, the possible saving of 900 units per annum is benefited, if the operator of PCs and Photostat machines are not preferring sleep mode (Figs. 7 and 8).

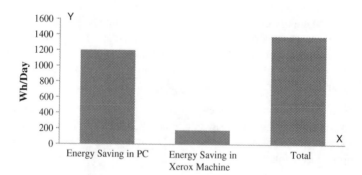

Fig. 6 Energy saved with PC and Xerox machine

Fig. 7 Energy saved with PC and Xerox machine

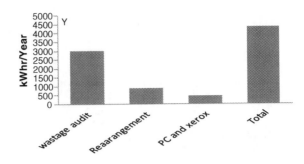

Fig. 8 CFL and LED comparison

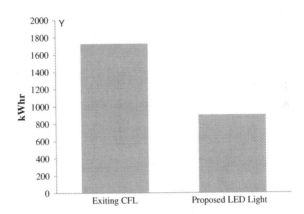

5 Conclusion

A famous quote "Energy saved is Energy generated". This shows that apart from increasing the generation capacity with investment, one must go for the energy audit to save the electricity at lower cost. The outcome result of the work shows the possibility of energy savings in the University Office. The wastage and electrical energy audit was conducted in the University office. The wastage audit recommendation was implemented in the office from Jan 2016. The suggested other recommendation also agreed to implement within 6 months of time span.

Acknowledgments The authors' expressed their valuable gratitude to Er. A.C.S. Arunkumar, President of Dr.M.G.R Educational and Research Institute, who provide constant support to the MGR Vision 10 MW initiative. We convey our special thanks to the Registrar and Dean E&T for their valuable suggestions in the present work.

References

1. Bobdubley, S.: BP Stastitical Review Report of World Energy Statistical Yearbook, 64th edn (2015)
2. Singh, M., Singh, G., Singh, H.: Energy audit: a case study to reduce lighting cost. Asian J. Comput. Sci. Inf. Technol. 2(5), 119–122 (2012)
3. Pramanik, S., Chakraborty, T., Alam, K., Mal, S., Sarddar, D.: Reduction of energy consumption using modern electronic system. Int. J. Adv. Res. Electr. Electron. Instrum. Eng. 2(6), 2642–2646 (2013)
4. Kumar, A., et al.: Energy Audit of IIT-Roorkee Campus. Energy Audit Team, IIT Roorkee (2010)
5. Jain, K., Kumar, N.K., Ramesh, L., Raju, M.: Comparative analysis of residential houses for effective reduction in power demand. Int. J. Appl. Eng. Res. 10(6), 5489–5494 (2015)
6. Kumar, A., Raj, A., Yadav, A.K., Ramesh, L.: Energy audit for a residential house with considerable recommentation for implementation. Int. J. Appl. Eng. Res. 10(20), 15537–15541 (2015)
7. Raju, M., Ramesh, L., Balamurugan, S.: Residential house energy conservation analysis through proposed package. Int. J. Appl. Eng. Res. 10(6), 5526–5531 (2015)
8. Keerthi Jain, K., Kishore Kumar, N., Ramesh, L., Madhusudhana Raju, M.: An analysis to save electrical energy in a residential house. Int. J. Eng. Sci. 6(2), 59–66 (2014)
9. Kumar, A., Thanigivelu, M., Yogaraj, R., Ramesh, L.: The impact of ETAP in residential house electrical energy audit. In: Proceedings of International Conference on Smart Grid Technologies (2015)
10. Keerthi Jain, K., Kishore Kumar, N., Muralikrishanan, S., Ramesh, L.: An analysis to save electrical energy in a residential house using ETAP. In: Proceedings of International Conference on Communication, Computing and Power Technologies (ICCCPT—2015), April 2015, Chennai (2015)
11. Washington State University Energy Program Energy Audit Workbook
12. Code of practice of interior illumination – IS 3646-1 (1992)
13. Lynne, C.: Writing User-Friendly Energy Audit Reports. University of Florida Energy Analysis and Diagnostic Center, Department of Industrial and Systems Engineering, University of Florida (2009)

Design and Analysis of Grid Connected PV Generation System

T.D. Sudhakar, K.N. Srinivas, M. Mohana Krishnan
and R. Raja Prabu

Abstract As there is a power shortage across any country, usage of renewable energy sources is being encouraged nowadays. This energy produced has to be properly synchronized with the grid to ensure safety and energy continuity. In this paper, the standard procedure for grid interconnection is discussed and the impact of not following the procedure is shown using the voltage and current waveforms. Later a step by step procedure for designing a MPPT algorithm based PV generation system connecting to the grid is developed. The developed system feeds a common load, as the micro grid systems feeds the local loads. The system takes care of all the specified guidelines while feeding the load, which is proved by using the voltage and current waveforms. Therefore this paper acts as guide in developing renewable energy based grid connected system.

Keywords Renewable energy · Grid interconnection · MPPT · PV generation

1 Introduction

The basic thing needed for grid synchronization is that the voltage magnitude, phase sequence and frequency of the interconnecting power source must be same as that of the grid. A basic integration of solar panel based distribution generator (DG) with grid is shown in Fig. 1.

T.D. Sudhakar (✉)
St. Joseph's College of Engineering, Semmencherry, Chennai 600 119
Tamil Nadu, India
e-mail: t.d.sudhakar@gmail.com

K.N. Srinivas · R. Raja Prabu
BS Abdur Rahman University, Vandalur, Chennai 600 048, Tamil Nadu, India

M. Mohana Krishnan
SSN College of Engineering, Semmencherry, Chennai 603 110, Tamil Nadu, India
e-mail: mohana2621994@gmail.com

© Springer Science+Business Media Singapore 2017
P. Deiva Sundari et al. (eds.), *Proceedings of 2nd International Conference on Intelligent Computing and Applications*, Advances in Intelligent Systems and Computing 467, DOI 10.1007/978-981-10-1645-5_35

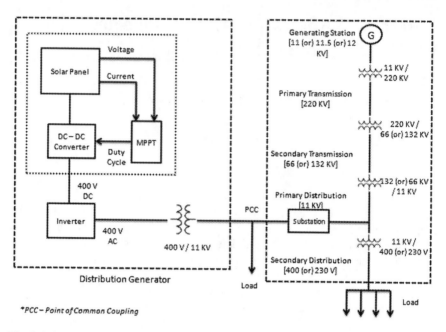

Fig. 1 DG interconnected with grid

Consider a DG [1] connected to the primary distribution network of the electrical grid [2] as shown in Fig. 1. The load is connected at the Point of Common Coupling (PCC). PCC is a point in the system where the connected load could draw power from either DG or grid or both as shown in Fig. 1. The DG shown in Fig. 1, comprises of solar panel modules with a DC-DC converter to track the maximum power point.

The following parameters of the DG must be same as that of the primary distribution substation:

- Voltage magnitude
- Phase sequence and
- Frequency.

If these synchronization parameters are equal, then according to Kirchhoff's current law (KCL), Eq. 1 will be satisfied.

$$I_{DG} + I_{SS} = I_L \qquad (1)$$

where I_{DG} is the magnitude of current from DG, I_{SS} is the magnitude of current from the interconnected substation and I_L is the magnitude of current drawn by the load at the point of common coupling as shown in Fig. 1. I_{DG}, I_{SS} and I_L should be in phase with each other for a properly synchronized DG.

2 Survey

The various types of synchronization techniques used for grid interconnection are zero cross detector [3], Phased Locked Oscillator (PLO), Phase Locked Loop (PLL) and Space Vector based phase detection [5]. Of these techniques PLL proves to track the grid frequency and phase with great robustness.

The various types of PLL commonly used for grid synchronization are analog PLL, digital PLL, Synchronous Reference Frame (SRF) PLL [5], SRF based Moving Average Filter (MAF) [6], high bandwidth PLL [7], Active and reactive power based PLL [4, 5, 8], double synchronous frame (DSF) based PLL [5, 8], Sinusoidal signal integrator (SSI) [5], Double second order generalized integrator (DSOGI) [5, 8], enhanced PLL (EPLL) [5], 3 phase magnitude PLL (3MPLL) [5, 9] and Quadrature PLL [5]. Though there is large variety of implementing a PLL, SRF-PLL (which is implemented in this paper) is the basic and the most suitable synchronization technique for grid interconnection.

If the DG is powered through a DC supply as shown in Fig. 1, similar to a solar power, then inverters are required to convert it into AC. Inverters used for such purpose may be divided as central, string and module inverters [10]. In this paper, a central inverter based interconnection is opted due to less number of inverter control components.

3 Problems Faced in Grid Connected Distribution Generator

Consider the following cases to understand the concept of DG interconnection at the primary distribution level of grid,

- Stand alone substation
- DG + Substation
- With same parameters
- With different voltage magnitude
- With different frequency
- With different phase sequence
- Stand alone inverter based DG with closed loop control
- Substation + Inverter based DG without any synchronization control

The load resistance used for the entire simulation process is maintained same in order to compare the results on the same aspect. The simulation parameters based on which the grid interconnection is analyzed is given in the upcoming sections with results at the end. The result of the simulation is tabulated under Table 1. It shows the phase to phase voltage and current of the system at the point of common coupling (PCC).

Table 1 Parameters of substation + inverter based DG (without synchronization control)

Power source	Phase to phase voltage (KV)	Current (A)		
		Substation	DG	Load
Standalone substation	11	6.35	–	6.35
Substation + DG (with equal parameters)	11	3.175	3.175	6.351
Substation + DG (with different voltage magnitude)	10.45	516.7	517.5	6.035
Substation + DG (with different phase sequence)	5.601	8999	8996	3.176
Standalone inverter based DG with closed loop control	11	–	6.317	6.317
Substation + inverter based DG (without synchronization control)	11	35.65	32.68	6.332

From Table 1, it is seen that the magnitude of current from DG and substation add up to give load current only for substation interconnected to DG with equal parameters. It can be clearly seen from Table 1 that if the voltage magnitude, phase sequence and frequency of the interconnecting sources are not the same, then current drawn from DG and substation is large for same amount of load. Thus, it can be concluded that the usage of inverter with only a closed loop voltage control is not sufficient for grid synchronization. The control method needed for proper grid synchronization is studied in detail in the next section.

4 Proposed Methodology

The inverter based DG is interconnected to grid via substation with a synchronization control as shown in Fig. 2. The DC source of inverter is replaced with solar panel along with a buck converter for Maximum Power Point Tracking (MPPT) with Perturb and Observe (PO) algorithm.

The 400 V DC is converted to phase to phase 400 V AC using a voltage source inverter which is controlled using the reference signal from synchronization control block as shown in Fig. 2. This 400 V AC is then stepped up to 11 kV using a power transformer and then is interconnected to an 11 kV substation with load at PCC.

4.1 Phase Locked Loop (PLL)

Phase locked loop (PLL) is used to synchronize the output voltage of an inverter with a reference frequency and phase of the grid voltage [4]. Synchronous

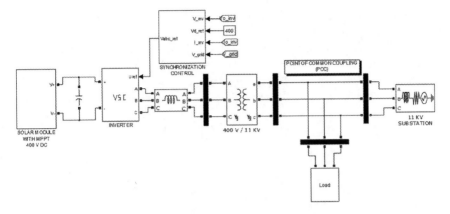

Fig. 2 Substation + inverter based DG (with synchronization control)

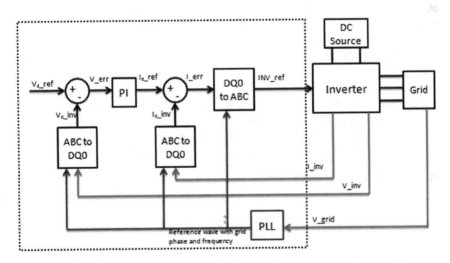

Fig. 3 Block diagram of SRF PLL

Reference Frame based Phase Locked Loop (SRF-PLL) is one of the synchronization control used to interconnect an inverter based DG with grid as shown in Fig. 3.

The measured 3 phase AC voltage and current waveforms are converted to DC components using Park's transformation (ABC to DQ0) [11]. Figure 3 shows that there is an outer voltage control loop used to find the voltage error and an inner current control loop used to find the current error. The voltage error is converted into current reference using PI controller. The direct and quadrature axis current error, which is represented as I_err in Fig. 3 is converted to 3 phase wave using Inverse Park's transformation (DQ0 to ABC) [11] and then it is fed as a reference to Voltage Source Inverter (VSI). It is to be noted that Fig. 3 shows the voltage and current components for direct axis only but the same control could be used for quadrature components also. Equations 2–5 are used to perform SRF PLL control.

$$V_{err} = V_{d(ref)} - V_{d(inv)} \qquad (2)$$

$$I_{d(ref)} = [P + (I*T_s)/(z-1)]*V_{err} \qquad (3)$$

$$I_{d(err)} = I_{d(ref)} - I_{d(inv)} \qquad (4)$$

$$I_{q(err)} = I_{q(ref)} - I_{q(inv)} \qquad (5)$$

where $V_{d(ref)}$ is the direct axis reference voltage, $V_{d(inv)}$ is the direct axis output voltage of inverter, V_{err} is the direct axis voltage error component, P is the proportional constant of PI controller, I is the integral constant of PI controller, Ts is the sampling time, $I_{d(ref)}$ is the direct axis current reference, $I_{d(inv)}$ is the direct axis output current of inverter, $I_{d(err)}$ is the direct axis current error, $I_{q(ref)}$ is the quadrature axis current reference, $I_{q(inv)}$ is the quadrature axis output current of inverter, $I_{q(err)}$ is the quadrature axis current error.

The control mechanism shown in Fig. 3 produces a constant voltage and constant current output from inverter in a grid connected system irrespective of load value. If the load consumes less current than the set value of inverter, then the excess current flows in the reverse direction. So, the load value must be specified such that it consumes a greater current than the constant current value of inverter. The maximum value of load resistance for which the system works properly is given in Eq. 6

$$R_{max} = K^2*(V\_inv/I\_inv) \qquad (6)$$

where K is the transformer ratio, V_inv is the constant phase to ground voltage output of inverter and I_inv is the constant current output of inverter.

4.2 Solar Panel with MPPT

Solar panel is modeled using the values of solarex_msx_60 [12]. Perturb and Observe (P&O) algorithm [13] is used to extract maximum power from solar panel via DC-DC buck converter. The characteristic parameters of solar panel used for simulation is given in Table 2.

Table 2 Solar panel parameters

Parameter	Value
Open circuit voltage (V_{oc})	21.1 V
Short circuit current (I_{sc})	3.8 A
Number of series cell (N_s)	20
Number of parallel cell (N_p)	1
Module voltage	422 V
Module current	3.8 A
Maximum power (P_{MPP})	1137 W

5 Simulation Results

The PI controller for simulation shown in Fig. 3 is tuned such that a constant current of 2.9508 A is produced at the inverter output. The maximum load resistance for which the system works properly is found using Eq. 6 as follows,

$$R_{max} = (11e3/400)^2 * [(400/1.732)/2.9508]$$
$$= 59,189 \, Ohms = 59 \, K \, Ohms \, (Approx.)$$

The value of load resistance used for simulation is 1 K Ohms, which is less than R_{max}. The characteristic of the modeled solar panel is shown in Fig. 4. Figure 5 shows the output power from solar panel with MPPT in a grid connected system.

The maximum power obtainable from the solar panel characteristic shown in Fig. 4 is 1137 W. From Fig. 5, it can be seen that the power injected by the solar panel to the grid is approximately equal to its maximum power.

A constant supply of 400 V and 2.9508 A is produced at inverter output which is fed to a power transformer to get a output of 11 kV and 0.1073 A. The current waveform at PCC for the substation integrated with solar based DG is shown in Fig. 6.

It can be seen from Fig. 6 that the current from solar based DG and substation adds up to provide the load current satisfying Eq. 1. Moreover, all the three currents are in phase with each other.

Table 3 clearly shows that the current from solar panel based DG is constant for any value of load resistance. The additional current required by the load is supplied by the substation.

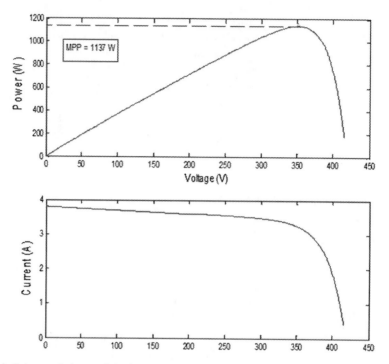

Fig. 4 Solar panel characteristics

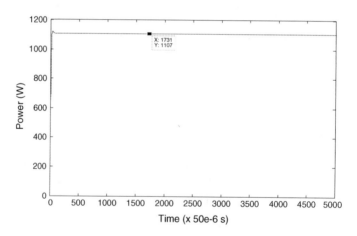

Fig. 5 Output power of solar panel integrated with substation

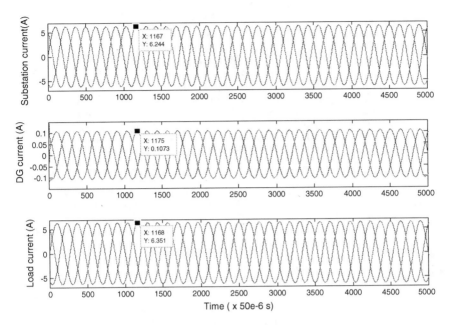

Fig. 6 Current waveform of substation + solar based DG (with synchronization control)

Table 3 Magnitude of current at PCC

Load resistance (Ohms)	Solar panel based DG current (A)	Substation current (A)	Load current (A)
1 K	0.1073	6.244	6.351
5 K	0.1073	1.1629	1.2702
10 K	0.1073	0.5278	0.6351
15 K	0.1073	0.3161	0.4234
20 K	0.1073	0.2102	0.3175
25 K	0.1073	0.1467	0.2540

6 Conclusion

From the simulation results shown in this paper, it can be clearly understood that if an inverter based source is interconnected with grid then proper synchronization control is needed. In this paper, the solar panel provides a constant current irrespective of any load condition. The extra amount of current needed by load is taken from the grid.. The current output from both grid and solar inverter will be in phase with each other only with proper synchronization control. The SRF PLL used in this paper proves to be working well with solar power integration. The PI controller could be tuned to provide any value of reference current making it possible to integrate with solar panel of any rating.

References

1. Blaabjerg, F., Teodorescu, R., Liserre, M., Timbus, A.V.: Overview of control and grid synchronization for distributed power generation systems. IEEE Trans. Ind. Electron. **53**(5), 1398–1409 (2006)
2. Saadat, H.: Power System Analysis. Tata McGraw-Hill, New York (2002)
3. Azrik, M., Ahmed, K.H., Finney, S.J., Williams, B.W.: Voltage synchronization scheme based on zero crossing detection for parallel connected inverters in AC microgrids. In: Industrial Electronics (ISIE), IEEE International Symposium, pp. 588–593, May 2012
4. dos Santos Filho, R.M., Seixas, P.F., Cortizo, P.C.: A comparative study of three phase and single phase PLL algorithms for Grid connected systems. In: Proceedings of INDUSCON Conference, pp. 1–7 (2006)
5. Guo, X.-Q., Wu, W.-Y., Gu, H.-R.: Phase locked loop and synchronization methods for grid-interfaced converters: a review. PRZEGLĄD ELEKTROTECHNICZNY (Electrical Review), pp. 182–187 (2011). ISSN 0033-2097, R. 87 NR 4/2011
6. da Silva, S.A.O., Modesto, R.A.: A comparative analysis of SRF-based controllers applied to active power line conditioners. In: 34th Annual Conference of IEEE: Industrial Electronics, IECON 2008, pp. 405–410, Nov 2008
7. Furlan, I., Balemi, S.: High-bandwidth three-phase phase-locked loop. In: IEEE International Symposium on Industrial Electronics (ISIE) Bari, pp. 1627–1632, July 2010
8. Bobrowska-Rafal, M., Rafal, K., Jasinski, M., Kazmierkowski, M.P.: Grid synchronization and symmetrical components extraction with PLL algorithm for grid connected power electronic converters—a review. Bull. Pol. Acad. Sci. Tech. Sci. **59**(4), 485–497 (2011)
9. Zhang, X., Jiang, Z., Ding, N.: Three-phase magnitude-phase detection based on T/4 time-lapse elimination method. J. Softw. **9**(2), 523–529 (2014)
10. Haeberlin, H.: Evolution of inverters for grid connected PV-systems from 1989 to 2000. In: 17th European Photovoltaic Solar Conference, Munich, Germany, 22–26 Oct 2001, pp. 1–5
11. Krause, P.C., Wasynczuk, O., Sudhoff, S.D.: Analysis of electric machinery and drive systems. In: IEEE Series on Power Engineering, 2nd edn. Wiley (2002)
12. Khlifi, Y.: Mathematical modelling and simulation of photovoltaic solar module in matlab-mathworks environment. Int. J. Sci. Eng. Res. **5**(2), 448–454 (2014)
13. Alsadi, S., Alsayid, B.: Maximum power point tracking simulation for photovoltaic systems using perturb and observe algorithm. Int. J. Eng. Innov. Technol. **2**(6), 80–85 (2012)

A Robust Energy Management System for Smart Grid

N. Loganathan, K. Lakshmi and J. Arun Venkatesh

Abstract This paper presents an energy management system with reduced energy consumption as well as to look for alternative sources of energy which are cheaper to minimize the total cost of energy consumption. A cluster of interconnected price-responsive demands (e.g., a college campus) that is supplied by the main grid and a stochastic distributed energy resources (DER) e.g., a wind and solar power plants with energy storage facilities is considered. An energy management system (EMS) arranges the value responsive requests inside of the bunch and gives the interface to vitality exchanging between the requests and the suppliers, primary lattice and DER. Vitality administration calculation permits the bunch of requests to purchase, store and offer vitality at suitable times. To solve this EMS problem, an optimization algorithm base on linear programming (LP) approach has been implemented. Toward estimate the performance of the planned algorithm an IEEE 14 bus system was consider. The outcome show with the purpose of the group of load of energy management system with the planned approach increases the effectiveness by minimizing the losses while compared to existing method. Improvement in the method is the optimization problem having two sources vulnerability identified with both the generation level of the DER and the cost of the vitality acquired from/sold to the fundamental network, which is demonstrated utilizing robust optimization (RO) procedures. Shrewd grid (SG) innovation is utilized to acknowledge 2-route correspondence between the EMS and the primary lattice and between the EMS and DER.

N. Loganathan (✉) · J. Arun Venkatesh
Department of Electrical and Electronics Engineering, KCG College of Technology, Chennai 600097, India
e-mail: logukirsh@gmail.com

J. Arun Venkatesh
e-mail: arunvenkatesh1991@gmail.com

K. Lakshmi
PSG Institute of Technology and Applied Research, Coimbatore, India
e-mail: lakshmik@psgitech.ac.in

© Springer Science+Business Media Singapore 2017
P. Deiva Sundari et al. (eds.), *Proceedings of 2nd International Conference on Intelligent Computing and Applications*, Advances in Intelligent Systems and Computing 467, DOI 10.1007/978-981-10-1645-5_36

Keywords Energy management system · Demand response · Distributed energy resources · Real time pricing

1 Introduction

Energy management incorporates arranging and operation of vitality related creation and utilization power. Goals are supply preservation, atmosphere insurance and expense reserve funds, while the clients have perpetual induction to the vitality they require. It is associated nearly to green administration, creation administration. Generation is the region with the biggest vitality utilization inside of an association.

In this paper propose a demand reaction model for a group of cost-responsive load organized through a Small Size of Electric Energy Management System (SSEEMS). Demands supply consumption in order toward the Demand Side Energy Management System (DSEMS) to be in give away for their power deliver. Base on the energy interest scope of utility and vitality cost data, the EMS ideally chooses the hourly power utilization for every interest and decides the aggregate force utilization to the vitality sources. It has considered three vitality sources, particularly, the primary lattice, photovoltaic and a wind force plant structure. The gathering of requests possesses a vitality stockpiling capacity to store vitality and to utilize it at suitable times when fancied. It supports market contribution of wind generation sizing and control of power flow batteries [1]. Through high saturation of wind energy, the information of uncertainties ahead can be extremely valuable to a number of determinations [2]. There exists a two-way communications system connecting each consumer to the energy supplier [3]. In the direction of address this problem [4] proposed a innovative billing approach, where each consumer is charged based on his/her direct load in each time period during the next operation period, synchronous and asynchronous algorithms were respectively developed in [4, 5] for the consumers to achieve their best possible strategies in a distributed method.

The demand, the storage space, space unit, and the distributed energy resources (DERs) occupation as a virtual power plant that buys and stores energy in time of low electricity costs, and sells energy in time of high prices [6]. Additional to demand response problems, robust optimization (RO) [7] and energy storage space process [8]. In addition [9] RO advance developed a conventional energy producer. Demand response [10, 11], the impact of price-based DR on voltage summary and losses of a distribution network was explored in [12]. It have been special aspects of the system operation, including network peak load, power losses, voltage profiles, and service dependability, are to be considered [13, 14] has been used in unlike power system optimization problems, management of distributed generation [15].

Renewable energy is the simply sustainable solution of secure energy which is environmental approachable [16], during current decade with advancement in the growing supplies of various sectors of the power industry for monitoring to grid as a systematic and practical solution to the utility grid [17]. Wind force determining systems can be extensively isolated into two gatherings: physical strategies and information driven techniques [18]. PI budgetary articulation for additional wellsprings of uncertainly, including model misspecification and clamor fluctuation [19, 20]. Proposed a straightforward charging methodology, where the shoppers were charged in corresponding to their aggregate vitality utilization for the following operation period.

Toward take care of this EMS issue, this paper proposes a calculation taking into account a linear programming (LP) model has been executed to amplify the utility for the gathering of interest with reverence to an arrangement of limitations, for example, least every day vitality usage, most extreme and least hourly load levels, vitality storage room breaking points, and power availability from the primary framework and the DERs. Determined the above point of view, the commitments of this paper is fourfold:

(1) To representation the indeterminate parameters in the calculation in through certainty interims that permits utilizing RO strategies.
(2) To data results from a handy contextual investigation that demonstrates the adequacy of the proposed neighborhood EMS calculation.

The paper is structured as follows: Sect. 2 provides the energy management system. Section 3 provides the energy management system algorithm. Section 4 presents the simulation results. Section 5 provides the conclusion.

2 Energy Management System

Energy administration is the positive, arranged and proficient coordination of obtainment, change, supply and utilization of vitality to get together the supplies, appealing into record ecological and budgetary goals. Vitality administration is a nonstop capacity of vitality chiefs. In the proposed EMS, PV exhibit's primary capacity is to create 500 kW vitality to supply the stores. The square diagram of EMS is showed up in Fig. 1. Wind farm produces 20 kW essentialness supply to the pile. The converter goes about as a buck bolster converter to manufacture the yield voltage range. From the DC converter the data indication of voltage and current is given to the MPPT controller. This controller discovers the bumble hail and prompts the IGBT entryway drive; this significantly diminishes the amount of emphases in the MPPT technique.

Wind homestead pitch controller has a dynamic control framework that can shift the pitch edge of the turbine cutting edges to diminish the torque delivered by the sharp edges in a settled rate turbine and to diminish the rotational pace in variable rate turbines. The variable rate operation of wind electric frameworks yields higher yields for both low and high wind speeds. Battery bank is utilized to store the DC vitality at the voltage of 115 kW. From the battery AC burdens are joined.

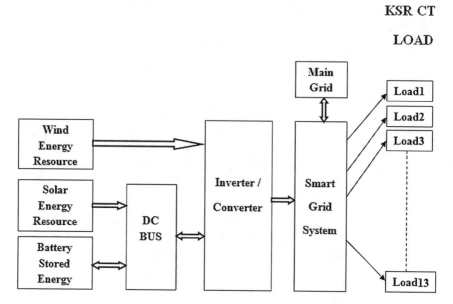

Fig. 1 Architecture of EMS

These AC burdens are organized as lighting burden which considered light and fan for the building considered. In the event that the heap interest is lower than the DERs yield, the abundance vitality will be utilized to implicate the battery. Battery will be totally charged, the force can be inverter from DC to AC for the usage of AC burdens or abundance vitality may be encouraged to the matrix. On the off chance that the DC supply does not exist or be just in part accessible and the interest is on the dc stacks, the battery will supply the force straightforwardly. On the off chance that the heap require be upper more noteworthy than the battery yield, the AC framework will give vitality when it is accessible.

3 Energy Management System Algorithm

3.1 LP Model

The orderly approach for numerical detailing of direct programming technique to take care of vitality administration issue is as per the following.

Step 1: Information the interest variables for ongoing information and pre decided information utilizing Neural Network (NN) of the vitality administration framework.

Step 2: Plan the interest capacity to be improved (greatest or least) as a straight capacity of the diverse variables.

Step 3: Plan the requirements of vitality administration framework, for example, asset constraints, business sector requests, between connections between diverse interest variables.

Step 4: From the considered contextual investigation thirteen unique sorts of requests accessible and three distinct sorts of vitality sources accessible. Let b_{kl} signify the quantity of units of vitality sources l in the unit of requests k, l = 1, 2, 3: k = 1, 2, 3, 4, 5, 6, 7, 8, 9, 10, 11, 12, 13. Let xj be the quantity of units devoured for interest. At that point the aggregate number of units of requests I in the favored source.

$$\sum_{k=1}^{13}\sum_{l=1}^{3} b_{kl}y_l \tag{1}$$

Step 5: Let c_k be the number of units of minimum daily requirement of the demand k and it can be expressed as follows

$$\sum_{k=1}^{13}\sum_{l=1}^{3} b_{kl}y_l \geq c_k \tag{2}$$

where k = 1, 2, 3, ..., 13.

Step 6: For every source l, yl must be either positive or zero.

$$Y_l \geq 0 \tag{3}$$

where l = 1, 2, 3

Step 7: Let d_l be the vitality administration framework yield of vitality source l. along these lines the aggregate yield of vitality administration framework is given underneath

$$m = d_1y_1 + d_2y_2 + \cdots + d_{13}y_{13} \tag{4}$$

Step 8: The most imperative normal for Prediction Intervals (PIs) is their scope likelihood. PI scope likelihood (PICP) is measured by tallying the quantity of target qualities secured by the built PIs.

$$PICP = 1/p \sum_{k=1}^{n} d_k \tag{5}$$

PICP is a measure of legitimacy of PIs developed with a related certainty level.

Step 9: PI standardized found the middle value of width (PINAW) evaluates PIs from this viewpoint and measures how wide they are

$$PINAW = 1/pS \sum_{k=1}^{p} (V_k - M_k) \qquad (6)$$

where V_k, M_k furthest utmost and lower cutoff of interest, S is the scope of the basic target characterized as the distinction between its base and greatest qualities. PINAW is the normal width of PIs as a rate of the fundamental target range.

3.2 Robust Model

The robust counterpart of model (7) is presented as the equivalent linear model below

$$\text{Min} \left[\lambda_t^s e_t^s + \lambda_t^w e_t^{AW} + \lambda_t^{PV} e_t^{APV} - \sum_{i=1}^{Nc} u_{i,t}(e_{i,t}) \right] \sum_{h=1}^{24-t} [\lambda_{t+h}^s \{e_{t+h}^s\} + \lambda_{t+h}^w \{e_{t+h}^{AW}\}$$
$$+ \lambda_{t+h}^{PV} \{e_{t+h}^{APV}\} - \sum_{i=1}^{Nc} u_{i,t+h}(e_{i,t+h})] \qquad (7)$$

The vulnerabilities in vitality value, primary framework, wind force and sun oriented force, separately, in a strong way. The variables of the hearty model are the variables of the introductory model its double variables, and some assistant variables. The double variables $u\_(i,t + h)(e\_(i,t + h))$ are identified with the instability in vitality value, while the double variables $\sum\_(t + h+1)^{\wedge}W$ and primary lattices are identified with the vulnerability in wind and sun oriented force. The helper variablesy$\_(t + h)^{\wedge}s$, y$\_(t + h+1)^{\wedge}w$ and y$\_(t + h+1)^{\wedge}pv$ are utilized to accomplish a straight streamlining issue.

The parameters permit controlling the level of strength of the arrangement regarding the vulnerabilities in vitality cost from the fundamental matrix and in the sun oriented power and wind power creation, individually. In the event that we select zero, the effects of the value deviations on the goal capacity. The most extreme worth results in a preservationist choice that considers the synchronous effects of all value deviations on the goal capacity. Not as a matter of course number, the control parameters of wind uncertainty$\_$ (t + h+1)$^{\wedge}$W, h = 1...24-t, can be chosen in the interim.

The parameters decide the insurance levels for limitation of the starting model against the instability in the accessible wind power production P$\_$(t + h+1)$^{\wedge}$AW, h = 1...24-t. the control parameters of sun powered illumination level uncertainty T$\_$(t + h+1)$^{\wedge}$PV, h = 1...24-t, can be chosen in the interim. These parameters decide the insurance levels for limitation of the starting model against the vulnerability in the accessible sun based force production P$\_$(t + h+1)$^{\wedge}$APV, h = 1...24-t

For straightforwardness, take note of that the proposed vigorous model is not versatile, i.e., it has no plan of action. Accept that the primary framework and the vitality stockpiling give final resort adjusting vitality.

3.3   Calculation

Given the vigorous issue (7) in the past subsection, propose a vitality administration calculation, whose working is portrayed underneath:

Step 1: Instate t = 1.

Step 2: The cost of the vitality from the fundamental network for current hour is sent to the EMS administrator, e.g., 10 min before hour t.

Step 3: The wind maker sends the wind energy to be delivered in hour to the EMS administrator, e.g., 10 min before hour t.

Step 4: The sun oriented maker sends the sunlight based energy to be delivered in hour to the EMS administrator, e.g., 10 min before hour t.

Step 5: Value responsive requests send to the EMS administrator their heap levels and utilities for current hour t, and the remaining hours of the day, e.g., 10 min before hour t.

Step 6: Utilizing verifiable time arrangement of vitality costs from the primary matrix and wind power preparations from the DER, the EMS administrator registers limits at both vitality costs and wind and sun oriented creations for the remaining hours of the day. The EMS administrator is capable of the danger connected with the vulnerability demonstrating.

Step 7: The EMS administrator understands the powerful model (5.3) and acquires the divisions of vitality to be supplied from the fundamental network, the wind maker, the sun powered maker and the stockpiling unit, and in addition the vitality to be supplied to every specific request, the part of vitality to be put away, and the portion of vitality to be sold to the primary matrix in hour t.

Step 8: The EMS administrator conveys the ideal choices got in Step (7) above to both vitality suppliers and requests, e.g., 5 min before hour t.

Step 9: Redesign the hour counter. In the event that, go to Step (2). On the off chance that, the calculation finish up for the present day and moves to the following day.

In Step (6), the EMS fuses the new data it gets from hour, redesigning the determining of the certainty limits of the dubious information for the remaining hours of the day.

4 Result and Discussions

The proposed LP technique reproduction be developed with MATLAB 7.10 software package as well as the structure arrangement is Intel Core i5-2410M Processor by way of 2.90 GHz speed and 4 GB RAM. In proposed exertion three power sources, 13 load and IEEE 14 bus organization considered as container study, over specified moment in time intervals. The computational consequences of energy management system trouble attained by the projected LP technique for the three power sources analyzed.

4.1 Case Study—IEEE 14 Bus System

This revision is conventional absent at the situation of preparation, procedure, be in charge of with money-spinning estimate. They continue living of make use of in significant the importance in addition to stage point of observation of consignment buses, and real and reactive control flow larger than transmission lines, and real and

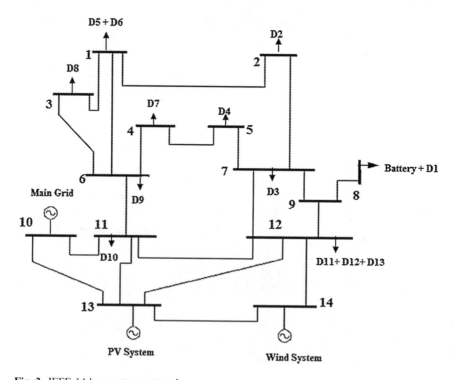

Fig. 2 IEEE 14 bus systems network

reactive energy by means of the function of be injected at the buses. For this occupation the linear programming method is use for arithmetic investigation.

The function of this project is to enlarge a MATLAB program make best use of the consumption of the group of load while it is subjected to a located of constraint. Figure 2 shows the whole load and the power sources. This LP algorithm allows the group of demand to purchase, stock up as well as put up for sale energy at appropriate times to regulate the hourly consignment level to analyze voltages, active and reactive power control on each buses second-hand for IEEE 14 bus systems. With principal IEEE 5 bus structure is calculated by using supply calculation as well as compared through MATLAB plan consequences and subsequently IEEE 14 bus system MATLAB program is executed with the part data. This category of study is positive used for solving the organize flow problem in diverse power systems which will practical to calculate the strange quantities. The energy management system considers 13 load positioned 14-bus scheme. The DERs with a power storage capacity are situated at bus 14.

4.2 Load Demand Data

This paper considers the EMS load situated within the K.S.Rangasamy College of Technology (KSRCT) campus. The main grid is associated to bus3. The whole control supplied beginning the DERs, main grid and power storage space facility. The accessible DSM capability is in use as a division of the listed require of the consequent hour. In adding, the devoted DSM capabilities include to exist put reverse to the demand during the same day in order to the behavior of power charge sensitive smart appliance. Condition an elevated degree of DERs generation is considered; the major giving of DSM is leveling the demand and dropping the DERs inconsistency.

4.3 Simulation Model for Proposed EMS

The representations have been elaborated and it works but the simulation times make it impossible to be used. One simulation with a very simple case take additional than 8 h to be run. However, taking into account that it has been part of this project, it is going to be explained. Demand side management toward the stand representation. Since it is a linear programming model, the mutual study just requires all the factor to facilitate be additional previously for each possibility. This means that if the reader has understood properly the mathematical formulation of previous chapter's equations should be enough to understand the whole model. The new factors in the system constraints of the base model have been highlighted in bold. This feature allows the model to curtail DERs while the overload of it avoids

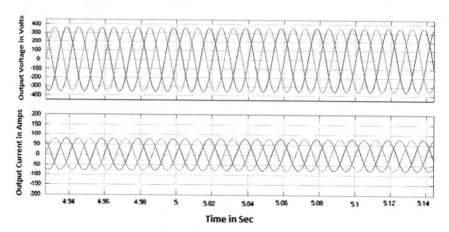

Fig. 3 Output voltages and current—wind system

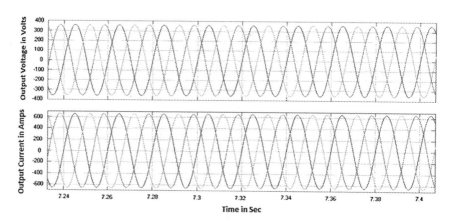

Fig. 4 Output voltages and current—PV system

achieve a practicable resolution (production larger that load) however at the same time wind curtailment is avoided.

If modeling of flexibility in the generation and load balances that evaluates requirement considering operational with capability price. The change contact simply on condensed require because the simply restriction happening shifted order that affects more than a few instants of instance is used for every day, the smallest size so as to has been calculated to produce the clusters. Within order to estimate the presentation of the mold and comprehend the contact of shift DSM on the demand-supply stability, this part presents a easy case learn for 24 h. The demand is modeled in a sinusoidal approach to afford various type of difference. The demand

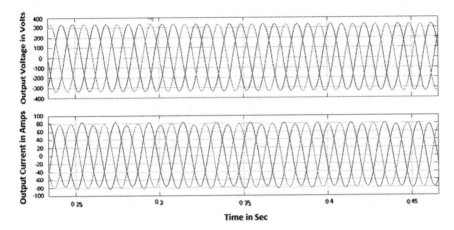

Fig. 5 Output voltages and current—academic block

Table 1 Consumption of power

Energy source	Rated capacity in KW	Output power in KW
Solar	500	252
Wind	42.5	28
Main grid	900	193
Total	1442.5	473

Total energy consumption in demand 469.55 kW

Fig. 6 Peak energy consumption at KSRCT

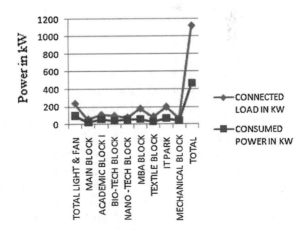

would not be sufficient to reach this product in a realistic submission, peripheral manage techniques would be required. DMS is that there are not efficiency losses in its utilization and it is a resource that would require just a change in the policies to be used.

The wind produces 28 kW from generating station. The wind power plant produces 28 kW and the waveforms for output voltage and current in the wind energy system are shown in Fig. 3. The photo voltaic power plant produces 252 kW and the waveforms for output voltage and current in photo voltaic energy system are shown in Fig. 4. The power consumed in academic block is about 60.4 kW and the waveforms for output voltage and current in the academic block are shown in Fig. 5. Consumption of power the various is shown in Table 1.

The Fig. 4 shows the entire connected demand and whole consumed power for the possible LP method of the point in time stage is shown in Fig. 6.

5 Conclusion

The projected method provides the actual time monitoring as well as control of load side management system. It improves the presentation of structure load to the stage of distributed energy resources infiltration. The photovoltaic and wind representation was planned using MATLAB. The peak demand of proposed system was satisfied from beginning to end in all the buses and also the protection system for photovoltaic and wind power place was implemented. The distributed energy resources, demand and the power from the main grid were related toward the IEEE fourteen bus system. While connecting all these energy sources to the bus system, the losses can be reduced by using the inductance in each bus. The proposed approaches satisfy the group of demands within the energy management system and in addition get better the system effectiveness and curtail the fatalities. The overload energy from distributed energy resources preserve also is store in the succession moreover it possibly will be utilize by the demand when in attendance is a demand of energy. The distributed energy resources used in the real time system improves the generation side flexibility.

Synchronization of value responsive requests through an EMS is imperative to encourage the correspondence of the requests with the suppliers and to perceive the full planned of SG innovation. The availability of 2-way correspondence empowers viable interest reaction that outcomes in most extreme utility for the requests. Instability demonstrating is imperative to stay away from interest confinement while minimizing supply cost (or expanding interest utility). A strong streamlining strategy gives an adaptable instability administration instrument that permits changing the favored level of conservatism. The proposed EMS calculation is most helpful at a group of interconnected cost responsive burden, for example, a KSRCT grounds. The proposed EMS is effectively executed at a little cost gave that SG innovation is sent, and along these lines, 2-way correspondence is accessible.

References

1. Brekken, T.K.A., Yokochi, A., von Jouanne, A., Yen, Z.Z., Hapke, H.M., Halamay, D.A.: Optimal energy storage sizing and control for wind energy applications. IEEE Trans. Sustain. Energy **2**, 69–77 (2011)
2. Matos, M.A., Bessa, R.J.: Setting the operating reserve using probabilistic wind energy forecasts. IEEE Trans. Energy Syst. **26**(2), 594–603 (2011)
3. Samadi, P., Mohsenian-Rad, H., Schober, R., Wong, V.W.S.: Advanced demand side management for the future smart grid using mechanism design. IEEE Trans. Smart Grid **3**(3), 1170–1180 (2012)
4. Atzeni, I., Ordonez, L.G., Scutari, G., Palomar, D.P., Fonollosa, J.R.: Demand-side management via distributed energy generation and storage optimization. IEEE Trans. Smart Grid **4**(2), 866–876 (2013)
5. Atzeni, I., Ordonez, L.G., Scutari, G., Palomar, D.P., Fonollosa, J.R.: Noncooperative and cooperative optimization of distributed energy generation and storage in the demand-side of the smart grid. IEEE Trans. Signal Process. **61**(10), 2454–2472 (2013)
6. Pandic, H., Morales, J.M., Conejo, A.J., Kuzle, I.: Offering model for a virtual power plant based on stochastic programming. Appl. Energy **105**, 282–292 (2013)
7. Baric, M., Borrelli, F.: Decentralized robust control invariance for a network of storage devices. IEEE Trans. Power Syst. **57**(4), 1018–1024 (2003)
8. Thatte, A.A., Xie, L.: Towards a unified operational value index of energy storage in smart grid environment. IEEE Trans. Smart Grid **3**(3), 1418–1426 (2012)
9. Baringo, L., Conejo, A.J.: Offering strategy via robust optimization. IEEE Trans. Power Syst. **26**(3), 1418–1425 (2011)
10. Ferreira, R.S., Barroso, L.A., Carvalho, M.M.: Demand response models with correlated price data: a robust optimization approach. Appl. Energy **96**, 133–149 (2012)
11. Shao, S., Pipattanasomporn, M., Rahman, S.: Grid integration of electric vehicles and demand response with customer choice. IEEE Trans. Smart Grid **3**, 543–550 (2012)
12. Safdarian, A., Fotuhi-Firuzabad, M., Lehtonen, M.: A stochastic framework for short-term operation of a distribution company. IEEE Trans. Energy Syst. **28**(4), 4712–4721 (2013)
13. Venkatesan, N., Solanki, J., Solanki, S.K.: Residential demand response model and impact on voltage profile and losses of an electric distribution network. Appl. Energy **96**, 84–91 (2012)
14. Mejia-Giraldo, D.: Adjustable decisions for reducing the price of robustness of capacity expansion planning. IEEE Trans. Power Syst. **29**(4), 1573–1582 (2004)
15. Sioshansi, R., Short, W.: Evaluating the impacts of real-time pricing on the usage of wind generation. IEEE Trans. Power Syst. **24**(2), 516–524 (2009)
16. Hossain, M.A., Hossain, M.Z., Rahman, M.M., Rahman, M.A.: Perspective and challenge of tidal power in Bangladesh. TELKOMINIKA Indones. J. Electr. Eng. **12**(11), 1127-1130 (2014)
17. Shahinzadeh, H., Hasanalizadeh-Khosroshahi, A.: Implementation of smart metering systems: challenges and solutions. TELKOMINIKA Indones. J. Electr. Eng. **12**(7), 5104–5109 (2014)
18. Foley, A.M., Leahy, P.G., McKeogh, E.J.: Current methods and advances in forecasting of wind power generation. IEEE Trans. Renew. Energy **37**(1), 1–8 (2012)
19. Khosravi, E., Mazloumi, S., Nahavandi, D.C., Van Lint, J.: A genetic algorithm-based method for improving quality of travel time prediction intervals. Transp. Res. Part C Emerg. Technol. **19**(6), 1364–1376 (2011)
20. Mohsenian-Rad, V.W.S., Wong, J., Jatskevich, R.S., Leon-Garcia, A.: Autonomous demand-side management based on game theoretic energy consumption scheduling for the future smart grid. IEEE Trans. Smart Grid **1**(3), 320–331 (2010)

An Electric Vehicle Powertrain Design

**G.N. Ripujit, K. Avinash, P. Kranthi Kumar Reddy
and M. Santhosh Rani**

Abstract The powertrain is an integral part in any vehicle and this article addresses powertrain design for an Electric Vehicle emphasising on cost effective range extension by coupling a gearbox with the electric motor thereby, reducing the torque requirement on the motor in turn reducing the battery capacity.

Keywords Transmission Gearbox · Vehicular Torque Requirements · Bearings · Couplings · Electric Vehicular Range · Electric Motor · Bending Stress · Crankshaft (Powershaft)

1 Introduction

Powertrain is the mechanism which transmits the drive from the power source to the axle. The power source may be an IC engine (Gasoline or Diesel) or an electric motor [Electric Vehicle (EV)] or a combination of both (Hybrid Vehicle). Generally, an IC engine powertrain consists of the engine, gearbox, a transmission shaft which ends at a differential axle.

A gearbox acts as a torque multiplier as the torque provided by the IC Engine may not be sufficient to carry forward the respective payload. This is because an IC Engine's torque output is variable i.e., it is different at different speeds. Whereas, when it comes to an Electric motor, the motor can give its highest torque output from zero rpm itself. Therefore, the power train of an EV comprises of an electric

G.N. Ripujit (✉) · K. Avinash · P. Kranthi Kumar Reddy · M. Santhosh Rani
Mechatronics Department, SRM University, Kattankulathur, Chennai, India
e-mail: ripujit.gindam@gmail.com

K. Avinash
e-mail: avinashkalluri@gmail.com

P. Kranthi Kumar Reddy
e-mail: kranthikumarreddys1994@gmail.com

M. Santhosh Rani
e-mail: santhoshrani.m@ktr.srmuniv.ac.in

© Springer Science+Business Media Singapore 2017
P. Deiva Sundari et al. (eds.), *Proceedings of 2nd International Conference on Intelligent Computing and Applications*, Advances in Intelligent Systems and Computing 467, DOI 10.1007/978-981-10-1645-5_37

motor and an optional gearbox which is either used or left out depending upon the manufacturer's priorities.

Range of an EV mainly depends on the capacity of the battery pack driving the motor. The range can be increased in two methods.

- Employing a larger battery pack
- Coupling a gearbox to the motor

The torque generated by an electric motor is directly proportional to the current absorbed. So, the greater the torque requirement, more current is absorbed thereby, depleting the capacity of the battery pack faster. By employing a gearbox, the torque requirement at the motor end can be restricted hence, the capacity of the battery pack is restrained for a longer duration which leads to extended range and is also economical. Keeping this in consideration, the powertrain for this electric vehicle was designed and fabricated.

2 Motor

A machine which converts electrical energy to mechanical energy is called a motor and the reverse of it is a generator. EVs generally employ self commutated motors (ground vehicles). There are many motors which are suitable and are distinguished according to the motor commutation techniques which are as follows (Fig. 1).

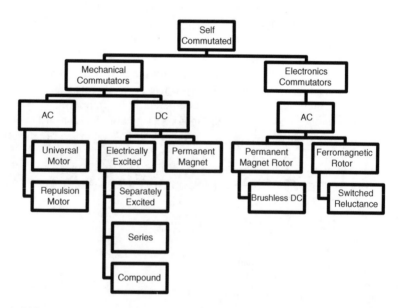

Fig. 1 Different types of self commutated motors

The selection of a motor for a specific EV depends on various parameters of which the most important and precedential one is the torque requirement. Its calculation is shown below.

3 Vehicle Dynamics—Torque Requirement Calculation [1].

Torque of a vehicle can be defined as the amount of load a vehicle is capable of pulling. The torque required for a vehicle is calculated from four different variable force components the vehicle has to overcome which are:

(i)   Inertial Force ($\mathbf{F_i}$)
(ii) Aerodynamic Resistance ($\mathbf{F_a}$)
(iii) Gradient Force ($\mathbf{F_g}$)
(iv) Road Resistance ($\mathbf{F_r}$)

$$F_i = \left[m_b + J_c \cdot \left(\frac{i_d}{r_{\omega st}} \right)^2 + J_e \cdot \left(\frac{i_d \cdot i_g}{r_{\omega st}} \right)^2 \right] \cdot a^2$$

where
m_b vehicle gross weight
J_c moment of inertia of Cardan Shaft (not present)
i_d final drive ratio (differential to wheel)
$r_{\omega st}$ radius of the vehicle wheel
J_e moment of inertia of Crankshaft
i_g gearbox ratio (connecting rod to power shaft)
a desired vehicle acceleration

$$F_a = C_d \cdot A \cdot \delta_{air} \cdot v_r^2$$

where
C_d Coefficient of drag
A Frontal Area of the vehicle
δ_{air} density of air
v_r velocity of the vehicle

$$F_g = m_b \cdot g \sin \alpha$$

where
m_b vehicle gross weight
g acceleration due to gravity
α angle of inclination

F_r depends on terrain, it is different for off road conditions, track conditions, asphalt conditions. For highway roads, it is 150 N

Torque Required,

$$T = (F_i + F_a + F_g + F_r) \cdot r$$

where
r radius of the vehicle wheel

4 Motor and Gearbox Selection

After calculating the torque required for the vehicle to possess the desired characteristics, the motor has to be selected with emphasis made on few reasons.

- Current rating at max torque
- Max rpm of the motor
- Efficiency of the motor

Considering these constraints, a thorough search for a suitable motor was conducted and finally a 48 V 5KW Golden motor was selected to serve the purpose coupled with a stock gearbox of Bajaj RE Compact. The stock gearbox was chosen to be coupled with the motor as it was already designed to be coupled with an IC Engine whose output characteristics such as torque and power were similar to that of the motor. The motor characteristics are shown below and the calculations pertaining for the selection of this report are given further in the report (Fig. 2; Table 1).

4.1 Calculations

$$C_d = 0.3$$
$$A = (1.3) \times (1.69) = 2.2 \text{ m}^2$$
$$\delta_{air} = 1.3$$
$$v_r = 35 \text{ Kmph}$$
$$= 9.72 \text{ m/s (average velocity taken)}$$
$$F_a = (0.3) \times (2.2) \times (1.3) \times (9.72)^2$$
$$= 40.53 \text{ N}$$
$$m_b = 820 \text{ kg}$$

(chassis, weight of 4 passengers with an average of 75 kg and batteries)

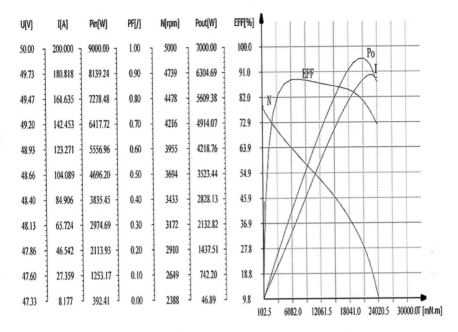

U[V]	I[A]	Pin[W]	PF[/]	N[rpm]	Pout[W]	EFF[%]
50.00	200.000	9000.00	1.00	5000	7000.00	100.0
49.73	180.818	8139.24	0.90	4739	6304.69	91.0
49.47	161.635	7278.48	0.80	4478	5609.38	82.0
49.20	142.453	6417.72	0.70	4216	4914.07	72.9
48.93	123.271	5556.96	0.60	3955	4218.76	63.9
48.66	104.089	4696.20	0.50	3694	3523.44	54.9
48.40	84.906	3835.45	0.40	3433	2828.13	45.9
48.13	65.724	2974.69	0.30	3172	2132.82	36.9
47.86	46.542	2113.93	0.20	2910	1437.51	27.8
47.60	27.359	1253.17	0.10	2649	742.20	18.8
47.33	8.177	392.41	0.00	2388	46.89	9.8

Fig. 2 Golden motor HP5000 data curve (*Source* www.goldenmotor.com)

$$\alpha - 18° \text{ (maximum scenario)}$$
$$g - 9.81 \text{ m/s}^2$$
$$F_g = (820) \times (9.81) \times \sin 18°$$
$$= 2483.26 \text{ N}$$
$$J_e - 1.68 \times 10^{-4}$$

A gearbox and differential of an existing autorickshaw are used in the vehicle. Hence, the ratios in the gearbox are

gear 1—5:1
gear 2—2.9:1
gear 3—1.84:1
gear 4—1.12:1
Final drive—4.125:1
i_d—4.125
i_g—5 (starting condition)
a—0.555 m/s^2
r_{ost}—0.15 m

Table 1 Golden motor HP5000 statistics

Company: GOLDEN MOTOR
Type: HPM48-5000
No.: G20130514008
Operator: 001
Date: 2013-5-14

rated U: 48 V
rated I: 120 A
rated P.: 5000 W
rated N: 3500 RPM

Items No.	Voltage (V)	Current (A)	P. input (W)	P. factor (PF)	Frequency (Hz)	Torque (mN m)	Rotate (rpm)	P. output (W)	Efficiency (%)
1	47.99	8.177	392.41	1.000	0.00	360.0	4389	165.45	42.2
2	47.98	8.538	409.70	1.000	0.00	242.5	4384	111.32	27.2
3	47.98	9.967	478.17	1.000	0.00	102.5	4369	46.89	9.8
4	47.95	13.222	633.99	1.000	0.00	577.5	4335	262.14	41.4
5	47.91	18.686	895.30	1.000	0.00	1412.5	4279	632.89	70.7
6	47.86	26.320	1259.60	1.000	0.00	2415.0	4204	1063.11	84.4
7	47.80	35.715	1707.06	1.000	0.00	3552.5	4116	1531.11	89.7
8	47.72	46.523	2219.96	1.000	0.00	4812.5	4021	2026.29	91.3
9	47.63	58.475	2785.48	1.000	0.00	6182.5	3923	2559.68	91.2
10	47.55	71.460	3397.57	1.000	0.00	7680.0	3826	3076.83	90.6
11	47.46	85.414	4053.55	1.000	0.00	9262.5	3734	3621.59	89.3
12	47.38	100.283	4751.16	1.000	0.00	10,920.0	3647	4170.18	87.8
13	47.40	116.273	5511.32	1.000	0.00	12,647.5	3573	4731.89	85.9
14	47.41	132.690	6291.16	1.000	0.00	14,387.5	3501	5274.41	83.8
15	47.39	149.915	7104.47	1.000	0.00	16,157.5	3429	5801.47	81.7
16	47.37	167.085	7915.23	1.000	0.00	17,950.0	3350	6296.60	79.5
17	47.33	174.525	8260.27	1.000	0.00	19,495.0	3206	6544.60	79.2
18	47.33	174.870	8277.47	1.000	0.00	20,797.5	2991	6520.18	78.8

$$F_i = \left[m_b + J_c \cdot \left(\frac{i_d}{r_{\omega st}} \right)^2 + J_e \cdot \left(\frac{i_d i_g}{r_{\omega st}} \right)^2 \right] \cdot a^2$$

$$= \left[(820) + (1.68 \times 10^{-4}) \cdot \left(\frac{4.125 \times 5}{0.15} \right)^2 \right] (0.555)^2$$

$$= (820.0231) \times (0.308)$$

$$= 252.59 \text{ N}$$

$$T = (F_i + F_a + F_g + F_r) \cdot r$$
$$= (252.59 + 40.53 + 2483.26 + 150) \times (0.15)$$
$$= (2926.38).(0.15) = 438.96 \text{ N-m.}$$

With these gear ratios, the maximum torque required at the motor $((T_M)_{max})$ would be

$$(T_M)_{max} = \frac{T}{i_d \cdot i_g}$$
$$= \frac{438.96}{(5 \times 4.125)}$$
$$= 21.2 \text{ N-m.}$$

The top speed of the vehicle can be calculated using the motor datasheet and the gearbox ratios.

$$rpm_W = \frac{rpm_M}{i_d \cdot i_g}$$

where

rpm_W—rpm at wheel
rpm_M—rpm at motor
maximum speed of the motor—4389 rpm
gear ratio, for the top speed, highest gear is used—1.12
differential ratio—4.125

$$rpm_W = \frac{4389}{1.12 \times 4.125}$$
$$= 950 \text{ rpm}$$
$$= 14 \text{ m/s}$$
$$= 53.72 \text{ Kmph}$$

Therefore, the vehicle when mounted with this motor coupled with the specified gearbox can deliver a similar performance as that of its IC Engine counterparts.

5 Motor and Gearbox Coupling

The motor shaft and gearbox shaft are coupled using a rigid coupling with a clutch mechanism. The purpose of a clutch in transmission is to increase the life of gear teeth by reducing the grinding effect in dynamic mating of two rotating shafts. This is required when there is a need to mate two gears rotating at different speeds. Theoretically, clutch is not necessary in EVs as there is not much difference in the speeds of the motor shaft and gear shaft. But it is used in this vehicle to taking into consideration the hypothetical situations where there is a need to shift to a lower gear in high speeds.

The rigid coupling was designed considering the following calculations.

The gearbox assembly used is same as the IC engine vehicle, so the crankshaft should be coupled with the motor shaft in order to transmit the torque. A set screw rigid coupling is used for this purpose as it is easy to be fabricated and also due to the space constraints for the motor mounting.

The shaft diameter is calculated by using the torque it transmits. But in this case, the torque is transmitted by the motor shaft and this dimension is fixed. So all the further calculations are done, taking this shaft diameter into the consideration for the coupling.

diameter of the motor shaft, d—22 mm
shear stress in yield, τ_y—230 N/mm^2
σ_c, normal yield strength—450 N/mm^2
torque transmitted by the coupling,

Hub dimensions [2],
Hub diameter,

$$d = 1.75d + 6.5 \text{ mm}$$
$$= 1.75(22) + 6.5 = 45 \text{ mm}.$$

Hub length,

$$l = 1.5d = 1.5 \times 22 = 33 \text{ mm}.$$

Key dimensions [2],
Length of the key,

$$L_k = \frac{8T}{d^2 \cdot T_y} = \frac{8 \times 9000}{22^2 \times 230} = \frac{72000}{111320} = 6.467 \text{ mm}.$$

For a square key,

$$b = \frac{d}{4} = \frac{22}{4} = 5.5 \text{ mm}.$$

Check for crushing of the coupling,

$$
\begin{aligned}
\sigma &= \frac{16 \times T}{L_k \cdot d^2} \\
&= \frac{16 \times 9000}{22^2 \times 6.467} \\
&= \frac{144000}{3130} \\
&= 46 \text{ N/mm}^2.
\end{aligned}
$$

As $\sigma < \sigma_c$ hence the key design is safe from crushing.

In the design of the coupling, hub diameter and the lengths of the key are taken more than the values acquired from the calculations, taking a factor of safety into consideration. The fabricated coupling is as follows (Figs. 3 and 4):

Fig. 3 Square keys

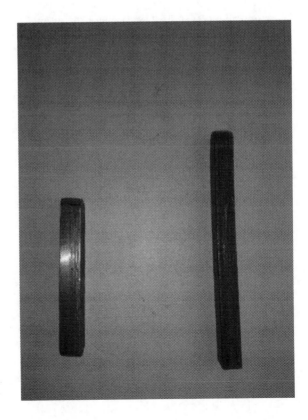

Fig. 4 Motor gearbox coupling

6 Motor Mount

As an old working auto rickshaw is used for the purpose, the mounting of the motor needs to be studied and designed for this purpose. The engine is removed and the crankshaft is connected to the motor using a coupling. The motor cannot stand by on its own on this coupling as its weight is pretty high and would result in the bending of the power shaft, so a motor mount is designed to bear its weight [1].

Designing the mounting using bending,

$$T = \frac{My}{I}$$

where

T   bending stress
M   bending moment
y   distance from neutral axis
I   moment of inertia [1].

$$I = \frac{b.d^3}{12}.$$

Central loading is on the lower beam, therefore for bending moment,

$$M = 11 \times 76.2$$
$$= 838.2 \text{ Kg mm}$$

b—154.2 mm

d—6 mm

$$I = 6^3 \times \frac{154.2}{12}$$
$$= \frac{33307.2}{12}$$
$$= 2775.2 \ \text{mm}^4.$$

y—3 mm

$$T = \frac{838.2 \times 3}{2775.2}$$
$$= 0.911 \ \text{kg/mm}^2$$
$$= 8.93 \ \text{N/mm}^2.$$

The selected material is mild steel, which has yield strength of 250 N/mm², [3]. so this steel provides the enough bearing capacity for the motor mount for the 11 kg motor. This is taken a bit high not only accounting the factor of safety, but also to take the sudden loads which might be created because of the vibrations caused by the motor in the starting condition.

Auto rickshaw has design constraints for the mount, so the design strategy followed is to find the material using these constraints instead of finding the thickness of the material using its properties. The final mount was designed using DS Solidworks and its model drawing is presented (Fig. 5):

Fig. 5 Motor mount model drawing

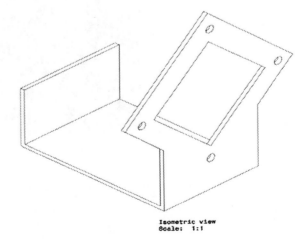

Isometric view
Scale: 1:1

7 Conclusion

A powertrain for an electric vehicle was designed and fabricated to meet the design constraint of cost effective range extension. Few illustrations are shown (Figs. 6, 7, 8, 9, 10 and 11).

Fig. 6 Original crankshaft

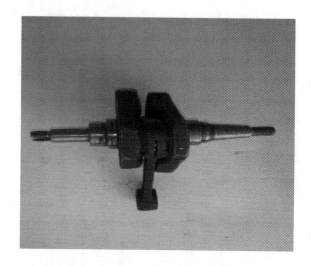

Fig. 7 Modified crankshaft (referred as powershaft)

Fig. 8 Gearbox

Fig. 9 Powershaft introduced into gearbox

Fig. 10 The gearbox and powershaft fixed in the coupling

Fig. 11 Complete functional assembly of gearbox, powershaft and electric motor coupled

References

1. Sharma, P.C., Agarwal, D.K.: A Textbook of Machine Design. S.K.Kataria and Sons, New Delhi (1998)
2. Shigley, J.E.: Mechanical Engineering Design. McGraw Hill, New York City (1986)
3. Dhawan, R.K.: A textbook of Machine Drawing. S. Chand and Co., Coimbatore (1996)

Innovation in the Operation of UAV Sensors

B.S.M. Augustine, M. Mohankumar and T.M. Yoga Anandh

Abstract The paper is aimed at reviewing the existing trends of UAV sensors and brings out innovation in the operation of each UAV sensor for increased capability in performance. It is well known that unmanned aerial vehicles (UAVs) play a major role in defense as replacements for manned aircraft tasks such as surveillance, Electronic warfare, reconnaissance and evaluation of the targets of enemy. Besides the defense application, they are pleasingly in demand in civilian application of disaster management. This paper brings out increase in current growth of UAV by research specially in performing different operational missions by the inclusion of imaging and non-imaging sensors. The objective of this research is to prevent the modified versions of the imaging sensors of TV and infrared sensors located in precision painting systems and synthetic aperture radar and non imaging sensor like electronic surveillance and increase their operational capabilities.

Keywords Sensors · Surveillance · Reconnaissance · Radar · Operational missions · Unmanned aerial vehicle · Synthetic aperture

1 Introduction

The unmanned aerial vehicle (UAVs) is increasingly in demand in the armed military forces. It has been proved that UAVs could perform more risky and dangerous activities successfully. In addition, UAVs get valuable information from the enemy's territories and also provide excellent communication to plan strategy and effectively enhance the tactics for out seating the planning of enemy with counter challenge. Modern UAVs platform consists of various items such as

B.S.M. Augustine (✉) · M. Mohankumar · T.M. Yoga Anandh
Department of Aeronautical Engineering, KCG College of Technology,
Chennai, India
e-mail: augustinesuvi@gmail.com

M. Mohankumar
e-mail: mohank.mk49@gmail.com

© Springer Science+Business Media Singapore 2017   451
P. Deiva Sundari et al. (eds.), *Proceedings of 2nd International Conference
on Intelligent Computing and Applications*, Advances in Intelligent Systems
and Computing 467, DOI 10.1007/978-981-10-1645-5_38

gyroplane, glider, fixed wing, single or multi-engine aircraft and solar powered. Further, they are capable of performing at low altitude as well as high altitude and having appreciably long endurance. Our study of research is based on to develop better operation of UAVs sensors to enhance the standard of their performance in turn advance the payload technologies. The present growth in Indian and global UAV development is thoroughly examined and introduced all the possible means to improve their performance. Besides, an in-depth study on imaging and non-imaging sensors would deeply give the account of the emerging trends of their growth and study the ways [1] to enhance their performance.

The desirable attributes of a UAV payload are small size, low weight, low power consumption, simple interface and autonomous operation capability. The degree of autonomy refers to self correction, error reporting and operation during command link breaks [2].

2 Imaging Sensors

Imaging sensors being the UAV payload can be operated in operational modes such as stereo imaging, spot light imaging and wide area search. Imaging sensors consists of visible light (EO) imager, infrared (IR) imager and synthetic aperture radar (SAR).

The flying operation is useful only during the morning and afternoon sessions as the climate aspects is more suitable as the cloud would be bright for effective at short ranges. As the imaging sensors being operative on thermal effects, they are very effective in assisting to differentiate between movable and non-movable transporters and centers. In addition, the high range imaging sensors are highly capable than the visible waves. Further, with a higher wavelength Infrared imager are less effective. When compared to visible light imager for a specified aperture and Infrared imagers are analyzed in minimum way [3].

2.1 Visible Light Imager (EO)

EO sensor functions with higher range being poor in resolution and in addition with simple being easy to work in it. They are very effectively during operation but can be effective only during daytime clear weather conditions at short ranges.

Over the years, today (EO) sensors have developed into dimensional Focal Planes Array (FPA) where in silicon used as doctor element and also as the read out circuitry. The camera becomes function able only up to long range or high speed imaging application [4]. It is due to the fact that there is a time delay between the reading of odd and even fields. It consists of a progressive scan camera image having the following advantages:

- Blur free image as pixels and lines are read sequentially.
- Flicker free image through high frame rate.

The EO sensor for future imaging payload configuration would use progressive scan camera, CCD camera will have lower noise at faster readout speeds thereby providing better resolutions. These cameras will have digital interfaces and analog to digital conversion will increasingly take place within the camera, freezing the image processor to handle more complex tasks processing functions [5].

2.2 Infrared Sensors

IR sensors operate exactly like EO sensors but can be used to works with thermal effects. They are very effective in assisting to differentiate between movable and non-movable transporters and centers. Further, Infrared waves go through atmospheric frame very easily and easily than the other imaging waves do. But they have the same limitations as EO sensors in connection with the aspects of clear weather as well as having relative small distance operation. In addition, when compared to visible imager. Infrared imager has high resolution for a given aperture and IR images are less easily interprets IR imaging in dependent on the variation of thermal emissivity. Two important area of infrared spectrum used for thermal imaging are 3–5 and 8–12 μm. These wavelengths are used as they are commonly generated by both natural and man-made objects and are not completely absorbed by the atmosphere. Both regions offer advantages and disadvantages for thermal imaging. Hence, the thermal imagers designed for applications which involve viewing the natural scene, or objects at near ambient temperature, are designed to operate in the 8–13 μm atmospheric window, which hot efflux tracking applications will more usually operate in the 3–5 μm band [6, 7].

In the Indian scenario, as a part of tactical UAV development, ADE has developed an imaging payload consisting of an EO and IR imager. The EO sensor (Fig. 1) is based on PULINIX CCD camera (570HX420V) and FUJINON H14 X 10.5 B–Y41 zoom lens. This sensor provides continuous zoom from 2.8° to 35°. The IR imager (Fig. 2) is an 8–12 μm MCT sensor from M/S ELOP Israel. This imager is built using a 128 element linear array. This system has three discrete fields of view.

ADE has developed a turret with two gimbals configuration (Fig. 3) having 100 micro radians (RMS) stabilization accuracy for use in its tactical UAV Nishant. This system is utilized for medium range imaging up to 4–6 km. For imaging up to 10–12 km, ADE is presently developing a turret with four gimbals configuration aiming at 15–30 micro radian stabilization accuracy. The proposed gimbals

Fig. 1 Nishant EO sensor

Fig. 2 Nishant IR sensors

Fig. 3 Nishant gimbal

Fig. 4 Four gimbal turret

configuration is shown in Fig. 4. Outer gimbals are used for coarse correction, while inner gimbals are for stabilization. The turret can house DTV, ICCD, thermal imager and laser range finder/designator combinations of payloads [2].

2.3 Synthetic Aperture Radar

The synthetic aperture radar is the most versatile imaging sensor that can be used in both day and night and in adverse weather conditions. Recent advances in technology have yielded much compact and powerful imaging radar as a viable payload for medium and high altitude UAVs with reasonable payload capability. Presently SARs weighing about 80 kg are available. The approaches followed in the development of SARs and UAVs are: development of UAV specific system, use of standard SARs developed for other commercial programs. In medium incorporate larger SAR system operating in X band.

The block diagram of simple synthetic aperture radar is given in Fig. 5. It consists of radar system which creates coherent radar pulses, which are transmitted through an antenna. The antenna sends the radar energy to illuminate the surface target area to be imaged and receive the target returns. Aircraft motions are captured by the Inertial Navigation System (INS) and by an optional inertial measurement unit. The sensed motions are used to compensate the radar returns remain phase coherent during collection. The radar returns are converted into concentrated imagery by imagery by an Image Formulation Processor (IFP). While early versions of the UAV SAR systems were based on a fixed antenna, the current versions of SARs incorporate gimbaled antenna that provide imaging while turning and MTI technologies [8].

Fig. 5 Synthetic aperture
radar

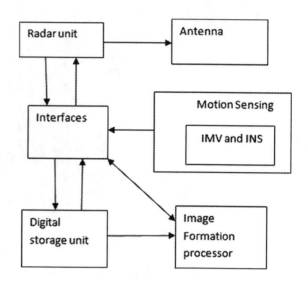

The modern SAR used in the UAVs provides a number of operating modes that can be exploited to get excellent results [9].

2.4 Hardware

The SAR hardware is typically configured as two different types of generic packages. They are Radar Electronic Assembly (REA) and the sensor front end or Gimbals Assembly (GA).

The REA contains radar control, wave form generation, up conversion, receiver, video ADE and signal processing functions. The Gimbals Assembly contains antenna, motion measurement hardware and front end microwave components, Figs. 6 and 7 gives a view of typical REA and Gimbals Assembly of a SAR system respectively [10]. Figure 8 shows am installation of a SAR system in UAV. Figures 7 and 8 provide an installation of images obtained from a UAV SAR system [10].

2.5 Hyper Spectral Imaging

The way of sub dividing the ultraviolet light, visible light and infrared spectra light into different types for imaging has been known. The Multi-spectral Imaging sensor

Fig. 6 Radar electronic
assembly

Fig. 7 SAR gimbal assembly

(MSI), uses multiple pictures of area or target that is produced with light from various places of the spectrum. If the proper wavelengths are selected, multi-spectral images can be used to detect many military important items such as camouflage, thermal emissions and hazardous wastes. The basic use and need of using multi-spectral/hyper spectral remote sensing image data is to differentiate, separate and to know the amount of objects found in the picture [11] (Figs. 9, 10 and 11).

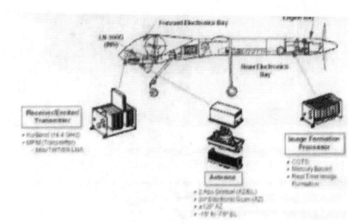

Fig. 8 Typical SAR system installations in a UAV

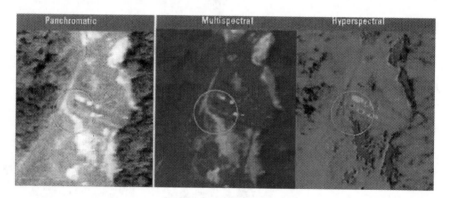

Fig. 9 Examples of spectral images

Fig. 10 Example figure of
UAV SAR system

Fig. 11 Example figure of
UAV SAR system

MILITARY EQUIPMENT
Resolution = 4 in. (0.1 m)

3 Non Imaging Payloads

The suite of non imaging payloads UAV includes ESM payload communication
relay payload and chemical sensor payload.

3.1 ESM Payloads

Electronic Support Measures (ESM) is the section of Electronic Warfare that
searches, locates and identifies radiated electromagnetic energy which can be used
to recognize threats and direct counter actions. In the respect of UAVs,

ESM mission are largely concerned with locating defense suppression weapon
systems and sites and location of enemy forces. This capability is achieved by
integrating a payload that can interrupt and locate by Direction Finding (DF) both
radar and communication signals. By virtue of being able to intercepts those signals
at fairly long distances, the ESM payloads can reduce the recon noising effort by
isolating the areas of interest quickly. The efface of the system in a target acqui-
sitions role is somewhat limited due to the limitations in DF accuracy. Hence the
ESM pay load are used in conjunction with payload to locate the target precisely.
When used with an EO payload the ESM payload can significantly enhance the
reconnaissance mission capability with a DF system of reasonable accuracy (10°).
Most of the MALE and HALE UAVs payloads must be capable of intercepting
signals that are emitted from a wide variety and communication equipment and
radars.

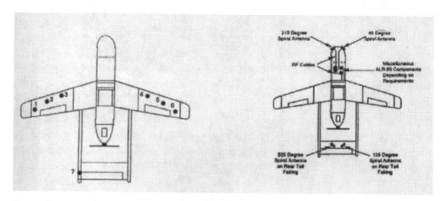

Fig. 12 COMINT and ELINT antenna installation

Due to the difference in the typical communication and radar frequencies to the ESM types of signals are classified as COMINT and ELINT payloads. The main issues in these two types of signals (Fig. 12) are illustrated of a practical antenna installation for COMINT and ELINT systems.

If this juncture the chemical and biological warfare is a distinct possibility and the UAVs capability to be deployed in the "dirty" mission makes the chemical sensors payload one of the interesting future prospects. A number of chemical sensors have been designed using surface acoustic wave SAW devices and flexural plate wave (FPW) devices coated with a thin layer of chemo selective material. These coatings, when exposed to specific chemical agents, when exposed to specific chemical agents resulting in increasing the mass of the polymer. When these devices are used as a resonator in an oscillator circuit, the change in mass due to absorption of the chemical agent results in change in the frequency of the oscillations. By detecting the change in frequency the presence of a particular agent could be detected.

Over the years, developments on the sensors include the use of MEMS (Micro Electro Mechanical Systems) technology based chemical sensors. It is expected that such a sensor would provide low cost miniaturized sensitive sensors with fast response time. Design of dual resonator systems capable of measuring humidity and specific chemical agents are being attempted [11].

4 Communication Relay Payload

The communication relay application demands the airborne antenna to be simple wide beam antenna providing uniform density across the coverage area. Another constraints is that the antenna of the aircraft in at least 1500 ft or more to provide a reasonable footprint on ground [5]. Stabilization of the airborne antenna to compensate for maneuvering and ensuring downward pointing of the antenna at all time is a desirable feature keeping this payload as small, compact and light weight and low power as possible to be within the limitations of the UAV platform is of the utmost importance. Another describe attribute of the communication payload as that of the systems should be as flexible as possible to be able to work with a variety different ground terminal equipments.

5 Conclusion

UAV are being used to play their major roles of surveillance, target identification and reconnaissance. Imaging sensors have developed a solid base as the major payloads. The growth of EO/IR technologies add and aid to achieve improved reliability and better sensitivity besides higher perception. Synthetic aperture radars have become the main imaging sensors for medium altitude and higher altitude UAVs with appreciable payload capability. Combined with the moving power of UAV, multi and hyper spectral imagers can assist to view a target or area under varying conditions of illumination and activity.

ESM payloads are pleasingly in demand in using as secondary payloads. They are further acknowledged as special categories of UAV sensors. Chemical sensor has a drastic growth in development to be used as remote detector. It is clearly evident from above that the sensors are able to achieve the performance required in most practical systems and attain a significant growth.

ESM payloads are special categories of UAV sensors and they are increasingly used as secondary payloads in MALE and HALE UAVs. The development of chemical sensor for remote detection is gaining momentum. Availability of MEMS based chemical sensors is a distinct possibility. The need for integrated operation with other aerial platforms has stimulated the development of a standard in UAV data link such as Tactical Common Data link (TCDL). SATCOM link providing communication with UAV over a long range has been proven.

The cost effectiveness of the payload system is an important issue to be addresses in the context of UAV payloads. The system designers resort to the following approaches to reduce the cost and size of the payloads systems [2].

- Use of commercially available and proven sensors that have been designed for other applications.
- Use of COTS systems with suitable environmental control schemes to extend the operating environments—pressurized and temperature controlled bay.
- Limiting the EO/IR system to either EO or IR imaging but not both simultaneously [13].
- Design the EO, IR and SAR/MTI sensors as an integrated sensor suite that shared hardware and software elements to avoid duplication of function and simplify the interfaces between the sensors, the aircraft power and control systems and the data link. The Global hawk's integrated sensor suite

(ISS) incorporates common reflector optical for EO and IR sensors and a common processor for processing the information [13].

The sensors, communication and other payloads, which are already able to achieve the performance required in most practical systems, will continue to decrease in weight and cost.

References

1. Major Brown, D.A.: MEDUSA's mirror: stepping forward to look back "Future UAV design implications from 21st centaury battlefield"; Monograph, School of Advanced Military studies, United States Army command and General staff College, Fort Leavenworth, Kansas (1997)
2. Proceedings of Aerospace Technologies: Developments and Strategies International Seminar, 05–08 Feb 2003
3. Borky, J.M.: Payload technologies and applications for uninhabited air vehicles (UAVs). Aerosp. Conf. Proc. 3, 267–283 (1997)
4. Stauart, D.M.: Sensor design for unmanned aerial vehicles. Aerosp. Conf. Proc. 3, 285–293 (1997)
5. Struz, R.: Airborne applications of stabilized video cameras. SPIE 1342, 46–51 (1990)
6. Bouchard, Y.: Low light level imaging. Photon. Spectra 102–103 (2001)
7. Haystead, J.: Thermal imaging technology has versatile and bright future. Def. Electron. 48–52 (1991)
8. Ruthbertson, G.M.: Thermal imaging for avionic application, advances in sensors and their integration into aircraft guidance and control system AGARD report no. 272, pp. 2.1–2.12 (1983)
9. Stockton, W.: Miniature Synthetic Aperture Radar System (MSAR), Unmanned Systems, pp. 3–10 (1999)
10. Tsunoda, S.I., et al.: A high resolution synthetic aperture radar. SPIE Aerosense 3704, 3–10 (1999)
11. www.ga.com/atg/atg.html#airborne
12. Jackson, A.W., Rose, Jr, E.A.: Electronic support measure for unmanned aerial vehicles. In: RPVs 11th International Conference, Bristol UK, 12–14 Sept 1994
13. Stuart, D.M.: Sensor design for unmanned aerial vehicle. In: IEEE Aerospace Conference (1997)

Lighting Electrical Energy Audit and Management in a Commercial Building

K. Keerthi Jain, N. Kishore Kumar, K. Senthil Kumar, P. Thangappan, K. Manikandan, P. Magesh, L. Ramesh and K. Sujatha

Abstract In the present scenario, the world is dependent upon the ways to conserve electrical energy in an effective manner with less cost investment in India. The demand that is lagging in the year 2016–2017 is 300 GW. It is seen that day by day the demand is increasing the government is behind the generation part but conservation is very much essential to reduce the demand in an effective manner, looking over this scenario an initiative has been taken in our University to conduct electrical energy audit and management in an effective manner to reduce the demand and save 10 MW generation in 10 years. The initial work was started under the vision **MGR-VISION 10 MW** which was inaugurated in our University. The team has completed audit in 25 residential flats, 2 commercial building and 2 industries so far. This paper delivers a lighting layout for a commercial building in which it consist of six floor. Lighting layout of one particular floor is done with electrical energy audit and energy management. The recommendations for the benefits of implementation with breakeven chart are given to reduce the consumption in an effective manner. Recommendation for usage of renewable energy is given so as to reduce the consumption to reduce demand and save electrical utilization bills.

Keywords Electrical · Energy audit · Renewable energy

K. Keerthi Jain (✉) · N. Kishore Kumar · K. Senthil Kumar · P. Thangappan · K. Manikandan · P. Magesh · L. Ramesh
Vision 10 MW, Dr. M.G.R Education and Research Institute, Chennai, India
e-mail: keerthijain12@gmail.com

L. Ramesh
e-mail: raameshl@rediffmail.com

K. Sujatha
EEE Department, Dr. M.G.R Education and Research Institute, Chennai, India

© Springer Science+Business Media Singapore 2017
P. Deiva Sundari et al. (eds.), *Proceedings of 2nd International Conference on Intelligent Computing and Applications*, Advances in Intelligent Systems and Computing 467, DOI 10.1007/978-981-10-1645-5_39

1 Introduction

The word scenario is dependent upon the conservation activities as the demand in all forms are increasing day by day management for this is become very much essential. According to this the electricity demand is also increased so for this conservation is a better source to reduce the demand. In India the total demand is increasing day by the day and it is seen that in the year 2016–2017 there will be a demand of 298 GW to overcome this, An initiative is taken to reduce the demand by conducting electrical energy auditing and management due to which the conservation of the demand will get low and also the saving of money for each client will be carried out in an effective manner. Why electrical auditing and management is essential? The answer to the question is that electrical energy auditing and management program is necessary for its originating within one division of saving, motivating people in all forms to undergo conservation activities. This conservation of electrical energy by auditing and management is started by VISION 10 MW which was inaugurated in our University so as to reduce 10 MW generation 10 years. A good program has been instigated who has recognized the potential and is willing to put forth the effort in addition to regular duties-will take the risk of pushing a new concept and is motivated by seemingly higher calling to save energy. This saving electrical energy of VISION 10 MW team has completed auditing in 25 residential flats, 2 commercial building and 2 industries till date.

2 Literature Review

Electrical energy management and audit has become very essential in the society as the demand is increasing day by day in the world scenario. In India it is seen that the total capacity generation from all the forms is 1102.9 TWh the total demand that is forecasted in the year 2016–2017 would be 298 GW and to minimize this demand an initiative is taken to conduct electrical energy audit [1]. In the book titled "Performance Contracting: Expanded Horizons," Shirley Hansen has described an effective procedure is given for the electrical management study. In this for an effective energy management following requirement is given i.e. set up plan for energy management, establishing energy audit plan, data collection for future assistance, evaluate for effectiveness [2]. In the book of Energy Audit Manual Energy Management Centre Kerala, Department of Power Government of Kerala. An Execution procedure is given in which is very much useful to conduct electrical energy audit in a industries, commercial building etc. [3]. The article entitled "The strategy of adjusting and optimizing energy, using systems and procedures so as to reduce energy requirements per unit of output while holding constant or reducing total costs of producing the output from these systems" for effective energy management in the book of ENERGY CONSERVATION HANDBOOK [4].

3 Procedure

How to conduct electrical energy audit and management? The answer to the question is that for electrical energy audit and management the first work is to design the layout to conduct the audit as shown in Fig. 1 in which the procedure how to conduct an audit and to the manage it in an effective manner. In this the layout described first is decided to conduct the audit and the electrical energy audit data are collected for which single line diagram with suitable recommendations are coated with appropriate results. Energy Management is the Techno-Managerial activity to achieve judicious and effective energy consumption pattern to ensure maximum profit and survival in this Competitive World. The government is behind the various ways to generate power and to reduce the demand but we are taking initiative to reduce the demand by electrical energy audit and this is our vision of 10 MW. The motto behind it is **SAVE ENERGY TO REDUCE DEMAND**. An Executive activity to conduct this electrical energy audit and management as given below;

- Collect all the Load details with maximum demand of the Electrical Equipment.
- Calculate the Load usage, Construct the single line diagram enter the values in ETAP (Electrical Transient and Analysis Program).
- Calculate the connected load with respective to single Line diagram.
- Plot a graph between day time load and night time for load analysis.
- Identify and calculate the unnecessary usage of power wastage in the layout with graph.
- Draw the Power Utilization Chart with respect to the Layout.

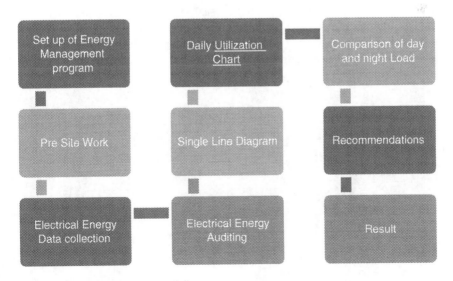

Fig. 1 Methodology to conduct audit

- Calculate the daily utilization of Power by all the equipment's and convert the same to pie chart.
- Data Collection for all the major equipment's and find out the performance.
- Interaction about the energy usage with suitable survey.
- Identify the Energy Saving and Conservations Opportunity.
- Report on suitable recommendation with existing and implementation suggestions.
- Plot Cost Benefit Analysis with Breakeven Chart.
- Check the earth resistance and report on the status of earthling in that house.
- Provide Awareness' on Electrical Safety to the people residing there.

Submission of Suitable Energy Audit Report with Breakeven Analysis and taking the benefits of renewable energy and simulating it in the ETAP software and provide them the best recommendation to reduce electrical consumption by renewable sources.

4 Electrical Energy Audit Survey

In this section, the distributed energy sources like wind and solar are modeled and analyzed.

4.1 Comparison of Day and Night Time Load

According to the pre-site survey taken an analysis for the energy management is done and the loads are compared as shown in Fig. 2. In this it is seen that the loads are compared according to the usage at day time and night time, and the categorized based on usage of that lighting (LDB, ELDB, PDB, UDB) are the loads which are

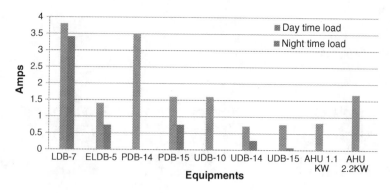

Fig. 2 Comparison of day and night load

used both in day time as well as night time. LDB (light distribution board), PDB (Power Distribution Board), UDB (UPS Distribution Board), ELDB (Electrical Distribution Board).

4.2 Daily Utilization Chart

The layout of the commercial building management is done by taking the survey of each appliances that are used in one single day and the graph is given in Fig. 3 which it is seen that lighting load are used at the maximum.

4.3 Wattage of Appliances

The wattage of the each appliances are taken at time of pre-site auditing and a graph is obtained as shown in Fig. 4 which the maximum wattage is taken by the AHU, FCU and sockets as the wattage of this appliance is very high so that the consumption for this will also be high according to their usage.

4.4 Lighting Single Line Diagram of Commercial Building

The lightning layout of the commercial building is taken and drawn in the ETAP simulation software in which the load analysis test is done and load flow report is generated (Fig. 5).

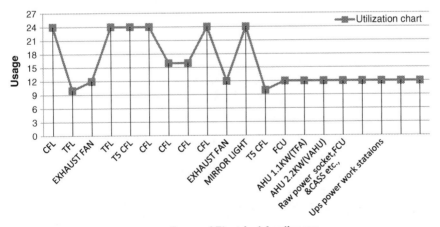

Fig. 3 Daily utilization chart

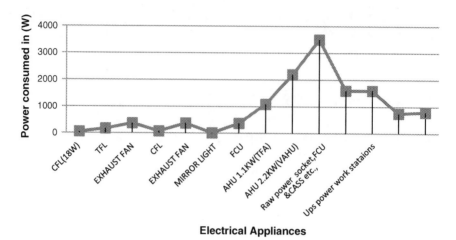

Fig. 4 Wattage of appliance

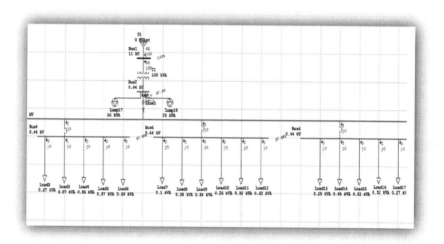

Fig. 5 Lighting single line diagram of commercial building

4.5 ETAP Voltage Load Analysis

The lighting layout of the commercial building is drawn in the ETAP simulation software through which the load analysis is done, according to the load flow analysis the bus voltage and no buses is drawn in which the voltage is draw (Fig. 6).

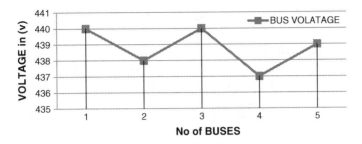

Fig. 6 ETAP voltage analysis

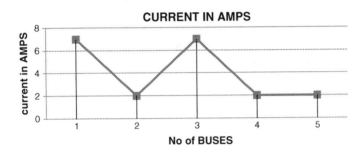

Fig. 7 ETAP current analysis

4.6 ETAP Current Load Analysis

The lighting layout of the commercial building is drawn in the ETAP simulation software through which the load analysis is done, according to the load flow analysis the bus current and no buses is drawn in which the current is draw (Fig. 7).

5 Recommendation

For an Managerial activity for the conservation of electrical energy audit and management in an effective manner certain recommendation are coated below according to the layout of the commercial building it is seen above that the lighting in this layout, consumption of lighting is very high as this is the maximum amount that the client is paying looking over the scenario the recommendation are given below for an effective reduction in power and also by saving money.

5.1 Recommendation with PID Sensors

It is a kind of proximity sensor in which it detects the human heat of the body by infrared signals it is used as a sensor in which the presence of the human is detected according to it the on/off function takes place hence to minimize the utilization of electrical loads, fans, AC etc.

In this layout the three phase's current is unbalanced in the following DB.

PDB-14

- The current R = 0.4, Y = 2.3, B = 1.7 this should be corrected it will increase the voltage regulation and of reduce power consumption.

PDB-15

- The current R = 1.3, Y = 2.2, B = 4 this should be corrected it will increase the voltage regulation and of reduce power consumption.

UDB-10

- The current R = 1.8, Y = 4.7, B = 3.6 this should be corrected it will increase the voltage regulation and of reduce power consumption.

UDB-14

- The current R = 1.4, Y = 3.7, B = 0.9 this should be corrected it will increase the voltage regulation and of reduce power consumption (Fig. 8).

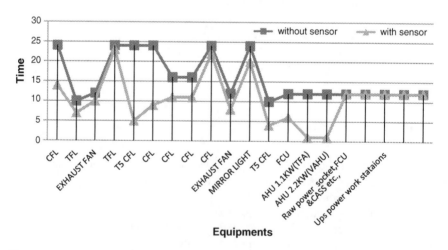

Fig. 8 Comparison with sensor and without sensor of lighting layout

Table 1 Recommendation with LED

Present energy usage	
Total number of tube light	45
Total number of watts	45*56 = 2520 W
Number of hours in a year	12 h*365 days = 4380
Total number of watts annually	2520*12*365 = 11,037,600 W
Total units consumed	11,037,600/1000 = 11,037.6 units/year
Cost annually	Rs. 77,263.2
If (20) number of tube light are replaced by LEDs wall light	
Total number of LED	20
Total number of watts	20*18 = 360 W
Total number of watts annually	360*12*365 = 1,576,800 W
Total number of units consumed	1,576,800/1000 = 1576.8 units per year
Cost annually	Rs. 11038
Saving	
Energy saved	11,037.6 − 1576.8 = 9460.8 units per year
Money saved	Rs. 77,263 − Rs. 11,038 = Rs. 66,225 per year
Payback time	
LEDs	Rs. 2400 (Philips)
Total investment	20*2400 = Rs. 48,000
Payback time	(Investment cost/annual saving)*12 months (48,000/66,225)*12 = 8.5 months

5.2 Recommendation with LED

According to the layout of the commercial building it seen that total 45 tube light are used which are of 56 W for the same lumens that is provided by the incandescent tube light it is recommended to go for 15 W LED tube light which will reduce the consumption as well as an effective reduction in the money and tariff can be saved (Table 1).

5.3 Replacement of Corridor Lighting into LED

According to the layout of the commercial building the corridor lights can be replaced into LED for an effective reduction in cost as well as electrical energy. So for betterment in lighting and also with same lumens and cost reduction LED is recommended in an effective manner. The layout of the corridor as 36 W CFL of 40 no's it is good to replace with LED of same lumens of 8 W of 20 no's for a better reduction of both cost and units (Table 2).

The graph that is given in Fig. 9 gives the comparative analysis of the CFL to LED in which the cost benefits and units consumed is given in an graphical manner.

Table 2 Replacement of corridor lighting into LED

Present energy usage	
Total number of CFL	40
Total number of watts	40*36 = 1440 W
Number of hours in a year	12 h*365 days = 4380
Total number of watts annually	1440*12*365 = 6,307,200 W
Total units consumed	6,307,200/1000 = 6307 units/year
Cost annually	Rs. 44,150
If (20) number of tube light are replaced by LEDs wall light	
Total number of LED	20
Total number of watts	20*8 = 160 W
Total number of watts annually	160*12*365 = 700,800 W
Total number of units consumed	700,800/1000 = 700 units per year
Cost annually	Rs. 4900
Saving	
Energy saved	6307 − 700 = 5607 units per year
Money saved	Rs. 44,150 − Rs. 4900 = Rs. 39,250 per year
Payback time	
LEDs	Rs. 650 (Philips)
Total investment	20*650 = Rs. 13,000
Payback time	(Investment cost/annual saving)*12 months (13,000/39,250)*12 = 4 months

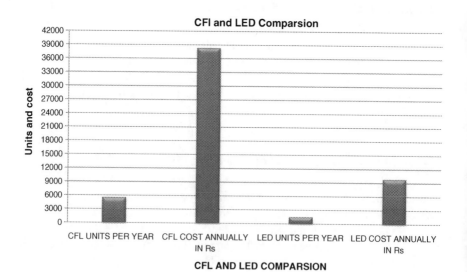

Fig. 9 Comparison with CFL to LED

Table 3 Recommendation LED with solar

LED lighting wattage for whole layout is	900 W
Solar panel capacity	1 KWp
Cost for 1 KWp solar panel	Rs. 85,000
Subsidy (30 %)	Rs. 20,000
Final investment cost	Rs. 85,000 − 20,000 = Rs. 65,000
Solar power generation	
Solar power generation Chennai	4–5 KW per day for 1 KW panel
Number of working days	300 per years
Total energy production	4*1*300 units per year
	5*1*300 units per year
Cost and benefits	
Number of units of grid power substituted	1200 units per year
	1500 units per year
Cost of grid power	Rs. 7 per unit
Power saving per year	1200*7 = Rs. 8400
	1500*7 = Rs. 10,500
Payback time	
Payback time	(Investment cost/annual saving)*12 months (65,000/8400)*12 = 92(7 years)

5.4 Recommendation LED with Solar

According to the layout the LDB loads are converted into LED and connected to the solar panel to get the desired output, as LED is a DC source so direct source is taken from the sunlight and an automatic control unit is set as at the time when there is no sunlight and at the night this automatic control unit will take the supply from the electrical board (Table 3).

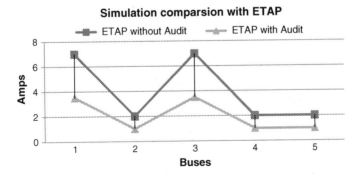

Fig. 10 ETAP output with and without audit

5.5 *ETAP Output with and Without Audit*

The ETAP output with and without audit graph is given in Fig. 10 which it is seen that if the implementation of all the recommendation is done as per given the amps will be reduced as the amps are reduced the consumption and the cost will be reduced in the effective manner.

6 Conclusion

This paper delivers electrical energy audit i.e. is done in an commercial building and this building is of six floor in which the audit is done and according to it the one floor lighting audit study is done and suitable recommendations are coated in an beneficial manner using the ETAP simulation software by which the consumption is reduced and there is a gradual decrement in the tariff bill.

References

1. Thumann, A., Niehus, T., Younger, W.J.: Hand Book of Energy Audit. CRC Press Taylor and Francis Group, Boca Raton
2. Energy Conservation Hand Book—Prepared by Uttarakhand Renewable Energy Development Agency (Ureda)
3. 2011 2nd International Conference on Advances in Energy Engineering (ICAEE 2011) Energy Conservation Measures in a Technical Institutional Building in Tamilnadu in India
4. www.Greencleanguide.Com
5. www.Cea.Nic

Advance Control Strategies for a Conical Process

Parashivappagol Praveen Kumar and Thirunavukkarasu Indira

Abstract In this paper a Dynamic Matric Control (DMC) and Receding Horizon Control (RHC) are designed for unconstrained SISO process. These MPC variants are then simulated for a conical tank system to control the level. A first order process with dead time model for conical tank is used for simulation. The response obtained for both methods are then compared and discussed in this paper.

Keywords Conical process · Dynamic matrix · Receding horizon · Unconstrained MPC

1 Introduction

Model Predictive Control (MPC) is optimization based control strategy widely used in process industries. Its efficiency lies in the fact that it handles constraints imposed on manipulated variable and control variable, also it is best suited for multivariable process control [1]. MPC uses explicit model for the process and the control signal is obtained by optimizing the objective function [2]. The process model for a linear process can be impulse response model, step response model or state space model. In this work Dynamic Matrix Control (DMC) and Receding Horizon Control (RHC) techniques are realized for a conical tank process.

RHC is based on optimal control and is developed as an alternative for LQ control [3]. At each current instant, the optimal control is obtained over a finite horizon and only the first one is adopted as current control law. Since the horizon recedes as time proceeds the name 'Receding Horizon'. RHC uses state space model of the process [4].

The DMC algorithm was first developed for petroleum refinery at Shell oil by Cutler [5]. The control problem for linear model boils down to simple matrix

P. Praveen Kumar · T. Indira (✉)
Department of Instrumentation and Control Engineering,
Manipal Institute of Technology, Manipal University, Manipal 576 104, India
e-mail: it.arasu@manipal.edu

© Springer Science+Business Media Singapore 2017
P. Deiva Sundari et al. (eds.), *Proceedings of 2nd International Conference on Intelligent Computing and Applications*, Advances in Intelligent Systems and Computing 467, DOI 10.1007/978-981-10-1645-5_40

algebra calculation, hence the name dynamic matric. DMC uses step response model of the process which can be obtained straightaway. To find the model we simply need to record the output response for step excitation.

In this work, DMC and RHC are simulated for conical tank process. The linear model is obtained from first principle method and FOPDT model is obtained [6]. This model is used for realizing both control algorithms.

In Sect. 2, we discuss the formulation of augmented matrices for RHC using state space model and dynamic matrix for DMC using step response model. In Sect. 3, the algorithms are simulated for conical tank SISO process and Sect. 4 makes concluding remarks.

2 Model Predictive Control

All the MPC algorithms possess common elements, and different option can be given to each element giving rise to different control algorithm.

Prediction model: A model that fully captures the process dynamics and allows prediction.
Objective function: The control law is formulated by minimizing the cost function to achieve certain aim, tracking the reference signal is most common.
Control law: The method of obtaining solution for cost function is not easy, hence certain structure is imposed on the control law.

The various MPC algorithms are DMC, GPC, MAC, RHC and so on.

2.1 Receding Horizon Control

RHC is usually represented by state feedback form, if all the states are available otherwise we estimate the states from observers like Luenberger [3]. The state space representation of the model with n states is given in Eq. 1. The output and state prediction is carried out using current states variables and hence we need to change the model to suit our purpose.

$$\begin{aligned} x(k+1) &= A_d x(k) + B_d u(k) \\ y(k) &= C_d x(k) \end{aligned} \tag{1}$$

The new state variable chosen for finding suitable model are $x(k) = [\,\Delta x' \quad y\,]'$. Thus the new state space model will of the form given in Eq. (2)

$$\begin{bmatrix} \Delta x(k+1) \\ y(k+1) \end{bmatrix} = \begin{bmatrix} A_d & o'_d \\ C_d A_d & 1 \end{bmatrix} \begin{bmatrix} \Delta x(k) \\ y(k) \end{bmatrix} + \begin{bmatrix} B_d \\ C_d B_d \end{bmatrix} \Delta u(k)$$

$$y(k) = \begin{bmatrix} o_d & 1 \end{bmatrix} \begin{bmatrix} \Delta x(k) \\ y(k) \end{bmatrix} \tag{2}$$

where, $o_d = zeros(1,n)$. These new matrix (A, B, C) forms the augmented model for the system. Due to this augmentation the system matrix will have additional eigen values along with that of the original system.

The next step in MPC is predicting the output and states. From a given state information we can predict up to p number of future state variables, where p is called prediction horizon. The future control trajectory are considered up to m, called control horizon. Let use define vectors Y and ΔU. The future prediction can be written in a compact matric form as in Eq. (3)

$$Y = [y(k_i + 1|k_i) \quad y(k_i + 2|k_i) \quad y(k_i + 3|k_i) \quad \cdots \quad y(k_i + p|k_i)]'$$

$$\Delta U = [\Delta u(k_i) \quad \Delta u(k_i + 1) \quad \Delta u(k_i + 2) \quad \cdots \quad \Delta u(k_i + m - 1)]' \tag{3}$$

$$Y = Fx(k_i) + \phi \Delta U$$

where,

$$F = \begin{bmatrix} CA \\ CA^2 \\ CA^3 \\ \vdots \\ CA^p \end{bmatrix} \quad \phi = \begin{bmatrix} CB & 0 & 0 & \cdots & 0 \\ CAB & CB & 0 & \cdots & 0 \\ CA^2B & CAB & CB & \cdots & 0 \\ \vdots & \vdots & \vdots & \vdots & \vdots \\ CA^{p-1}B & CA^{p-2}B & CA^{p-3}B & \cdots & CA^{p-m}B \end{bmatrix}$$

We define the cost function J that reflects the control objective as given in Eq. (4). $\bar{R}$ is a diagonal matrix of the form $\bar{R} = r_w I_{m \times m}$ where r_w is used as tuning parameter for desired closed loop performance. R_s is data vector that contain data set point information.

$$J = (R_s - Y)'(R_s - Y) + \Delta U^T \bar{R} U \tag{4}$$

The optimal solution ΔU found after minimizing J is given as control law given in Eq. (5) with assumption that inverse exists for hessian matrix.

$$\Delta U = (\phi^T \phi + \bar{R})^{-1} \phi^T (R_s - Fx(k_i)) \Delta U = (\phi^T \phi + \bar{R})^{-1} \phi^T (R_s - Fx(k_i)) \tag{5}$$

Because of receding horizon control principle, we take only the first element of ΔU at time k_i as the incremental control, thus $\Delta u(k_i) = [1 \quad 0 \quad 0 \quad \cdots \quad 0] \Delta U$.

2.2 Dynamic Matrix Control

As it was stated earlier DMC uses step response model taking first N terms as process is assumed to be stable. The Eq. (6) represents step response of the system where g_i are the sampled output values from step input and $\Delta u = \Delta u(t) - \Delta u(t-1)$. The output is predicted at kth instant of time.

$$\hat{y}(t+k|t) = \sum_{i=1}^{N} g_i \Delta u(t+k-i|t) \qquad (6)$$

$$\hat{y}(t+k|t) = \sum_{i=1}^{k} g_i \Delta u(t+k-i|t) + \sum_{i=k+1}^{N} g_i \Delta u(t+k-i|t) \qquad (7)$$

The first term in the Eq. (7) represents the future control action to be calculated and the second terms contains the past control action and called as free response. The process model can finally written in matrix form as in Eq. (8). G and F matrix are formed using the coefficients g_i

$$\hat{Y} = Gu + f$$

$$f(t+k) = y_m(t) + \sum_{i=1}^{N} (g_{k+i} - g_i)f(t+k) = y_m(t) + \sum_{i=1}^{N} (g_{k+i} - g_i) \qquad (8)$$

where,

$$G = \begin{bmatrix} g_1 & 0 & \cdots & 0 & 0 \\ g_2 & g_1 & \cdots & 0 & 0 \\ \vdots & \vdots & \ddots & \vdots & \vdots \\ g_m & g_{m-1} & \cdots & g_2 & g_1 \\ \vdots & \vdots & \cdots & \vdots & \vdots \\ g_p & g_{p-1} & \cdots & g_{p-m} & g_{p-m+1} \end{bmatrix}$$

The matrix G, takes the form of a special type of matrix called toeplitz matrix in which the step response coefficients are shifted orderly. f is a free response vector, as after N steps the coefficients g_i remain same the summation is only up till N. We need our output y to track the reference signal w in least square sense and hence the objective function chosen is of the form given in Eq. (9)

$$J = \sum_{j=1}^{p} [\hat{y}(t+j|k) - w(t+j)]^2 + \lambda \sum_{j=1}^{m} [\Delta u(t+j-1)]^2 \qquad (9)$$

The reference trajectory $w(t+k)$ is normally a smooth approximation from current output $y(t)$ towards the known reference by means of a first order system given in Eq. (10). α is the parameter between 0 and 1, larger the value smoother the approximation [7].

$$w(t+k) = r(t+k) - \alpha^k(y(t) - r(t)) \tag{10}$$

For unconstrained case, the solution of the objective function will be of the form

$$\Delta u = (G^T G + \lambda I)^{-1} G^T (w - f) \tag{11}$$

In all the predictive strategies only the first element of the control input is sent to the plant. It is because of the fact that we cannot perfectly estimate or anticipate the disturbance that causes the changes in actual output from the predicted output.

3   Simulation on Conical Process

First step in MPC is to obtain a model for a process, conical tank considered is a nonlinear process and a transfer function model is obtained is from two point method [6]. This transfer function model given in Eq. (12) is used to obtain step response and state space models analytically.

$$G(s) = \frac{0.925e^{-1.09s}}{25.05s + 1} \tag{12}$$

3.1   RHC for Conical Process

The state space model for FOPDT process is obtained by pade approximation. The augmented matrix is calculated using Eq. (2).

$$A = \begin{bmatrix} 0.538 & -0.780 & -3.030 & 0 \\ 0.075 & 0.957 & -0.001 & 0 \\ 0.004 & 0.098 & 0.999 & 0 \\ 0.006 & -0.186 & 0.372 & 1.0 \end{bmatrix}, \quad B = \begin{bmatrix} 0.075 \\ 0.004 \\ 0 \\ 0.002 \end{bmatrix}, \quad C = [0 \ \ 0 \ \ 0 \ \ 1]$$

Using the above given augmented matrix, we need to find matrix F and ϕ to predict the future output, take prediction horizon m = 10 and control horizon n = 4. The control law give for RHC in Eq. (5) is then realized with control weight $r_w = 0.1$.

Fig. 1 Controlled variable
and manipulated variable for
DMC and RHC

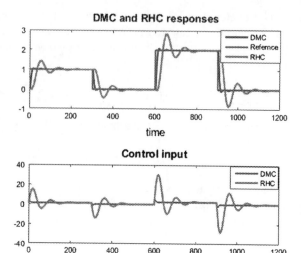

3.2 DMC for Conical Process

The step response model is found analytically and the coefficients are used to obtain the toeplitz matrix for m = 10 and n = 4

$$g = \begin{bmatrix} 0 & 0.035 & 0.103 & 0.166 & 0.225 & 0.278 & 0.328 & 0.374 & 0.416 & 0.455 \end{bmatrix}$$

From the coefficients free response can also be found. The tuning parameter α is taken as 0.4 and weight λ as 1. Finally the control law for DMC given is realized.

The graph shown in Fig. 1. shows the actual output (controlled variable) and the input variable (manipulated variable) obtained from control laws.

4 Conclusion

The control law obtained from both the algorithms are simulated for a conical tank process, the suitable models are derived from a given FOPDT transfer function. It can be seen that DMC tracks the reference fast while RHC whose model is approximated (pade) has some overshoot. As both cases are unconstrained, the controlled variable is varying freely. The weights used in objective function indicates the aggression of a controller and hence is a tuning parameter. The smaller value of control weight indicated less suppression of controlled variable and thus a more aggressive controller. The tuning parameter α in DMC will damp the controlled variable for larger value as it will generate a smoother reference signal.

Acknowledgments The authors would like to thank Dept. of Instrumentation and Control Engineering, MIT, Manipal for providing the facility towards the hardware and software, e-journal access for the work.

References

1. Camacho, E.F., Bordons, A.C.: Model Predictive Control. Springer, Carlos (2007)
2. Liuping, W.: Model Predictive Control System Design and Implementation Using MATLAB. Springer, Berlin (2009)
3. Kwon, W.H., Han, S.H.: Receding Horizon Control Model Predictive Control for State Models. Springer, Berlin (2005)
4. Mayne, D.Q.: Receding horizon control of nonlinear systems. IEEE Trans. Autom. Control. **35**, 814–824 (2002)
5. Garcia, C.E., Morari, M.: Model predictive control: theory and practice—a survey. Automatica **25**, 335–348 (1989)
6. Rakesh, M.K., Satheesh babu, R., Thirunavukkarasu, I.: Sliding model with dead time compensation for conical tank level process. IOSR J. Electr. Electron. Eng. **9**, 59–64 (2014)
7. Rhinehart, R.: CV damping and MV suppression for MPC tuning. In: Proceedings of American Control Conference (2002)
8. Muske, K.R., Rawling, J.B.: Model predictive control with linear models. AIChE J. **39**, 262–287 (1993)

ZVC Based Bidirectional DC-DC Converter for Multiport RES Applications

V. Jagadeesh Babu and K. Iyswarya Annapoorani

Abstract This article proposes an advanced converter technique that integrates all major ports in that double unidirectional source port, one battery port for (charging and discharging), and a single ideal power consumption port (load port). A new multi-port DC-DC power converter is developed when added to a pair of semi conducting switches and a pair of PN junction diodes to a basic half bridge (HB) converter circuit. Zero voltage conduction (ZVC) is performed on every major switch. This technique is implemented for savings huge number of component count and to reduced losses obtained from renewable energy applications.

Keywords HBC (half-bridge converter) rectifier · Zero-voltage conduction (ZVC) · Multiple-input single-output (MISO) · Multiple ports · Advanced DC-DC boost converter (chopper)

1 Introduction

Main purpose of advanced power converters is to integrate, and control many power ports with less costing small circuit designs [1]. Renewable energy resources are fluctuating in nature because of that a normal storage is needed if there is no ac main. This paper introduced an advanced multi-port DC-DC converter technique; this is also applicable for several other renewable energy resources applications.

Zero voltage conduction (ZVC) is implemented for all major switches to get optimal efficiency in the system at lower/higher conducting frequencies [2]; this gives better design for multiport power converter. The control circuit is developed based on the remodeling of the proposed modified half-bridge (HB) terminology.

V. Jagadeesh Babu (✉) · K. Iyswarya Annapoorani
VIT University, Chennai, India
e-mail: jagadeeshbabu.v2015@vit.ac.in

K. Iyswarya Annapoorani
e-mail: iyswarya.annapoorani@vit.ac.in

© Springer Science+Business Media Singapore 2017
P. Deiva Sundari et al. (eds.), *Proceedings of 2nd International Conference on Intelligent Computing and Applications*, Advances in Intelligent Systems and Computing 467, DOI 10.1007/978-981-10-1645-5_41

The new power converter technique is suited best for obtaining low-power in renewable energy resources.

2 Advanced Converter Technique with Circuit Description

The advanced four-port technique is derived from basic two port HBC (half-bridge converter), it is made with pair of main thyristor switches indicated by S_1 and S_2. Which is shown in Fig. 1, another source port is added with diode D_3 and a thyristor switch of S_3, last port is made up of bidirectional power path with freewheeling diode D_4 across primary side of the transformer, this mode consists of diode D_4 and another operating switch S_4. And the system is made with four main operating switches (S_1, S_2, S_3, S_4) and a pair of diodes (D_1, D_2), transformer and a rectifier circuit. The advanced power converter techniques were best suited for 'n' issues related to power obtained from RES, this paper will focus the integration of RES systems and their application.

2.1 Operational Methods

The gating sequences for switches S_1, S_2, S_3, and S_4 and two SR signals should give better operational outputs. We should not gate it ON S_2, S_3, S_4 all at the same time, until there is a block due to the parallel diode (D_1) of switch S_1. S_3 alone needs to be turned ON earlier before the S_2 switch is turned OFF, and S_4 is also like S$_3$ switch which has to be turned ON before S$_3$ is turned OFF as shown in Fig. 2 No resting time gap is required in-between switches S_2 and S_4, as the presence of diodes prevents zero voltage state issues. But the sleep time interval of S_1, S_2 and S_1, S_4 is a must to prevent zero voltage state.

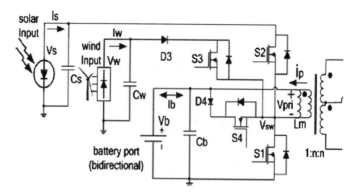

Fig. 1 Multi-port half-bridge converter terminology

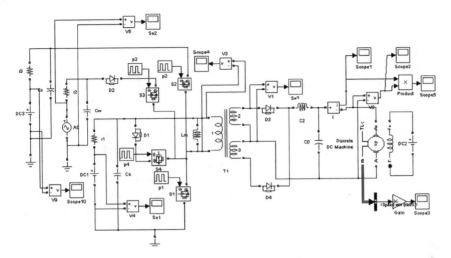

Fig. 2 Multi-port design in matlab

2.2 Circuit Diagrammatical Explanation

The switching, operational and output wavelets of advanced multi-port power converters are provided [3, 4] in Fig. 2. For simpler operational analysis, some elements are considered ideal, except measuring and indicating components. The important switching states were explained below.

State 1: at time (t_0–t_1): Before starting of this time interval, the parallel diode D_1 of switch S_1 has to be turned ON to discharge the stored energy in electro-magnetic property of transformer and leakage inductor, and at output of time t_0 it will turn ON S_1 with ZVC.

State 2: at interval (t_1–t_2): During t_1, primary current of the transformer increases so that can be reflected to inductor current (I_{Lo}) on secondary side of the transformer, the parallel diode of output side switch SR_2 becomes open circuited, and the power from DC/DC power converter delivered to output circuit.

State 3: at (t_2–t_3): During t_2 interval, the switch S_1 is turned OFF, this can reflect the leakage current L_{ip}, which helps the S_1 capacitor to charge and dissipating the energy in the S_2, S_3, S_4 switch connected capacitors.

Stage 4: at (t_3–t_4): During t_3, voltage in S_2 capacitor is discharged to resting (zero), and switch S_2 parallel diode conducts, reset to zero and makes reverse-charged in the battery port to carry the charging current, which induces to ZVC for switch S_2. At this period, the output flows through output switch SR_1 and SR_2 parallel diodes.

Stage 5: while (t_4-t_5) *period:* During t_4 time period, switch S_2 is turned ON with ZVC, the leakage current in inductor is made zero and also reverse-charging is enabled in the battery port.

Stage 6: at (t_5-t_6): During t_5 time interval, the I_p (primary current) of the transformer increases to the wasted inductor current (I_{Lo}), on secondary side of the transformer, the parallel diode of switch SR_1 is turned ON, and the output power from converter is delivered power to output circuit.

Stage 7: when (t_6-t_7): During t_6 time interval, main switch S_2 is tuned OFF, this will cause the large leakage current *ip* to charge the S_2 parallel capacitor and discharge through D_3 diode combinational capacitors.

Stage 8: at (t_7-t_8): During time t_7, the voltage across diode D_3 is discharged to zero, and D_3 diode starts conducting. S_3 switch is turned ON earlier to this period itself, since switch S_3 is naturally ZVC. Output inductor current (I_L) returned back via switch SR_2.

Stage 9: during (t_8-t_9): When t_8 time interval, switch S_3 is turned OFF, this induces the leakage current *ip* to charge S_2, S_3 storage capacitors and dissipates main switch S_1 and D_4 storage capacitors.

Stage 10: at time (t_9-t_{10}): Here during t_9 time interval, the potential across diode D_4 becomes zero and D_4 diode starts to conduct. Since main switch $S4$ is getting turned ON in time period t_8 and t_9 itself, the leakage current is discharged via diode D_4 and thyristor S_2, whereas the output inductor current gets discharged in SR_1 and SR_2.

Stage 11: in $(t_{10}-t_{11})$: At t_{10} time interval, the main switch S_4 is turned OFF, causing center tapped leakage current to be discharged through S_1 switch storage capacitor and starts charging the main S_2, S_3, S_4 stray capacitors.

Stage 12: during $(t_{11}-t_{12})$: At t_{11} time interval, the voltage across main Switch (S_1) reaches to zero voltage level, and the freewheeling diode across S_1 can carry the current to provide ZVC condition for main switch S_1. During this time period, the

Table 1 Switching sequences of advanced four port power converter	Switching time	Conducting switching
	t_0-t_1	S_1, SR_1
	t_1-t_2	S_1, SR_1
	t_2-t_3	–
	t_3-t_4	–
	t_4-t_5	S_2, S_3, S_4, SR_2
	t_5-t_6	S_2, S_3, S_4, SR_2
	t_6-t_7	S_3, S_4, SR_2
	t_7-t_8	S_3, S_4, SR_2
	t_8-t_9	S_4, SR_2
	t_9-t_{10}	S_4, SR_2, SR_1
	$t_{10}-t_{11}$	SR_1
	$t_{11}-t_{12}$	–

output reaches to load. This is the end process of the switching cycles which is shown in Tabulation (Table 1).

2.3 Zero Voltage Switching Algorithm

ZVC for the main switches S_1 and S_2 is easily implemented with stored energy in transformer and its leakage inductance, when ZVC for Switch S_3, S_4 is made constant for better continuity of current, the modernized driving scheme will maintain shunt connected diodes of S_3, S_4 and has to be forced to turn ON before two main switches S3, S4 getting turned ON. Once switch S_4 is OFF, the excess energy is dissipated through switch S_1's storage capacitance, and energizes S_2, S_3, S_4's capacitors too, to obtain ZVC condition of switch S_1. This leads to satisfy the following expression [5]:

$$1/2L_t(I_m + nI_t)2 > 2\,Cos\,Vb2 + Cos\,Vs\,Vb + Cos\,Vw\,Vb, I_m + nI_l > 0 \quad (1)$$

where L_t is leakage inductance of the transformer,

MOSFET capacitances of C_1, C_2, C_3 and C_4 are assumed to be equal to Cos, and I_m is magnetizing current of the transformer, which is representing in the above Eq. (1) accordingly.

Which obeys the following expression number (2):

$$I_b = D_1(I_m - nIo) + D_2(I_m + nI_t) + D_3(I_m + nIo) \quad (2)$$

$$I_m = I_b + (D_1 - D_2 - D_3)nI_t/(D_1 + D_2 + D_3) \quad (3)$$

when magnetizing current I_m of the transformer is larger than load current I_t, then it persists and the ZVC of switch S_2 is unavailable, so we should not allow the condition of $I_m - nI_t < 0$.

The ZVC of S_3, S_4 is obtained when the potential difference relation of $V_b < V_w < V_s$ is fulfilled to ensure that the shunt diodes will be gated ON earlier than turning ON of switches S3, S4. If $V_w < V_s$ is easier to meet from Eq. (3), then the wind and solar port can be altered if wind voltage V_w is higher than the voltage V_s obtained from solar port. If incase wind voltage V_w is at times higher than source voltage V_s in the full potential difference range, the power converter should turn ON always, but switching loss for a period of S_2 switch depends on gating ON of switches S_2, S_3. While it is happening we need to change switching sequence of S_2, S_3 to avoid overlap of switching condition [3]. On the contrary side buck conversion is introduced from PV to battery and wind to battery side, since PV and wind voltage is always higher then battery voltage. Summarizing ZVC of all the main major switches is maintained to obtain better efficiency with the DC-DC power converter when operated at high conducting frequency.

2.4   Circuit Model Considerations

While considering power semiconductor switches of modified DC-DC power converter [6], it is similar to the conventional half-bridge converter terminology. The difference between conventional and modified DC-DC converter is that the CT (current transformer) design for this multi-port advanced DC-DC power converter needs to pass unidirectional current flow, and therefore, it is as like that of an electromagnetic field or a multi terminal transformer. The DC excitation is must to determine the actual quantity of air gap required to be maintained to fulfill the reactive power injection in the transformer. Except transformer model all other circuit and evolutionary techniques are used in convention (HB) half-bridge terminology which is used for advanced multi-port topology that allows good understanding and implementation of various power stages model [7].

3   Simulation Circuit

Simulated output waveforms (Figs. 3, 4, 5, 6, 7, 8, 9, 10 and 11).

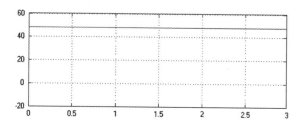

Fig. 3 Solar voltage

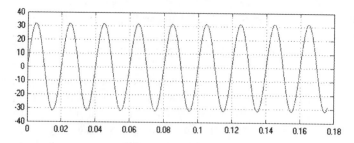

Fig. 4 Wind voltage

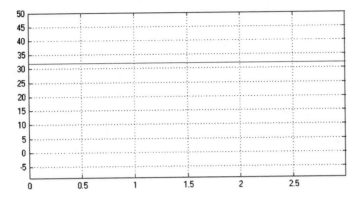

Fig. 5 Battery voltage

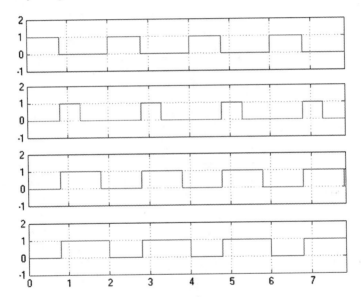

Fig. 6 Switching sequence in s1, s2, s3, s4

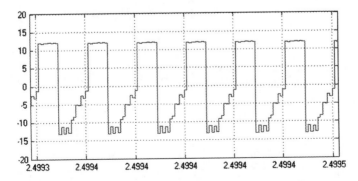

Fig. 7 Transformer primary voltage

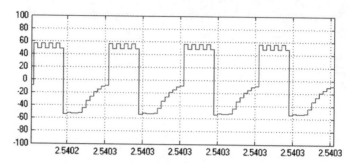

Fig. 8 Transformer secondary voltage

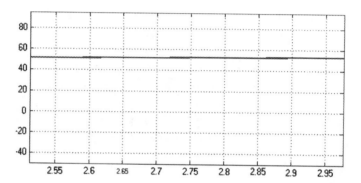

Fig. 9 Output DC voltage

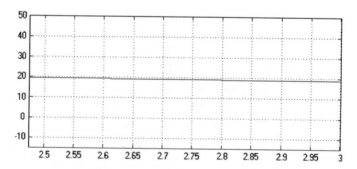

Fig. 10 Output DC current

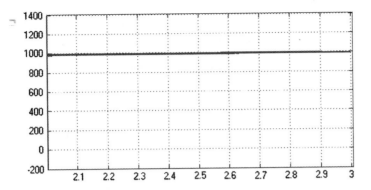

Fig. 11 Output DC power

4 Conclusion

This article covered an advanced DC-DC power converter technique which can interface four DC power ports using zero voltage conduction (ZVC) analysis in that dual source inputs, a two directional battery port, and a separate load balancing port. This multi-port converter deals with less switching elements and ZVC implementation for main semi conducting switches. Alteration is done based on conventional half-bridge converter terminology which makes this reliable for all budding engineers in order to understand the power circuit modeling. The advanced multi-port power converter suites best for systems which use renewable energy resources. In this system the energy storage system (battery) is most prominent which provides continues load regulation [2]. The conventional half-bridge terminology is suited for low power applications, but the multiport converter based on full-bridge terminology is suitable for high-power applications too.

References

1. Khaligh, A., Cao, J., Lee, Y.: A multiple-input DC–DC converter topology. IEEE Trans. Power Electron. **24**(3), 862–868 (2009)
2. Peng, F.Z., Li, H., Su, G.J., Lawler, J.S.: A new ZVS bidirectional DC-DC converter for fuel cell and battery applications. IEEE Trans. Power Electron. **19**(1), 54–65 (2004)
3. Qian, Z., Abdel-Rahman, O., Al-Atrash, H., Batarseh, I.: Modeling and control of three-port DC/DC converter interface for satellite applications. IEEE Trans. Power Electron. **25**(3), 637–649 (2010)
4. Zhao, C., Round, S.D., Kolar, J.W.: An isolated three-port bidirectional DC-DC converter with decoupled power flow management. IEEE Trans. Power Electron. **21**(5), 2443–2453 (2008)
5. Jiang, W., Fahimi, B.: Multi-port power electric interface for renewable energy sources. In: Proceedings of IEEE Applied Power Electronics Conference, pp. 347–352 (2009)

6. Krishnaswami, H., Mohan, N.: Three-port series-resonant DC–DC converter to interface renewable energy sources with bidirectional load and energy storage ports. IEEE Trans. Power Electron. **24**(9–10), 2289–2297 (2009)
7. Chen, Y.M., Liu, Y.C., Wu, F.Y.: Multi-input DC/DC converter based on the multiwinding transformer for renewable energy applications. IEEE Trans. Ind. Appl. **38**(4), 1096–1104 (2002)
8. Tao, H., Duarte, J.L., Hendrix, M.A.M.: Three-port triple-half-bridge bidirectional converter with zero-voltage switching. IEEE Trans. Power Electron. **23**(2), 782–792 (2008)
9. Matsuo, H., Lin, W., Kurokawa, F., Shigemizu, T., Watanabe, N.: Characteristics of the multiple-input DC–DC converter. IEEE Trans. Ind. Appl. **51**(3), 625–631 (2004)
10. Benavides, N.D., Chapman, P.L.: Power budgeting of a multipleinput buck-boost converter. IEEE Trans. Power Electron. **20**(6), 1303–1309 (2005)

Power Electronic Drives and Control Technology Status: Brief Review

B. Gunapriya, M. Sabrigiriraj, M. Karthik, B. Deepa
and R. Nithya Devi

Abstract In the human race, 70 % of energy is devoured by electric motors. This rate may be expanded because of the growth of power electronic devices and the fast advancement of automation technology. Most assembling units overall depend on electric motors for their generation, therefore highlighting the requirement for a viable speed control motors to build creation. It is underlined power electronic devices innovation has encountered a dynamic improvement in the previous four decades. As of late, its applications are quick growing in modern, business, private, transportation, utility, aviation and military situations, principally because of the lessening of expense, size, and performance enhancement. Soft computing techniques, especially the neural systems are having as of late huge effect on electrical drives and power devices. Neural systems have created another new edge and development power devices, that is currently a fancy and multidisciplinary innovation that goes through the dynamic improvement as lately. In this article, the significance of power hardware, the late advances in power semiconductor devices, converters, AC motors with variable frequency, the dawn of microprocessors/microcontrollers/microcomputers permitted to actualize and these control methods will be discussed briefly.

Keywords DC motors · AC motors · Control · Variable speed electrical drives · PI controller

B. Gunapriya (✉) · M. Karthik · B. Deepa · R.N. Devi
Department of EEE, Coimbatore Institute of Engineering and Technology, Coimbatore, Tamil Nadu, India
e-mail: gunapriya78@yahoo.in

M. Sabrigiriraj
Department of ECE, SVS College of Engineering, Coimbatore, Tamil Nadu, India

© Springer Science+Business Media Singapore 2017
P. Deiva Sundari et al. (eds.), *Proceedings of 2nd International Conference on Intelligent Computing and Applications*, Advances in Intelligent Systems and Computing 467, DOI 10.1007/978-981-10-1645-5_42

1 Introduction

Electric drives could be a multidisciplinary field of study, requiring a total combination of various electrical equipment data and are taking place with advancements inside of the field of electrical drives [1–4]. Electronic commutated motors utilize the DC motor principle however, put back the mechanical switch through inverter-based commutations. Induction motors are referred to as the workhorse of business on account of their across the prevalent use in mechanical drives [5–12]. PMSMs were created when the development of alloy, a static magnet material, in 1930 [5, 6, 13–18]. Nowadays for creating cost-effective power electronic converters huge power semiconductor devices are on hand. The converter will be utilized as various power modulators. The simulation tool could be an essential stride for acting advanced control for industry [10, 11, 15, 19]. These control stages give adaptability of control and alter the usage of advanced control algorithms, related to field oriented control (FOC) techniques, direct torque control (DTC) methods, nonlinear control strategies, and artificial intelligence technique. Intelligence techniques are at present being increasingly utilized for electrical drive speed control applications [20, 21]. A changeover of electrical drive controls, given by learning through brain emotional, is according to the literature [22]. A sensorless drive have the features of a simple drive with decreased upkeep, lessened the cost, and its capacity to oppose unforgiving natural conditions.

1.1 High Performance Refer Capacity Drives

It provides precise control, in addition to a fast dynamic response and steady state response. High performance drives are considered for the vital security applications due to the precise control [23–25]. For high accuracy and a quick movement, but as a response to the steady state vector control methods should be used with closed loop feedback control [19].

1.2 The Electrical Drive Challenges and Needs for Industrial Applications

Industrial automation needs precisely controlled electrical drive systems. The problems and requirements for electric drive systems based on the exact applications are used. Among the totally typical classes of electric drives, medium voltage drives are used as a part of the industry, such as within the various drive systems. To conquer the limitations of assessments electric converters, arrangement and parallel mixtures power devices are educated [26, 27].

1.3　Switching Frequency of Inverter

The use of high-frequency switching devices in power converters causes rapid transitions and current tensions. Switching losses can be a vital issue that must be taken under the thinking in the organization of electric drives; accordingly, they cause the most distant point of the switching frequency and thereby the quantity produced influences the level converters. The problems are compounded because of the great length of the cables between the converters and motors, in addition to generating power line due to switching transients [28, 29].

1.4　Signal Manifestation and Large dv/dt

The high frequency switching power devices causes a high dv/dt at rising and falling edges of the output voltage wave electric converter. This can cause the failure of the motor winding insulation due to partial discharges and high stress. High dv/dt further delivers voltages of the rotor shaft which make the current flow in the shaft carrying through the parasitic coupling capacitors finally cause the motor bearing failures. This can be a disadvantage in the typical variable speed drive systems in the industry [30].

1.5　Variable Speed Drives and Their Application

Variable speed drives (VSDs) will develop the ability of those motors by relating to in a few application. Introduction of semiconductor devices like diodes and transistors, VSDs became very trendy over traditional control strategies. But modifying the speed of an electrical machine, VSDs might keep speed at a proceeding level wherever the load is variable. Alternatively, modifiable speed drives will change the speed of the machine once the load changes, then deteriorating the connected energy to the system.

1.6　The Consequence of Electrical Motor Drives

Electric motors impact substantially on either side of the recent life. Coolers, vacuum cleaners, ventilation systems, fans, readers of notebook PCs, programmable logic controllers, and many alternative engines, all use electric motors convert the energy in support of tension. Peak motor drives are suitable for greater accuracy, therefore, they use a large number of DSP chips or microcontroller controllers to observe and regulate the power output. They also provide greater efficiency by using a large number of economic topologies.

1.7 The Motor Efficiency Impact

Nowadays the energy devoured by electrical motors represents almost a large portion of all power created inside of the world. The effect of wasted energy inside of the assortment of motor loss is frequently balanced into two principle classifications: economical and ecological. Industrial drives are utilized in various fields like mining and atmosphere control of enormous business structures. Residential motors are utilized in applications in the areas of cooling units, iceboxes and dishwashers. However, it's even additionally possible to introduce additional economical motor drives to residential applications. So there are extra chances to augment the proficiency of motor drives utilized in private applications. There are a few drawbacks to ASDs, similar to harmonics, voltage notching, and ripples, however these is mounted mistreatment harmonic filters, voltage regulators, and switch management mechanisms. Advancements in the computing power of computerized PCs and improvement systems have prompted the development of new methodologies. In this manner, evolutionary algorithms have turned into a prominent and helpful field of exploration and application in a long time [26–30].

2 Survey of Electrical Machinery

2.1 Direct Current Motor

The direct current machine overwhelmed the field of variable speed drives for over a century; they are still the most well-known decision if a controlled electric drive working over a wide speed range is indicated. This is because of the superb properties and the control of working qualities; the main crucial downside is the mechanical switch, which restricts the power and speed of the motor, expands the inertia and the axial length of the request and maintenance. In any case, it is very much perceived that DC motors have various limitations: Their energy is restricted at high speeds because of the switching. The collector and brush require maintenance. DC motors are not suitable for risky situations, for example, mines, chemical plants, and so on. The power to weight DC motor ratio is not good for AC motors, particularly the induction motor [31–33].

2.2 Permanent Magnet Motor

The number of utilizations for motor drives without DC brush has altogether expanded in the course of the most recent 20 years. This is basically because of various essential advances in material innovation and brushless drive created by the improvement of power semiconductor conducting devices and permanent magnet materials. The utilization of permanent magnets (PM) in electrical hardware apparatus set up of electromagnetic excitation results numerous favourable circumstances such that no excitation losses, simplified construction, improved productivity, implementation of the rapid component, and high torque or power per unit volume [34–36]. Brushless motors with permanent magnets can be divided into two subclasses. The main classification uses a product with the rotor position feedback to provide sinusoidal voltages and currents of the motor. The EMF is a perfect sine wave so enthusiastic that cooperation with sinusoidal currents creates a constant torque low torque ripple. Of the two types of PMBL motors, PMSM is the privileged way for applications where accuracy is desired, for example, robotics, CNC motors, solar tracking, etc. Moreover, PMBLDCM can be used as part of general and low cost applications. These engines are preferred for some applications due to their high level of competence, quiet operation, lesser size and low down maintenance [37–41].

2.3 Induction Motors

These engines were used for over a 100 years prior to their tough structure, low maintenance cost, high efficiency, high reliability, and noiseless operation. Despite the fact that there are many points of interest, sliding and power losses are to test the control parameters of the induction machine. Three phase induction motor stator and rotor. The stator carries a stator three-phase winding and the rotor can be short-circuited copper rod or three-phase rotor winding. Only the stator winding is powered by the 3-step feeding. The winding rotor obtains its voltage and the power voltage of the outer stator winding through electromagnetic induction and thus the name. The synchronous rate is squarely with respect to the proportion of the supply frequency and the number of shafts in the engine. Rotor flux rotates slower than the synchronous speed of the slip speed.

2.4 Switched Reluctance Motor

In switching reluctance motor torque is produced due to the tendency of the magnetic circuit to achieve the minimum reluctance to say, the rotor moves in conformity with the stator pole amplification therefore the inductance of the voltage. When the coil of a rotor pole is adjusted to a stator pole, torque because the fact that the field lines are orthogonal to the surfaces. In the event that one moves the rotor of the uncommitted position, and after that several tends to move the rotor to the following adjusted position. SRM magnetic behavior is profoundly nonlinear have changed the advantages because of their ease, simple powerful structure, and generally report high torque-volume and low maintenance costs [40, 42]. These drives can work in an extensive variety of rate with no significant diminishment in effectiveness. Along these lines, among all the diverse types of electric motors, SRM drives are for the exploitation of the profound saturation to the growth of the output power density. Following the impact of immersion and the variation of the reluctance attractive, each normal to the dependent model of the machine are very non-direct ability of the phase current and two rotor position [41]. Therefore, in spite of the uncomplicated mechanical arrangement, they require difficult calculations of speed control and switching, and has high torque ripple and acoustic noise when the stator and rotor poles are stimulated further speed changes especially in a limited low speed area probability for direct drive of their modern applications. Due to problems above have an unpredictable structure of control and demonstrate critical thinking is so annoying with PI controllers [42]. Therefore, many techniques can be embraced eager to take care of electrical control engine problems for applications above [43–46]. Switched reluctance motor, when used as a generator is truly different option for customary inverters in many applications, especially in the extraction of the most extreme energy in the wind power generation system to the fluctuating wind speeds [47–51].

2.5 Stepper Motor

A wide range of stepper motor based on a variety of operating principles have been established for industrial applications. According to their rules of operation, stepper motor can be controlled in three types: the reluctance, permanent magnet and hybrid. Stepper motors with variable reluctance are remarkable trees on the stator and rotor excitation poles with the testator loop. At reluctance stepper motors variable, the reluctance torque is a torque which is created by the inclination of the rotor shafts of the stator and to adjust, when the stator poles are energized. Stepper reluctantly position control variable are exceptionally suitable for many applications. Stepping motors with permanent magnets have salient poles drive coil on the stator. The pair of stepper motors having permanent magnet is the electromagnetic torque produced by the interaction of the stator currents and the flux formed by the rotor magnets. Stepper hybrid (HSM) has salient poles of the stator and the rotor. The stator poles are teeth and tooth wear of the rotor excitation coils are magnetized by a permanent magnet and type of different shaft assemblies. In hybrid stepper motors, torque is provided both by electromagnetic effects and reticence. Among the various types of stepping motors, HSM is most often used in light of the fact that they have the advantages of greater efficiency and the torque capacity on the other stepper motor [48, 49].

2.6 Synchronous Reluctance Motors

One of the oldest and least complex electric motor is synchronous reluctance motor (SynRM) [50, 51]. A SynRM exceeds an induction motor due to the absence of losses in the copper rotor, brushless following the structure of reasonable and rotor switched reluctance for a much lower torque ripple and low noise. In SynRM while torque output is delivered by the saliency of the rotor in light of the fact that SynRMs lack of a permanent magnet or coil power in the rotor. In addition, there is no risk of demagnetization, so that the stator current is essential to create a large torque. These reasons inductance of the stator and the stator resistance changes with the variety of inductance caused by the attractive immersion is particularly critical [51]. Robustness, economy and reliability features, but the yield is generally reduced at least compared to the PM machine [50, 51]. The idea of mounting rectangular pieces in each pole of the phase of the machine for self-excitation was exposed and implemented the same three ways [51].

3 Concise Review of Controllers for Electric Drives

Traditional controllers require accurate mathematical models and systems are very sensitive to parameter variations [52]. Soft Computing methods have proved their excellence by giving better results in improving the condition of the characteristics of balance and performance indices. Conventional proportional integral derivative controller is widely used in many industrial applications due to its simple structure and ease of design [53, 54]. In proportional integral derivative controller (PID) controller has different control settings, or proportional derivative, proportional and integral values determines the reaction to the current error, determines the value of the integral of the reaction based on the sum of recent errors and determines the value of the derivative on the basis of the reaction rate at which the error was changing and the weighted sum of these three actions is used to adjust the process via the latest controllers control elements are widely used in industrial facilities because it is simple and robust. Industrial processes are subject to change parameters and settings disturbances which when high makes the system unstable. Tuning is important for best performance of PID controllers [52]. Soft Computing techniques as methods GA, PSO and EP have demonstrated their excellence by giving better results by improving the characteristics of the balance indices of the condition and performance. Optimization techniques such as genetic algorithm (GA), evolutionary programming (EP) belonging to the family of algorithms for calculating the change have been widely used in many automotive control applications. The computer system dynamic flexibly tuned to performance on that achieves the same monitor system with ZN for the following reasons. ZN method provides only the initial adjustment parameters. Fine tuning to improve the response depends on the experience and intuition of control. Performance improved focus can be achieved if the AG and EP are organized in a greater number of iterations. This comes at a cost of increased load and processing delay in the calculation. GA and EP depends on the genetic operators. This means that even weak solutions could contribute to the future composition of candidate solutions. This ability to "remember" its previous best solution means that PSO can converge much faster than PE and GA on an optimal solution [53]. Objective function that measures the quality of control performance is optimized by a series of small adjustments to the three PID control parameters, because optimization is driven by the actual performance and no assumptions are made on the process dynamics and behavior disruption of this technique enables optimum setting for almost any loops that can be controlled by PID [54]. For the design of the PID controller, it is ensured that the controller settings provide estimated results in a closed loop system stable. This is the hardest part of creating a genetic algorithm written the objective function and it could be created to find a PID controller that gives the smallest amount, the rise time faster or time faster stabilization [55]. Genetic algorithm uses the fitness value of the chromosome to create a new population composed of the most capable members. Each chromosome is made up of three separate chains constituting a P, I and D in the long term, as defined by the Declaration of 3 rows limits in the creation of the population. The newly formed PID controller is placed in a feedback loop of the apparatus with the system transfer function. This will result in a reduction of the duration of the construction program. The system transfer function is defined in a different file and imported as a global variable. Forced system is then given an entry level and the error is evaluated using an error performance criterion and is assigned an overall value fitness depending on the size of the error and the error on the fitness value. Many clever methods can be adopted to solve control problems electric drives for high-performance applications [56–59]. From the perspective of industrial applications divided into four main categories,

1. Modeling and identification [53].
2. Optimization and classification [54].
3. Process Control [55, 60].
4. Recognition [56, 57].

Various controls based heuristic AI showed a good prospect of bringing the robustness and adaptive nature of the variable speed constant torque applications or constant speed drive torque variables [53–56, 60]. Thus, the superior performance of artificial intelligence (AI) urged controllers supply system and power electronic engines on the base to replace the conventional speed control circuit with Intelligent Cruise control.

A simple FLC consists of four main elements: fuzzy logic, a rule base, the fuzzy inference system and defuzzifier. The basic rule expressed in IF-THEN rules is used by the inference unit. The defuzzifier takes the fuzzy reasoning results and produces a new real control action. It is understood from the results that the outputs of the fuzzy controller are less overruns compared to the PID controller [58, 59, 61–63]. Fuzzy Logic has emerged as one of the areas of active research activity especially in control applications. It is very powerful reasoning process when mathematical models are not available and the input data is imprecise. Fuzzy logic works better compared to conventional controls such as PID [63–66].

Despite impressive progress in the drive automation, there are still a number of challenges, including a very low near zero speed, zero speed operation with full load condition and too high speed operation [1]. Control network based on remote drive systems are still ongoing. Types of electric motors are an important area that can serve applications that influence the quality of life, such as non-conventional energy, automotive and biomedical applications. More diversity in the design of the machine with free rare earth motors is being investigated in the electric drive systems [1]. Several attempts have been made to model the emotional behavior of the human brain [62]. Behavioral perspective were introduced that were then named Emotional Learning (BEL) brain model. The model was presented solely for descriptive purposes and was not motivated by a motor or an industrial application that could push the model away from the simple reflection of cognitive functioning bio mid brain to facilitate this particular application. Based on the open-loop model on motivated cognitive, brain emotional learning has been introduced for the first time in 2004 [63], and in recent years, this regulator was used, with minimal modifications in the devices order for multiple industrial applications [64–66]. BELBIC was designed and implemented in the field programmable gate arrays (FPGAs) [65]. For the first time, implementation of the method BELBIC for the electric drive control has been introduced [66]. The speed control the type of highly non-linear permanent magnet synchronous motor inside (IPMSM) using BELBIC method was compared with a PID controller. The results show the control of superior characteristics, especially, setting out simple, fast and robust to disturbances and parameter variations. Based on the above evidence of emotional control approaches in computing and automotive control, it can be concluded that the application of emotion in systems could by its simple design and unique control, to overcome the problems of unacceptable [67–70].

4 Structure of Various Motors

Figures 1, 2, 3, 4, 5, 6 and 7 show the structure of the different electric motors that are reviewed.

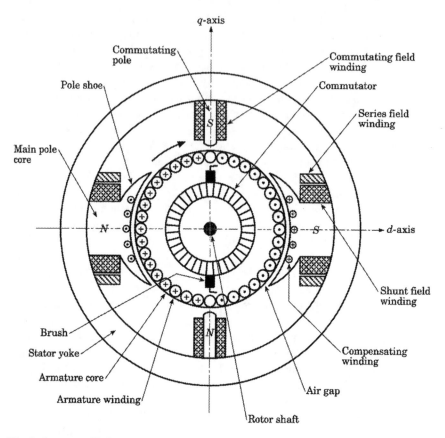

Fig. 1 Structure of DC machine

Fig. 2 Induction motor

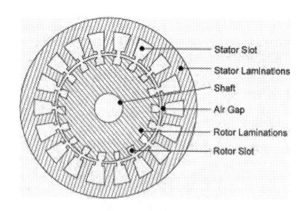

Fig. 3 Permanent magnet
brushless motor

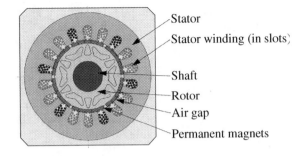

Fig. 4 Stepper motor

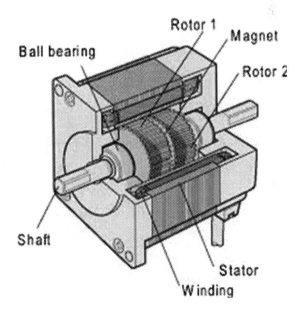

Fig. 5 Permanent magnet synchronous motor

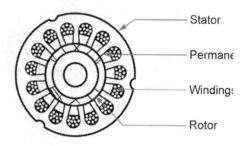

Fig. 6 Switched reluctance motor

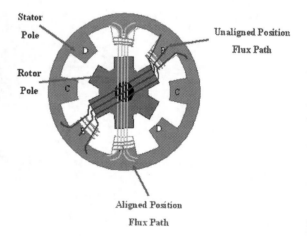

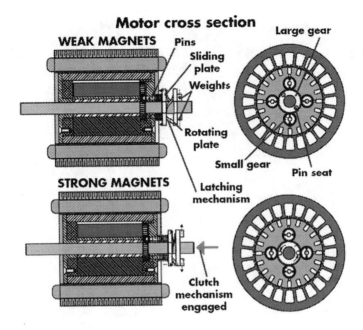

Fig. 7 Synchronous reluctance machine

5 Overview of Control Blocks of Electric Drives

Figures 8, 9, 10, 11, 12, 13, 14 and 15 show the control structure of the different electric drives that are reviewed.

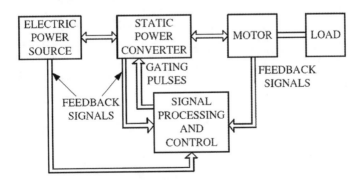

Fig. 8 Block diagram of electric motor drive

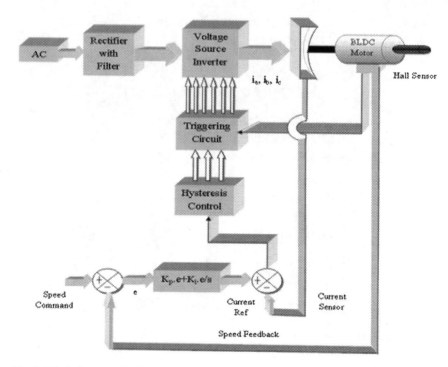

Fig. 9 Block diagram of BLDC motor control

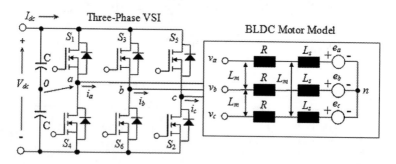

Fig. 10 Dynamic model of BLDC motor

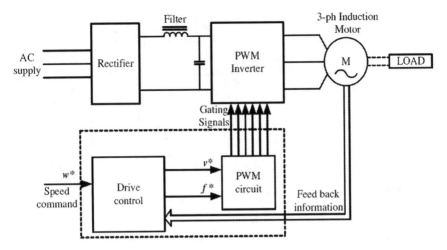

Fig. 11 Block diagram of IM motor control

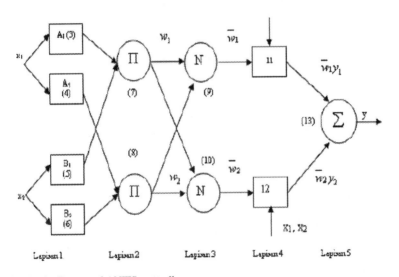

Fig. 12 Basic diagram of ANFIS controller

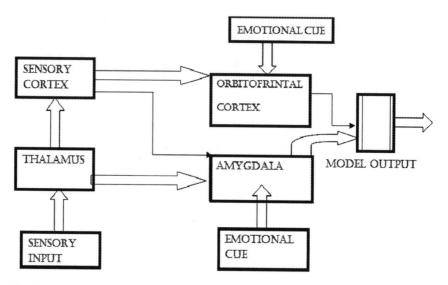

Fig. 13 Basic blocks of BELBIC

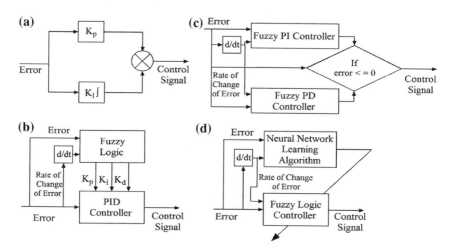

Fig. 14 **a** Proportional integral controller, **b** fuzzy tuned PID controller, **c** fuzzy variable structure controller and **d** ANFIS controller

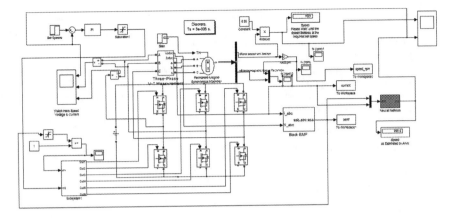

Fig. 15 Matlab simulink model of BLDC motor drive

6 Conclusion

Adjustable Speed drives and their applications have been quickly growing all over the world in most recent couple of years. This paper has presented a review of power devices and drives and its applications. Recent researches are included in the improvement of intelligent controllers to use the greatest capability of electric drives for more extensive applications. Different specialized parts of modelling, design, analysis and control in power electronics and drives have been reviewed comprehensively likewise talks about the grouping of electrical motors and their control techniques of field oriented control and direct torque control, and exhibits a review on High-Performance Drives. The variable speed drives are faster to develop innovations in electric motors and their future will undoubtedly be dynamic. The continuing difficulties for the industrial utilization of electric drives are further outlined. At last structure of the electric motors, control logic diagrams and controllers are surveyed from recent research in drives area.

References

1. http://www.sparkmuseum.com/MOTORS.HTM
2. Mecrow, B.C., Jack, A.G.: Efficiency trends in electric motors and drives. Energy Policy **36**, 4336–4341 (2008)
3. Rahman, M.A.: Modern electric motors in electronic world. 0-7803-0891-3/93, pp. 644–648 (1993)
4. Lorenz, R.D.: Advances in electric drive control. In: Proceedings of the International Conference on Electric Machines and Drives IEMD, pp. 9–16 (1999)
5. Finch, J.W., Giaouris, D.: Controlled AC electrical drives. IEEE Trans. Ind. Electron. **55**(2), 481–491 (2008)

6. Toliyat, H.A.: Recent advances and applications of power electronics and motor drives: electric motors and motor drives. In: Proceedings of the 34th IEEE Industrial Electronics Conference, IECON, Orlando, pp. 34–36 (2008)

7. Capilino, G.A.: Recent advances and applications of power electronics and motor drives: advanced and intelligent control techniques. In: Proceedings of the IEEE Industrial Electronics Conference, IECON, pp. 37–39 (2008)

8. Muller, S., Deicke, M., De Doncker, R.W.: Doubly-fed induction generator systems for wind turbines. IEEE IAS Mag. **8**(3), 26–33 (2002)

9. Krishnan, R.: Electric Motor Drives: Modeling, Analysis, and Control. Prentice-Hall, Upper Saddle River (2001)

10. El-Sharkawi, A.: Fundamentals of Electric Drives. Brooks/Cole Publishing, PacificGrove (2000)

11. Mohan, N., Undeland, T.M., Robbins, W.P.: Power Electronics: Converters, Applications, and Design. Wiley, New York (2003)

12. Bose, B.K.: Modern Power Electronics and AC Drives. Prentice Hall PTR, Upper Saddle (2002)

13. Bose, B.K.: Advances in power electronics and drives: their impact on energy and environment. In: Proceedings of the International Conference on Power Electronic Drives Energy Systems for Industrial Growth, PEDES, vol. 1 (1998)

14. Skvarenina, T.L.: The Power Electronics Handbook. CRC Press, Boca Raton (2002)

15. Iwanski, G., Koczara, W.: DFIG-based power generation system with UPS function for variable-speed applications. IEEE Trans. Ind. Electron. **55**(8), 3047–3054 (2008)

16. Pena, R., Clare, J.C., Asher, G.M.: Doubly fed induction generator using back-to-back PWM converters and its application to variable-speed wind energy generation. IEEE Proc. Electr. Power Appl. **143**(3), 231–241 (1996)

17. Forchetti, D., Garcia, G., Valla, M.I.: Vector control strategy for a doubly-fed stand-alone induction generator. In: Proceedings of the 28th IEEE International Conference, IECON, vol. 2, pp. 991–995 (2002)

18. Rich, N.: Xilinx puts ARM core into its FPGAs. EE Times. http://www.eetimes.com/electronicsproducts/processors/4115523/Xilinx-puts-ARM-core-into-its-FPGAs. Accessed 27 Apr 2010

19. Leonhard, W.: Control of Electrical Drives, 2nd edn. Springer, New York (1996)

20. Vas, P.: Artificial Intelligence Based Electric Machine and Drives: Application of Fuzzy, Neural, Fuzzy-Neural and Genetic Algorithm Based Techniques. Oxford University Press, Oxford (1999)

21. Daryabeigi, E., Markadeh, G.R.A., Lucas. C.: Emotional controller (BELBIC) for electric drives: a review. In: Proceedings of the IEEE IECON-2010, pp. 2901–2907 (2010)

22. Krzeminski, Z.: Non-linear control of induction motor. In: IFAC 10th World Congress Automotive Content, Munich, pp. 349–354 (1987)

23. Abu-Rub, H., Krzemisnki, Z., Guzinski, J.: Non-linear control of induction motor: idea and application. In: proceedings of the 9th International Conference on Power Electronics and Motion Control, EPE–PEMC, Kosice, vol. 6, pp. 213–218 (2000)

24. Levi, E.: Multiphase electric motors for variable-speed applications. IEEE Trans. Ind. Electron. **55**(5), 1893–1909 (2008)

25. Balkenius, C., Moren, J.: Emotional learning: a computational model of the amygdala. Cybern. Syst. **32**(6), 611–636 (2000)

26. Gopakumar, K.: Power Electronics and Electrical Drives, Video Lectures 1–25, vol. 55(3), pp. 157–164, Centre for Electronics and Technology, Indian Institute of Science, Bangalore (2000)

27. Bimbhra, P.S.: Power Electronics. Khanna Publishers, New Delhi (2006)

28. Dubey, G.K.: Fundamentals of Electrical Drives. Narosa Publishing House, New Delhi (2009)

29. Bose, B.K.: Power electronics, and motor drives recent technological advances. In: Proceedings of the IEEE International Symposium on Industrial Electronics, IEEE, pp. 22–25 (2002)

30. Kenjo, T., Nagamori, S.: Permanent Magnet DC Brushless Motors. Clarendon Press, Oxford (1985)
31. Bose, B.K.: Power Electronics and Variable Frequency Readers, Technology and Applications. IEEE Press, Piscataway (1997)
32. Puttaswamy, C.L.: Analysis, Design, and Control Permanent Magnet Brushless Motors. Ph.D. Thesis, IIT Delhi (1996)
33. Singh, B., Murthy, S.S., Reddy, A.H.N.: A micro speed controller for permanent magnet brushless DC motor. IETE Tech. Rev. 17(5), 299–310 (2000)
34. Singh, B., Reddy, A.H.N., Murthy, S.: Gain planning control brushless permanent magnet DC motor. IE (I) EL-J. 84, 52–62 (2003)
35. Kim, T., Lee, H.W., Parsa, L., Ehsani, M.: Optimal power and torque control of brushless DC (BLDC) motor drive/electric and hybrid electric generator vehicles. In: Proceedings of the IEEE Industry Applications Conference, Flight. 3, pp. 1276–1281 (2006)
36. Dwivedi, S.K.: Power quality improvements and sensor reductions permanent magnet synchronous drives. Ph.D. Thesis, IIT Delhi (2006)
37. Murphy, J.M.D., Turnbull, F.G.: Electronic Power Control of AC Motors. Pergamon Press, Oxford (1988)
38. Miller, T.J.E.: Brushless Permanent Magnet and Reluctance Motor Drive. Clarendon Press, Oxford (1989)
39. Singh, B., Singh, B.P., Dwivedi, S.K.: A state of the art of various permanent magnet brush configurations motors. IE (I) J.-EL, 78, 63–73 (2006)
40. Soares, F., Costa Branco, P.J.: Simulation of a switched 6/4 reluctance based on matlab/simulink environment. IEEE Trans. Aerosp. Electron. Syst. Flight 37(3), 989–1009 (2001)
41. Comparison of design and FET Harris Miller: Performance parameters in induction motors and switched reluctance. In: IEE in 1989 Electrical Motors and Drives Conference, London (1989)
42. Staton, D.A., Deodhar, R.P., Soong, W.L., Miller, F.E.T.: Torque forecasting using the flowchart MMF AC, DC, and reluctance motors. IEEE Trans. Ind. Appl. 32(1), 180–188 (1996)
43. Lawrenson, P.J.: Switched reluctance drives: a point of view. In: Proceedings of the 1992 International Conference on Electrical Motors, Manchester, pp. 12–21 (1992)
44. Slemon, G.R.: Chapter 2. In: Bose, B.K. (ed.) Power Electronics and Frequency Variation. IEEE Press, Piscataway (1997)
45. Miller, T.J.E.: Electronic Control of Switched Reluctance Motors. Newnes, Oxford (2001)
46. Chirila, A., Deacon, I., Navrapescu, V., Albu, M., Ghita, C.: On the model of a hybrid step motor. In: Proceedings of the IEEE International Conference on Industrial Electronics, pp. 496–501 (2008)
47. Huy, H., Brunelle, P., Sybille, G.: Design and implementation of a multi-motor model step by step Sim Power Systems Simulink. In: Proceedings of the IEEE International Conference on Industrial Electronics, pp. 437–442 (2008)
48. Nassereddine, M., Rizk, J., Nagrial, M.: Switched reluctance generator for wind energy applications. In: Proceedings of the World Academy Science, Motoring and Technology, vol. 31, pp. 126–130 (2008)
49. Nedic, V., Lipo, T.A.: Experimental verification voltage induced self-excitation of a switched reluctance generator. In: Proceedings of the Industrial Applications Conference, pp. 51–56 (2000)
50. Allawa, B., Gasbaoui, B., Mebarki, B.: PID configuration DC motor speed changing the control parameters using particle swarm optimization strategy. Leonardo Electron. J. Pract. Technol. 14, 19–32 (2009). ISSN1583-1078
51. Ellis, G.: Design Guide Control Systems. Academic Press, London (1991)
52. Gaing, Z.L.: A particle swarm optimization approach for optimum design of PID controller in the system. IEEE Trans. Energy Convers. Method 19(2), 284–291 (2004)

53. Karaboga, D., Kahuh, A.: PID controller parameters tuning using the tabu search algorithm. In: Proceedings of the IEEE Conference on Systems, Man, and Cybernetics, pp. 134–136 (1996)

54. Astrom, K.J., Hagglund, T.: PID controllers: theory, design, and tuning. ISA, Triangle Park (1995)

55. Daryabeigi, E., Arabic Markade, G., Lucas, C.: Interior permanent magnet synchronous motor (IPMSM), with an emotional learning developed brain based intelligent controller (BELBIC). In: Proceedings of the IEEE, IEMDC, pp. 1633–1640 (2009)

56. Rouhani, H., Jalili, M., Arabi, B., Eppler, W., Lucas, C.: Brain intelligent controller based on emotional learning applied to a neuro-fuzzy model of the micro-heat exchanger. Expert Syst. Appl. 32(3), 911–918 (2007)

57. Wang, Y., Xia, C., Zhang, M., Liu, D.: Adaptive control for speed brushless DC motors after the genetic algorithm and RBF neural network. In: IEEE Proceedings of the CICA, pp. 1219–1222 (2007)

58. Bose, B.K.: Neural network applications in power electronics and drives a motor-introduction and perspective. IEEE Trans. Ind. Electron. 54(1), 14–33 (2007)

59. Jamaly, M.R., Armani, A., Dehyadegari, M., Lucas, C., Navabi, Z.: Emotion FPGA: model-oriented approach. Expert Syst. Appl. 36(4), 7369–7378 (2009)

60. Meireles, M.R.G., Almeida, P.E.M., Simões, M.G.: A complete review on industrial application of artificial neurons networks. IEEE Trans. Ind. Electron. 50(3), 585–601 (2003)

61. Moren, J.: Emotions and learning: a model for calculating the amygdala. Ph.D. thesis, Lund University Cognitive 93 studies. University of Lund, Sweden (2002). ISSN 1101-8453

62. Lucas, C., Shahmirzadi, D., Sheikholeslami, N.: BELBIC presentation: intelligent control based on the brain emotional learning. Int. J. Intell. Autom. Soft Comput. 10(1), 11–22 (2004)

63. Jamali, M.R., Arami, A., Hosseini, B., Moshiri, B., Lucas, C.: Real emotional control time for anti-tilt and positioning control SIMO crane. Int. J. Innov Comput Inform Control 4(9), 2333–2344 (2008)

64. Jamali, M.R., Dehyadegari, M.R., Arami, A., Lucas, C., Navabi, Z.: Real time emotional embedded controller. J. Neural Comput. Appl. 19(1), 13–19 (2009)

65. Rahman, A., Milasi, R.M., Lucas, C., Arrabi, B.N., Radwan, T.S.: Implementation of emotional controller interior permanent magnet synchronous motor drive. IEEE Trans. Ind. Appl. 44(5), 1466–1476 (2008)

66. Kassakian, J.G., Schlecht, M.F., Verghese, G.C.: Principles of Power Electronics. Addison-Wesley, Upper Saddle River (1991)

67. Krein, P.T.: Elements of Power Electronics. Oxford University Press, New York (1998)

68. Rashid, M.H.: Power Electronics, 3rd edn. Prentice Hall, Upper Saddle River (2003)

69. Hart, D.H.: Introduction to Power Electronics. Prentice-Hall, Upper Saddle River (1997)

70. Vas, P.: Sensorless Vector and Direct Torque Control. Oxford University Press, London (1998)

Modelling and Simulation of Transformer Less Dynamic Voltage Restorer for Power Quality Improvement Using Combined Clark's and Park's Transformation Technique

Mohanasundaram Ravi and R. Chendur Kumaran

Abstract This paper presents a series connected custom power device (CPD) called transformer less dynamic voltage restorer (DVR) is used to protect sensitive loads in the load side by mitigating power quality problems present in the source side or at the point of common coupling (PCC). Series connected three single phase voltage source inverter (VSI) with filter design makes transformer less DVR to achieve desired load voltage magnitude with smooth sinusoidal waveform. Hysteresis controller is used for generating pulses for the VSI present in the Transformer less DVR and the error signal is generated by using combined Clark's and Park's transformation technique. The performance analysis of the transformer less DVR is verified using the MATLAB/SIMULINK software.

Keywords Transformer less dynamic voltage restorer (DVR) · Hysteresis controller · Voltage source inverter (VSI) · Clark's and Park's transformation

1 Introduction

Due to increasing usage of the nonlinear applications in day to day life leads to harmonics in the system. Other than harmonics some switching conditions and faults affects the quality of the power delivered to the load. This quality less power delivery makes the sensitive loads to malfunction or even shutdown. Thus these problems are to be mitigated before it affects the sensitive loads [1]. Power quality problems are voltage sag, voltage unbalance, voltage swell, voltage fluctuation, voltage interruption and waveform distortion. Voltage sag defined as decrease of

M. Ravi (✉) · R. Chendur Kumaran
School of Electrical Engineering, VIT University, Chennai, India
e-mail: Mohana.sundaramr2014@vit.ac.in

R. Chendur Kumaran
e-mail: chendurkumaran.r@vit.ac.in

© Springer Science+Business Media Singapore 2017
P. Deiva Sundari et al. (eds.), *Proceedings of 2nd International Conference on Intelligent Computing and Applications*, Advances in Intelligent Systems and Computing 467, DOI 10.1007/978-981-10-1645-5_43

fundamental rms voltage for duration 0.5 cycle to 1 min. Voltage sag ranges from 10 to 90 % of nominal voltage. Voltage sag is caused mainly due to energization of heavy loads. Voltage swell is defined as an increase in rms voltage for durations 0.5 cycles to 1 min and magnitudes are between 1.1 and 1.8 p.u. Voltage swell is caused mainly due to de-energization of heavy loads. Voltage unbalance is defined as the magnitude of three phase supply is not in equal. This is caused due to using of single phase loads in three phase circuit. Voltage unbalance is restricted to within five percentages. Voltage interruption is defined as the supply voltage decreased less than 0.1 p.u caused due to system faults. Waveform distortion is defined as steady state deviation in the voltage or current waveform from an ideal sine wave.

   To protect sensitive loads from these power quality problems custom power devices (CPD) [2, 3] are used in between the source side or point of common coupling (PCC) and the load side [4]. Dynamic voltage restorer (DVR) [5, 6] is one of the custom power device used to protect sensitive loads effectively by injecting voltage directly into the line. A conventional DVR has an injection transformer which is used to inject voltage into the line. This makes the DVR bulky, require more space and costly. Instead of that series connected VSI with filter circuit can able to do same function of conventional DVR without transformer called transformer less DVR. This result in reduce bulkiness of DVR, less cost and require less space [7, 8].

2 Conventional Dynamic Voltage Restorer

A conventional DVR consist of a series injection transformer, filter, voltage source inverter (VSI), DC energy storage device and control system. Figure 1 shows block diagram of conventional dynamic voltage restorer [9].

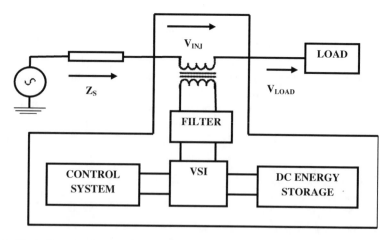

Fig. 1 Block diagram of conventional dynamic voltage restorer (DVR)

Series injection transformer is connected in series with the line such that it injects voltage directly into the system to make load voltage into desired level. Harmonics filter are used to convert the inverted PWM waveform into a sinusoidal waveform by eliminating the harmonic components generated by VSI action. It consists of an inductor in series with the output of the inverter and capacitor in parallel to that. Voltage source inverter is a power electronics converter which converters DC to AC by means of switching devices action which generates any required voltage magnitude, frequency and phase angle. DC energy storage is the place where DC energy is stored by means of storing elements like capacitors and batteries used for the input of the VSI. Control circuit generates pluses for the switches placed in the VSI for generating AC which is injected into the line. DC link controller circuit is used to maintain constant DC link voltage even on the time of voltage injection.

3 Transformer Less Dynamic Voltage Restorer (DVR)

Block diagram of the transformer less dynamic voltage restorer (DVR) is shown in the Fig. 2.

Voltage from the source is represented by v_s. Load in the feeder 2 is considered as a sensitive load and it is protected from the power quality problems by installing

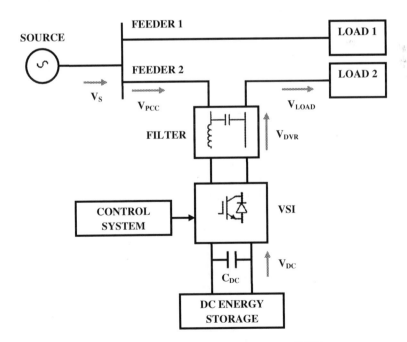

Fig. 2 Block diagram of transformer less dynamic voltage restorer (DVR)

a series connected transformer less DVR. Voltage at the point of common coupling (PCC) is represented as V_{pcc}. Load voltage in the feeder 2 is represented as V_{load}. Voltage injected by the DVR is represented by V_{dvr}. Series capacitor and filter inductor represented as C_{dvr} and L_{dvr} respectively A fault in feeder 1 affect all the feeders connected in the same system. DC link capacitor is represented as C_{bus}. The pulses for the voltage source inverter (VSI) is generated by using hysteresis controller and the reference signal is generated by using combined Clark's and Park's transformation technique.

4 Controller for Transformer Less Dynamic Voltage Restorer

There are various types of control techniques used to control custom power devices, among them in this paper controller based on combination of Clark's and Park's transformation control technique is used [10, 11]. The controller block diagram for DVR is shown in Fig. 3. The three phase PCC voltage V_{pcc} is converted into alpha and beta by using Clark's transformation and alpha and beta is converted into V_d and V_q by using Park's transformation as given in the Eq. 1. V_d and V_q compared with the reference value or desired voltage magnitude to get the error signal. This error signal is again converted into three phase voltage signal by means of inverse Clark's and Park's transformation given in the Eq. 2 [12].

$$\begin{pmatrix} v_d \\ v_q \end{pmatrix} = \sqrt{2} * \begin{pmatrix} \sin \omega t & -\cos \omega t \\ \cos \omega t & \sin \omega t \end{pmatrix} * \begin{pmatrix} 1 & \frac{-1}{2} & \frac{-1}{2} \\ 0 & \frac{\sqrt{3}}{2} & \frac{\sqrt{3}}{2} \end{pmatrix} * \begin{pmatrix} v_a \\ v_b \\ v_c \end{pmatrix} \quad (1)$$

$$\begin{pmatrix} v_a \\ v_b \\ v_c \end{pmatrix} = \sqrt{2} * \begin{pmatrix} 1 & 0 \\ \frac{-1}{2} & \frac{\sqrt{3}}{2} \\ \frac{-1}{2} & \frac{\sqrt{3}}{2} \end{pmatrix} * \begin{pmatrix} \sin \omega t & \cos \omega t \\ -\cos \omega t & \sin \omega t \end{pmatrix} * \begin{pmatrix} v_d \\ v_q \end{pmatrix} \quad (2)$$

This three phase reference voltage signal is converted into pulses for the switches in the VSI by using hysteresis controller. Thus the series connected inverter with filter circuit of transformer less DVR injects voltage in such a way that

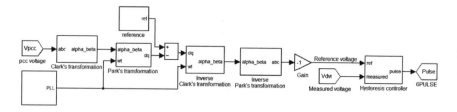

Fig. 3 Controller block diagram of transformer less dynamic voltage restorer

load voltage maintained desired constant voltage magnitude, cancel out distortion present in the supply voltage [12].

5　Simulation and Results of Transformer Less Dynamic Voltage Restorer

The simulation diagram of transformer less DVR is shown in Fig. 4.

Simulation done for transformer less dynamic voltage restorer based on combined Clark's and Park's transformation technique. The pulse for the VSI is generated by using hysteresis controller. Loads in the feeder 2 is considered as a sensitive loads and protected by using transformer less dynamic voltage restorer. By switching ON rectifier fed rl loads in feeder 1 produce voltage sag and harmonics at the point of common coupling in feeder 2. By connecting same load only in two phases creates voltage unbalance at the PCC point. Figure 5 shows output voltage waveform of voltage sag at PCC point, injection voltage and at load point. Figures 6 and 7 show THD value of voltage sag before and after compensation. Figure 8 shows output voltage waveform of voltage unbalance at PCC point, injection voltage and at load point. Figures 9 and 10 shows THD value of voltage unbalance before and after compensation. During voltage sag nominal voltage reduces from 1 Per Unit (P.U) to 0.77 P.U. After compensation load voltage is maintained 1 P.U constant by injecting 0.23 P.U voltage through series connected transformer less DVR. During voltage unbalance voltages in two phases got decreased to 0.77 P.U and other phase remain constant 1 P.U. After compensation load voltage at load point is maintained 1 P.U by injection 0.23 P.U on both phases and zero injection in unaffected phase. Sag and unbalance duration from 0.05 to 1 s and 0.15 to 2 s. Table 1 shows THD comparison of transformer less DVR [13–15].

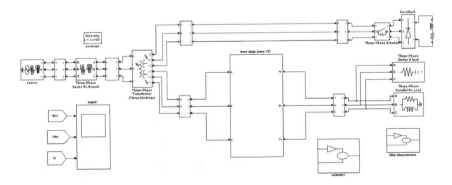

Fig. 4 Simulation diagram of transformer less DVR

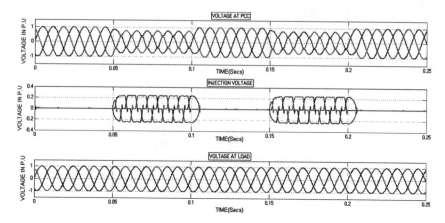

Fig. 5 Waveform of voltage sag at PCC, injection voltage and at load point

Fig. 6 THD value of voltage sag before compensation

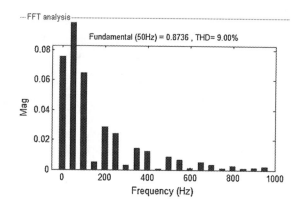

Fig. 7 THD value of voltage sag after compensation

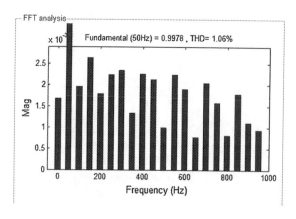

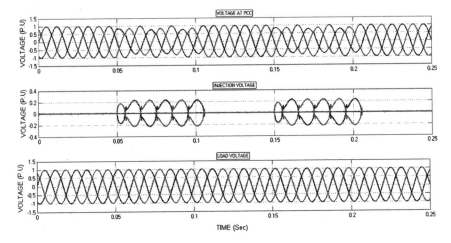

Fig. 8 Waveform of voltage unbalance at PCC, injection voltage and at load point

Fig. 9 THD value of voltage unbalance before compensation

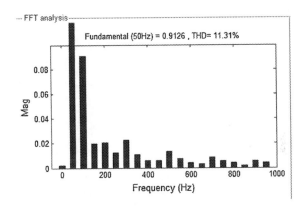

Fig. 10 THD value of voltage unbalance after compensation

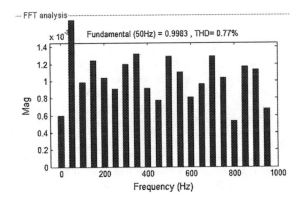

Table 1 THD comparison of transformer less DVR

Conditions	THD	
	Before compensation (%)	After compensation (%)
Voltage sag	9	1.06
Voltage unbalance	11.31	0.77

6 Conclusion

Transformer less dynamic voltage restorer based on combination of Clark's and Park's transformation technique is used for improving power quality is presented. The above simulation results show voltage sag, voltage unbalance and harmonics present in PCC is mitigated and protected the sensitive loads present in the feeder 2 by using series connected transformer less DVR.

References

1. Ghosh, A., Ledwich, G.: Power quality enhancement using custom power devices. Kluwer Academic Publishers, Dordrecht (2002). http://link.springer.com/book/10.1007%2F978-1-4615-1153-3
2. Hingorani, H.: Introducing custom power. IEEE Spectr. **32**(6), 41–48 (1995)
3. Akagi, H.: New trends in active filters for power conditioning. IEEE Trans. Ind. Appl. **32**(6), 1312–1322 (1996)
4. Moreno-Munoz, A.: Power Quality: Mitigation Technologies in a Distributed Environment. Springer, London (2007)
5. Chung, I.-Y., Won, D.-J., Park, S.-Y., Moon, S.-I., Park, J.-K.: The DC link energy control method in dynamic voltage restorer system. Electr. Power Energy Syst. **25**, 525–531 (2003)
6. Babaei, E., Kangarlu, M.F.: Sensitive load voltage compensation against voltage sags/swells and harmonics in the grid voltage and limit downstream fault currents using DVR. Electr. Power Syst. Res. **83**, 80–90 (2012)
7. Lu, Y., Xiao, G., Lei, B., Wu, X., Zhu, S.: A transformerless active voltage quality regulator with the parasitic boost circuit. IEEE Trans. Power Electron. **29**(4), 1746–1756 (2014)
8. Marei, M.I., Eltantawy, A.B.: An energy optimized control scheme for a transformerless DVR. Electr. Power Syst. Res. **83**, 110–118 (2012)
9. Kumar, C., Mishra, M.K.: Predictive voltage control of transformerless dynamic voltage restorer. IEEE Trans. Ind. Electron. **62**(5), 2693–2697 (2015)
10. Jayaprakash, P., Singh, B., Kothari, D.P., Chandra, A., Al-Haddad, K.: Control of reduced rating dynamic voltage restorer with battery energy storage system. In: Power System Technology and IEEE Power India Conference, POWERCON (2008)
11. Singh, B., Adya, A., Gupta, J.: Power quality enhancement with DSTATCOM for small isolated alternator feeding distribution system. In: Power Electronics and Drive Systems 2005 (PEDS 2005), vol. 1, 16–18 Jan 2005
12. Barros, J.D., Silva, J.F.: Multilevel optimal predictive dynamic voltage restorer. IEEE Trans. Ind. Electron. **57**, 2747–2760 (2010)
13. Ravi, M.: Comparision of PV Supported DVR and DSTATCOM with multiple feeders in stand alone WECS by mitigating power quality problems. Indian J. Sci. Technol. **8**(15), 1–10 (2015)

14. Dharmalingam, R., Dash, S.S., Senthilnathan, K., Mayilvaganan, A.B., Chinnamuthu, S.: Power quality improvement by unified power quality conditioner based on CSC topology using synchronous reference frame theory. Sci. World J. (2014). doi:10.1155/2014/391975
15. Khadkikar, V.: Enhancing electric power quality using UPQC: a comprehensive overview. IEEE Trans. Power Electron. **27**(5), 2284–2297 (2012)

Computational Study of Coil Helical Spring: Automobile Clutch

V.C. Sathish Gandhi, R. Kumaravelan, S. Ramesh
and M. Venkatesan

Abstract An open coil and closed coil helical springs works usually under critical conditions due to the continuous variations of load acting on the top surface. The analysis of an open coil and closed coil helical springs of an automobile clutches are carried out in 'ANSYS' software and the experimental study has been conducted for different load conditions. The various parameters like total deformation, stress intensity, strain energy and equivalent von-misses stress are considered for study. An experiment has been carried out for total deformation of an open and closed coil helical springs at various loads like 500, 1000 and 1500 N. The experimental results shows a good agreement for the simulation results. The results shows that the performance of an open coil helical spring is good compared with closed coil helical spring.

Keywords Helical spring · Total deformation · Stress intensity · Von-misses stress

1 Introduction

A spring may be defined as an elastic member whose primary function is to deflect or distort under the action of applied load; it recovers its original shape when load is released. Coil springs are manufactured to very tight tolerances to allow the coils

V.C. Sathish Gandhi (✉) · M. Venkatesan
Department of Mechanical Engineering, University College of Engineering Nagercoil
(A Constituent College of Anna University, Chennai), Konam, Nagercoil 629 004,
Tamilnadu, India
e-mail: vcsgandhi@gmail.com

R. Kumaravelan
Department of Mechanical Engineering, Velalar College of Engineering
and Technology, Erode 638 012, Tamilnadu, India

S. Ramesh
Department of Mechanical Engineering, KCG College of Technology, Karapakkam,
Chennai 600 097, Tamilnadu, India

© Springer Science+Business Media Singapore 2017
P. Deiva Sundari et al. (eds.), *Proceedings of 2nd International Conference
on Intelligent Computing and Applications*, Advances in Intelligent Systems
and Computing 467, DOI 10.1007/978-981-10-1645-5_44

spring to precisely fit in a hole or around a shaft. Most structures are designed will undergo acceptable deformation under specified loading conditions, but their main requirement is to remain rigid. A spring, however, will store energy elastically due to its relatively large displacement. The various application of springs are used to absorber shocks, applying forces in brakes and clutches, to control the motion between cams and followers and store energy in watches, toys etc.,. To ensure the application of springs the study of performance of coil springs are very important.

2 Literature Review

The literature of the coil springs are composed based on the information from the simulation techniques and its applications. Gaikwad and Kachare [1] presented the static analysis of helical compression spring used in two-wheeler horn. In this work the safe load of spring is calculated analytically and compared with Simulation results. The simulation is performed in the NASTRAN solver. Patil et al. [2] studied the comparison of cylindrical and conical helical springs for their buckling load and deflection. Based on the existing theories and analytical buckling equation they performed the analysis and verified with an experimental results. These results shows that under the given operating conditions it is to decided the suitability of conical springs against buckling failure of cylindrical springs. Chavan et al. [3] studied the fatigue life of suspension spring by Finite element analysis. The stiffness of the spring is evaluated by the experimental studied and it is compared with the simulation results. Salwinski and Michalczyk [4] discussed an accurate calculation of helical springs with a rectangular cross-section wire. The helical spring is machined from the tubular blanks is considered for the study. It is pointed out that the FEA analysis of spring with open ended is not shown the accurate stress values. Dakhore and Bissa [5] discussed the failure analysis of locomotive suspension coil spring through FEA. The stress distribution, material characteristic, manufacturing and common failures are discussed. The analytical and simulation results are presented for a failed coil springs. Harale and Elango [6] presented a design of helical coil suspension system by combination of conventional Steel and composite material. It is pointed out that the conventional steel helical coil spring is very bulky and costly for the required stiffness. So that, the E-Glass fiber/Epoxy reinforced with conventional steel material. The simulation has been carried out for different design in COMSOL software. Rathore and Joshi [7] reported a review of fatigue stress analysis of helical compression spring. It is identified that for better estimation of numerical solution, fatigue stress, shear stress and life cycle the finite element simulation is the best one. Mulla et al. [8] presented finite element analysis of helical coil compression spring for three wheeler auto-rickshaw front suspension system. For this analysis the elastic characteristics and fatigue strength are considered. The simulation results of stress distribution of the springs are reported. Tati et al. [9] discussed the design and analysis for coiled spring for application subjected to cyclic loading. This design and analysis procedure provided the methods of analysis of springs in the FEA. Pyttel et al. [10] has been investigated the probable failure position in helical compression springs used in fuel injection system of diesel engines. The simulation is performed in the ABAQUS 6.10 software. The results shown that an oscillatory behaviour of stresses along the length at inner side of the spring. Budan and Manjunatha [11] is investigated on the feasibility of composite coil spring for automotive applications. The three different composition of springs are fabricated such as glass fiber, carbon fiber and combination of glass fiber with carbon fiber. The experiments are carried out for the weight comparison with each other.

From the literature studied it is identified that the helical spring applications and its performances is studied. It is identified that the performance of the type of helical coil springs is not clearly disused. Therefore the present work represent the performance of the open and closed coil helical springs for the same input and identified the suitability of application.

3 Materials and Methods

The helical springs are used in the clutches due to reduce the shock to the pressure plate. In this work the open coil and closed coil helical spring's performance has been studied. The various parameters like total deformation, stress intensity, strain energy and equivalent von misses stress are considered for this study. An experimental study has been made to determine the total deformation of closed and opened coil springs at various loads. The spring models are created as per the dimensions of the spring which is used in the clutches in modeling software 'Pro-E'. The various loads of 500, 1000 and 1500 N are considered for this study. The attempt has been made to analysis the open coil and closed coil helical spring using simulation software "ANSYS V-12" under static conditions. The Table 1 shows the design parameters of an open coil and closed coil helical spring taken from an automobile clutches.

Table 1 shows the various dimensions of an open coil and closed coil helical springs. These values must be considered while designing and modeling of open coil and closed coil helical springs. Figure 1 shows the open coil helical spring and Fig. 2 shows the proto type of the closed coil helical spring used in the automobile clutches.

Table 1 Design parameters of helical springs

S. No.	Parameters	Units	Open coil	Closed coil
1	Wire diameter	mm	3.5	2.5
2	Mean coil diameter	mm	30	30
3	Spring index	–	8.57	12
4	Number of coils	–	7	12
5	Solid length	mm	24.5	32.5
6	Pitch	mm	8	3
7	Free length	mm	47	38.5
8	Poisson's ratio	–	0.3	0.3
9	Modulus of rigidity	N/mm^2	4.967×10^5	4.967×10^5

Fig. 1 Open coil spring

Fig. 2 Closed coil spring

3.1 Finite Element Analysis of Coil Springs

The following are the boundary conditions considered for the finite element analysis of both open and closed coil helical springs.

(a) Bottom fixed and load on top surface 500 N
(b) Bottom fixed and load on top surface 1000 N
(c) Bottom fixed and load on top surface 1500 N

3.1.1 Meshing Model—Open and Closed Coil

Mesh generation is one of the most critical aspects of engineering simulation. Too many cells may result in long solver runs, and too few may lead to inaccurate results. ANSYS meshing technology provides a means to balance these requirements and obtain the right mesh for each simulation in the most automated way possible. The meshing of the open coil and closed coil helical spring is done in ANSYS using mesh tool of size 2.9 mm. The suitable combination type spring-damper 14 element is selected for both open coil and closed coil helical springs. Mesh scale factor is 1000 mm.

Figure 3 shows the meshed model of the open coil helical spring. The mesh scale of this spring is 2.9 mm. Mesh scale factor is 1000 mm. The number of nodes are 6026. The number of elements are 19,212. Figure 4 shows the meshed model of the closed coil helical spring. The mesh scale of this spring is 2.4 mm. Mesh scale factor is 1000 mm. The number of nodes is 20,717. The number of elements is 55,899.

3.1.2 Analysis of Open Coil Spring—1500 N Load

Figure 5 bottom of the open coil helical spring is fixed and the load has been applied on the top surface of the helical spring. The analysis is carried out for different loads such as 500, 1000 and 1500 N. Here the simulation plots are presented for the maximum applied load of 1500 N. The 1500 N force has been

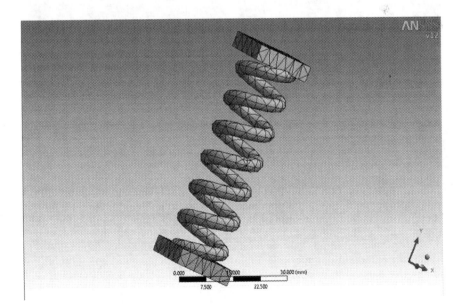

Fig. 3 Meshing model of open coil helical spring

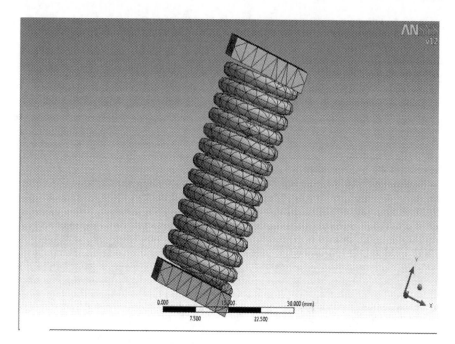

Fig. 4 Meshing model of closed coil helical spring

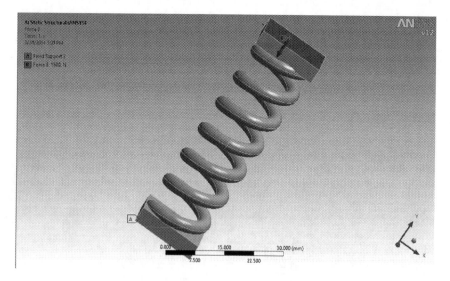

Fig. 5 Boundary condition—open coil

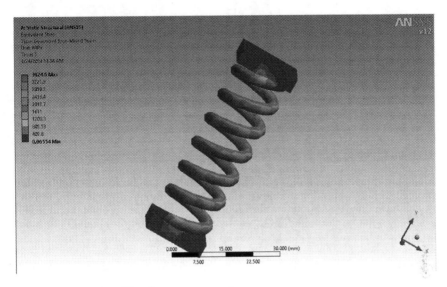

Fig. 6 Equivalent (von-Misses) stress—open coil

applied on the top side and various results of the stress, strain energy, total deformation has been taken and analyzed. In this case boundary conditions, equivalent stress, stress intensity, strain energy and the total deformation has been taken and analyzed.

Figure 5 shows the boundary condition of open coil helical spring in the static condition. The bottom side of the spring is fixed in all directions. The 1500 N load is applied on top surface of a spring. Figure 6 shows the equivalent von-misses stress of the spring. The maximum stress distribution is on the side and the value is 3.6246×10^3 N/mm^2.

Figure 7 shows the strain energy distribution of spring. The maximum strain energy distribution is on the body of the spring is 44.309×10^6 J. Figure 8 shows the total deformation of the open coil helical spring under 1500 N force applied. The maximum deformation is occurred on the top contact side of the spring of 15.225 mm. Figure 9 shows the stress intensity of the open coil helical spring. The maximum stress intensity acting on the spring is 4.1648×10^3 N/mm^2.

3.1.3 Analysis of Closed Coil Spring—1500 N Load

Figure 10 bottom of the closed coil helical spring is fixed and the load has been applied on the top surface of the helical spring. The analysis is carried out for different loads such as 500, 1000 and 1500 N. Here the simulation plots are presented for the maximum applied load of 1500 N.

The 1500 N force has been applied on the top side and various results of the stress, strain energy, total deformation has been taken and analyzed. In this case

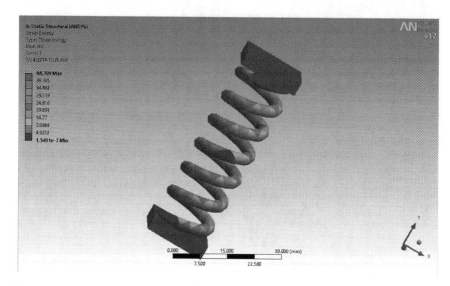

Fig. 7 Strain energy—open coil

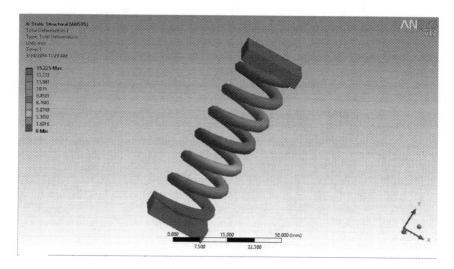

Fig. 8 Total deformation—open coil

boundary conditions, equivalent stress, stress intensity, strain energy and the total deformation has been taken and analyzed. Figure 10 shows the boundary condition of closed coil helical spring in the static condition. The bottom side of the spring is fixed in all directions. The 1500 N load is applied on top surface of spring.

Figure 11 shows the equivalent von-misses stress of the spring. The maximum stress distribution is on the side and the value is 644 N/mm^2. Figure 12 shows the

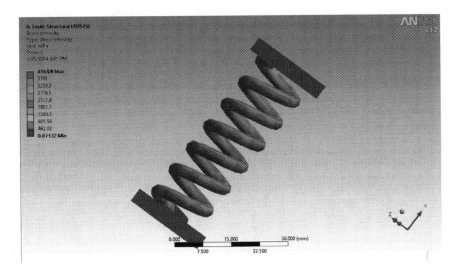

Fig. 9 Stress intensity—open coil

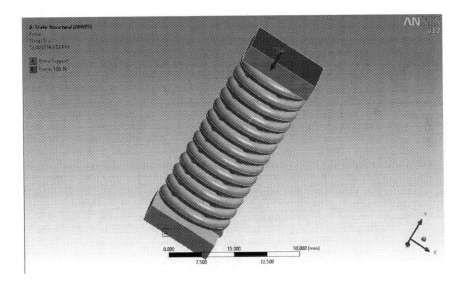

Fig. 10 Boundary condition—closed coil

strain energy distribution of spring. The maximum strain energy distribution is on the body of the spring is 0.75174×10^6 J.

Figure 13 shows the total deformation of the closed coil helical spring under 1500 N force applied. The maximum deformation is occurred on the top contact side of the spring of 11.228 mm. Figure 14 shows the stress intensity of the closed coil helical spring. The maximum stress intensity acting on the spring is 742.66 N/mm^2.

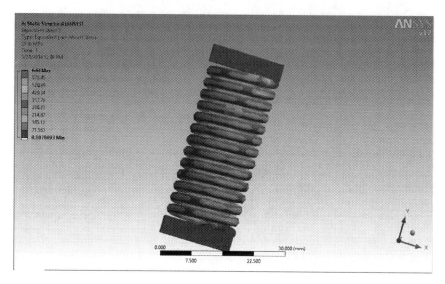

Fig. 11 Equivalent (von-Misses) stress—closed coil

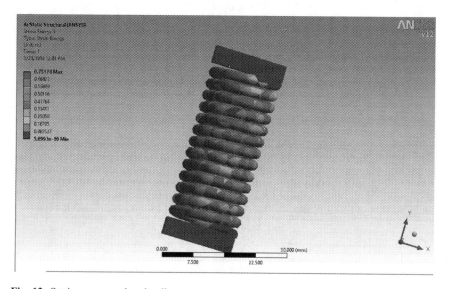

Fig. 12 Strain energy—closed coil

3.2 Experimental Study of Helical Springs

Experimental study shows the total deformation of open coil and closed coil helical springs are calculated experimentally using a manually operated spring testing machine. The total deformation is calculated under varying loads like 500, 1000

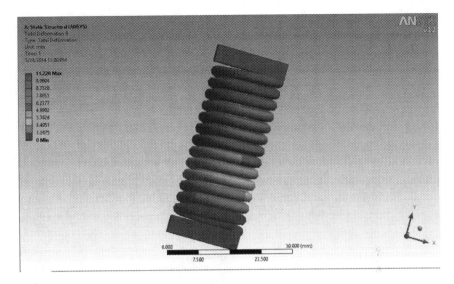

Fig. 13 Total deformation—closed coil

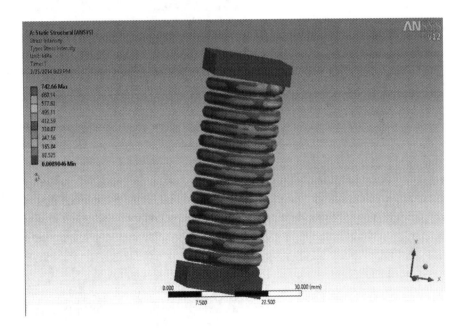

Fig. 14 Stress intensity—closed coil

and 1500 N. Figure 15 shows the testing of open coil helical spring is under compressive stress by apply the loads of 500, 1000 and 1500 N for testing the spring. The total deformation values for loads of 500, 1000 and 1500 N are 5, 10

534

V.C. Sathish Gandhi et al.

Fig. 15 Testing of open coil
helical spring

and 15 mm respectively. Figure 16 shows the testing of closed coil helical spring is under compressive stress by apply the load of 1000 N for testing the spring. The total deformation values for loads of 500, 1000 and 1500 N are 5, 9 and 11 mm respectively.

Fig. 16 Testing of closed
coil helical spring

4 Result and Discussion

The static analysis of open and closed coil helical springs are studied for three various loads of 500, 1000 and 1500 N. The study has been made for the various parameters like strain energy, total deformation, equivalent von-misses stress and stress intensity are analyzed. The results of open and closed coil helical springs were obtained as discuss below.

4.1 Results of 500 N Load

Table 2 Shows the results of static analysis of an open and closed coil helical springs. The load of 500 N is applied for analysis the helical spring. The various parameters have been evaluated in the open and closed coil helical springs.

4.2 Results of 1000 N Load

Table 3 Shows the results of static analysis of an open and closed coil helical springs. The load of 1000 N is applied for analysis the helical spring. The various parameters have been evaluated in the open and closed coil helical springs.

4.3 Results of 1500 N Load

Table 4 Shows the results of static analysis of an open and closed coil helical springs. The load of 1500 N is applied for analysis the helical spring. The various parameters have been evaluated in the open and closed coil helical springs.

Table 2 Results of 500 N load for open and closed coil helical spring

S. No.	Parameters	Value (open coil)	Value (closed coil)
1	Total deformation	5.0794 mm	5.6139 mm
2	Strain energy	4.9232 MJ	0.18794 MJ
3	Stress intensity	1388.3 MPa	371.33 MPa
4	Equivalent von-misses stress	1208.2 MPa	322 MPa

Table 3 Results of 1000 N load for open and closed coil helical spring

S. No.	Parameters	Value (open coil)	Value (closed coil)
1	Total deformation	10.15 mm	9.3192 mm
2	Strain energy	19.693 MJ	0.51788 MJ
3	Stress intensity	2776.5 MPa	616.41 MPa
4	Equivalent von-misses stress	2416.4 MPa	534.52 MPa

Table 4 Results of 1500 N load for open and closed coil helical spring

S. No.	Parameter	Value (open coil)	Value (closed coil)
1	Total deformation	15.225 mm	11.228 mm
2	Strain energy	44.309 MJ	0.75174 MJ
3	Stress intensity	4164.8 MPa	742.66 MPa
4	Equivalent von-misses stress	3624.6 MPa	644 MPa

4.4 Comparison of Results

The comparison of results of static analysis for open and closed coil helical springs is discussed. The parameters such us total deformation, stress intensity and von-misses stress are compared for the materials like Spring steel (Grade 1) for various loads of 500, 1000 and 1500 N.

Figure 17 shows the value of total deformation for the various loads of 500, 1000 and 1500 N like the material are spring steel (Grade 1). It is observed that the closed coil helical spring has low value of total deformation as compared with the various loads. Figure 17 shows the relation between load and deflection of open and closed coil helical springs. The linear deformation is occurring on open coil helical spring. But closed coil helical spring gives linear output till the load of 500 N.

Figure 18 shows the value of stress intensity for the various loads of 500, 1000 and 1500 N. It is observed that the closed coil helical springs has low value of stress intensity as compared with the loads. Figure 18 shows the comparison of stress intensity in graphical representation. In this graph the intensity variation is plotted linearly for open coil. For closed coil the variation of stress intensity is very minimum. The load is above 500 N the stress intensity is low for closed coil helical spring.

Figure 19 shows the comparison of von-misses stress in graphical representation. In this graph the von-misses stress variation is plotted linearly for open coil. For closed coil the variation of stress is very minimum. The load is above 500N the von-misses stress is low for closed coil helical spring. The von-misses stress is maximum in the top edge of the both open coil and closed coil helical springs.

Figure 20 shows the relationship between strain energy and load for closed and open coil helical spring. It is observed that in an open coil the strain energy rate is high as compared with closed coil.

Fig. 17 Load versus total deformation of coil springs

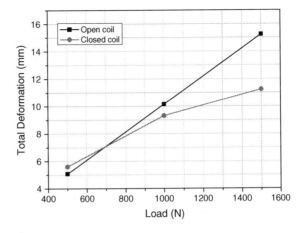

Fig. 18 Load versus stress intensity of coil springs

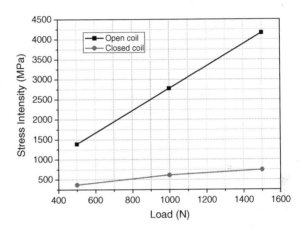

Fig. 19 Load versus equivalent Von-Misses stress of coil springs

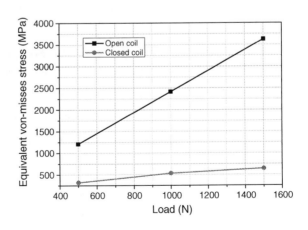

Fig. 20 Load versus strain
energy of coil springs

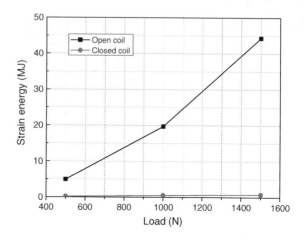

4.5 Experimental Study

The spring testing machine is used for determine the total deformation in open coil
and closed coil helical springs. The spring which is used in the clutches is taken for
this study. The total deformation in the spring has been estimated for three different
loads of 500, 1000, 1500 N are considered for this study. Tables 5 and 6 shows the
experimental value of total deformation for an open and closed coil helical springs
respectively.

Figure 21 shows the relationship between load and total deformation for an open
coil helical spring. An experimental and simulation results has been compared and
it is very close to each other. It is observed that when the load increases the total
deformation of an open coil helical spring is increased. It shows that the load and
deformation are directly proportional to each other.

Figure 22 shows the relationship between load and total deformation for closed
coil helical spring. An experimental and simulation results has been compared and
it is very close to each other. It is observed that when the load increases the total
deformation of closed coil helical spring is increased. The deformation of the closed
coil helical spring is less with respective to load as compared with open coil helical
spring.

Table 5 Total deformation
—Open coil spring
(Experiment)

S. No.	Parameters	Unit	Open coil Helical spring—load (N)		
			500	1000	1500
1	Total deformation	mm	5	10	15

Table 6 Total deformation
—closed coil spring
(experiment)

S. No.	Parameters	Unit	Closed coil Helical spring—load (N)		
			500	1000	1500
1	Total deformation	mm	5	9	11

Fig. 21 Load versus total deformation (open coil—experiment and simulation)

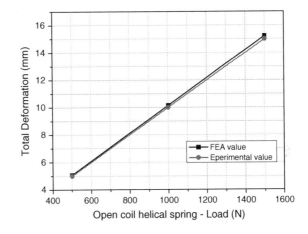

Fig. 22 Load versus total deformation (closed coil—experiment and simulation)

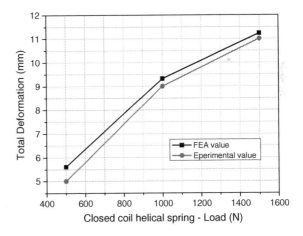

5 Conclusion

The open coil and closed coil helical spring are used in a automobile clutches has been Studied. The open and closed coil helical springs models are developed as per the dimensions of the springs which are used in clutches in modeling software 'Pro-E' and the analysis through simulation software "ANSYS V-12". The study has been made for three different loads like 500, 1000 and 1500 N. For analyzing the open and closed coil springs the bottom side of the spring is fixed and the load is applied on the top side. The parameters like total deformation, equivalent von-misses stress, strain energy and stress intensity are analyzed. These results are compared and it is observed that, the open coil helical spring has high stress, strain energy and deformation. The total deformation of the open and closed coil springs are estimated in a spring testing machine and compared with simulation results. It shows a good agreement for the simulation results. From this study it is concluded that performance of an open coil helical spring is good compared with closed coil helical spring.

References

1. Gaikwad, S.S., Kachare, P.S.: Static analysis of helical compression spring used in two-wheeler horn. Int. J. Eng. Adv. Technol. 2(3), 161–165 (2013)
2. Patil, V.R., Reddy, P.R., Laxminarayana, P.: Comparison of cylindrical and conical helical springs for their buckling load and deflection. Int. J. Adv. Sci. Technol. 73, 33–50 (2014)
3. Chavan, C., Kakandikar, G.M., Kulkarni, S.S.: Analysis for suspension spring to determine and improve its fatigue life using finite element methodology. Int. J. Sci. Res. Manag. Stud. 1 (11), 352–362 (2015)
4. Salwinski, J., Michalczyk, K.: Stress analysis in helical springs with closed end coils machined from cylindrical sleeves. Mechanics 25(4), 169–172 (2006)
5. Dakhore, M., Bissa, B.: Failure analysis of locomotive suspension coil spring using finite element analysis. Int. Mon. Refereed J. Res. Manag. Technol. II, 96–104 (2013)
6. Harale, G.S., Elango, M.: Design of helical coil suspension system by combination of conventional steel and composite material. Int. J. Innov. Res. Sci. Eng. Technol. 3(8), 15144–15150 (2014)
7. Rathore, S.G., Joshi, U.K.: Fatigue stress analysis of helical compression spring: a review. Int. J. Emerg. Trends Eng. Dev. 2(3), 512–520 (2013)
8. Mulla, M.T., Kadam, J.S., Kengar, S.V.: Finite element analysis of helical coil compression spring for three wheeler automotive front suspension. Int. J. Mech. Ind. Eng. 2(3), 74–77 (2012)
9. Tati, N., Shinde, S.L., Kulkarni, S.S.: Design and analysis for coiled spring for application subjected to cyclic loading. Int. J. Adv. Eng. Res. Stud. 3(3), 23–24 (2014)
10. Pyttel, B., Ray, K.K., Brunner, I., Tiwari, A., Kaoua, S.A.: Investigation of probable failure position in helical compression springs used in fuel injection system of diesel engines. IOSR J. Mech. Civil Eng. 2(3), 24–29 (2012)
11. Budan, D.A., Manjunatha, T.S.: Investigation on the feasibility of composite coil spring for automotive applications. World Acad. Sci. Eng. Technol. 4(10), 1035–1039 (2010)

Adaptive Control Technique for Generator Side Power System Voltage Stability at Wind Power Station

M. Presh Nave, K. Priyatharshini, N. Nijandhan and S. Pradeep

Abstract Voltage instability and over voltage are one of the main problems in today wind industry. In wind power production system the output is depends upon the nature of source called wind. But that source is not constant one. It may be varying depending upon the climate. Due to that oscillation the output voltage from the generator side is instability. These problems are becoming a more serious concern with the ever-increasing utilization and higher loading of existing transmission systems, particularly with increasing energy wastage, energy demands, and competitive generation and supply requirements. Our aim is to improve power system voltage stability by enhancing generator reactive, active power control and voltage control. Adaptive feedback system along with the HSVC ways to improve power system voltage stability by enhancing generator controls in the wind power station. To solve this problem adaptive exciter system are used which will adjust the load voltage and the system voltage with that of the reference voltage. The generator output voltage is applied to the adaptive exciter controller (AEC) which updates its stability weight value on demand. The design of modules can be done in Xilinx system generator (XSG). The modules that are designed in system generator can be implemented in FPGA.

Keywords Advanced over-excitation limiters · High side voltage control · Adjustable speed machines · Adaptive linear network · Adaptive exciter controller · Xilinx system generator · Field programmable gate array

M. Presh Nave (✉) · K. Priyatharshini · N. Nijandhan
Department of EEE, KCG College of Technology, Chennai, India
e-mail: preshnavem@gmail.com

S. Pradeep
Department of ECE, P.A College of Engineering, Pollachi, India

1 Introduction

Vitality is the fundamental need for the conservative advancement of the nation. Vitality exists in distinctive structures in nature however the most imperative structure is the electrical vitality. The current society is such a great amount of ward upon the utilization of electrical vitality that it has turned into an integral part of our life. We should center our consideration on the general parts of electrical vitality. Vitality may be required as warmth, as light, as thought process power and so on. The present day headway in science and innovation has made conceivable to change over electrical vitality into any wanted structure. This has given electrical vitality a position of pride in the current world.

The survival of mechanical undertaking and our social structure depends principally upon ease and continuous supply of electrical vitality. Truth be told, the progression of a nation in measured regarding per capita utilization of electrical vitality. Electrical vitality is created from vitality accessible in different structures in nature; it is attractive look into the different wellsprings of vitality that is sun, wind, water, energizes atomic vitality. Out of these sources, vitality because of sun and wind not been used on vast scale because of number of restriction. At present, the other three sources water, energizes, atomic are principally utilized for the era of electrical vitality.

The general fundamental structure of wind station as appeared in Fig. 1. Prime mover is mechanical gadget. The prime wellspring of electrical supplied by utilities or the active vitality of water. Prime mover changes over these wellsprings of vitality into mechanical vitality that is thus, changed over to electrical vitality by synchronous generator [10]. Power frameworks are by and large nonlinear and the

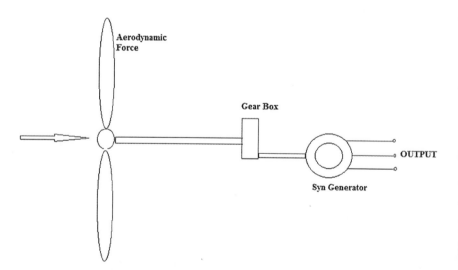

Fig. 1 General structure of wind station

working conditions can differ over a wide range. As of late, little flag steadiness has gotten much consideration. The expanding size of producing units, the stacking of the transmission lines and the fast excitation frameworks are the fundamental driver influencing the little flag strength.

Voltage shakiness issues are drawing in expanding consideration in today's energy frameworks, with colossal accentuation on force framework operation, arranging, and control [7]. These issues are turning into a more genuine worry with the always expanding use and higher stacking of existing transmission frameworks, especially with expanding vitality requests, and focused era and supply necessities.

It is settled that upgrading the responsive force applying so as to supply capacity of a force framework synchronous condensers, shunt and arrangement capacitor banks, static compensators (STATCOM), static VAR compensators (SVCs), and other flexible AC transmission system (FACTS) controllers is viable in enhancing voltage strength. For instance, systems, for example, high side voltage control (HSVC), progressed over-excitation limiters (OEL), and the utilization of cus-tomizable velocity machines (ASM) have all been produced and connected for the change of voltage security. This paper portrays different approaches to enhance power framework voltage dependability by upgrading generator controls [8].

2 Outline of Field Verification Test

For the field check tests, the U.S. Armed force Corps of Engineers. "The Dulles" hydro force station situated on the Co-lumbia River was chosen. The Dallas had the accompanying attractive attributes.

- Two generators those are straightforwardly associated with the same stride up transformer. This was valuable for confirming the cross-current concealment capacity on the low-voltage side of a stage up transformer. (The current AVR utilizes receptive hang pay that is "in reverse looking.")
- There are extra generators past the two utilized for the tests. This was helpful for checking the impact of the "high side" slant (hang).
- The AVRs of the test units were advanced sort.
- There was a huge shunt capacitor bank in the transmission switchyard. Capacitor bank exchanging brought on a stage change in system voltage, and the reaction of both ordinary control and HSVC could be observed.

The receptive hang remuneration capacity of the current AVR lessens voltage security. In the HSVC mode, after the capacitor bank is exchanged off, the terminal voltage is expanded, and the drop of the high side voltage is additionally dimin-ished by the increment of receptive force. In addition, the connection between generators is totally suppressed [5]. Voltage insecurity happens in its immaculate structure the edge and voltage precariousness go as an inseparable unit. One may prompt other and refinement may not be clear. On the other hand, a qualification between edge soundness and voltage security critical for comprehension of the under laying instances of the issues keeping in mind the end goal to create suitable plan and working strategy.

The change of the responsive force in the HSVC mode is bigger than that in the AVR mode. Regardless of high hang of 0.06 p.u. Around four times the responsive force of the AVR mode can be supplied to the force framework by the HSVC [4]. By applying littler hang rates, more impact can be normal. With DC framework associated with feeble framework especially on inverter side and in addition direct voltage is exceptionally touchy to change in stacking. An increment in direct current is joined by a fall of rotating voltage thus; the genuine increment in force is little or insignificant [3]. Control of voltage and recuperation from unsettling influence get to be troublesome. The DC framework reaction may even add to fall of AC framework. The affectability increments with vast measure of shunt capacitors.

The way of the framework reaction to little unsettling influence relies on upon the quantity of variables including the introductory operation, the transmission quality, and the sorts of generator excitation control utilized. For a generator joined profoundly to a huge force framework without programmed voltage controller. The precariousness is because of lack of adequate synchronizing torque.

The outcome is precariousness to a non oscillatory mode with consistently acting voltage controller; the little unsettling influence soundness is one of the guaranteeing adequate damping of framework motions. Insecurity is typically motions of expanded regularly. Nearby modes or machine framework mode are connected with the swinging of units at the producing station as for rest of the force framework. The term neighborhood is utilized on the grounds that the motions are confined at one station or the little piece of the force framework.

3   Outcome

Channels are connected in sound frameworks. The bass control on a sound preamplifier applies pretty much pick up at lower frequencies than at higher frequencies [2]. The finished result is that lower frequencies (bass) are stressed or lessened. Note that for this situation, recurrence parts were not being uprooted, but rather basically moulded. Sifting is especially simple to watch the impact of separating in the recurrence area. For instance, the high-pass channel appeared on the graph, weakens (opposes) the low-recurrence parts of the sign, while the high recurrence segments of the sign are gone through without observable adjustment (Fig. 2).

Figure 3 shows the 8 tap FIR channel plan for voltage steadiness process [1, 9]. The square chart comprises of info voltage from the prime mover with reference voltage. It is looked at by utilizing comparator framework and that info sign given

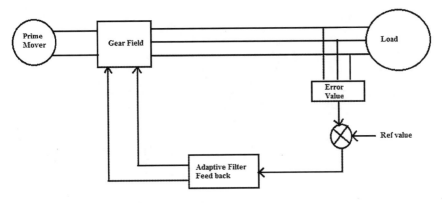

Fig. 2 Block diagram of voltage stability system

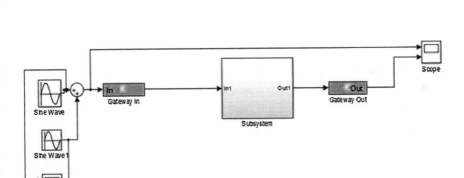

Fig. 3 Xilinx system generator block

to the 'door in'. At that point it is given to the versatile channel area. That versatile channel contains 8 tap Fir framework. It contains the Add/Sub, Multiplier, postponement and consistent component. Those consistent qualities in light of the prime mover voltage worth taking into account the supply framework and nature of the source.

The 8 tap FIR adaptive filter as shown in Fig. 4. It contains the Add/Sub, Multiplier, delay and constant element.

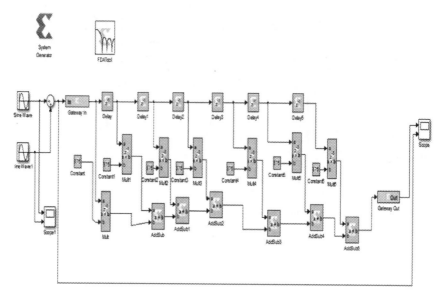

Fig. 4 Adaptive filter block

4 Simulation Result

The yield of the recreation comprises of versatile channel and terminal voltage and rotor point deviation as appeared in Figs. 5, 6, 7 and 8. The capacity of force framework to keep up synchronism under little unsettling influence such aggravation happens consistently on the framework in view of little variety in burden and era [6]. This unsettling influence is considered adequately for linearization of framework mathematical statement to be passable for reason for investigation.

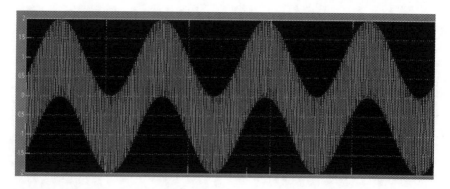

Fig. 5 Input voltage of adaptive filter

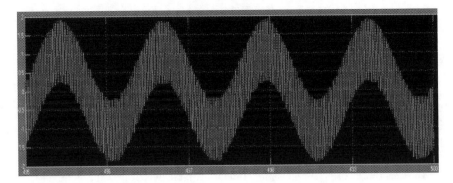

Fig. 6 Output voltage of adaptive filter

Fig. 7 Terminal voltage
wave form

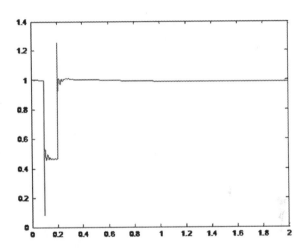

Fig. 8 Rotor angle deviation
wave form

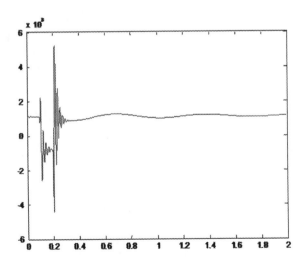

4.1 Adaptive Filter Simulation Result

4.2 Synchronous Generator Simulation Result

In the exciter mode, the terminal voltage Vg is altered in the converse course because of the receptive hang remuneration capacity of the current AVR. The voltage solidness is worry with the framework's capacity to control voltages for little incremental changes in framework load. This type of strength is normal for burden, constant control and discrete control at a given moment of time.

The reactive droop compensation function given by (1) and the compensated target voltage V_{gref} of the AVR control is, (Figs. 9 and 10)

$$\mathbf{V}'_{\mathbf{gref}} = \mathbf{V}_{\mathbf{gref}} - \mathbf{X}_{\mathbf{c}} \mathbf{I}_{\mathbf{q}}$$ (1)

Fig. 9 Generator side input voltage

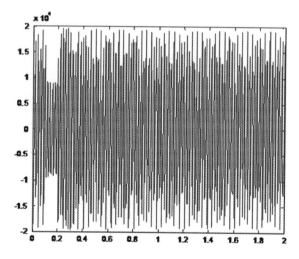

Fig. 10 Generator side input voltage

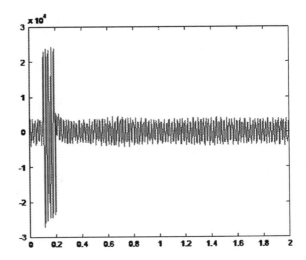

5 Conclusion

This paper presents Generator side field check and versatile voltage adjustment after-effects of a propelled System Generator Adaptive procedure. Voltage precariousness issues are drawing in expanding consideration in today's energy frameworks, with colossal accentuation on force framework operation and control. It is entrenched that upgrading the receptive force supplying capacity of a force framework is a successful intends to enhance voltage soundness. The initializing so as to learn method of AEC begins the weight quality to an irregular worth and after that computing the AEC yield from which the mistake sign is figured. The learning is done in a manner that the blunder ought to be lessened by altering the benefit of learning coefficient. This AEC framework has been demonstrated in Simulink where the prime mover and burden are displayed as necessary components. The modules are outlined in Xilinx System Generator are mimicked in MATLAB/Simulink R2013a.

References

1. Davies, J.B., Midford, L.E.: High side voltage control at Manitoba hydro. In: Proceedings of IEEE PES Summer Meeting, Seattle, WA, 16–20 July 2000
2. Noguchi, S., Shimomura, M., Paserba, J.: Improvement to an advanced high side voltage control. IEEE Trans. Power Syst. 21(2), 693–702 (2006)
3. Taylor, C.W.: Line drop compensation, high side voltage control, secondary voltage control why not control a generator like a static var compensator? In: Proceedings of IEEE PES Summer Meeting, Seattle, WA, USA, 16–20 July 2000
4. Kaminski, M., Orlowska-Kowalska, T.: FPGA implementation of ADALINE based speed controller in a two-mass system. IEEE Trans. Ind. Inf. 9(3), 1301–1311 (2013)

5. Orlowska-Kowalska, T., Szabat, K.: Optimization of fuzzy-logic speed controller for DC drive system with elastic joints. IEEE Trans. Ind. Appl. **40**(4), 1138–1144 (2004)
6. Orlowska-Kowalska, T., Kaminski, M., Szabat, K.: Implementation of a sliding mode controller with an integral function and fuzzy gain value for the electrical drive with an elastic joint. IEEE Trans. Ind. Electron. **57**(4), 1309–1317 (2010)
7. Fallahi, M., Azadi S.: Adaptive control of a DC motor using neural network sliding mode control. In: Proceedings of the International Multiconference on Engineers and Computer Scientists, vol. 2 (2009)
8. Memon, A.P., Uqaili, M.A., Memon, Z.: Design of AVR for enhancement of power system stability using matlab/simulink. Mehran Univ. Res. J Eng. Technol. **31**(03), 25–31 (2012)
9. Maurya, L., Srivastava, V. K., Mehra, R.: Simulink based design simulations of band pass FIR filter. IJRET: Int. J. Res. Eng. Technol. **3**(2), eISSN: 2319-1163, pISSN: 2321-7308 (2014)
10. Hsu, Y.-Y., Liou, K.L.: Design of self-tuning PID power system stabilizers for synchronous generators. IEEE Trans. Energy Convers. **2**, 343–348 (1987)

Analysis and Implementation of MPPT Algorithm for a PV System with High Efficiency Interleaved Isolated Converter

T. Anuradha, V. Senthil Kumar and P. Deiva Sundari

Abstract In this paper, a high efficiency interleaved flyback (IFB) converter is proposed for photovoltaic (PV) electric power generation. An incremental conductance (INC) based maximum power point tracking (MPPT) algorithm is implemented for the proposed PV system to maximize the output power extracted from the PV array. Among the isolated converter topologies, flyback converter is selected for its simple configuration and low cost. It is also the most popular isolated converter used in low power, electronic applications. Interleaving technique is applied for the flyback converters in order to reduce the input current and output voltage ripples, and to increase the efficiency. A 100 W prototype is developed and tested. The simulation and experimental results obtained for different irradiations validate the effectiveness of proposed converter for the PV applications.

Keywords PV system · Incremental conductance (INC) MPPT method · Flyback (FB) converter · Interleaving technique

1 Introduction

In recent times, there is an increase in the development of renewable and alternative energy sources to satisfy the exponentially growing energy demand. The Photovoltaic (PV) energy is a good choice among the renewable energy sources as

T. Anuradha (✉)
Department of Electrical and Electronics Engineering, Anand Institute
of Higher Technology, Anna University, Chennai, India
e-mail: anuradha.tn@gmail.com

V. Senthil Kumar
Department of Electrical and Electronics Engineering, College of Engineering,
Guindy, Anna University, Chennai, India

P. Deiva Sundari
Department of Electrical and Electronics Engineering, KCG College of Technology,
Anna University, Chennai, India

© Springer Science+Business Media Singapore 2017
P. Deiva Sundari et al. (eds.), *Proceedings of 2nd International Conference
on Intelligent Computing and Applications*, Advances in Intelligent Systems
and Computing 467, DOI 10.1007/978-981-10-1645-5_46

it is clean, environmental friendly and abundant. However, PV cells have very low efficiency and power output varies with the changing insolation level, temperature and load. One important method to increase the efficiency of the PV system is to employ MPPT algorithms which maximize the power extracted from PV modules by tracking MPP at all operating conditions [1–3].

Recently, several MPPT algorithms and control schemes are developed to track MPP. MPPT algorithms are ranging from simple, indirect techniques viz open circuit voltage and short circuit current methods to complex, direct techniques which include Hill climbing, Perturb and Observation (P&O), Incremental conductance (INC), fuzzy logic and neural network based methods [4–11].

In order to implement the MPPT algorithm, a DC-DC converter interface is required. Depending on the applications and voltage requirements, many popular converter topologies viz boost, buck, flyback, SEPIC, push pull, forward, etc. are used for the implementation of MPPT controller. Fly-back converter is the most popular, low cost, simple, isolated DC-DC converter and is widely used in electronic applications. An interleaving technique is employed to reduce the ripple content and peak currents [12–16].

This Paper proposes a PV system with a new high efficiency interleaved flyback (IFB) converter for implementing the INC method based MPPT algorithm. The overall system is described in Sect. 2. The simulation and experimental results are discussed in Sect. 3.

2 System Overview

The block diagram in Fig. 1 shows the major components of proposed PV system which include the PV panels, IFB converter module, MPPT controller with INC algorithm, PWM generator and Resistive load.

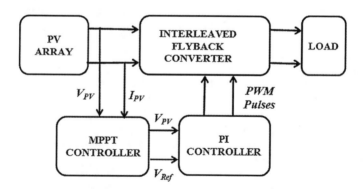

Fig. 1 Block diagram of proposed PV system with INC MPPT controller

2.1 PV Module

A PV cell is generally symbolized by a PN junction diode working on photovoltaic effect. PV Array of required power and voltage is formed by connecting several PV cells and PV modules in series and parallel. The single diode equivalent circuit is used for modeling the PV panel [17]. The mathematical equation for plotting the nonlinear I–V and P–V characteristics is given by (1) where I_{PV} is the PV panel output current, I_{Ph} is light generated current and I_D is the current generated in the PN junction diode [18].

$$I_{PV} = I_{Ph} - I_D \qquad (1)$$

2.2 Proposed IFB Converter

The arrangement of circuit components in the proposed IFB converter is shown in Fig. 2. It consists of MOSFET switches S1 and S2, flyback transformers Tr-1 & Tr-2 and diodes. High frequency transformers with large magnetizing inductance and negligible leakage inductance are considered for isolating the input and output side of converter. Interleaving is achieved by connecting two identical FB converters in parallel. The gating pulses for switches S1 and S2 are complimentary to

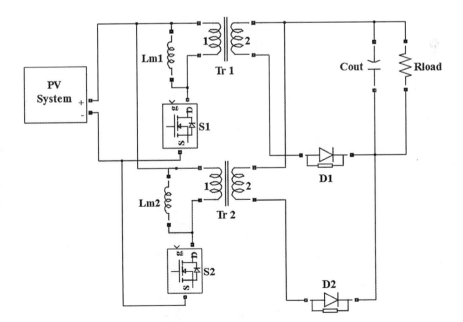

Fig. 2 Proposed interleaved flyback (IFB) converter topology

one another and any instant only one switch is on. The duty ratio (D) of the proposed converter is calculated using the Eq. (2) where V_{PV} and V_O are the input PV voltage and output converter voltage respectively and transformer turns ratio is taken as 1:1. The output capacitor C_{OUT} is selected such that the ripples in the output voltage are within ± 2 %.

$$D = \frac{V_O}{V_{PV} + V_O} \tag{2}$$

2.3   INC MPPT Algorithm

The Maximum Power Point (MPP) of the PV panel is dependent on irradiation and temperature. It changes dynamically when there is a change in the environment. In this paper, an INC MPPT algorithm is used to track the MPP of proposed PV system. INC MPPT algorithm is based on the comparison of instantaneous conductance (I/V) and change in conductance (dI/dV) at given switching period. The error (Ve) between (I/V) and (dI/dV) is used to calculate the V_{REF} and it is

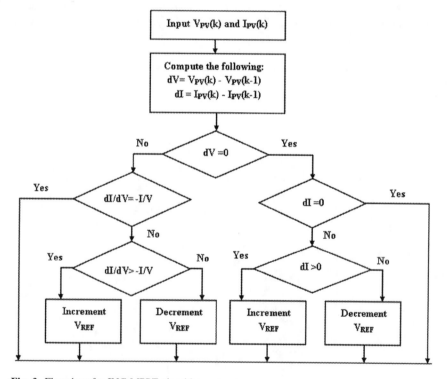

Fig. 3 Flowchart for INC MPPT algorithm

increased or decreased depending on positive or negative error (Ve). V_{REF} is updated until it reaches the voltage corresponding to MPP. The flow chart for the INC algorithm is shown in Fig. 3

3 Simulation and Experimental Results Discussions

To analyze and compare the performances characteristics, the PV system with conventional FB converter and PV system with proposed IFB converter were simulated using MATLAB/SIMULINK software. The simulation results obtained for the proposed IFB converter PV system were validated by building a 100 W prototype. The circuit parameters used for simulation and hardware are listed as below:

PV panel: Open circuit voltage $(V_{OC}) = 21.6$ V; Short circuit current $(I_{SC}) = 1.30$ Å; Maximum Power $(P_M) = 20$ W; $V_M = 17.1$ V; $I_M = 1.17$ Å; n = no of panels connected in series = 5;
Flyback converter: Power rating (P) = 100 W, Input voltage (Vi) = 75 V, and Switching frequency $f_{SW} = 100$ kHz; FB transformer turns ratio = 1:1; load resistance $(R_{load}) = 125$ Ω.

The control scheme and hardware setup of the proposed PV system with IFB converter are shown in Figs. 4 and 5. The control scheme was implemented using PIC microcontroller. The voltage (V_{REF}) corresponding to maximum power (P_M)

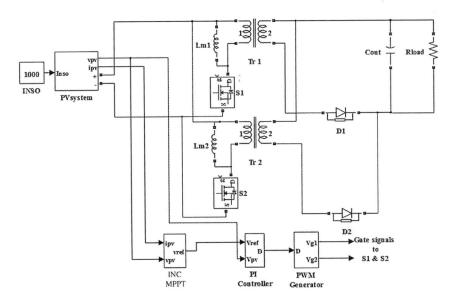

Fig. 4 Control schematic diagram of proposed PV system

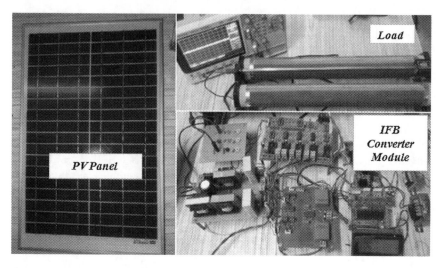

Fig. 5 Hardware setup of proposed PV system

was tracked by the INC MPPT algorithm implemented on the MPPT controller. The PV voltage (V_{PV}) is measured and compared with the reference voltage (V_{REF}) and the error signal obtained was given to the PI controller to generate gate signals for MOSFET switches S1 and S2.

Figures 6 and 7 show the simulated waveforms of input and output side of the PV system with conventional FB converter and PV system with proposed IFB converter for irradiation of 900 W/m^2. The ripples in the input current (i_{PV}), output current (i_O) and output voltage (V_O) and corresponding duty ratio (D) for irradiation 900 W/m^2 are shown in Figs. 8 and 9 respectively. It has been demonstrated that the percentage of ripples for the PV system with IFB converter are far lesser than that of PV system with conventional FB converter. Table 1 presents the comparative analysis of ripple

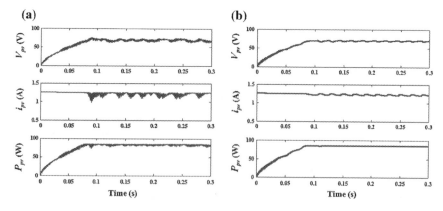

Fig. 6 Simulation waveforms of V_{PV}, i_{PV}, and P_{PV} for irradiation of 900 W/m^2. **a** Conventional single switch flyback converter. **b** Proposed IFB converter

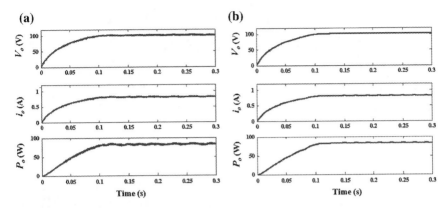

Fig. 7 Simulation waveforms of V_O, i_O, and P_O for irradiation of 900 W/m^2. **a** Conventional single switch flyback converter. **b** Proposed IFB converter

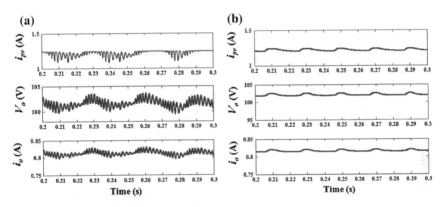

Fig. 8 Comparison of ripples in i_{PV}, V_O and i_O for irradiation of 900 W/m^2. **a** Conventional single switch flyback converter. **b** Proposed IFB converter

Table 1 Comparison of ripples in PV system with conventional single switch FB converter and proposed IFB converter

Irradiation (W/m^2)	i_{PV} (A)	i_O (A)	V_O (V)	Conventional flyback converter			Proposed interleaved flyback converter		
				i_{PV} Ripple (%)	i_O Ripple (%)	V_O Ripple (%)	i_{PV} Ripple (%)	i_O Ripple (%)	V_O Ripple (%)
900	1.235	0.820	102.5	11.9	5.9	4.5	1.83	0.61	0.68
750	1.037	0.682	58.30	13.59	6.2	5.2	1.81	0.65	0.68
500	0.675	0.456	57.01	15.11	6.9	5.75	1.85	0.68	0.682
250	0.343	0.228	28.50	15.8	7.1	6.14	1.88	0.73	0.685

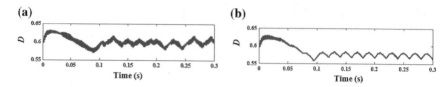

Fig. 9 Comparison of duty ratio for irradiation of 900 W/m². **a** Conventional single switch flyback converter. **b** Proposed IFB converter

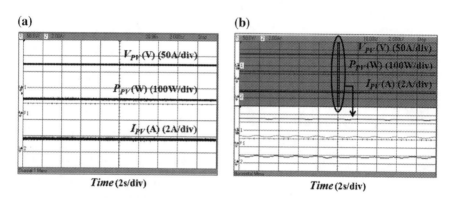

Fig. 10 a Experimental waveforms of V_{PV}, i_{PV}, and P_{PV} for irradiation of 900 W/m². **b** Closer look for ripple content

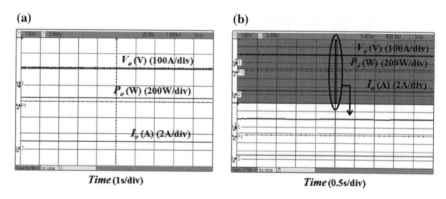

Fig. 11 a Experimental waveforms of converter output V_O, i_O, and P_O for irradiation of 900 W/m². **b** Closer look for ripple analysis

percentages in conventional and proposed system under various irradiation conditions. Figures 10 and 11 show the hardware waveforms for proposed IFB converter for irradiation 900 W/m². The verification of simulation and hardware results of MPPT efficiency of the proposed system at low, medium and high irradiations are presented in Table 2.

Table 2 Comparison of simulation and hardware results for proposed IFB converter PV system

Irradiation (W/m^2)	P_{MPP} (W)	INC method			
		Simulation results		Hardware results	
		P_{MPP} tracked (W)	η_{MPPT} (%)	P_{MPP} tracked (W)	η_{MPPT} (%)
900	86.65	85.20	98.32	84.15	97.11
750	60.19	58.78	97.66	58.22	96.72
500	27.1	25.91	95.61	25.82	95.27
250	6.88	5.52	80.23	5.48	79.65

4 Conclusion

In this paper, an interleaved flyback (IFB) converter is introduced for PV power generation system. The traditional Incremental Conductance (INC) MPPT algorithm is implemented for the proposed converter system and its performance characteristics and ripple content were analyzed for various irradiation levels. The simulation results have been compared with that of the PV system with conventional flyback (FB) converter and presented. A 100 W prototype of proposed PV system was developed and tested. The hardware results have been verified with the simulation results. The proposed topology has smaller current ripples (<2 %) and voltage ripples (<1 %). It has been also demonstrated that proposed IFB converter topology deliver higher efficiency of about 97 % for medium and high irradiation. Hence it is proved that the selected IFB converter is suitable for implementing INC MPPT algorithm and shows the excellent performance characteristics for medium and higher irradiations.

References

1. Koutroulis, E., Kalaitzakis, K., Voulgaris, N.C.: Development of a microcontroller-based, photovoltaic maximum power point tracking control system. IEEE Trans. Power Electron. **16**(1), 46–54 (2001)
2. Esram, T., Chapman, P.L.: Comparison of photovoltaic array maximum power point tracking techniques. IEEE Trans. Energy Convers. **22**(2), 439–449 (2007)
3. Salas, V., Olias, E., Barrado, A., Lazaro, A.: Review of the maximum power point tracking algorithms for stand-alone photovoltaic systems. Sol. Energy Mater. Sol. Cells **90**(2), 1555–1578 (2006)
4. Piegari, L., Rizzo, R.: Adaptive perturb and observe algorithm for photovoltaic maximum power point tracking. IET Renew. Power Gener. **4**(4), 317–328 (2010)
5. Kim, T.-Y., Ahn, H.-G., Park, S.K., Lee, Y.-K.: A novel maximum power point tracking control for photovoltaic power system under rapidly changing solar radiation. In: IEEE International Symposium on Industrial Electronics, pp. 1011–1014 (2001)
6. Kumar, M., Ansari, F., Jha, A.K.: Maximum power point tracking using perturbation and observation as well as incremental conductance algorithm. Int. J. Res. Eng. Appl. Sci. **1**(4), 19–31 (2011)

7. Veerachary, M., Senjyu, T., Uezato, K.: Neural-network-based maximum-power-point tracking of coupled-inductor interleaved-boost converter-supplied PV system using fuzzy controller. IEEE Trans. Ind. Electron. **50**(4), 749–758 (2003)
8. Khaehintung, N., Pramotung, K., Tuvirat, B., Sirisuk, P.: RISC microcontroller built-in fuzzy logic controller of maximum power point tracking for solar-powered light-flasher applications. In: Proceedings of the 30th Annual Conference on IEEE Industrial Electronics Society, pp. 2673–2678 (2004)
9. Wilamowski, B.M., et al.: Microprocessor implementation of fuzzy system and neural networks. In: International Joint Conference on Neural Networks, Washington, DC, vol. 1, pp. 234–239 (2001)
10. Hiyama, T., et al.: Evaluation of neural network based real time maximum power tracking controller for PV system. IEEE Trans. Energy Convers. **10**(3), 543–548 (1995)
11. Hiyama, T., et al.: Neural network based estimation of maximum power generation. IEEE Trans. Energy Convers. **12**, 241–247 (1997)
12. Abidi, H., Ben Abdelghani A.B., Montesinos-Miracle, D.: MPPT algorithm and photovoltaic array emulator using DC/DC converters. In: 16th IEEE Mediterranean Electrotechnical Conference (MELECON), pp. 567–572 (2012)
13. Veerachary, M., Senjyu, T., Uezato, K.: Maximum power point tracking control of IDB converter supplied PV system. Proc. Inst. Electr. Eng. Electr. Power Appl. **148**(6), 494–502 (2001)
14. Chang, Y.C., Liaw, C.M.: Design and control for a charge regulated flyback switch-mode rectifier. IEEE Trans. Power Electron. **24**(1), 59–74 (2009)
15. Kasa, N., Iida, T., Chen, L.: Flyback inverter controlled by sensorless current MPPT for photovoltaic power system. IEEE Trans. Ind. Electron. **52**(4), 1145–1152 (2005)
16. Hsieh, Y.C., Chen, M.R., Cheng, H.L.: An interleaved flyback converter with zero voltage transition. IEEE Trans. Power Electron. **26**(1), 79–84 (2011)
17. De Soto, W., Klein, S.A., Beckman, W.A.: Improvement and validation of a model for photovoltaic array performance. Sol. Energy **80**(1), 78–88 (2006)
18. Villalva, M.G., Gazoli, J.R., Filho, E.R.: Comprehensive approach to modeling and simulation of photovoltaic arrays. IEEE Trans. Power Electron. **24**(5), 1198–1208 (2009)

Analysis of Optimum THD in Asymmetrical H-Bridge Multilevel Inverter Using HPSO Algorithm

M. Ammal Dhanalakshmi, M. Parani Ganesh and Keerthana Paul

Abstract This paper analyses an optimum Total Harmonic Distortion (THD) value for multilevel inverters using Hybrid Particle Swarm Optimization Algorithm (HPSO). The proposed paper instigates an algorithm which supersedes the conventional algorithms in finding out an optimum global value. The prime objective of this paper is to reduce the THD value of the Asymmetric H-bridge multilevel inverters by choosing appropriate values from the inequality constraints. The proposed algorithm is simulated for an 11-level H bridge inverter using matlab 2014 version software. The results interpreted seemed to have a satisfactory higher order harmonic elimination compared to the conventional methods.

Keywords Total harmonic distortion (THD) · Asymmetrical H-bridge multilevel inverter · Hybrid particle swarm optimization algorithm (HPSO)

1 Introduction

Power electronics is considered as a boon for various power system applications such as standalone and grid connected systems. The conventional inverter output possess deviations in voltage and frequency from its fundamental value, which is in turn injected into the power system lines leading to power quality issues. To overcome this problem, many researches are focused on inverters. One eminent solution to the above problem is the improvisation of multilevel inverters in particular the H-bridge inverters [1]. The multilevel inverters eliminate maximum level of harmonics in the output waveform, thus providing an optimum result.

The H-bridge multilevel inverters structure operates by utilizing individual DC sources at each bridge [2, 3]. An asymmetric H-bridge multilevel inverter is

M. Ammal Dhanalakshmi (✉) · M. Parani Ganesh · K. Paul
Department of EEE, KCG College of Technology, Chennai, India
e-mail: ammal.eee@kcgcollege.com

M. Parani Ganesh
e-mail: parani.eee@kcgcollege.com

© Springer Science+Business Media Singapore 2017
P. Deiva Sundari et al. (eds.), *Proceedings of 2nd International Conference on Intelligent Computing and Applications*, Advances in Intelligent Systems and Computing 467, DOI 10.1007/978-981-10-1645-5_47

561

powered by different DC sources at each bridge level. Peculiar characteristic feature of this type is that, it can generate higher levels with the same number of DC sources compared to the symmetrical inverter.

Particle Swarm Optimization (PSO) algorithm is a pertinent approach to complex optimization problems due to the fact that the convergence time is faster compared to the conventional methods. The Hybrid PSO (HPSO) emerges to be a much more rapid convergent method in identifying the global best value [4–8]. Owing to the above factors, HPSO algorithm is effectively used for the H bridge inverter, wherein the particles swarm with a velocity so as to obtain efficient, harmonic free output.

2 Asymmetrical Cascaded Multilevel Inverter

Multilevel has laid its strong foundation in fields of industrial locomotives, custom power devices, distributed generations and grids [9]. There are three main divisions of multilevel inverters namely diode clamped, flying capacitors and cascaded or H-bridge inverters. Out of the three types, the H-bridge inverter excludes the use of large transformers, clamping diodes and the flying capacitors thus extending its applications towards medium and high power systems [10–14].

H-bridges can be made to operate in symmetrical and asymmetrical mode based on the input DC sources. Asymmetrical mode functions with dissimilar DC supply [15, 16]. Figure 1 shows the asymmetrical 11 level H-bridge inverter.

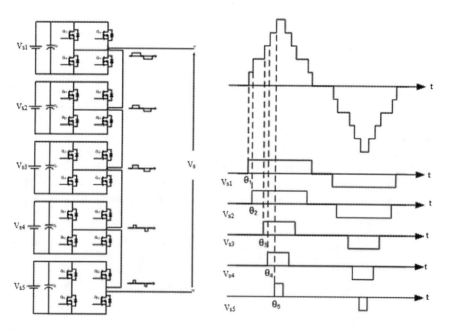

Fig. 1 Asymmetrical 11 level H-bridge inverter

Each bridge of multilevel inverter consists of 4 IGBT switches powered by a DC source. Five such bridges are connected in series to obtain an 11 level output waveform. The main advantage of asymmetric mode is that, many levels can be obtained by using the same number of cells as that of symmetric mode. By combining the output AC voltage waveforms of all the bridges we can obtain an output nearly similar to the fundamental value [17].

For an 11 level H-bridge inverter, the cascaded output voltage is given by,

$$V_{RN} = V_{S1} + V_{S2} + V_{S3} + V_{S4} + V_{S5}$$

where,

V_{RN} Cascaded output AC voltage
$V_{s1}, V_{s2}, V_{s3}, V_{s4}, V_{s5}$ unequal DC voltages across each bridge

2.1 Constraints for Choosing DC Voltages

- The DC voltages should be such that the voltage at the second bridge should be higher than the first voltage. Similarly choose for the succeeding bridges.
- The values chosen for the DC voltage must be a natural number.
- The bridge possessing maximum voltage must be the same or slightly less than the given load.

3 Total Harmonic Distortion (THD)

THD is the measurement of fundamental voltage with respect to its harmonic voltages.

It is given by the expression,

$$\%THD = \frac{\sqrt{V_0^2 - V_1^2}}{V_1} * 100$$

where,

V_0 Output AC voltage
V_1 Fundamental AC voltage

3.1 Consequences of THD

Presence of harmonic content in the power system will lead to enormous current in the loads. This owes to exponential increase in temperature of the conductors. It also affects the transformers in the nearby vicinity [18–21]. High frequency harmonics results in core loss in transformers and motors. Hence it becomes mandatory to keep the THD value as minimum as possible. As per the standards, safer limits for THD are 5 %.

4 Hybrid Particle Swarm Optimization (HPSO)

HPSO is a powerful tool to solve any complicated optimization problem. It has comparatively faster convergence compared to other conventional evolutionary algorithms. The special feature of HPSO algorithm is to accelerate the velocity in case the particle gets stagnated inside the local optima. This is achieved by creating a turbulence effect to the particle. As a result the particle gets shifted to a new optimum location. This makes the convergence faster and effective.

Steps to code HPSO algorithm:

- Assume $X = [x1, x2, ..., xs]$ is a solution within a boundary of 's'.
- Choose a fitness function $f(X)$ which has to be minimized.
- Set boundary for the inequality constraints. The values inside the boundaries are reformed according to the swarm search area.
- Let $X' = [x1', x2', ..., xs']$ be the solution in the reformed search area.
- The reformed solution x' is given by,

$$x' = k(m+n) - x$$

  where $x \, \varepsilon \, R$ within the boundary of $[m, n]$ and k can be randomly chosen between $[0, 1]$
- If $f(X')$ is better than $f(X)$, replace X with X'
  Else retain the same X
  Here the optimization problem travels towards the optimum direction
- The updated X' is

$$X'_{ij} = k\left(m_j(t) + n_j(t)\right) - X_{ij}$$

  where, $m_j(t) = min(X_{ij}(t))$ and $n_j(t) = max(X_{ij}(t))$; $i = 1,2, ...$ Population size and $j = 1, 2, ... s$
- If observed $pbest_i > T_1$, HPSO subjects to

$$c_1 r_1 \left(p_{id}^t - d_1 x_{id}^t\right)$$

  where, observed $pbest_i$ gives the data of repetition of stagnated pbest,
  T_1 predefined threshold
  d_1 random number $[0,1]$
- If observed $gbest_i > T_2$, HPSO subjects to

$$c_2 r_2 (p_{gd}^t - d_2 x_{id}^t)$$

where, observed gbest$_i$ gives the data of repetition of stagnated gbest,
T$_2$ predefined threshold
d$_2$ random number [0,1]
- Now the velocity of HPSO becomes

$$v_{id}^{t+1} = \omega v_{id}^t + c_1 r_1 (p_{id}^t - d_1 x_{id}^t) + c_2 r_2 (p_{gd}^t - d_2 x_{id}^t)$$

- Update particle position using

$$x_{id}^{t+1} = x_{id}^t + v_{id}^{t+1}$$

- Calculate fitness at this position
- Find new pbest and gbest
- If new gbest is better than the previous, update the gbest with the new value
- Extract the corresponding pbest
- Repeat the procedure until optimum fitness is obtained.

5   Problem Formulation

Population Size = 500
Maximum no. of iterations = 100
No. of Dimensions = 5

The main objective of this paper is to minimize the total harmonic distortion (THD) which is given by,

$$\%THD = \frac{\sqrt{\sum_{n=5,7,11,13} \left(\frac{1}{n} \sum_{k=1}^{5} \cos(n\alpha_k)\right)^2}}{\sum_{k=1}^{5} \cos(\alpha_k)} * 100$$

α Firing angles in deg
n   harmonic levels
k   No. of Bridges
Subject to the equality constraints:

$$\cos(\alpha_1) + \cos(\alpha_2) + \cos(\alpha_3) + \cos(\alpha_4) + \cos(\alpha_5) = 5M$$
$$\cos(5\alpha_1) + \cos(5\alpha_2) + \cos(5\alpha_3) + \cos(5\alpha_4) + \cos(5\alpha_5) = 0$$
$$\cos(7\alpha_1) + \cos(7\alpha_2) + \cos(7\alpha_3) + \cos(7\alpha_4) + \cos(7\alpha_5) = 0$$
$$\cos(11\alpha_1) + \cos(11\alpha_2) + \cos(11\alpha_3) + \cos(11\alpha_4) + \cos(11\alpha_5) = 0$$
$$\cos(13\alpha_1) + \cos(13\alpha_2) + \cos(13\alpha_3) + \cos(13\alpha_4) + \cos(13\alpha_5) = 0$$

The inequality constraints are

$$\alpha_i = [0\ \pi/2]$$

where i = 1, 2, ...5.

6 Simulation Results

The simulation is carried for a 11 level asymmetric H-bridge multilevel inverter. For a 11 level inverter we require,

$$No.\,of\,Bridges = \frac{(No.\,of\,levels - 1)}{2}$$

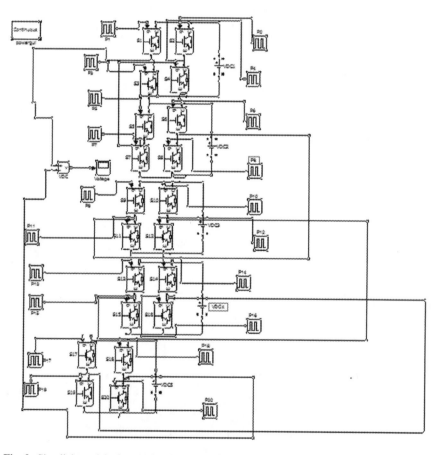

Fig. 2 Simulink model of an 11 level asymmetric H-bridge multilevel inverter

The asymmetric voltages given are,

V_s = 100 V, V_{s1} = 108 V, V_{s2} = 98 V, V_{s3} = 90 V, V_{s4} = 86 V, V_{s5} = 80 V

The firing angles for the IGBT switches are determined from the HPSO algorithm. Figure 2 shows the simulink model of an 11 level asymmetric H-bridge multilevel inverter. Figure 3 shows the output waveform of the inverter.

Figure 4 shows the FFT analysis of the 11 level asymmetric H-bridge multilevel inverter. The THD value obtained for these values of voltage is 5.21 %.

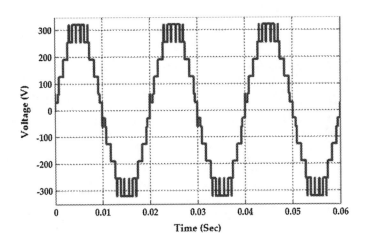

Fig. 3 Output waveform of an 11 level asymmetric H-bridge multilevel inverter

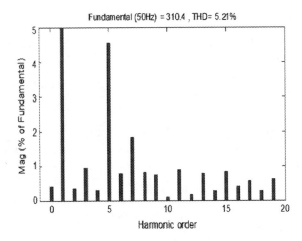

Fig. 4 FFT analysis of the 11 level asymmetric H-bridge multilevel inverter

7 Conclusion

An 11 level H-bridge multilevel inverter is simulated with five dissimilar DC source voltages. The suitable firing angles for triggering the IGBT switches are found out using Hybrid Particle Swarm Optimization Algorithm (HPSO) algorithm. The AC voltage waveform thus obtained is observed to have minimum THD value of 5.21 %. This result is comparatively better than the conventional algorithms.

References

1. Peng, F.Z., Lai, J.S.: Multilevel converters—a new breed of power converters. IEEE Trans. Ind. Applicant **32**, 509–517 (1996)
2. Rodriguez, J., Lai, J.S., Peng, F.Z.: Multilevel inverters: a survey of topologies, controls and applications. IEEE Trans. Ind. Electr. **49**(4), 724–738 (2002)
3. Rech, C., Hey, H.L., Gründling, H.A., Pinheiro, H., Pinheiro, J.R.: Analysis and comparison of hybrid multilevel voltage source inverters. In: Proceedings of PESC, pp. 491–496 (2002)
4. Yu, S., Wu, Z., Wang, H., Chen, Z., He, Z.: A hybrid particle swarm optimization algorithm based on space transformation search and a modified velocity model. Int. J. Numer. Anal. Model. **9**(2), 371–377 (2012)
5. Kennedy, J., Eberhart, R.C.: Particle swarm optimization. In: Proceedings of IEEE International Conference on Neural Networks, pp. 1942–1948 (1995)
6. Eberhart, R.C., Shi, Y.: Comparison between genetic algorithms and particle swarm optimization. In: Proceedings of the 7th Annual Conference on Evolutionary Programming, pp. 69–73 (1998)
7. Shi, Y., Eberhart, R.C.: A modified particle swarm optimization. In: Proceedings of IEEE Congress Evolutionary Computation, pp. 69–73 (1998)
8. Wang, H., Wu, Z.J., Liu, Y.: Space transformation search: a new evolutionary technique. In: Genetic and Evolutionary Computation, pp. 537–544 (2009)
9. Zhang, J.: Analysis of a cascade voltage source multilevel converter for a high power motor drive. Master Thesis, KTH-EME (2008)
10. Peng, F.Z., Lai, J.S., McKeever, J., Van Coevering J.: A multilevel voltage-source converter system with balanced DC voltages. Power-Electronics Specialists Conference, pp. 1144–1150 (1995)
11. Lai, J.S., Peng, F.Z.: Power converter options for power system compatible mass transit systems. PCIM/Power Quality and Mass Transit System Compatibility Conference, Dallas, Texas, pp. 285–294 (1994)
12. Kim, I.D., Nho, E.C., Kim, H.G., Ko, J.S.: A generalized undeland snubber for flying capacitor multilevel inverter and converter. IEEE Tran. Ind. Electr. **51**(6), 1290–1296 (2004)
13. Peng, F.Z., Lai, J.S.: Multilevel cascade voltage-source inverter with separate DC source. US Patent 5,642,275, June 24, 1997, (1991)
14. Choi, N.S., Cho, J.G., Cho, G.H.: A general circuit topology of multilevel inverter. In: Proceedings of IEEE PESC'91, pp. 96–103 (1991)
15. Patel, D., Chaudhari, H.N., Chandwani, H., Damle, A.: Analysis and simulation of asymmetrical type multilevel inverter using optimization angle control technique. Int. J. Adv. Electr. Electron. Eng. **1**(3), 78–82 (2012)
16. Kumar, P.V., Kishore, P.M., Nema, R.K.: Simulation of cascaded H-bridge multilevel inverters for PV applications. Int. J. Chem. Tech. Res. **5**(2), 918–924 (2013)
17. Prathiba, T., Renuga, P.: Performance analysis of symmetrical and asymmetrical cascaded H-bridge inverter. J. Electr. Eng. **13** (2013)

18. Lundquist, J.: On harmonic distortion in power systems. Chalmers University of Technology, Department of Electrical Power Engineering (2001)
19. Gosbell, V.: Harmonic distortion in the electrical supply system. PQC Tech Note No. 3 (Power Quality Centre), Elliot Sound Products. http://sound.westhost.com/lamps/technote3.pdf
20. Harmonics (electrical power). Wikipedia, The Free Encyclopedia. Wikimedia Foundation, Inc. 4 April 2011. Web. 5 April 2011
21. IEEE Std 519-1992.: IEEE Recommended Practices and Requirements for Harmonic Control in Electrical Power Systems. IEEE, New York, NY

A Survey on Challenges in Integrating Big Data

Akula V.S. Siva Rama Rao and R. Dhana Lakshmi

Abstract The Big Data is a Buzzword, which is being generated from various sources in and around in our daily life. Big data is the conjunction of big transactional data i.e. relational data base system, users activities huge data e.g. face book, twitter, LinkedIn, web logs, scanned, sensor devices, mails, and big data processing. The four striking characteristics of Big Data are volume, variety, velocity and veracity. Big data analytics refers to the process of gathering, arranging and analyzing huge data set to uncover the hidden knowledge that enables us to take effective and efficient decision making. The source data mostly may contain heterogeneity, noise, outliers, missing values and inconsistency. The poor source data can produce poor quality of analytical results. Traditional data processing system does not resolve these problems. The proposed data integration frame work with NoSQL technology could resolve integration, transformation, inconsistencies, noise challenges in big data.

Keywords Big data · Pre-processing · NoSQL

1 Introduction

Today we find digital data in and around in all fields of the world. The digital data is being generated from various sources such transactions, inventory, Web information and documents, Web logs, online marketing, user activity data e.g. face book twitter, youtube, scanned, satellite, aeroplane, health, insurance etc. Big data

A.V.S. Siva Rama Rao (✉)
CSE-E-JULY.14-PH24, Department of CSE, Hindustan University, Chennai, India
e-mail: shiva.akula@gmail.com

A.V.S. Siva Rama Rao
Department of CSE, SITE, Tadepalligudem 534101, Andhra Pradesh, India

R. Dhana Lakshmi
Department of CSE, KCG College of Technology, Chennai 600097, India
e-mail: dhanalakshmi.cse@kcgcollege.com

© Springer Science+Business Media Singapore 2017
P. Deiva Sundari et al. (eds.), *Proceedings of 2nd International Conference on Intelligent Computing and Applications*, Advances in Intelligent Systems and Computing 467, DOI 10.1007/978-981-10-1645-5_48

Integration differs from traditional data integration in many ways. The Act of increasing data sources, data types and managing and storing is the major hurdle of integrating data make into use full form. Many companies putting their efforts in storing and maintaining data than that what actually they do work with it. The challenges involved with data integration are increasing day by day. Overcoming Big Data Integration Challenges will examine the key big data integration issues facing enterprises today, real-world solutions and best practices.

1.1 4 Vs Characteristics of Big Data

- **Volume**: Unlike traditional data integration sources, big data sources are complex, enormous and huge in number has been growing in lakhs (Fig. 1).
- **Velocity**: Velocity refers to the speed at which new data can created and data transfer. The technology should support to analyse while it being generated from the multiple sources.
- **Variety**: Variety is the data, which is in the form of different types and from different data sources. Data type includes text, relation, images, videos, audio i.e. both structured, semi-structured and unstructured data/.
- **Veracity**: The quality of huge data being in correct state, which is collected from different sources and in different data types.

1.2 Classification of Big Data

The huge complex data can not be understood easily, for better understanding feature of big data, Big data is classified into various categories, which are data sources, content format, data stores, data staging, and data processing. Each of category is unique from the other in characteristics wise and complexity wise. Data

Fig. 1 4 V's

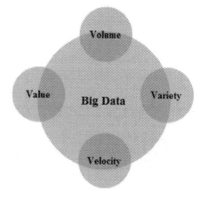

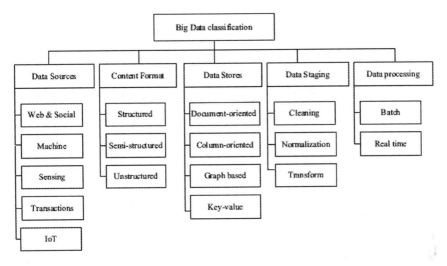

Fig. 2 Data classification

source include structured, unstructured and semi-structured which may contain inconsistencies, noise, outliers and redundant data (Fig. 2).

1.3 Big Data Analytics

Big data analytics is part of the database searching, pattern recognition, and analysis, which play an important roll to improve firm performance. Even though some leading companies are actively adopting big data analytics to strengthen market competition and to open up new business opportunities, many firms are still in the early stage of the adoption curve due to lack of understanding and experience with big data.

We define big data analytical technologies and techniques that an organization can be used to analyze huge and complex data for various decision makings in their organizations. With that usage innovative data storage, organization, analytical capability and presentation technologies are required in the big data analytics.

Any company expect to attain long term benefits, it is mandatory to maintain data consistency, veracity. Once data inconsistency or incompleteness becomes an issue due to some intentional or accidental faults such as flawed system design, data input errors, and data operator's subjective judgment trust in data quality can be significantly impaired. To make the matter worse, the extent of business damages due to missing and inconsistent data is difficult to assess as it becomes virtually impossible to restore the data quality. This undoubtedly discourages data-driven decision making.

2 Research Challenges

2.1 Scalability

The major challenge of cloud computing infrastructure is scalability, the scalability is the functionality of storing data while growing it.

2.2 Transformation

The data which is gathered from various sources is not suitable for data analysis because it have variety of data formats. Before placing data in relational database it should undergone for the preprocessing that should meet schema constraints.

2.3 Data Integrity

The Data which is stored in database is can be accessed by the authorized persons only, unauthorized persons have to prevent from accessing the data is data integrity. Big data will be stored in cloud data centers which will be accessed by various applications programs in an organization ensuring data integrity become major challenge. The cloud should provide a mechanism to ensure data integrity for data analytics by storing data in distributed manner such as HBase.

2.4 Data Quality

In the past, data processing was typically performed on clean datasets from well-known and limited sources. Therefore the results were accurate. However, with the emergence of big data, data originate from many different sources; not all of these sources are well-known or verifiable. Poor data quality has become a serious problem for many cloud service providers because data are often collected from different sources. For example, huge amounts of data are generated from smartphones, where inconsistent data formats can be produced as a result of heterogeneous sources. Therefore, obtaining high-quality data from vast collections of data sources is a challenge. High-quality data in the cloud is characterized by data consistency. If data from new sources are consistent with data from other sources, then the new data are of high quality.

2.5 Heterogeneity

Variety, one of the major aspects of big data characterization, is the result of the growth of virtually unlimited different sources of data. This growth leads to the heterogeneous nature of big data. Data from multiple sources are generally of different types and representation forms and significantly interconnected; they have incompatible formats and are inconsistently represented in a cloud environment, users can store data in structured, semi-structured, or unstructured format. Structured data formats are appropriate for today's database systems, whereas semi-structured data formats are appropriate only to some extent.

3 Related Work on Big Data Integration

3.1 Intelligent Similarity Joins for Big Data Integration

With the growing amount of data, the record joining has become a challenge for big data integration [1]. Similarity join is an efficient approach to address the record linkage, but it is difficult to achieve by the single server environment. Intelligent Similarity Joins for Big Data Integration propose a framework based on MapReduce for set similarity join. The techniques of framework improve the efficiency by reducing candidate pairs and load balance. In reducing candidate pairs, the propose algorithms that combines multiple filtering techniques to reduce the amount of candidate pairs. Which includes length filter, prefix filter and position filter. The techniques for load balance are used to address the skew data and decrease the replication transfer volume. Examined results on real time datasets are in encouraging the approaches and can achieve the fastness than the previous algorithms on big data.

3.2 Semantic ETL Framework for Big Data Integration

Semantic ETL frame for Big data Integration [2] uses of semantic technologies in the Transform phase of an ETL process to create a semantic data model and generate semantic linked data to be stored in a data mart or warehouse. The transform phase will still continue to perform other activities such as normalizing and cleansing of data. Extract and Load phases of the ETL process would remain the same (Fig. 3).

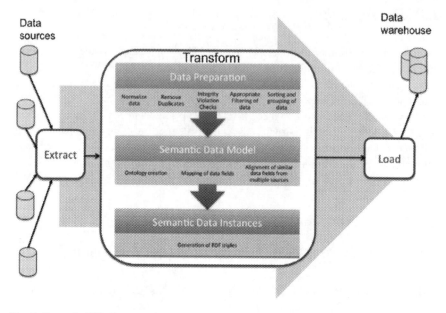

Fig. 3 Semantic ETL frame work

Limitation: Transform phase will involve a manual process of analyzing the datasets, the schema and their purpose. Based on the findings, the schema will have to be mapped to an existing domain-specific ontology or an ontology will have to be created from scratch. And it is a time taking process. The project was not implemented on real-time data.

3.3 NoSQL Database

The characteristics, data model, Classification of databases are defined according to the CAP theorem [3]. The major advantages of NoSQL database are reading and writing data can be performed quickly, Supporting huge storage, it can be easily expanded and cost effective.

Limitations: The major disadvantage of NoSQL is, that it does not support industry SQL standards and can not generate the reports.

3.4 Inconsistencies in Big Data

Inconsistency is same data will be stored differently in different places [4]. They are different types of inconsistencies in big data such temporal inconsistencies and spatial inconsistencies.

Limitations: Only a few inconsistencies were addressed briefly, but there is need to address other frequently encountered types of inconsistencies in big data and their impact on big data analysis.

3.5 A Federated Approach on Heterogeneous NoSQL Data Stores

The main goal of A Federated Approach on Heterogeneous NoSQL [5]. Data Stores was to examine the feasibility of having a federated NoSQL solution. This research has managed to achieve this task by implementing a federated system between MongoDB, CouchDB and Cassandra. The choice of the NoSQL systems contain two document stores and a column family.

Limitation: This research an assumption was made that the federation is used instead of accessing multiple data stores from the end user application itself. Therefore the consistency differences can be ignored.

3.6 Schema Conversion Model of SQL Database to NoSQL

Schema Conversion Model of SQL Database to NoSQL relational database [6] to NoSQL database schema conversion model, which uses nested idea to improve query speed of NoSQL database. The experiment achieves a MySQL-MongoDB migration system, and use real-world data to verity the correctness of migration and compare the query performance.

Limitations: The model is the expense of a certain amount of space to exchange query efficiency.

3.7 Challenges of Data Integration and Interoperability in Big Data

The companies or research organization, who generate huge data must be accustomed to new infrastructure [7].

The integration and interoperability play an important roll in installing big data architectures because of heterogeneity of data. In directly data integration and interoperability effect will be on the performance of the organization. To overcome integration and interoperability a comprehensive method required.

This paper address the challenges of data integration and data interoperability in big data which includes accommodate scope of data, data Inconsistency, query optimization, inadequate Resources, scalability, implementing support system and ETL Process in big data.

Limitations: They is no big data integration architecture that can handle the challenges.

3.8 Big-Data Integration Methodologies for Effective Management and Data Mining of Petroleum Digital Ecosystems

Big-data Integration Methodologies for Effective Management and Data Mining of Petroleum Digital Ecosystems uses the integrated method for building, semantic, syntactic and schematic relations a cross multidimensional and heterogeneous data sources [8]. A general form of Petroleum ontology is required to resolve the heterogeneity among petroleum data sources. Petroleum ontology, which allows multiple levels of information stored in the ontology for information query and retrieval.

Limitations: Fine-grained ontologies are effective for mining South East Petroleum Ecosystems.

3.9 Cloud Data Service Frame Work

Cloud Data Service Center which provides data integration, data sharing and data management, can share data between server and client in the distributed manner [9]. A set of cloud data service algorithms can resolve elastic, security and distribute issues. Many service provides need various API and increase the problem of managing services, this problem can be resolved by Infrastructure as a service by using the DMTF common information model meta-model. Infrastructure as a service proxy model provide by different Infrastructure as a service providers for various environments. Many services provider require different APIs offered by corresponding providers, and increases the problem of managing services.

3.10 Modeling and Querying Data in NoSQL Databases

Still today importance of flexibility and scalability of face book, twitter, youtube and mobile applications can't be ruled out [10]. The technologies like NoSQL can support for incremental data and non-relational database which provide scalable model. NoSQL technologies has been used by the prominent organizations like face book, e-commerce giant Amazon and Google in their applications.

3.11   Data Capture and Integration

Still today there is need of Big data ETL tool that can load data from existing data sources into a big data. Hadoop Distributed File system, Cassandra Mongo DB, GraphDB provide solutions for big data storage [11] (Fig. 4).

Even there is no such ETL tool that can transfer mass data from traditional data sources to big data. But there are some technologies like Key-value store i.e. Couch Base, Redis, Cassnadra, Riak, DocumentStore i.e. Hbase, MongoDB, graph data-bases etc.

Big Data integration support from ETL tools: Some of big data solutions are Hadoop frame work and Hive data base.

A seldom requirement in data integration is to upload data into Hadoop frame work from the online Relational data base Management system this can done by using export capability into flat files by using informatica. The technologies like PIG is used to load mass data into Hadoop frame work. Once it is loaded in HDFS then by using MapReduce aggregation operations performed then after loaded into NoSQL database.

Today the general technologies like Chukwa, Scribe, Flume are used to gathered data from various sources.

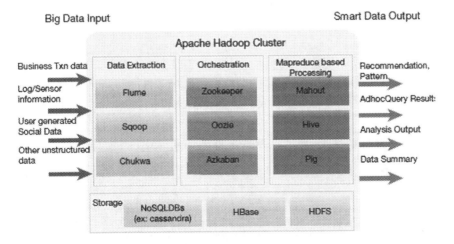

Fig. 4 Architecture of big data processing system

4 Proposed Work

Because of the 4Vs (i.e. volume, variety, velocity and veracity) characteristics of Big Data traditional data processing technologies can not perform tasks like Big data scalability, integration, transformation, noise, missing, redundant.

Traditional data processing methods can't be resolved continuously growing needs of Big data. Hadoop MapReduce parallel processing frame work can resolve big data problems. A NoSQL database technology can assure in resolving storage and access of huge data problems. The important features of NoSQL databases include simple API, easy replication support, schema-free and consistent.

4.1 Cassandra

Cassandra is one of the open source database owned by facebook. The key features of Cassandra includes high range of queries, easily managed schema, provide high performance with huge volume data.

4.2 MongoDB

MongoDB is a database between relational databases and non-relational database, its features includes non-relational database, Support complex data types, Powerful query language, High-speed access to mass data.

5 Conclusion

The proposed data integration frame work with NoSQL technology can resolve scalability, integration, transformation, noise, missing, redundant challenges in big data.

References

1. Wang, M., Nie, T., Shen, D., Kou, Y., Yu, G.: Intelligent similarity joins for big data integration. In: 10th Web Information System and Application Conference (2013)
2. Bansal, S.K.: Towards a semantic extract-transform-load (ETL) framework for big data integration. In: IEEE International Congress on Big Data (2014)
3. Han, J., Haihong, E., Le, G.: Survey on NoSQL Database, 978-1-4577-0208-2/11/$26.00 ©2011 IEEE

4. Zhang, D., Hsu, D.F., Wang, Y., Rao, A.R., Zhang, D., Kinsner, W., Pedrycz, W., Berwick, R. C., Zadeh, L.A. (eds.): Inconsistencies in Big Data. 978-1-4799-0783-0/13/$31.00 ©2013 IEEE
5. Dharmasiri, H.M.L., Goonetillake, M.D.J.S.: A federated approach on heterogeneous NoSQL data stores. In: International Conference on Advances in ICT for Emerging Regions (ICTer), pp. 234–239 (2013)
6. Zhao, G., Lin, Q., Li, L., Li, Z.: Schema conversion model of SQL database to NoSQL. In: Ninth International Conference on P2P, Parallel, Grid, Cloud and Internet Computing (2014)
7. Kadadi, A., Agrawal, R., Nyamful, C., Atiq, R.: Challenges of data integration and interoperability in big data. In: IEEE International Conference on Big Data (2014)
8. Nimmagadda, S.L., Dreher, H.V.: Big-data integration methodologies for effective management and data mining of petroleum digital ecosystems. 978-1-4799-0786-1/13/$31.00 ©2013 IEEE
9. Hong, X., Rong, C.M.: Cloud Data Integration Sharing and Service. 978-1-4799-3351-8/14/$31.00 ©2014 IEEE
10. Kaur, K., Rani, R.: Modeling and Querying Data in NoSQL Databases. 978-1-4799-1293-3/13/$31.00 ©2013 IEEE
11. Gopala Krishnan, S.: Integration of Big Data Technologies into Enterprise Landscape. Co-Chairman, Infosys limited, Bangalore, Big data Spectrum (2012)

RETRACTED CHAPTER:
Coronary Heart Disease Detection from Variation of Speech and Voice

Suman Mishra, S. Balakrishnan and M. Babitha

Abstract A large percentage of ailment related death today is attributed to coronary heart disease (CHD). Due to lack of physical activity, life style and increased stress level, heart disease are common now-a-days. Complications in the heart affect the larynx, breathing process and consequently the quality of speech. This work is primarily aimed to understand the variation of voice in the coronary heart disease patients for detection of CHD. Computerized Speech Lab (CSL) model 4500 is used for processing the voice signal. CSL contains Multi Dimensional Voice Program (MDVP) that breaks down and shows up to 32 voice parameters from a voice test. Voice samples of a group of 100 coronary heart disease patients (males and females) are compared with that of a group of 100 normal people (males and females). The study reveals variations in voice parameters like spectrogram, long term average spectrum (LTAS), jitter, shimmer, amplitude perturbation quotient (APQ) of the coronary heart disease patients in comparison with the normal people.

Keywords Coronary heart disease · Larynx · Computerized Speech Lab · Multi-Dimensional Voice Program · Spectrogram · Long term average spectrum · Jitter · Shimmer · Amplitude perturbation quotient

---

The original version of this chapter was revised: The plagiarized chapter has been retracted. The erratum to this chapter is available at https://doi.org/10.1007/978-981-10-1645-5_58

S. Mishra (✉) · S. Balakrishnan
Sri Venkateswara College of Engineering and Technology (Autonomous), Chittoor, AP, India
e-mail: emailssuman@gmail.com

S. Balakrishnan
e-mail: balkiparu@gmail.com

M. Babitha
Adhiyamaan College of Engineering, Hosur, Tamilnadu, India
e-mail: mageshbabitha@yahoo.co.in

© Springer Science+Business Media Singapore 2017
P. Deiva Sundari et al. (eds.), *Proceedings of 2nd International Conference on Intelligent Computing and Applications*, Advances in Intelligent Systems and Computing 467, DOI 10.1007/978-981-10-1645-5_49

1 Introduction

Heart is the indispensable organ in the human body and is a standout amongst the most essential muscle that pumps blood all through the body by method for composed compression [1]. The measure of the heart is about the gripped clench hand, arranged in the mid-section cavity encompassed by two lungs and partitioned into two sections. Each part contains one chamber called atrium and ventricle. The atrium gather blood and the ventricles contract to push blood out of the heart. The right portion of the heart pumps oxygen to blood and the new oxygenated blood streams from the lungs into the left atrium and the left ventricle. The left ventricle pumps blood to organs and tissues of the body. The oxygen content in the blood provides energy to the body and keeps it healthy. Heart disease is a pathological condition that affects the heart [2].

Coronary illness for the most part alludes to condition that includes limited or blocked veins. The problems that come under the heart diseases are atherosclerosis, heart attack, ischemic stroke, heart failure, heart valve problems, arrhythmia, coronary heart disease etc., [3].

Most normal issue is coronary heart disease (CHD) because of which a waxy substance rang plaque assembles inside the coronary arteries. These arteries supply oxygen-rich blood to the heart muscle. At the point when plaque develops in the arteries, the condition is called atherosclerosis. The development of plaque happens over numerous years. After some time, plaque can solidify and crack (tear open). Solidified plaque contracts the coronary arteries and lessens the stream of oxygen-rich blood to the heart. In the event that the plaque bursts, a blood coagulation can shape on its surface. A substantial blood cluster can form the most part or totally square blood course through a coronary conduit. After some time, cracked plaque additionally solidifies and narrows the coronary arteries [4].

Several factors that attribute CHD includes high blood pressure, high blood cholesterol, physical inactiveness, overweight, smoking, diabetes, family history of heart disease, etc. Chronic stress is also one among the factors that could maximize the possibility for causing coronary heart disease. Asthenic voice can be associated with various cardiac related disorders.

The vocal tract includes the larynx and the pharynx (the area above the larynx often thought of as the throat) and has a rich and complex circulatory system. Controlled movement of the vocal tract makes various activities such as singing or speaking possible. The automatic nervous system (ANS) that regulates the function of the internal organs such as blood pressure, heart rate etc. is also responsible for controlled movement of the vocal tract [5].

Heart is dynamically related to the variation in vocal cord parameters and also related to the acoustic parameters of speech. The outcomes of coronary heart complications incorporate respiratory changes, sporadic breathing, and expanded muscle pressure of vocal strings and vocal tract that might straightforwardly or in a roundabout way influence the nature of discourse [6]. By performing careful analysis of human speech, these variations in voice can be detected for diagnosis of coronary heart disease non-invasively.

2 Data and Method

This study comprises 100 patients diagnosed with coronary heart disease and 100 normal and healthy males and females of same age group. Details about each patient like age, history, clinical examination and investigations were recorded. Voice samples were recorded with a Shure SM48 mouthpiece at a separation of around 15 cm from the mouth. The voice parameters were extracted with the Multi-Dimensional Voice Program (MDVP) model 5105 from Kay Pentax Corporation, actualized in a Computerized Speech Lab (CSL model 4500, Kay Pentax Corp.).

Only sustained vowels were analyzed on the grounds that they permit simple and compelling partition of typical voice from neurotic voice. In this way, the speakers were solicited to talk a managed phonation from the vowel a/for inexact 4 s at an agreeable pitch and clamor. Directions were given to keep up a relentless phonation however much as could be expected. In this concentrate all voice tests were recorded at 44.1 kHz and specifically put away in the hostPC. The samples were recorded in a soundproof studio. The number of people involved in this experiment and their age are shown in Table 1. MDVP acoustic parameters give objective and noninvasive measures of vocal capacity parameters [7].

3 MDVP Parameters

MDVP extracts parameters related to frequency, amplitude, noise and turbulence. These parameters provide complete information about the speech signal [7].

3.1 Fundamental Frequency

Average fundamental frequency (Hz): It is the average value of all extracted period-to-period fundamental frequency values. Voice break territories are barred. It is represented by F0.

Table 1 Characteristics of people involved in this study

Characteristics	Normal people		CHD patients	
M/F	M (60)	F (40)	M (60)	F (40)
Age (mean ± SD)	53.4 ± 7.2	50.9 ± 5.7	47.3 ± 8.3	48.75 ± 5.9

3.2 Frequency Related Parameters

(1) *Absolute Jitter (Jita)/µs/*: Absolute jitter is an assessment of the period-to-period variability of the pitch period inside of the broke down voice sample. Voice break areas are excluded in all frequency related parameters. Jita is computed from the extracted period to period pitch data using Eq. (1).

$$\text{Jita} = \frac{1}{N-1} \sum_{i=1}^{N-1} \left| T^{(i)} - T^{(i+1)} \right|.$$

(1)

where $T^{(i)}$, i = 1, 2..., N—extracted pitch period data
N = PER—Number of extracted pitch periods.
It is very sensitive to the pitch variations occurring between sequential pitch periods.

(2) *Jitter Percent (Jitt)/%/*: Jitt is the relative period-to-period variability of the pitch period. Jitt is computed using Eq. (2).

$$\text{Jitt} = \frac{\frac{1}{N-1}\sum_{i=1}^{N-1} \left| T^{(i)} - T^{(i+1)} \right|}{\frac{1}{N}\sum_{i=1}^{N} \left| T^{(i)} \right|}.$$

(2)

where $T^{(i)}$, i = 1, 2..., N—extracted pitch period data
N = PER—Number of extracted pitch periods.
It is very sensitive to the pitch variations occurring between consecutive pitch periods.

(3) *Relative Average Perturbation (RAP)/%/*: It gives the relative evaluation of the period-to-period variability of the pitch with a smoothing factor of 3 periods.
(4) *Pitch Perturbation Quotient (PPQ)/%/*: This gives the variability of the pitch period at a smoothing factor of 5 periods.
(5) *Smoothed Pitch Perturbation Quotient (SPPQ)/%/*: SPPQ is an assessment of the long haul variability of the pitch period inside of the dissected voice test, with a smoothing element of 55 periods.

3.3 Amplitude Related Parameters

(1) *Shimmer in dB (ShdB)/dB/*: ShdB is an evaluation in dB of the period-to-period variability of the peak-to-peak amplitude within the analyzed voice sample. As in other parameters, voice break areas are excluded. ShdB is computed from the extracted peak-to-peak amplitude data using Eq. (3).

$$ShdB = \frac{1}{N-1}\sum_{i=1}^{N-1}\left|20\log(A^{(i+1)}/A^{(i)})\right| \tag{3}$$

where; $A^{(i)}$, i = 1, 2... ,N—extracted peak to peak amplitude data
N = Number of extracted impulses

(2) *Shimmer Percent (Shim)/%/:* *Shim is the* relative evaluation of the period-to-period variability of the peak-to-peak amplitude. Shim is computed using Eq. (4).

$$Shim = \frac{\frac{1}{N-1}\sum_{i=1}^{N-1}\left|A^{(i)} - A^{(i+1)}\right|}{\frac{1}{N}\sum_{i=1}^{N}\left|A^{(i)}\right|}. \tag{4}$$

where; $A^{(i)}$, i = 1, 2..., N—extracted peak to peak amplitude data
N = Number of extracted impulses

(3) *Amplitude Perturbation Quotient (APQ)/%/:* This gives a relative evaluation of the variability of the peak-to-peak amplitude at a smoothing of 11 periods. The smoothing reduces the sensitivity of APQ to pitch extraction errors.

(4) *Smoothed Amplitude Perturbation Quotient (sAPQ)/%/:* It is the evaluation of the long-term period-to-period variability of the peak-to-peak amplitude at a smoothing of 55 periods.

3.4 Noise Parameters

(1) *Noise-to-Harmonic Ratio (NHR):* Noise-to-harmonic ratio is the average ratio of the inharmonic spectral energy in the frequency range 1500–4500 Hz to the harmonic spectral energy in the frequency range 70–4500 Hz. This is a general evaluation of noise present in the analyzed signal. Increased values of NHR are interpreted as increased spectral noise which can be due to amplitude and frequency variations (i.e., shimmer and jitter), turbulent noise, subharmonic components and/or voice breaks.

(2) *Voice Turbulence Index (VTI):* *Voice turbulence index* is the ratio of the non-harmonic energy in the range 2800–5800 Hz to the harmonic spectral energy in the range 70–4500 Hz.

(3) *Soft Phonation Index (SPI):* It is the ratio of the harmonic energy in the range 70–1600 Hz to the harmonic energy in the range 1600–4500 Hz. This is very sensitive to the vowel formant structure. SPI is computed using a pitch-synchronous frequency-domain method.

4 CSL Analysis

For analysis first voice signal is recorded with the sampling rate 44.1 kHz. The recorded voice signals are shown below in Fig. 1a, b for typical persons from the group of normal people and group of CHD patients.

The extensive acoustic analysis including spectrograms, long term average spectrum, pitch contour and many more is carried using CSL. Analyses show significant variations in relevant parameters of normal people group and CHD patients group.

4.1 Spectrogram

Speech signal are composed of acoustic patterns which vary in frequency, time and intensity. In order to show these variation simultaneously a display known as spectrogram is used [8]. The spectrogram is one of the most effective ways to observe the complete spectrum of speech.

Heart disease results in variable heart rate. ECG is method using to measure the heart rate and to detect any irregularity in heart function. In recent years, many techniques and algorithms are developed for analyzing the ECG signal. The classifying methods which have been proposed during these years are Data Mining technology [9], Dempster–Shafer theory [10], cost sensitive algorithm [11] and wavelet transforms [12] etc.

Various investigates demonstrate that speech signals contain linguistic, expressive, organic and biological information [13]. The human voice signal is also

Fig. 1 Recorded audio signal in CSL at 44.1 kHz. **a** normal person male, **b** coronary heart patient male

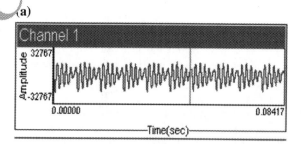

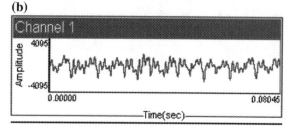

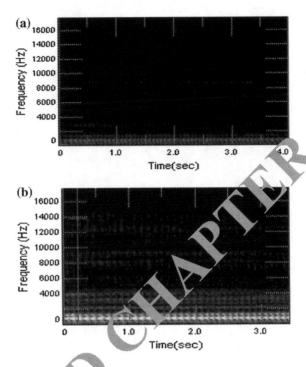

Fig. 2 Spectrogram with analysis size 512 points, Blackmann window weighting, and linear frequency scale of **a** normal male person, **b** CHD male patient

affected by the heart rhythms, so heart signal extraction using voice signal are possible. Many researchers used various techniques like pattern recognition technique [14], heart rate extraction using speech signal [15, 16]. Figure 2a, b show the spectrogram of one of the typical case in both CHD patients group and normal people group.

Spectrogram of a CHD patient voice displays greater noise and less energy in the harmonics of the signal compared to spectrogram of a normal voice sample. Wide band spectrogram is useful in identifying level of noise in the voice signal. Heavier resolution of individual harmonics is sacrificed in wide band analysis. Y. Anagihara in 1967 proposed that the regular harmonic components are mixed with the noise component chiefly in the formant region of vowel [17].

Long Term Average Spectrum

There are important properties of the glottal source spectrum that can only be observed in decibel spectrum [18].

According to the Nordemberg and Sundberg (2003) the LTAS particularly "reflects the contribution of the glottal source and the vocal tract for the vocal quality". The long term average spectrum for person from normal people group and person from CHD patients group is plotted in Fig. 3. The LTAS contains the

Fig. 3 LTAS for male
person from control group
(*red*) and person from CHD
group (*black*) with analysis
size 512 points and
Blackmann window
weighting

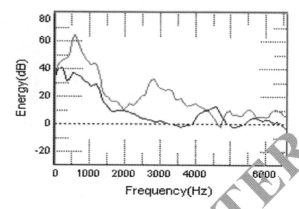

average of several spectra obtained. As the results shows, the control group's long
term average spectrums have higher sound level in 1–4 kHz.

Spectrogram as well as long term average spectrum shows a significant variation in normal male and CHD patient male. Almost similar variations have
observed in the female groups as well.

5   Results and Discussions

The primary goal of this paper is to extract, analyze, and classify the coronary heart
patient and normal persons on the basis of variations in their voice parameters. In
this work parameters have been extracted successfully by using MDVP—and
observing of spectrogram and long term average spectrum. Multi-Dimensional
Voice Program radial graphs for the normal person and the CHD patient are shown
in Fig. 4.

Parameters shown in the Fig. 4 are like this, APQ, amplitude perturbation
quotient; ATRI, amplitude tremor intensity index; DSH, degree of sub harmonics;
DUV, degree of voiceless; DVB, degree of voice breaks; FTRI, frequency tremor
intensity index; Jita, jitter absolute; Jitt, jitter percentage; NHR, noise to harmonic
ratio; PPQ, pitch perturbation quotient; RAP, relative amplitude perturbation;
sAPQ, smoothed amplitude perturbation quotient; ShdB, shimmer in dB; Shim,
shimmer percentage; SPI, soft phonation index; sPPQ, smoothed pitch perturbation
quotient; vAm, peak to peak variation in amplitude; vF0, fundamental frequency
variation; VTI, voice turbulence index [7]. The values for every parameter are
compared between two groups for both males and females.

Comparison of both the normal people group and CHD patients male group's
amplitude and frequency related parameters like Jita, Jitt, Shim, ShdB, RAP, PPQ,
APQ graphs are shown in the Fig. 5.

Fig. 4 MDVP radial graph
showing voice parameters
a for normal person and
b CHD patient with *green
colour* shows CSL normative
threshold and *olive-red
colours* person's personal
parameter

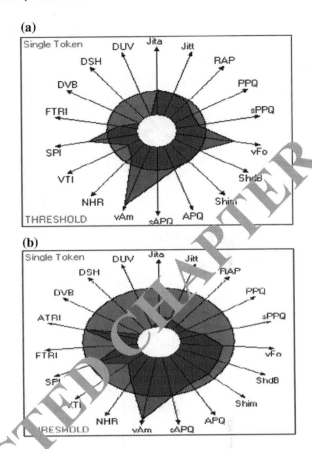

Table 2 shows the me and standard deviation for all the above parameters for
the normal people gr and CHD patients group.

According the graph shown in Fig. 5 and Table 2 the voice parameters
MDVP value ha ear and significant difference among normal people group and
CHD patients up for males. In statistics, the *p* value is a component of the
watched results that is used for testing a statistical hypothesis. The *p* value for
the parameters Jita, Jitt, RAP, PPQ, sPPQ, ShdB, Shim, APQ, sAPQ and NHR are
$<0.$ and for VTI and SPI are 0.2 and 0.06 respectively. Comparison between
emale CHD patients group and female normal people group on the basis of MDVP
parameters is summarized in Table 3.

Table 3 clearly shows that the normal females have significant difference in
parameters value compared to CHD. The p values in females for parameters Jita,
Jitt, RAP, PPQ, sPPQ, ShdB, APQ, and SPI are <0.001 and sPPQ, Shim, NHR and
VTI are 0.01–0.001. Hence, both males and females suffering from CHD have been
successfully differentiated from their normal counterparts.

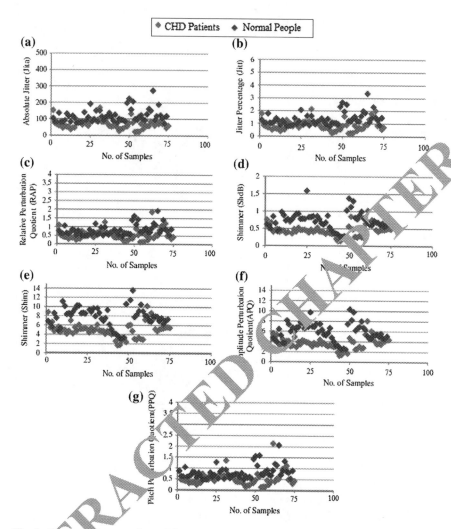

Fig. 5 Graph of the comparison of the parameters. **a** Jita, **b** Jitt, **c** RAP, **d** ShdB, **e** Shim, **f** APQ, and **g** PPQ, for normal persons and CHD patients (males)

Table 2 MDVP parameter values for males

Parameters	Normal people (mean ± SD)	CHD patients (mean ± SD)
Jita (μs)	116.56 ± 41.09	68.45 ± 27.88
Jitt (%)	1.35 ± 0.509	0.86 ± 0.41
RAP (%)	0.80 ± 0.30	0.54 ± 0.30
PPQ (%)	0.81 ± 0.30	0.53 ± 0.31
sPPQ (%)	1.19 ± 0.52	0.76 ± 0.28
ShdB (dB)	0.73 ± 0.25	0.45 ± 0.16
Shim (%)	8.056 ± 2.59	4.98 ± 1.59
APQ (%)	5.87 ± 1.76	3.70 ± 1.14
sAPQ (%)	8.69 ± 3.23	6.32 ± 2.30
NHR	0.19 ± 0.033	0.15 ± 0.02
VTI	0.046 ± 0.01	0.04 ± 0.01
SPI	19.02 ± 4.88	17.59 ± 5.13

Table 3 MDVP parameter values for females

Parameters	Normal people (mean ± SD)	CHD patients (mean ± SD)
Jita (μs)	97.77 ± 35.27	56.?? ± 16.37
Jitt (%)	1.80 ± 0.64	0.93 ± 0.25
RAP (%)	1.09 ± 0.37	0.55 ± 0.15
PPQ (%)	1.11 ± 0.42	0.54 ± 0.14
sPPQ (%)	2.34 ± 2.03	0.68 ± 0.11
ShdB (dB)	0.84 ± 0.27	0.59 ± 0.11
Shim (%)	9.02 ± 2.92	6.77 ± 1.21
APQ (%)	6.15 ± 1.7?	4.64 ± 0.76
sAPQ (%)	8.99 ± 2.07	5.52 ± 0.81
NHR	0.21 ± 0.07	0.17 ± 0.02
VTI	0.?? ± 0.014	0.05 ± 0.01
SPI	2?.25 ± 8.58	13.25 ± 4.38

6 Conclusion

The voice samples of coronary heart patients shows clear deviation from the normal people in both male and female cases of the same age group. The result shows that, coronary heart disease can be detected using MDVP and voice spectrum analysis non-invasively.

References

1. Cherry E.M., Flavio, H.: Fenton, Heart Structure, Function and Arrhythmias. Department of Biomedical Sciences, Collage of Veterinary Medicine, Cornell University, Itchaca, NY
2. https://www.cardiosmart.org/Heart-Basics/How-the-Heart-Works
3. Haskell, W.L., Lee, I.M., Pate R.R., Powell K.E, Blair, S.N.: Physical activity and public health updated recommendation for adults from the American College of Sports Medicine and Science in Sports and Exercise. ISSN: 0195-9131. **39**(8), 1423–1424 (2007)
4. Coronary Heart Disease. http://www.nhlbi.nih.gov/book/export/html/4847
5. Openstax College: Circulatory and Respiratory Systems—Concepts of Biology. Connexion Biology NM
6. Brown, R., DiMarco, A.F., Hoit, J.D., Garshick, E.: Respiratory Dysfunction and Management in Spinal Cord Injury. Respir Care. **51**, 853–870 (2006)
7. Software Instruction Manual. MDVP model 5105 KayPENTAX, Issue E, Jun 2008
8. Sonnleitner, R., Niedermayel, B., et. al.: A simple and effective spectral feature for speech detection in mixed audio signals. In: Proceedings of the 15th International Conference on Digital audio effects, York, UK, 17–21 September 2012
9. Shrinivas, K., Rao, G.R., Govardhan, A.: Analysis of coronary heart disease and prediction of heart attack in coal mining regions using data mining techniques. 5th International Conference on Computer Science and Education, Hefei, China, 24–27 August 2010
10. Khatibi, V., Montuzer, G.A.: Coronary heart disease risk assessment using Dempster–Shafer theory. In: Proceedings of the 14th IEEE International CSI Computer Conference
11. Alizadeshani, R., Hosseini, M.J.: Diagnosis of coronary artery disease using cost sensitive algorithm. In: IEEE 12th International Conference Data Mining Workshops (2012)
12. Li, C., Zheng, C., Tai, C.: Detection of ECG characteristic points using wavelet transforms. IEEE Trans. Biomed. Eng. **42**, 21–28 (1995)
13. Pickette, J.M.: Acoustics of Speech Communication, the Fundamentals Speech Perception Theory and Technology. Allyn and Bacon, Boston (1999). ISBN 0205198872
14. Ahamed, V.I.T., Dhanasekran, P., K. Tilek, N.G., Joseph, P.K.: A novel pattern recognition techniques for quantification of heart rate variability. In: International Conference on Biomedical and Pharmaceutical Engineering (2006)
15. Schuller, B, Friedmann, F., Eyaben, F.: Automatic recognition of physiological parameters in the human voice: heart rate and skin conductance. In: ICASSP 2013 978-1-4799-0356-6/13, IEEE (2013)
16. Masleh, A., Skopin, Baglicov, S., et al.: Heart rate extraction from vowel speech signal. J Comput. Sci. Technol. **27**(6), 1243–1251 (2012). doi:10.1007/s11390-012-300-6
17. Yumoto, E, Gould, W.J., Baer, T.: Harmonic-to-noise ratio as an index of the degree of hoarseness. Acoust. Soc. Am. **71**, 1544–1550 (1982)
18. Nordberg, M., Sundberg, J.: Effect on LTAS of Vocal Loudness Variation. TMH–QPSR, KTH. 4(1), 093–100 (2003)

Extensibility of File Set Over Encoded Cloud Data Through Empowered Fine Grained Multi Keyword Search

S. Balakrishnan, J. Janet and S. Spandana

Abstract By utilizing distributed computing, we can store the information on remote servers and permit information access to open inquiry clients through the cloud server's. The document set ought to ordinarily encoded before transferred to cloud server. These data is called outsourced data. It has certain limit to search on encrypted data. We are providing security for search data through cloud server. Encryption of data in the form of keywords. The search users can easily access the data from cloud servers by using empower of fine grained multi-keyword search. In this paper, we address the two issues with respect to multi-watchword seek. First issue as we consider the extensibility of document set and multiuser environment. By using these type of search users can easily access the data in the file format. These files consist of some indexes to search. The complete data can be stored in the form of file. And the total file will be stored in an order and encrypted by the information proprietor (data owner) and send to the cloud. Second issues as we think about creating as a scalable searchable encryption to get effective results on pragmatic databases.

Keywords Extensibility of file set · Fine grained · Multi keyword · Multiuser cloud environment · Scalable searchable encryption

S. Balakrishnan (✉) · J. Janet · S. Spandana
Sri Venkateswara College of Engineering and Technology (Autonomous),
Chittoor, Andra Pradesh, India
e-mail: balkiparu@gmail.com

J. Janet
e-mail: janetjude1@rediffmail.com

S. Spandana
e-mail: spandu181@gmail.com

P. Deiva Sundari et al. (eds.), *Proceedings of 2nd International Conference on Intelligent Computing and Applications*, Advances in Intelligent Systems and Computing 467, DOI 10.1007/978-981-10-1645-5_50

1   Introduction

Cloud computing [1] has a great demand resource, and internet based computing and which can be used to store the data and access. It is useful in business application. By using cloud we may get profits in business and it is very cheap at cost of service and get high performance. It is very helpful to the public users.

The outsourced data contain sensible information because of that we are providing security to the data and stored in the cloud server. In this we want to search for keywords [2] by using searchable encryption. The searchable encryption has been recently developed. In this we have a cloud server, a search user and data owner. The "data owner" generates keywords for (outsourced) a file set. In this file set, we have whole documents which can be kept in a file format which can be search by the search users. Data owner encrypt the total file set and send to the cloud server and search users can access the keywords by using trapdoor once we can search in one direction and get the results. In this we have multiple users have to search the data at a time. Previously we have ranked search, index search and scalable search are their. It can be done on predefined. Fine grained search is used to search in depth and detail and we get the fine details in results.

Extensibility [3] of file set is used to extend the file by storing the lot of data in an order. It has an index to match the keywords in a document. By using the hashing technique we may extend the index file. In this we have to increase the index size and the complexity. By using the searchable algorithm [4] we have to secure the index file. Dynamic approach is used to get secure index and efficient updates on practical databases.

Scalability is utilized to resize the size and size of a system. In this paper we need very scalable searchable encryption to get proficient results on pragmatic database. Searchable encryptions are essentially two sorts searchable symmetric encryption and searchable private key encryption. Searchable symmetric encryption contains three operations. Encryption [5] changes a watchword document pair using a mystery key as a part of to a figure content. Using the discharge key one can deliver a mission token for a specific watchword. Using this token one can look in a game plan of figure substance for those that match the watchword. The application for searchable encryption is distributed storage where the client outsources its stockpiling, yet encode it reports for mystery of catchphrases and holds the key.

2   Literature Survey

2.1   Searchable Symmetric Encryption

Song et al. [6] created an idea of Searchable Symmetric Encryption (SSE). Wang et al. [7, 8] are developed the "ranked keyword search scheme, which consider index file". Then again, the above plans can't productively bolster multikeyword seek which is generally used to give the better experience to the hunt client [9]. Later, Sun et al. [10] propose a "multikeyword seek plan which considers the list record of watchwords, and it can accomplish productive inquiry by using the multidimensional tree procedure".

To give a protected and proficient recovery of information [11], one can guarantee that the client can perform a pursuit over the scrambled information without uncovering the substance and the looked catchphrase to the server [12]. This can be known as searchable encryption (SE) [13]. This examination intends to concentrate on the searchable encryption plans in point of interest and actualize an answer that empowers extensibility of document set over encoded information and recovery framework in distributed computing. To empower the extensibility of inquiry, this plan will create an encoded catchphrase record which will be outsourced to the cloud server alongside the scrambled information set. The encoded watchword record rattles off the scrambled catchphrase and pointer to the relating report containing that catchphrase. To look a watchword, customer can just encode the catchphrase to create the pursuit token and send it to the remote cloud server. The server can recover pointer to the matching so as to compare report the pursuit token with the scrambled catchphrase file table. For our proposed arrangement, we have filed every one of the words from the record rather than a particular watchword set. This will permit the customer to hunt down any word in the record instead of a particular catchphrase. Additionally, indexing every one of the expressions of the record ensure that the client does not have to keep up a watchword list table at the customer side since it can seek any words. Nonetheless, indexing every one of the words will accompany the exchange off of a marginally bigger record size. One of the restrictions of the proposed arrangement is that it doesn't bolster the expansion of new records following the file redesign is static.

2.2 Searchable Public Key Encryption

Boneh et al. [14] proposed a Searchable Public key Encryption (SPE) which "bolsters single keyword search on encrypted data yet the calculation overhead is overwhelming". In the system of SPE, Boneh et al. [15] propose "conjunctive, subset, and range inquiries on encoded information". Hwang et al. [16] propose "a conjunctive catchphrase plan which underpins multi-watchword seek". Zhang et al. [17] propose a productive open key encryption with conjunctive subset catchphrases look. In any case, these conjunctive catchphrases plans can just give back the outcomes which coordinate every one of the watchwords all the while, and can't rank the returned results. Qin et al. propose a positioned question plan which utilizes a veil lattice to accomplish cost-viability. Yu et al. [18] propose a multi-catchphrase top-k recovery plan with completely encryption, which can return positioned comes about and accomplish high security. When all is said in done, in spite of the fact that SPE permits more expressive questions than SSE, it is less proficient, and in this way we embrace SPE in the work.

2.3 K-Nearest Neighbor

The most straightforward method for doing this is to utilize K-nearest Neighbor. K-nearest neighbor calculation (KNN) is a piece of administered learning and it has been utilized as a part of numerous applications in the field of information mining, measurable example acknowledgment and numerous others. KNN is a technique for grouping objects in view of nearest preparing. K is dependably a positive number. The neighbors are taken from an arrangement of articles for which the right order is known. It is common to utilize the Euclidean distance, however other distance measures, for example, the Manhattans distance could on a basic level be utilized.

3 Proposed System Model

The proposed system model diagram is given in Fig. 1. And mainly this model will be divided into three different phases namely: "data owner, cloud server and search user". It is elaborated this section.

The information proprietor (data owner) outsources the information to the cloud server which may make a bunches of catchphrases taking into account the outsourced information. By using keywords we are providing security to the data. Which can encrypted in a file set which may consist of indexes to choose the correct key word. It can be kept in a secure way by the data owner. This cannot be shown to others.

The cloud server is a moderate element which can stores scrambled archives and relating files records that are gotten from the information proprietor, and gives

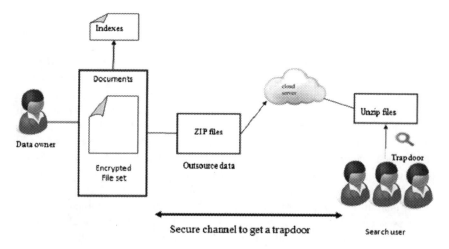

Fig. 1 System Model

information get to and seek administrations to pursuit clients, when a client's sends a watchword trapdoor to the cloud server, it would give back an accumulation of coordinating reports taking into account certain operations.

A pursuit client inquiries the outsourced archives from the cloud server with taking after three stages. By utilizing the trapdoor they can get results. In this hunt client send a trapdoor to cloud server and it sent the coordinating keywords to the server.

Step1: Initially the data owner will created some keywords in the form of a document.
Step2: The whole document can be encrypted as a file set in an order and the keywords has some index which can be orderly stored in a file.
Step3: Next the file set will be encrypted by data owner and sent to the cloud server.
Step4: The cloud server is an intermediate between data owner and search users. Search users will get output from cloud servers via trapdoor.
Step5: Total file set will be compressed into a zip file and then send to cloud server by using compression we reduce the size of a file. And we can easily transmit large amount of data through compression.
Step6: Then the search users types some input to cloud server to get some key from the data owner by using trapdoor to search.
Step7: The cloud will give out put in the form of unzip file that the data owner.
Step8: Search users known the symmetric key they will get the output through it. They has a secret key and a symmetric key which sent by the data owner. They can easily open the compression file.

4 Peril Model and Security Requirements

In Peril model, the cloud server is thought to be "straightforward yet inquisitive", it identified with secure cloud information look as well as takes after the standards. Cloud server examines information (file) in its stockpiling and message streams got amid the convention in order to learn extra data.

4.1 Known Cipher Text Model

The cloud server can simply know encoded report accumulation C and record gathering I, which are outsourced from data proprietor which are as archive (document) set.

4.2 Known Background Model

The cloud server can have more realizing than what can be gotten to in the relationship of trap entryways and the related accurate of other information [19]. In a solitary pursuit we can get vast measure of information.

5 Conclusion

By implementing empower fined grained multikeyword search we are using some extensibility of file set, scalability to increase the index file.

References

1. Cao, N., Wang, C., Li, M., Ren, K., Lou, W.: Privacy-preserving multi-keyword ranked search over encrypted cloud data. In: INFOCOM 2011, pp. 829–837 (2011)
2. https://support.google.com/websearch/answer/173733?hl=en
3. Li, H., Liu, D., Jia, K., Lin, X.: Achieving authorized and ranked multi keyword search over encrypted cloud data. In: Proceedings of IEEE (2015)
4. Yang, Y., Li, H., Liu, W., Yang, H., Wen, M.: Secure dynamic searchable symmetric encryption with constant document update cost. In: Proceedings of Globcom IEEE (2014)
5. Golle, P., Staddon, J., Waters, B.: Secure conjunctive keyword search over encrypted data. In: Golle, P. (ed.) Applied Cryptography and Network Security, pp. 31–45. Springer, Berlin (2004)
6. Song, D.X., Wagner, D., Perrig, A.: Practical techniques for searches on encrypted data. In: Proceedings of S&P, pp. 44–55. IEEE (2000)
7. Wang, C., Cao, N., Li, J., Ren, K., Lou, W.: Secure ranked keyword search over encrypted cloud data. In: Proceedings of ICDCS, pp. 253–262. IEEE (2010)
8. Wang, C., Cao, N., Ren, K., Lou, W.: Enabling secure and efficient ranked keyword search over outsourced cloud data. IEEE Trans. Parallel Distrib. Syst. **23**(8), 1467–1479 (2012)
9. Arvanitis, A., Koutrika, G.: Towards preference-aware relational databases. In: International Conference on Data Engineering (ICDE), pp. 426–437. IEEE (2012)
10. Sun, W., Wang, B., Cao, N., Li, M., Lou, W., Hou, Y.T., Li, H.: Verifiable privacy-preserving multi-keyword text search in the cloud supporting similarity-based ranking. IEEE Trans. Parallel Distrib. Syst. (2013). doi:10.1109/TPDS.2013.282
11. Liu, Q., Tan, C.C., Wu, J., Wang, G.: Efficient information retrieval for ranked queries in cost-effective cloud environments. In: Proceedings of INFOCOM, pp. 2581–2585. IEEE (2012)
12. Li, H., Liu, D., Dai, Y., Luan, T.H., Shen, X.: Enabling efficient multi-keyword ranked search over encrypted cloud data through blind storage. IEEE Trans. Emerg. Top. Comput. (2015). doi:10.1109/TETC.2014.2371239
13. Boldyreva, A., Chenette, N., Lee, Y., Oneill, A.: Order-preserving symmetric encryption. In: Advances in cryptology-EUROCRYPT, pp. 224–241. Springer, Berlin (2009)
14. Boneh, D., Di Crescenzo, G., Ostrovsky, R., Persiano, G.: Public key encryption with keyword search. In: Advances in Cryptology—EUROCRYPT, pp. 506–522. Springer (2004)
15. Boneh, D., Waters, B.: Conjunctive, subset, and range queries on encrypted data. In: Theory of Cryptography, pp. 535–554. Springer, Berlin (2007)
16. Hwang, Y., Lee, P.: Public key encryption with conjunctive keyword search and its extension to a multi-user system. In: Proceeding of Pairing, pp. 2–22. Springer, Berlin (2007)

17. Jung, T., Mao, X., Li, X., Tang, S.-J., Gong, W., Zhang, L.: Privacy preserving and aggregation without secure channel: multivariate polynomial evolution. IEEE (2013)
18. Yu, J., Lu, P., Zhu, Y., Xue, G., Li, M.: Towards secure multi keyword top-k retrieval over encrypted cloud data. IEEE Trans. Dependable Secure Comput. 10(4), 239–250 (2013)
19. Li, R., Xu, Z., Kang, W., Yow, K.C., Xu, C.-Z.: Efficient multi keyword ranked query over encrypted data in cloud computing. Future Gener. Comput. Syst. 30, 179–190 (2014)
20. Cao, N., Wang, C., Li, M., Ren, K., Lou, W.: Privacy-preserving multikeyword ranked search over encrypted cloud data. IEEE Trans. Parallel Distrib. Syst. 25(1), 222–233 (2014)

An Efficient On-demand Link Failure Local Recovery Multicast Routing Protocol

Deepika Vodnala, S. Phani Kumar and Srinivas Aluvala

Abstract Multicast routing is a crucial task in Mobile Ad-hoc Networks due to the wireless nature of the network. Route maintenance is challenging in MANETs due to frequent link failures which causes high packet loss and end-to-end delay. To overcome these problems many link recovery techniques have been proposed but every technique has its own limitations. The aim of this paper is to overcome the problem of link failures through local recovery technique in MANETs. In this study, we propose a novel route recovery algorithm which helps to find out an alternate path based on localization in case of a link failure in the network. Our proposed algorithm may reduce the overhead of control packet, delay and increases the throughput.

Keywords Dynamic topology · Link failure recovery · Localization · MANET · Multicast routing

1 Introduction

Group of wireless nodes get communicate through wireless links to form an infrastuctureless network which is referred as Mobile Ad-hoc Networks. MANET [1] is a self configuring network without any centralized administration. It has dynamic topology due to the random mobility of nodes. Because of which link

D. Vodnala (✉) · S. Phani Kumar
Department of CSE, GITAM University, Hyderabad, India
e-mail: deepuvodnala19@gmail.com

S. Phani Kumar
e-mail: phanikumar.s@gmail.com

S. Aluvala
Department of CSE, SR Engineering College, Warangal, India
e-mail: srinu.aluvala@gmail.com

© Springer Science+Business Media Singapore 2017
P. Deiva Sundari et al. (eds.), *Proceedings of 2nd International Conference on Intelligent Computing and Applications*, Advances in Intelligent Systems and Computing 467, DOI 10.1007/978-981-10-1645-5_51

failures takes place, results in packet loss, end-to-end delay and degradation of performance of network.

In MANET, routing is a crucial part as the topology is dynamic. With the different sorts of communication in MANETs, to a great extent multicasting [2] is advantageous and challenging operation. In time delivery of data without exploiting the resources available is most essential in important application scenarios sending to a group of destinations.

Efficiency, simplicity, controlling overhead, management of available resources, quality service, and robustness are the basic features of good multicasting routing protocols. Multicast protocols are categorized into tree based, mesh based and hybrid routing protocols. In tree based, protocol provides a single path from one node to another in the multicast group, bandwidth efficient. Mesh protocol multicast packets [3] are distributed in the mesh structures. Mesh multicast protocols are highly robust compared to tree based protocols in high robustness environment and creates redundant paths from source and destinations while sending data packets. Hybrid multicasting protocols combine with the best features of both tree and meshed based approaches.

Even though, with the existence of many routing protocols a crucial problem is finding an efficient and optimal route between source and destination [4]. As of now many link failure detection and recovery protocols have been proposed and each protocol has its own limitations. In this proposal, we come up with a solution in multicast routing for quick recovery from link failures, which is the primary and main aim of this research work. The remaining sections contain the cause of link failures, detection techniques, recovery mechanisms, proposed protocol, its simulation results and conclusion.

2 Cause of Link Failures

In MANETs particularly, a reliable connection could be achieved by protecting the links between each pair of nodes in the network. To achieve a reliable connection by protecting the links between each pair of nodes in the path link expiry is crucial. We study, in particular, the latter approach using multi-paths. In addition to link expiry, another major reason for a connection to break down is when an intermediate node or destination node becomes unreachable. The node can become unreachable due to several reasons such as running out of energy, node failure, or when a node becomes unresponsive. Due to overhearing at the mobile nodes may cause energy loss, because of shared nature of medium, flooding of packets [5]. In comparison to wired network in MANETs energy consumption is increased because of congestion and packet losses, which can also be a cause for link failures. To the best of our knowledge, there has been no efficient link failure recovery routing algorithms for MANETs.

3 Link Failure Detection Techniques

In MANETs, link failure detection plays a vital role. As of now various detection mechanisms [6] have been adopted such as timely detecting a failed link using cross layer model. Most of the routing protocols adopt neighbor discovery mechanisms to detect the link failures.

3.1 Neighbor Discovery Mechanism to Detect Link Failures

The process of neighbor discovery mechanism is based on the concept of HELLO packets. Every node of the network sends HELLO packet to the neighboring nodes [6] in its range of communication. After receiving HELLO packet, the neighboring nodes of received HELLO packets should send back a reply HELLO packet. Then the link is assumed to be in good condition for routing of packets. In case if the reply packet is not given within a specific interval of time it is assumed to be a link failure. Because of large amount of delay in link failure detection, this detection mechanism cannot be implemented in the real time applications.

3.2 Cross Layer Approach—Link Failure Detection

In this approach, if a MAC frame is sent then the acknowledgement is sent back for that frame. In case acknowledgement is not received then the frame is retransmitted. After a number of times of MAC frame retransmissions [6] are done without getting back a reply it is assumed that the link is failed and frames sent are lost. Link failure detection is carried out by considering the number of MAC frame delivery failures, in MAC layer.

In comparison with neighbor discovery technique, cross layer approach is advantageous for quick finding of link failure. Transmission errors to be considered to ensure the link failure detection precisely, which are categorized primarily into two types, they are transient error and permanent error [6]. By the retransmission of frames in MAC layer, transient errors can be removed. While different approach need to be adopted to handle the permanent errors. In case of permanent link failure, routing layer chooses an alternate path. The correct error type should be detected otherwise false alert occurs which causes misconception of assuming transient errors [6] as permanent errors. This leads to increased overhead of packets re-routing takes place. To improve the performance of MANETs, false alerts need to be avoided.

4 Link Failure Recovery Techniques

MANETs has dynamic topology due to the random mobility of nodes, link failures takes place, which results in packet loss, end-to-end delay and degradation of performance of network. There are various techniques used for link failure recovery [3]. Some of them are briefly summarized below.

4.1 On-demand Multicast Routing Protocol (ODMRP)

ODMRP is mesh-based multicast routing protocol rather than a conventional tree based multicast routing scheme and uses a forwarding group concept [7] where only a subset of nodes forward the packets by scoped flooding in multicast. To build and maintain routes dynamically and to maintain multicast group membership, it uses the on-demand routing methods. It improves scalability and reduces the channel overheads.

4.2 Enhanced On-demand Multicast Routing Protocol (EODMRP)

It's an enhancement of ODMRP [7]. This protocol is mainly used to reduce the control overhead and network congestion.

4.3 Multicast Zone Routing Protocol (MZRP)

MZRP proposed by Zhang and Jacob [8]. In MZRP multicasting is done using the shared tree concept. This protocol adopts the zone routing concept [8]. The zones will be created depending on the zone radius that is in terms of number of hops by the neighboring nodes. Coordinator node of the group sends the messages over the network to inform all the nodes of the multicast group and also other group coordinator nodes. In MZRP the group information can be tracked from the local routing of each node [8]. As MZRP uses shared tree multicasting there exist redundant paths between tree members, in case of link failure while it is repaired by the protocol temporarily it uses an IP tunnel for data packets delivery.

4.4 Ad-hoc On-demand Distance Vector Protocol (AODV)

For connecting multicast group members it constructs multicast trees as needed. There is no single point of failure as the multicast trees are distributed. Loop free routes [9] are available in AODV for both unicast and multicast, at the time of broken links repairing. To reduce the broken links due to high mobility of nodes and variations among the intermediate nodes including forwarding zone criteria.

4.5 Forwarding Group Multicast Protocol (FGMP)

Chiang et al. [10] proposed FGMP, which provides an efficient and simple way of multicasting in MANETs. The efficiency of multicasting is achieved by network forwarding the packets from source to destination without resorting the global flooding. It adopts the advantage of wireless broadcast transmissions to reduce overhead on the channel and storages along with improving the performance and scalability of the network.

4.6 Optimized Link State Routing Protocol (OLSR)

In OLSR Multipoint Relay nodes generate link state information, which is used for route calculation. In OSLR no central administration is needed to handle the process of routing and provides optimal routes [11]. One of the major advantages of the OLSR protocol is link status is known immediately and possibility to extend the quality of service information.

4.7 Efficient Geographic Multicast Protocol (EGMP)

Efficient Geographic Multicast Protocol proposed by Xiang et al. [12]. It provides scalable and efficient group membership management by adopting hierarchical structure. Construction of Zone based bidirectional tree achieves more efficiency in multicast delivery which effectively reduces the tree structure maintenance [12] and route searching overhead.

4.8 Core-Assisted Mesh Protocol (CAMP)

The connectivity of multicast group's is maintained by CAMP even the routers move frequently. In the multicast group of CAMP [7] within a finite time every receiver has a reversed shortest path to each source of the multicast group. From sources to receivers all multicast packets of a specific group are forwarded along the shortest paths defined within the group's mesh.

4.9 Dynamic Source Routing- Localization of Link Failures (DSR-LLF)

When a link failure occurs in MANET DSR-LLF algorithm [1, 13] works. Based on the location of link failure in the source route it takes decision. Source route will be clustered into three regions, and depending on the location of relay node a right mechanism for route maintenance is adopted by DSR-LLF [1, 13]. Intermediate nodes load will be varied and depends on the cluster in which the node is. DSR-LLF improves the solutions in route discovery, maintenance and the overall performance of DSR [1, 14].

Limitations of existing protocols:

- As multiple packets are forwarded on the same path causes flooding and maximum consumption of energy and bandwidth occurs.
- Packet delivery ratio is less
- As network size increases the performance decreases.
- Packet delay and overhead vice versa.

Protocol Analysis:
Table 1 shows the comparison of existing routing protocols.

Table 1 Comparison of existing routing protocols

Protocol	Multicast topology	Loop free	Periodic message	Delay	QoS support	Overhead	Link recovery support
ODMRP	Y	Y	Y	N	Y	Y	Y
EODMRP	Y	Y	Y	N	Y	Y	Y
MZRP	Y	Y	Y	Y	Y	Y	Y
AODV	Y	Y	Y	Y	N	Y	Y
FGMP	Y	Y	N	Y	Y	Y	Y
OLSR	Y	Y	Y	N	Y	Y	Y
EGMP	Y	Y	N	Y	Y	Y	Y
CAMP	Y	Y	Y	N	Y	Y	Y
DSR-LLF	Y	Y	Y	N	Y	Y	Y

5 Proposed Protocol

5.1 An Efficient On-demand Link Failure Local Recovery Multicast Routing Protocol [EOD-LLR]

Here we propose a mechanism of route recovery in multicasting that does the quick recovery from link failures through localization. In the process of source node intends to send data packets to the destination node, it initially refers into its routing cache for a path to the destination. If the source node founds the path it forwards the packets to the destination accordingly to the route found in the route cache. Else, a Route Request Packet (RREQ) is sent to all the neighbor nodes by the source node, which are in the range of its transmission. Every RREQ carry sender address, receiver address, request ID, and route record. In this process in receiving the route request packet by a node, it processes the request. If the requested route is already found with that node, in its recently seen requests then request is not processed further. In case, if in current node route cache, requested nodes address is already present then the route request packet is discarded.

Else, if the address in the route request packet matches with the nodes own address, the packets route record knows the route by which the request packet reached the destination from source node. And a route reply packet (RREP) is sent to the source with the route record it reached the destination. Else, the address of the current node is added in the route request packet and again it broadcasts the request. So the propagation of the route request is done over the network till it finds the intended destination node, then which results in replying route reply packet to the source with the route record.

5.2 Working of EOD-LLR

In the process of sending a packet from source to destination, route discovery is done with the part of implementing DSR (Dynamic Source Routing). In our proposed protocol, every node maintains neighboring nodes information which existed in one-hop distance. In case of a link failure, the upstream node sends route request packet to all the one-hop neighboring nodes to create an on-demand alternate path towards the destination.

In the below Fig. 1a, the route to destination from source is S-a-b-c-d-D, generated as per route discovery mechanism. Assuming link failed between nodes c and d. In Fig. 1b at the point of link failure, node c (upstream node) sends RREQ packet to its entire one-hop neighboring nodes towards the destination. In Fig. 1c the nodes which receive RREQ packet forwards it to its neighboring nodes which leads to destination. In Fig. 1d RREP packet is sent back to the upstream node from destination through the same path packet request is received. In some scenarios, there exist multiple alternate paths. As shown in Fig. 1d path1:c-h-i-j-D; path2: c-e-f-g-D. In Fig. 1e from the available multiple alternate paths one optimal alternate path is selected from c to D, path: c-e-f-g-D (assumed) based on route discovery mechanism.

Algorithm

- Once the route is discovered, every upstream node in the path periodically sends a control packet to check the link liveliness.
- If the upstream node doesn't receive a control reply packet within a stipulated time, it is assumed that the link is failed.
- In case it is found link is failed, upstream node sends RREQ packet to all the one-hop neighboring nodes towards the destination node.
- The RREQ is propagated over the one-hop neighboring nodes until the destination is found.
- As the destination is found, a route reply packet is sent back to the upstream node which has sent RREQ packet.
- Upon receiving the multiple alternate paths, an optimal alternate path is adopted based on route discovery mechanism.
- So that, an alternate path is generated from the point of failure link to the destination.

The below Fig. 2 illustrates flowchart of EOD-LLR Protocol.

Fig. 1 Representation of EOD-LLR technique. **a** An example of link failure, **b** upstream node sending RREQ packet to one—hop distance neighboring nodes, **c** forwarding RREQ packet to other neighboring nodes towards destination, **d** receiving of RREP from destination, **e** generation of an alternate link

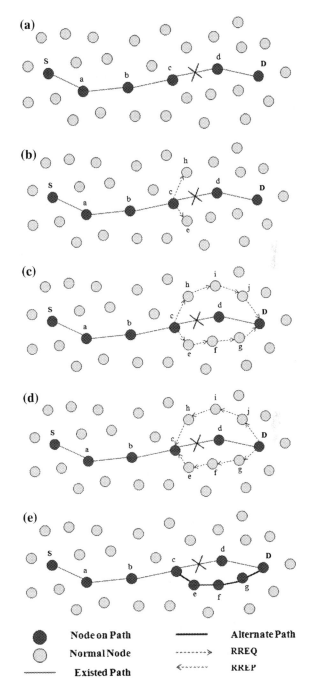

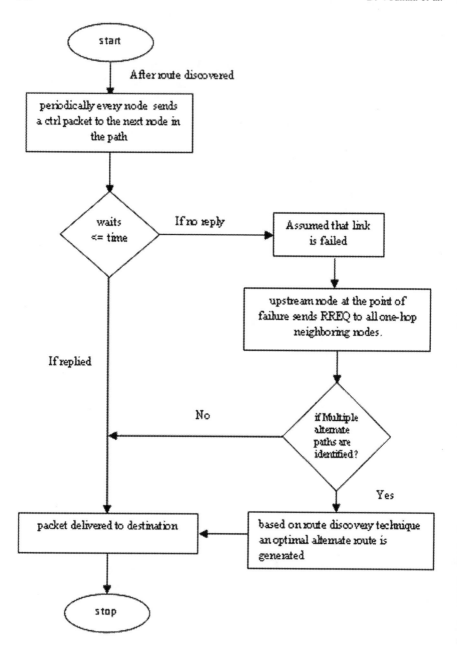

Fig. 2 Flowchart of EOD-LLR

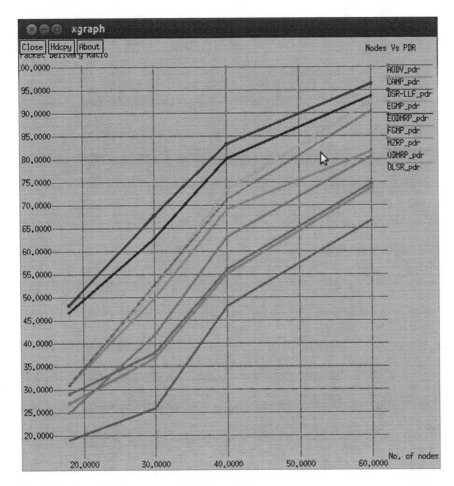

Fig. 3 Number of nodes versus packet delivery ratio

6 Simulation Results

Here we compared the performance of existing protocols which we discussed in Sect. 4, using Network Simulator 2 tool (NS2) by considering some parameters. Experimental scenario considered with a network of mobile nodes placed randomly within 800 m × 800 m. (Figs. 3, 4, 5)

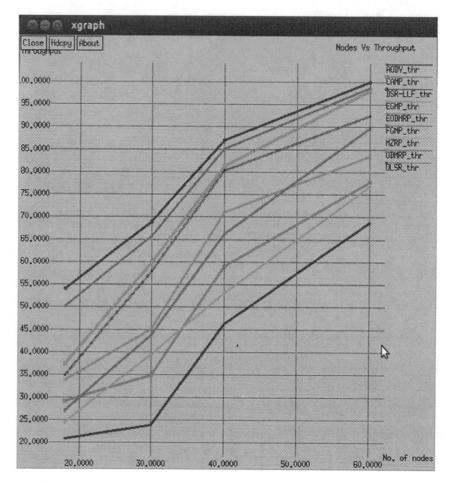

Fig. 4 Number of nodes versus throughput

7 Conclusion

EOD-LLR is a route maintenance algorithm for the quick recovery of link failures in MANETs by constructing alternate paths. It works when a link failure occurs during the transmission of the packet form source to the destination in the route discovered prior the packet multicasting. Upon finding the link failure by the mechanism adopted, by the protocol it constructs an alternate path from the point of link failure to the destination based on localization. Further this protocol can be modified in future to make it adaptable in localization of link failures where the source route can be clustered into parts to implement it more efficiently by reducing the length of constructing alternate paths. The proposed protocol EOD-LLR is helpful to overcome the effect of link failures and reduces overhead and improves the network overall performance.

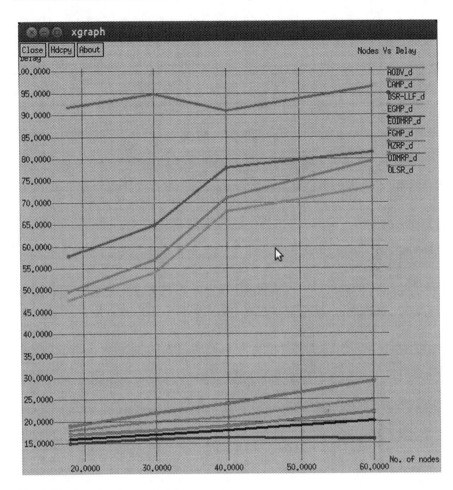

Fig. 5 Number of nodes versus delay

References

1. Yadav, P., Bhattacharjee, J., Rahguwanshi, K.S.: A novel routing algorithm based on link failure localization for MANET. Int. J. Eng. Res. Appl. **3**(4), 1133–1139 (2013). ISSN: 2248-9622
2. Huang, C.-C., Lo, S.-C.: A comprehensive survey of multicast routing protocols for mobile adhoc networks. J. Internet. Technol. (2008)
3. Selvi, M., Balakrishna, R.: Design and developing a multicast routing protocol for link failure and reliable data delivery. IOSR J. Comput. Eng. **16**(1), 6–10 (2014). E-ISSN: 2278-0661; P- ISSN: 2278-8727. doi: 10.14419/ijet.v3i2.2318
4. Zadin, A., Fevens, T.: Maintaining path stability with node failure in mobile ad hoc networks. Elsevier Procedia Comput. Sci. 1068–1073 (2013)

5. Kumar, A., Rafiq, M.Q., Bansal K.: A survey of link failure mechanism and overhead of routing protocols in MANET. Int. J. Comput. Sci. Inf. Technol. **2**(5), 2421–2425 (2011). ISSN: 0975-9646

6. Pendke, K., Nimbhorkar, S.U.: Study of various schemes for link recovery in wireless mesh network. Int. J. Adhoc Netw. Syst. (IJANS). **2**(4), (2011). doi: 10.1016/j.procs.2013.06.150

7. Lakshmi, S.M., Sikamani, K.T.: Quick recovery from link failures using enhanced on-demand multicast routing protocol. Res. J. Appl. Sci. Eng. Technol. **9**(7), 526–530 (2015). 2040-7459; E-ISSN: 2040-7467

8. Zhang, X., Jacob, L.: MZRP: an extension of the zone routing protocol for multicasting in MANETS. J. Inf. Sci. Eng. **20**, 535–551 (2004). doi: 10.5121/ijasuc.2011.2202

9. Anttila, A.: Multicast routing with AODV routing protocol Cygate Networks Vattuniemenkatu 21, P.O. Box 187, 00201 Helsinki, Finland

10. Chiang, C.-C., Gerla, M., Zhang, L.: Forwarding group multicast protocol (FGMP) for multihop, mobile wireless networks. J. Cluster Comput. **1**(2), 187–196 (1988)

11. Amnai, M., Fakhri, Y., Abouchabaka, J.: QOS routing and performance evaluation for mobile ad hoc networks using OLSR protocol. Int. J. Ad Hoc Sens. Ubiquitous Comput. (IJASUC). **2**(2) (2011)

12. Xiang, X., Wangy, X., Zhou, Z.: An efficient geographic multicast protocol for mobile ad hoc networks. In: IEEE International Symposium on a World of Wireless, Mobile and Multimedia Networks, (2006)

13. Aluvala, S., Vodnala, D., Yamsani, N., Phani Kumar, S.: A routing algorithm for localization of link failure in MANET. Int. J. Curr. Eng. Sci. Res. (IJCESR). **1**(3), 25–30 (2014) ISSN (Print): 2393-8374, (Online): 2394-0697

14. Jeni, P.R.J., Juliet, A.V., Bose, A.M.: An efficient quantum based routing protocol with local link failure recovery algorithm for MANET. Int. J. Eng. Technol. **3**(2), 237–244 (2014). doi: 10.5121/Ijans.2012.2405

Delay Analysis of ControlNet and DeviceNet in Distributed Control System

Smak Azad and K. Srinivasan

Abstract The DCS (Distributed Control System) have made a great impact on the process industries like manufacturing, refinery, textile allowing automatic industrial processes with multiple input–output arrangements in real time. The above applications and its better performance indirectly depends on the communication between devices, control stations where sharing of network for optimal communication is most important. The present paper deals with study of communication parameters which decide the performance of the network in network control system (NCS). The model developed is a network of nodes explains the time delay analysis of token passing in ControlNet and DeviceNet to get the performance of the NCS. Simulation of the model for 4 nodes using token passing protocol among DeviceNet for NCS has been simulated and analyzed.

Keywords Time delays · Network control system · DCS · Field devices network · Token passing · Scheduled data

1 Introduction

The advances in technologies made the process control leading and growing drastically towards the peak. It is developed by replacing the analog control system with control processors in which the distributed control systems were introduced decades ago. The necessity to communicate the devices and the controllers has increased to enhance flexible and uninterrupted control which drives towards digitalization of the communication. Several multinational organizations started researching and developed protocols like FIELDBUS [1]. The Distributed Control Systems consists of various devices arranged in a hierarchical pattern. Communication between the devices is of utmost importance for making automation possible in the various process industries. During communication, the delays in

S. Azad (✉) · K. Srinivasan
Department of ICE, NIT, Trichy, India
e-mail: freedom_smak@rediffmail.com

© Springer Science+Business Media Singapore 2017
P. Deiva Sundari et al. (eds.), *Proceedings of 2nd International Conference on Intelligent Computing and Applications*, Advances in Intelligent Systems and Computing 467, DOI 10.1007/978-981-10-1645-5_52

transmission time should always be taken into consideration in order to analyze the performance of the system. Feng Li-Lian had an analysis and modeling of controlNet and DeviceNet in which different time delays are studied [2]. The present paper deals with relative time delays with the average time delays and performance, efficiency of a proposed 4 node network of both ControlNet and DeviceNet. This model shows the simulation results of time delays inside the device and on the control network with the real time data propagation between the nodes.

2 Distributed Control System

Distributed Control System (DCS) is a complete solution for control system for a process Industry. A DCS consists of processors as controllers and uses both proprietary interconnections and open communications protocol for communication. Input and output modules complete the control loop interconnection to control the process variables in DCS. Field devices are the factory level control chain in DCS which connects controllers to the sensors and actuators of a control loop [1]. Besides of controllers for controlling the loop, DCS also has dedicated and redundant control network for data acquisition of a process control loop. This control loop supports and accesses the proprietary data through it and necessary diagnostics [3]. The other network of communication between the field devices called as DeviceNet also plays a vital role in processing the control loop. The ControlNet and DeviceNet of the network are discussed in coming chapters.

3 ControlNet

ControlNetis a general real-time, open communication protocol usually available in industrial automation applications for control data communication through input and output modules connected with the controller [4, 5]. A scheduled communication is a network where the control loop is configured through host system called DCS, designed for continuous loop data access. The parameters in a loop of the network will be updated with the scan period of each loop execution, known as Network period or Network Update Interval (NUI). Network update time consists of scheduled, unscheduled and guard band communication sections [2]. In scheduled communication the nodes connected can have a transmission whereas in unscheduled communication is used for data sharing and other maintenance purpose between the nodes. Figure 1 shows schedule, unscheduled and guard bands in a given period of data communication [2]. Token-passing bus topology used in ControlNet the maximum waiting time before sending a message frame can be characterized by the token rotation time [2]. The nodes in the token bus network are formed into a ring and with the address for consecutives nodes. In normal operation

Fig. 1 Scheduled,
Unscheduled and Guard band
access of a medium

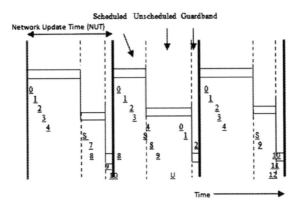

of the network, the node which has the token can transmits data packet till the end
of the data packets. In case, node has no data to transmit, it has to pass the token to
the next consecutive so that no collision of data packets occur [4].

4 DeviceNet

DeviceNet [6, 7] is a network among the field devices and remote operation with
interoperability among the vendors of industry supporting digital, multi-drop network connected with sensors and actuators. The DeviceNet is also a standard of
open automation communication as ControlNet. Categorization of communications
has been done as periodic between sensors and actuators and aperiodic used for
configuration and diagnostics purpose [8]. A periodic communication is to
exchange the data periodically between the field devices and the controller and a
Non-periodic communication is to configure and the diagnostics of field devices
with events management facility.

5 Time Delay Analysis in NCS

In ControlNet a node can send a message only after waiting till the token from
previous node is received. Blocking time is nothing but the sum of transmission
time and token rotation times of previous nodes.

$$T^B = T_{int} + \sum_{j \in Nn} Tt^{(j)} + \sum_{j \in Nq} \min\left(T_{txj}^{(j,n)}, T^n\right) + T^G$$

where,

T^B Blocking time,

T_{int} internal time,

T^G guardband time,

T_{tx} network delay time,

T^n token holding time. If the time of blocking, T^B, in DeviceNet is known value then following equation from k = 1 until it converges by iterative process

$$T_{(k)}^{B12} = T_{int} + \sum_{\forall j \in Nhp} \left[\left(T_{(k-1)}^B + T_{bit} \right) / T_{peri}^{(j)} \right] T_{tx}^{(j)}$$

T_{max} is the running time which depends on number of devices connected. The sum of delay time ($T^D{}_{sum}$) and average time delays ($T^D{}_{avg}$) for each node on a network can be calculated. It has been assumed that the messages are periodic and the total number of messages is equal to the time period of messages running time. The average delay can be computed for the entire network.

$$T_{sum}^D = \sum_{i \in Nn} \sum_{j=1}^{M(i)} \left[T^{D(i,j)} \right]$$

$$T_{avg}^D = \frac{1}{N} \sum_{i \in Nn} \sum_{j=1}^{M(j)} \left[T^{D(i,j)} / M^{(i)} \right]$$

The network efficiency, E^{net} is the ratio of the transmitting time to the sum of time required for sending messages and queuing time, blocking time. "$E^{net} = 1$" represents performance of the network is good with minimum time delay in the network and $E^{net} = 0$ represents delay time is utilized maximum for message interference [2].

$$P_{eff} = \left(\sum_{i \in Nn} \sum_{j=1}^{M(j)} [T_{tx}^{(i,j)}] \right) / \left(T_{sum}^D \right)$$

The message transmission delay of network control system is classified into three major delays located at the source node, on the network channel and at the destination node. Processing time (T_{pre}) and the waiting time (T_{wait}) can be observed from the Fig. 2 at source node. The pre-processing time at the source node is the sum of computation ($T^s{}_{comp}$) and encoding time ($T^s{}_{comp}$) times [2].

$$T_{pre} = T_{comp}^s + T_{code}^s$$

$$T_{wait} = T_{queue} + T_{block}$$

$$T_{post} = T_{code}^d + T_{comp}^d$$

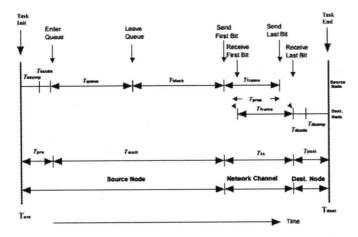

Fig. 2 Timing diagram of initial and ending of the task of sending a message over the network

where T_{queue} is queue time in which a message waits in queue at the source node as previous message in the queue is under progress. T_{wait} is wait time, T_{post} is post processing time and T^d_{code} and T^d_{comp} are decoding time and computation time respectively. Network delay time (T^X) covers the total time for transmission of a message T_{frame} and the propagation delay T^P of the network [2].

$$T^X = T_{frame} + T^P$$

The frame time T_{frame} varies with number of bits in the data, the overhead, padding, and the bit time T_{bit}.

Let N_{data} be the number of the bytes in data, N_{ovhd} be the number of bytes used as overhead, N_{pad} be the number of bytes used to pad the remaining part of the frame to meet the minimum frame size requirement, and N_{stuff} be the number of bytes used in a stuffing mechanism [2].

$$T_{frame} = \{ [[N_{data} + N_{ovhd} + N_{pad} + N_{stuff}]] 8 \} T_{bit}$$

The propagation time T^P varies with signal speed of the transmission and the distance between the source and destination. Least propagation delay from one end to the other of the network cable can be maintained as $T^P = 10$ μs for ControlNet with a distance of 1000 m, and $T^P = 1$ μs for DeviceNet for a distance of 100 m [9].

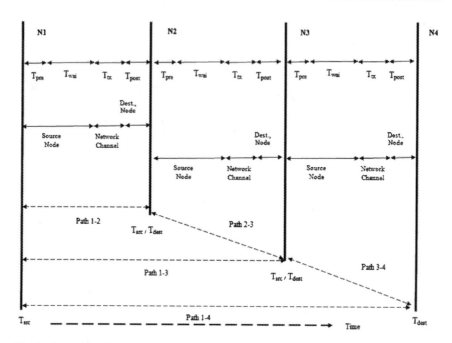

Fig. 3 Simulation of network model

6 Simulation Model

A network of 4 nodes of a control network system has been modeled in the present paper work shown in Fig. 3. In this paper, the complete time delays available in the network are considered and message transmission is based on token passing scheme [10]. There exist different kinds of time delay during the propagation of message in the network. There exists time delay when the token is passed from one node to the other.

There exists time

$$T^B = T^T + T_{tr}$$

where,

T^B blocking time
T^T time required for transmission
And, T_{tr} timetaken for a complete rotation by a token from the previous node

$$T_{TN} = T^T + T^P$$

where,

T_{TN} token time to be passed
T^P propagation time.

The Delays existing at node 1 are defined as

$$T^{B11} = T_{int} + \sum_{j \in Nn} Tt^{(j)} + \sum_{j \in Nq} \min\left(T_{txj}^{(j,n)}, T^n\right) + T^G T_{(k)}^{B12}$$

$$= T_{int} + \sum_{\forall j \in Nhp} \left[\left(T_{(k-1)}^B + T_{bit}\right) / T_{peri}^{(j)}\right] T_{tx}^{(j)}$$

$$T^{B1} = T^{B11} + T^{B12}$$

$$T^{D1} = T_{Tp} + T_{Twait} + T_{Tx} + T_{Tpost}$$

where,

T_{int}	internal time in device
T_t	token rotation time
T^n	node time
T^G	guard time
T_{tx}	network time delay
T_{Tp}	total preprocessing time
T_{bit}	bit time
T_{Tpost}	total post processing time
T_{Twait}	waiting time
T^{B11}	blocking time for the ControlNet in the first node
T^{B12}	blocking time for the DeviceNet in the node 1
T^{B1}	total blocking time at node1
T^{D1}	total time delay at node 1

The various network parameters are simulated and recorded.

7   Simulation Results

Simulation results for all the combinations from node 1 to 4 for blocking time, total and average time delays with efficiency of the network including both ControlNet and DeviceNet are recorded and tabulated. From the curve shown in Fig. 4, blocking time in the ControlNet for data accessing corresponding blocking times for both ControlNet and DeviceNet for different node combinations can be seen in the Table 1 Also from Fig. 4. The total blocking time of ControlNet and DeviceNet for data accessing and for different node combinationscan be seen in Table 1. The sum of the delay time in DeviceNet and ControlNet at all nodes in control network and the corresponding blocking time combinations are tabulated in Table 2.

The average of the delay time at the nodes in the present model of network for data accessing and the corresponding blocking times for different node combinations can be seen in the Table 2 and the efficiency of the network goes on

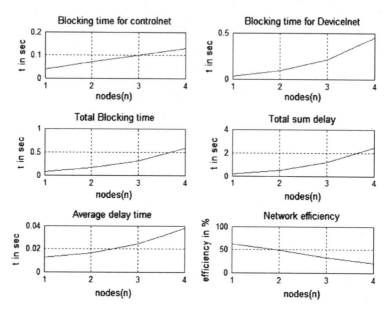

Fig. 4 Simulations of ControlNet and DeviceNet

Table 1 Values of the blocking time for ControlNet and DeviceNet

Blocking time	Nodes	N1	N2	N3	N4
ControlNet (in s)	N1	0.04*	0.03	0.06	0.09
	N2	0.03	0.04*	0.03	0.06
	N3	0.06	0.03	0.04*	0.03
	N4	0.09	0.06	0.03	0.04*
DeviceNet (in s)	N1	0.03*	0.06	0.18	0.42
	N2	0.06	0.03*	0.12	0.36
	N3	0.18	0.12	0.03*	0.24
	N4	0.42	0.36	0.24	0.03*
Total Blocking time (in s)	N1	0.07*	0.09	0.24	0.51
	N2	0.09	0.07*	0.15	0.42
	N3	0.24	0.15	0.07*	0.27
	N4	0.51	0.42	0.27	0.07*

*Blocking time (ms) for ControlNet and DeviceNet

decreasing from one node to the other due to the increase in the time delay. The efficiency at node 2 is less than node 1. Thus, we observed that network efficiency and time delay in the complete network containing 4 nodes are inversely proportional to each other.

Table 2 Total Sum delay and Avegare delay (ms)

Delay Time	Nodes	N1	N2	N3	N4
Total sum delay (in s)	N1	0.200*	0.32	0.96	2.24
	N2	0.32	0.200*	0.64	1.92
	N3	0.96	0.64	0.200*	1.28
	N4	2.24	1.92	1.28	0.200*
Average delay (in s)	N1	0.0125*	0.038	0.0117	0.0256
	N2	0.038	0.0125*	0.0079	0.0218
	N3	0.0117	0.0079	0.0125*	0.0139
	N4	0.0256	0.0218	0.0139	0.0125*

*Value of the internal time delay of field device

8 Conclusion and Future Work

The time required for each message to be transmitted from one node to the other 4 nodes in a network has been studied. Sharing of information in the network with delays like internal device time delay, net blocking time and waiting time are considered and modeling of a network of 4 node network system has developed. The time delays are calculated with the help of certain appropriate mathematical formulas simulated on MATLAB.

The proposed model of 4 nodes network has been simulated for performance parameters and found that the performance will be increased with the reduction of delays by selecting appropriate topology and communication techniques. The above model can be used for analysis and design of new communication protocol for process automation like DeviceNet, ControlNet and Ethernet in future work. Number of nodes can be increased for further hidden and complicated delays consideration in the network for better enhanced performance.

References

1. Segovia, V.R., Theorin, A.: History of control history of PLC and DCS. 2012-06-15 (minor revision 2013-07-26)
2. Li-Lian, F.: Analysis, Design, Modeling and Control of network Systems. University of Michigan, Michigan (2001)
3. Distributed Control Systems, http://www.honeywell.com
4. "ControlNet International", ControlNet Specifications, 2.0 edition (1998)
5. http://www.dia.uniroma3.it/autom/Reti_e_Sistemi_Automazione/PDF/ControlNetDetails.pdf
6. http://www.pcmag.com/encyclopedia/term/52962/token-passing
7. http://www.intechopen.com/books/factory-automation/token-passing-techniques-for-hard-real-time-communication
8. Lee, Y.H., Hong, S.H.: Dependency on prioritized data in the delay analysis of foundation fieldbus (2008)

9. Goktas, F., Smith, J.M., Bajcsy, R.: μ-Synthesis for distributed control systems with network induced delays. In: Proceedings of the 35th Conference on Decision and Control, pp. 813–814 (Dec 1996)
10. "Open DeviceNet Vendors Association". DeviceNet Specifications, 2.0 edn (1997)

A Novel Hybrid BAT Search for Security Confined Unit Commitment-3-Unit System

V. Lakshmidevi, P. Sujatha and K.S.R. Anjaneyulu

Abstract For the efficient and reliable operation of power systems, the most critical task is committing of generating units with respect to change in demand to supply. So, therefore as to control this challenging task named as security confined unit commitment (SCUC). This SCUC considered system's objective function as consisting of both equality and inequality constraints. By applying numerous optimization functions to the constraints as power balance, spinning reserve, operating limits of real power, minimum up and down times, emission etc., the ultimate conditions of the problem will be solved. These are handled to generate an answer to the problem by applying hybrid BAT search algorithm. So, for modern power systems, by implementing this most economical method of operation generates the optimal power generation for different units as by meeting all the considered system Conditions. The reliability of the recommended technique's computational results is assessed to a 3-unit testing system.

Keywords Unit commitment · Security constraints · BAT techniques · BAT-GA techniques

1 Introduction

The planning and the most favourable economic operation of generating power systems is an utmost concern in power sectors [1]. Unit commitment (UC) is a most concerned task referred to the power generating resource management in the

V. Lakshmidevi (✉) · P. Sujatha · K.S.R. Anjaneyulu
Department of Electrical Engineering, JNTUA College of Engineering,
Ananthapuramu, India
e-mail: vldevi.eee@gmail.com

P. Sujatha
e-mail: psujatha1993@gmail.com

K.S.R. Anjaneyulu
e-mail: ksralu@yahoo.co.uk

© Springer Science+Business Media Singapore 2017
P. Deiva Sundari et al. (eds.), *Proceedings of 2nd International Conference on Intelligent Computing and Applications*, Advances in Intelligent Systems and Computing 467, DOI 10.1007/978-981-10-1645-5_53

restricting power sector [2]. This operation of power systems is a centralised important task which is stable and most efficient [3]. The important objective seen in this unit commitment problem (UC) is definitely ascertaining the set of minimum cost turn on and off schedules of units which are meant to equalize the continuous changes of the load demand by accepting the operational constraints [4, 5]. Therefore the considered unit commitment (UC) problem reduces the power cost and start up cost as well which are due to generating units [6].

Evaluation of Unit Commitment problem (UC) is absolutely a detailed optimum result concern handled as two sub-optimization problems, that is as a connective concern even for generating units of the power system [7]. One of the problem named as unit scheduling is meant to attain, the start up and shut down operating costs of the units as reduced total operating cost by satisfying the generator Constraints and also the system constraints including economic dispatch problem [8, 9]. In electrical power sectors, the unit commitment (UC) which is an optimization concern, it determines the generating units both on and off states thereby minimizing the operational cost for considered deterministic levels of time [10].

There are several mathematical procedural programming methodologies for solving unit commitment like Genetic algorithm, Ant colony search algorithm, Evolutionary programming(EP) technique [11], Tabu search [12], Particle swarm (PSO) optimization and BAT optimisation [13–15]. Genetic algorithm (GA) deals with the population of desired solutions and these methods are developed on the conventions and procedures for eco selection and "survival of the fittest" [16]. The strength of every representative in the population is calculated by an interpretation function that measures wellness of an individual [17]. The Tabu search depresses the probability to converge locally within the foremost laps of the epochs. This enables the representatives to explore new clarification spaces to urge better solutions [18]. Particle swarm optimisation (PSO) has more potency in soluting integer programming dilemma. It is also worned as a pre-processor for generating good initial marks in a branch and bound technique of an integer programming problem [19]. In particle swam optimisation technique, the variables of unit commitment are coded as integers as they are formulated highly that minimises the decision variables and UC problems [20].

Ant colony search algorithm (ACSA) replicates the performance of ants [21]. Therefore this Ant colony optimization technique was used to determine the power system's problem of economic dispatch in large-scale systems [22]. For optimization problem the Ant colony optimization method is applied to achieve least total generation cost [23]. Chandrasekaran et al. [24] have explained a binary actual coded firefly (BRCFF) algorithm which is a new ecologically activated technique to determine the system's unit commitment (UC) under considered constraints [25]. The firefly (FF) algorithm was animated by the unique flash behaviour of fireflies and experiences bioluminescent communication. Zhao and Guan [26] have suggested a novel unified problematic and powerful unit commitment model which takes advantages of twain stochastic and problematic optimization paths of low familiar total cost by assuring the system robustness.

The rest of the paper is organized as follows: as Sect. 1 mentioned introduction, Sect. 2 deals with transmission system constraints related to unit commitment (UC) with security constraints formulation. The new eco-inspired BAT technique is expressed in Sect. 3. Introduction to GA is seen in Sect. 4. The proposed hybrid algorithm in Sect. 5. The proposed algorithm outcomes are proved with 3-unit system for 24 h load demand in Sect. 6 and at last Sect. 7 summarizes conclusion with a noted advantage of proposed methodology.

2 SCUC and Formulation of SCUC

Unit commitment (UC) is the critical and powerful procedure to minimize the costs required for the start up, shut down and fuel in generating units. At the time of power generation there is a contingent for problem existence. This complication existence leads to generating system's inaccurate unit line-ups and economic dispatch problem, since the UC minimises the costs of the start up and output power of the generating system [27]. Unless the dilemma is segregated perfectly the required system cost cannot be reduced in comparison to the power generated as output. The complication identification and distribution of the generating units is incredibly necessary in power systems. The most unbiased operate of the unit commitment (UC) is to resolve a reduced value for turn-on and turn-off roots for an electrical power generating units to satisfy a demanded load by satisfactory choices by opting operational constraints [28]. The two concepts related to reduce the standard unit commitment are the cost for power production and the start up cost by the generating units.

For N generating units and T hours, then the most objective operate of the unit commitment UC complication is often authored as follows:

$$\text{Min } F = \sum_{i-1}^{Ng} F_i(P_{gi}) = \sum_{i=1}^{Ng} (a_i + b_i P_{gi} + c_i P_{gi}^2) \tag{1}$$

where F is that the value for total generation (Rs/Mwh), F_i is that the 'ith' generator's input-output operate, Ng is that the total on-line generators count, P_{gi} is that the 'ith' generator's active power output (Mw) and a, b, c are the fuel value coefficients of 'ith' generator.

The fuel value operate is depicted as:

$$C_i(P_i) = a_i + b_i(P_{gi}) + c_i(P_{gi}^2) + |e_i \sin(f_i(P_i^{min} - P_i))| \tag{2}$$

2.1 The Constraints Subjected Are

2.1.1 Power Balance Constraint

The summation of total load demand and total real power loss within the transmission lines must be up to the whole power generated by the units. Hence the constraint is outlined as:

$$\sum_{i}^{Ng} P_{gi} = P_D + P_L \tag{3}$$

where P_D is that the total load of the considered system and P_L is that the loss of the transmission (Mw). The considered transmission losses are calculated by practising B-coefficients.

$$\text{Where } P_L = \sum_{m=1}^{Ng} \sum_{n=1}^{Ng} P_{Gm} B_{mn} P_{Gn} \tag{4}$$

$P_{Gm}, P_{Gn} = real\ power\ generation\ at\ m, nth\ plants$ B_{mm} = Loss coefficients that are constraints which are under simulated operating conditions.

2.1.2 Generation Capacity Constraints

With regard to their several lower and upper limits, the real output power of generating units must be restricted as follows: (For i = 1,...,N$_G$)

$$P_{gi}^{min} \leq P_{gi} \leq P_{gi}^{max} \tag{5}$$

where P_{gi}^{min} and P_{gi}^{max} are the minimum output power and maximum output power for the ith unit.

2.1.3 Spinning Reserve Constraints

It is the modification betwixt total maximum powers from all on-line units of generation with total demand in the required time. Spinning reserve constraint equation is expressed as,

$$\sum_{i-1}^{N} P_{i\,max} \geq P_D + R \tag{6}$$

2.1.4   Minimum Up and Down Time Constraints

It is the minimum time that the generating unit requires to turn on to travel back in rest methodology. Mean time of minimum down time during this unit commitment (UC) is to show on and to travel back in on-line mode as minimum time that generating unit had just turn on to go back in online mode. Minimum up and minimum down time are often denoted as in the following equation,

$$U_{ih} = 1 \quad for \sum_{t=h-up_i}^{h-1} U_{it} \leq up_i \tag{7}$$

$$U_{ih} = 0 \quad for \sum_{t=h-down_i}^{h-1} (1 - U_{it}) \leq down_i$$

3   BAT Algorithms

BAT algorithmic program could be a meta-heuristic optimisation formula developed by Xin-She Yang in 2010 [29]. The BAT formula is predicted on the localization behaviour of bats. Bats have the ability to search out their prey and discriminate differing kinds of insects even in complete darkness. The localization behaviour of micro-bats may be accustomed to optimize an objective operate [11].

3.1   Movement of Bats

The bat's movement hinge on the modification in velocity with relevance to step time. The new results x_i^t, and velocities v_i^t, at step time t are denoted as:

$$f_i = f_{min} + (f_{max} - f_{min}) * \beta \tag{8}$$

$$x_i^t = x_i^{t-1} + v_i^t \tag{9}$$

where, $\beta \in [0, 1]$ could be an uniformly distributed random vector. X^* is that the new global best location (solution) which is found by comparison to any or all the obtained results among all the 'n' bats. For the local search, once an answer is picked among the present best solutions, for each bat, using stochastic random walk condition, a new result is created internally as;

$$x_{new} = x_{old} + \varepsilon A^t \tag{10}$$

where, $\varepsilon \in [-1, 1]$ could be a random number, once $A^t = < A_i^t >$ which is the average loudness of all the bats at time t.

3.2 Loudness and Pulse Emission

The loudness A_i and therefore the rate of pulse emission r_i are modernised with various respective epochs go ahead.

$$A_i^{t+1} = \alpha A_i^t \tag{11}$$

$$r_i^{t+1} = r_i^0[1-\exp(-\gamma t)] \tag{12}$$

where α and γ are constants.

4 Genetic Algorithms

The Genetic algorithm is a simulating process where natural evolution system of Darwin's theory "fittest survival" is proposed by Holland in 1940s. This is method used for solving both constrained and unconstrained optimisation problem based on natural selection.

The steps involved in this algorithm are:

A. Selection

Depending on evolution function, chooses a chromosome in random fashion out of population.
B. Cross over

Mating of strings randomly in between two parents.
C. Mutation

An operation performed by random selection of chromosomes in pre-specified limit of probability.

Depending on complexity of the problem, the population size is to be considered.

5 Hybrid (BAT-GA) Algorithms

The BAT algorithm based Genetic algorithm is used for solving the unit commitment complication. By utilizing GA in BAT objective function helps as good search tool because of unplanned results and merging, this can be explained as a source to the entire generation is developing, but this may not be revealed as an individual within the entire generation considered. This hybrid technique reduces the speed of convergence range.

6 Simulation Results

BAT algorithm is comparatively a new optimization algorithm [1] but the hybrid BAT-GA algorithm approach is a new effective and reliable methodology for solving SCUC problem. In order to ensure the effective applicability of suggested combinatorial BAT-GA algorithm results are enclosed in terms of generating units cost. The procedural solution of considered unit commitment (UC) problem consisting the objective constraints is represented in given unit and time schedules. The hybridised BAT-GA algorithm was simulated to reveal the status of convergence and reliability of the selected system was tabulated and even compared with BAT algorithm's data and results.

In order to show the importance of the selected hybrid algorithm, here a 3-unit system with its system parameters are picked as shown in below tables.

The parameters are chosen for algorithmic procedure is as follows: length of the bat as 5, iterations considered up to 100 and time zone as for one day i.e. 24 h. Tables 1 and 2 represent the 3-unit system data. The results are tabulated for the

Table 1 Basic data for 3-unit system

Gen. no.	a	b	c	P_{max}	P_{min}	t_{on}	t_{off}	Strt-cost	C hour
1	176.9	13.5	0.1	220	100	4	4	100	1.2469
2	129.9	32.6	0.1	100	10	2	2	200	1.2469
3	137.4	17.6	0.1	20	10	2	2	0	1.2462

Table 2 Load demand for 24 hour

Hour	1	2	3	4	5	6	7	8	9	10	11	12
Demand	290	250	240	300	200	280	280	220	250	170	160	240
Hour	13	14	15	16	17	18	19	20	21	22	23	24
Demand	280	310	180	250	230	160	210	180	280	210	240	180

Table 3 Result for 3-unit system (using BAT)

Hour	Unit-1	Unit-2	Unit-3	Start up	Fuel cost
1	184.990	90.995	13.992	0	10,435.13
2	142.6551	90.6597	16.663	0	8512.12
3	146.6536	82.6615	10.667	0	8160.74
4	202.3165	85.3297	12.337	0	11,023.03
5	158.6466	22.6539	18.662	0	6265.46
6	170.9922	90.9858	17.990	0	9829.66
7	170.6583	92.6601	16.663	0	9871.10
8	156.1625	63.8389	0	0	7351.73
9	173.9937	75.9899	0	0	8748.25
10	151.9923	0	17.993	0	5034.54
11	110.6587	31.6541	17.658	200	4843.35
12	162.4885	77.4871	0	0	8276.90
13	195.9864	63.9902	19.993	0	9830.26
14	207.3191	85.3193	17.325	0	11,397.87
15	133.9892	45.9899	0	0	5629.78
16	193.9857	38.9897	16.999	0	8588.86
17	163.9967	65.9968	0	0	7807.14
18	140.9928	0	18.997	0	4584.50
19	193.9919	0	16.000	0	7015.33
20	164.4961	0	15.497	0	5547.54
21	199.6491	66.6511	13.672	200	10,213.82
22	144.4972	65.4974	0	0	6918.33
23	144.9929	76.9895	18.003	0	7964.46
24	126.9849	39.9866	13.001	0	5487.79
Total cost					189,337.7

considered system as III and IV. From Tables 3 and 4 the applicability of the selective method for solving UC problem is highlighted. The combinatorial method has minimised the operating value when compared to BAT technique [1]. Therefore, the total operating value is reduced there by representing its effectiveness. For unit-3 system, as in BAT approach [1] the total cost is seen as 189,337.7$, where as in proposed hybrid BAT-GA algorithm approach the minimised operating value is seen as 177,103.8301$. The comparison of BAT and HYBRID BAT-GA approach is represented with help of graphs as shown below in Fig. 1.

Table 4 Result for 3-unit system (using BAT-GA)

Hour	Unit-1	Unit-2	Unit-3	Start up	Fuel cost
1	185.237	100.7867	0	0	10,551.37
2	152.323	79.31469	18.32	0	8400.67
3	115.242	104.0906	0	0	7674.40
4	193.879	86.32343	0	0	10,254.03
5	180.729	0	0	0	5893.92
6	191.77	72.64643	0	200	9680.81
7	205.775	51.05751	0	0	9256.65
8	159.82	39.33063	0	0	6465.10
9	186.328	49.479	0	0	8163.10
10	146.998	0	22.997	0	4926.09
11	133.105	0	18.315	0	4246.79
12	129.469	87.19535	0	0	7341.50
13	166.666	94.38785	0	0	9312.51
14	203.331	94.08083	0	0	11,150.51
15	158.7	0	0	0	4847.46
16	181.649	48.65749	19.663	200	8614.75
17	167.704	40.99767	0	0	6897.94
18	137.44	0	0	0	3929.59
19	185.216	0	0	0	6118.93
20	155.006	0	13.49	0	5074.54
21	204.864	55.26853	0	0	9388.94
22	168.746	20.22655	0	0	6142.85
23	148.983	75.99006	14.993	0	8025.11
24	156.455	0	0	0	4746.28
Total cost					177,103.8

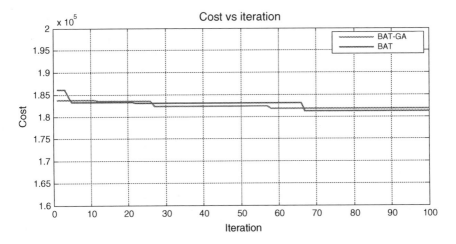

Fig. 1 Cost comparison of BAT and BAT-GA

7 Conclusions

This paper represents a methodological solution to security constrained unit commitment (SCUC) by availing all the objective operator's related to circuitry and units of an electrical power system. This hybrid algorithm generated optimal results under the specified constraints. When compared with conventional algorithms and nature inspired—BAT algorithm, this offspring approach has superior benefits with quality application, stable convergence and high computational heights. So, this proposed BAT-GA algorithm is a reliable technique for evaluating complex problems in electrical power sectors and reduces the uncertainties to implement optimal solution while using conventional techniques. This BAT-GA algorithm can be used to solve unit commitment problem and reduced the total production cost of power generation. The effective results are even compared with graphical representation in Fig. 1 as comparison of BAT and HYBRID BAT-GA approach in terms of optimal cost.

References

1. Lakshmidevi, V., Anjaneyulu, K.S.R., Sujatha, P.: A novel approach for security constrained unit commitment using BAT algorithm. IEEE Digital Explore, (2015) doi:10.1109/IADCC.2015.7154782
2. Roy, P.K.: Solution of unit commitment problem using gravitational search algorithm. Electr. Power Energy Syst. 53, 85–94 (2013)
3. Wang, L., Singh, C.: Unit commitment considering generator outages through a mixed-integer particle swarm optimization algorithm. In: Proceedings of IEEE Conference on Region 5, pp. 261–266 (2006) doi:10.1109/TPSD.2006.5507419
4. Caroe, C.C., Schultz, R.: A two-stage stochastic program for unit commitment under uncertainty in a hydro-thermal power system, pp 1–17 (1998) http://citeseerx.ist.psu.edu/viewdoc/summary?doi=10.1.1.28.7817
5. Sheble, G.B., Fahd, G.N.: Unit commitment literature synopsis. IEEE Trans. Power Syst. 9, 28–135 (1994)
6. Sum-Im, T., Ongsakul, W.: Ant colony search algorithm for unit commitment. In: Proceedings of International Conference on Industrial Technology, vol. 1, pp. 72–77 IEEE. (2003) doi:10.1109/ICIT.2003.1290244
7. Lenin, K., Reddy, B.R., Kalavathi, M.S.: Ant colony search algorithm for solving unit commitment problem. Int. J. Mechatron. Electr. Comput. Technol. 3, 193–207 (2013)
8. Christober, C, Rajan, A.: An evolutionary programming based simulated annealing method for unit commitment problem with cooling-banking constraints. In: Proceedings of IEEE India Annual Conference (INDICO 2004), pp. 435–440 (2004) doi:10.1109/INDICO.2004.1497790
9. Dhanapal, M, Lakshmi, K.: Co-ordination of thermal unit system with wind energy system for scheduling problem in restructured power system. In: "Proceedings of IEEE International Conference on Green Line Constraint", International Journal of Computer Science & Communication, vol. 1, no. 2, pp. 145–149, 2010
10. Liang, R.H., Kan, F.C.: Thermal generating unit commitment using an extended mean field annealing neural network. In: IEE Proceedings of Generation, Transmission and Distribution, vol. 147, no. 3 (2000) doi:10.1049/ip-gtd:20000303

11. Selvi, S.C., Devi, R.P.K., Rajan, C.C.A.: Hybrid evolutionary programming approach to multi-area unit commitment with import and export constraints. Int. J. Recent Trends Eng. 1(3), 223–228 (2009)

12. Mulyawan, A.B., Setiawan, A., Sudiarso, A.: Thermal unit commitment solution using genetic algorithm combined with the principle of tabu search and priority list method. In: Proceedings of International Conference on Information Technology and Electrical Engineering (ICITEE), pp. 414–419 (2013)

13. Kadam, D.P., Sonwan, P.M., Dhote, V.P., Kushare, B.E.: Fuzzy logic algorithm for unit commitment problem. In: Proceedings of International Conference on Control, Automation, Communication and Energy Conservation, pp. 1–4 (2009)

14. Biswal, S., Barisal, A.K., Behera, A., Prakash, T.: Optimal power dispatch using BAT algorithm. IEEE Trans. Power Syst. 978-1-4673-6150-7/13 IEEE (2013) doi:10.1109/ICEETS.2013.6533526

15. Gherbi, Y.A., Bouzeboudja, H., Lakdja, F.: Economic dispatch problem using BAT algorithm. Leonardo J. Sci. 24, pp. 75–84, ISSN 1583-0233

16. Swarup, K.S., Yamashiro, S.: Unit commitment solution methodology using genetic algorithm. IEEE Trans. Power Syst. 17(1), 87–91 (2002)

17. Dasgupta, D.: Unit commitment in thermal power generation using genetic algorithms. In: Proceedings of the Sixth International Conference on Industrial & Engineering Applications of Artificial Intelligence and Expert Systems (IEA/AIE-93) (1993) http://citeseerx.ist.psu.edu/viewdoc/summary?doi=10.1.1.48.9694

18. Kurahashi, S., Terano, T.: A genetic algorithm with tabu search for multimodal and multi objective function optimization, pp. 291–298 (2000) http://citeseerx.ist.psu.edu/viewdoc/summary?doi=10.1.1.592.4326

19. Gamot, R.M., Mesa, A.: Particle swarm optimization—tabu search approach to constrained engineering optimization problems. WSEAS Trans. Math. 7(11), 666–675 (2008)

20. Pappala, V.S., Erlich, I.: New approach for solving the unit commitment problem by adaptive particle swarm optimization. In: Proceedings of IEEE Conference on Power and Energy Society General Meeting, pp. 1–6 (2008) doi:10.1109/PES.2008.4596390

21. Sisworahardjo, N.S., El-Keib, A.A.: Unit commitment using the ant colony search algorithm. In: Proceedings of Conference on Power Engineering Large Engineering Systems (LESCOPE 02), pp. 2–6 (2002) doi:10.1109/LESCPE.2002.1020658

22. Song, Y.H., Chou, C.S.V.: Large scale economic dispatch by artificial ant colony search algorithms. Electr. Mach. Power Syst. 27, 679–690 (1999)

23. Ameli, A., Safari, A., Shayanfar, H.A.: Modified ant colony optimization technique for solving unit commitment problem. Int. J. Tech. Phys. Probl. Eng. 3(4), 29–35 (2011)

24. Chandrasekaran, K., Simon, S.P., Padhy, N.P.: Binary real coded firefly algorithm for solving unit commitment problem. Inf. Sci. 249, 67–84 (2013)

25. Jiang, Q., Zhou, B., Zhang, M.: Parallel augment Lagrangian relaxation method for transient stability constrained unit commitment. IEEE Trans. Power Syst. 28(2), 1140–1148 (2013)

26. Zhao, C., Guan, Y.: Unified stochastic and robust unit commitment. IEEE Trans. Power Syst. 28(3), 3353–3361 (2013)

27. Jiang, R., Wang, J., Zhang, M., Guan, Y.: Two-stage minimax regret robust unit commitment. IEEE Trans. Power Syst. 28(3), 2271–2282 (2013)

28. Bakirtzis, E.A., Biskas, P.N., Labridis, D.P., Bakirtzis, A.G.: Multiple time resolution unit commitment for short-term operations scheduling under high renewable penetration. IEEE Trans. Power Syst. 29(1), 149–159 (2014)

29. Rajan, C.C.A.: An evolutionary programming based tabu search method for unit commitment problem with cooling-banking constraints. In: Proceedings of Power India Conference, p. 8 (2006)

Application of Fuzzy Inference System to Estimate Perceived LOS Criteria of Urban Road Segments

Suprava Jena, Atmakuri Priyanka, Sambit Kumar Beura and P.K. Bhuyan

Abstract Perception based Level of service methodology depicts how well a transportation facility satisfy its road users. There are several engineering factors affecting driver's perception of service quality on urban arterials, which vary largely by individual user and not easy to be distinguished. Therefore, user-perceived service quality appraisals don't relate to the LOS assessed using Highway Capacity Manual (HCM) method. The main objective of this study is to speak to the variability of human discernment and to characterize level of satisfaction of road users using hierarchical fuzzy inference system (HFIS). Around 500 responses of road users were gathered from two emerging cities of India i.e. Bhubaneswar and Vishakhapatnam, where traffic facility is portrayed by distinctive sorts of roadway features and variation in number of pedestrians, motorists and each category of vehicles. So that our survey can include every circle of transportation and people belonging to all classes can be taken into account. The present approach, contains fuzzy reasoning experiences to combine important factors which reflect road user's perception of service quality. In order to decrease the intricacy of multivariable, the fuzzy inference system (FIS) is partitioned into two subsystems, together conducing to a hierarchical fuzzy inference system. First the four input variables namely roadway geometry, traffic facilities, pavement condition and safety are fuzzified. The resulting fuzzy values are entered into the fuzzy inference system, where number of fuzzy rules are generated by applying fuzzy operators. Applying max–min composition method all the outputs are aggregated and defuzzified to evaluate composite LOS of various road segments of both the cities. This methodology

S. Jena (✉) · A. Priyanka · S.K. Beura · P.K. Bhuyan
Department of Civil Engineering, National Institute of Technology Rourkela,
Rourkela 769008, Odisha, India
e-mail: suprava728@gmail.com

A. Priyanka
e-mail: priyanka.atmakuri@gmail.com

S.K. Beura
e-mail: sambit.beura@gmail.com

P.K. Bhuyan
e-mail: pkbtrans@gmail.com

© Springer Science+Business Media Singapore 2017
P. Deiva Sundari et al. (eds.), *Proceedings of 2nd International Conference
on Intelligent Computing and Applications*, Advances in Intelligent Systems
and Computing 467, DOI 10.1007/978-981-10-1645-5_54

639

offers unique and significant approach to define service quality provided by the transportation infrastructure and may defeat the constraints of customary delay-based methods to some extent.

Keywords Perceived level of service · Fuzzy inference system · Membership functions · LOS thresholds

1 Introduction

LOS is a quantitative way to describe the performance of transportation service, atonement level of all type of road users. LOS concept was initially presented in the 1965 Highway Capacity Manual (TRB, 1965) to assess the service quality and was portrayed by six classes from "A" (describes the best) to "F" (describes the worst) based upon "travel time" and "ratio of traffic flow rate to capacity". Nonetheless, each class was not characterized in a subjective way. In HCM 2010, LOS describes a range of operating conditions and individual's approach about those conditions. In real form HCM LOS thresholds were not determined based upon perception analysis, instead it is decided from a combination of capacity based results. Transportation facility is paid for and provided to the public. Transportation departments should focus on what attributes of the transportation system are vital to road users and how those influence their satisfaction level. However decision of investment does not encompass information about the views of road users. These are highly influenced by the results of LOS analysis procedures of HCM based upon the aggregate judgment of individuals from of the Highway Capacity and Quality of Service (HCQS) Committee. In a Similar manner the threshold values for different LOS criteria are determined.

Human thinking process and behavior is intricate, qualitative to judge and varies from person to person. So the user-perceived service quality appraisals don't relate to the LOS assessed using the Highway Capacity Manual (HCM) method. It is a tedious effort to assess service criteria and to get effective outcomes due to the attributes of individual behavior. But fuzzy logic will figure out erratic and insignificant aspects of public opinions. The fundamental goal of this research is the determination of LOS under heterogeneous traffic flow condition using HFIS. The proposed approach, contains fuzzy reasoning experiences to combine important factors which reflect road user's perception of service quality. With a specific goal to decrease the intricacy of multivariable fuzzy inference system (FIS), the suggested framework is partitioned into two subsystems: lower level FIS and upper level FIS. These two levels together put up to a hierarchical fuzzy inference system. The upper level FIS is brought about to deduce and evaluate the level of service (LOS) of roadways taking in account four input variables i.e. roadway geometry, traffic facilities, pavement condition and safety. Applying "max–min composition" method the concluding output of the fuzzy inference system corresponds to the composite satisfaction level of road users on various road segments of both the cities with respect to their operating conditions.

Sutaria and Haynes carried out a perception survey to figure out perception of LOS at signalized intersections. About 300 subjects had taken part and reveal necessary factors, influencing their satisfaction level. He concluded 'Delay' as the most important factor influencing LOS [1]. Pecheux concentrated on the Quality of Service (QOS) of urban streets contemplating of road user's perception. Information was gathered by in-vehicle approach. He had also carried out a written survey in which members gave their view of the driving experience considering roadway, operational and environmental conditions on urban roadways. This study delivered about 45 distinguished QOS components [2]. Hummer et al. created a model to assess LOS on shared use paths and developed a model relating user's view to operational as well as geometric variables. The outcomes of this research uncovers the strong correlation between path operation and overall satisfaction of the service quality [3].

Zhang and Prevedouros carried out a pilot questionnaire survey to reveal the variables adding to user ratings LOS at signalized intersection. Fuzzy weighted average technique was employed to aggregate all the variables in determining LOS [4]. Flannery et al. while relating quantitative to subjective LOS measuring techniques for urban roads found that LOS ascertained by HCM 2000 procedure, shows 35 % of the change in mean driver rating [5]. Lee et al. evaluated driver's perception of satisfaction by fuzzy weighted average method at signalized intersection. A consensus analysis has been applied to identify the weight of individual's perception and aggregate overall service quality of signalized intersection [6]. Xiaoming et al. suggested a method to anticipate user's opinion on signalized intersection using fuzzy neural network in mixed traffic condition. They complied to some degree with HCM LOS [7].

Fang and Kelly carried out video laboratory survey to review an intersection from bad to excellent on a 10 point scale without considering stopped and control delay. He applied fuzzy data mining approach and found out that subjects were not able to perceive more than 3–4 LOS categories as described in HCM and LOS F have a wide range of delays, that needs two or more levels to differentiate in between them [8]. Zhang and Prevedouros once more exhibited fuzzy logic technique to evaluate signalized intersection LOS taking into account the road user's opinion. The found out that road users consider multiple factors to evaluate signalized intersections [9].

The traffic in urban corridors of India is profoundly heterogeneous comprising of different kind of vehicles having diverse operational attributes. The geometric attributes of roadways and behaviour of individuals also differ from other countries. Hence the developed models for homogeneous traffic flow conditions in other countries can't be applied directly in Indian traffic conditions for the evaluation of the service quality provided by the transportation infrastructure. An endeavour has been made in this regard for the evaluation of roadway service quality using fuzzy logic to represent variability and complexity of human perceptions in urban Indian condition. This methodology offers unique and significant approach to define service quality provided by the transportation infrastructure and may hence conquer the constraints of conventional delay-based methods to some extent.

2 Methodology

L. A. Zadeh presented the concept of fuzzy sets in 1965, which is broadly utilized in numerous regions related to human perception. Among different applications of fuzzy set theory Fuzzy inference system was implemented in this study, which is the most common fuzzy technique. Public perceptions give erratic information which are represented through linguistic terms. Fuzzy inference is the deductive reasoning process of formulation from a given input to an output using fuzzy logic as shown in Fig. 1.

2.1 Fuzzification

The input data are crisp values, so they should be changed over to fuzzy sets. Figure 2 demonstrate the fuzzy sets for one input variable (i.e. roadway geometry) of one road segment. The membership functions (mf) considered in this approach are triangular in shape, having simple formulas and computational efficiency. It consists of three parameters to control the exact shape of the membership functions as well as the function value. Equation (1) shows the of triangular membership functions.

$$Triangle f(x, a, b, c) = \begin{bmatrix} 0 & x \leq a \\ \frac{x-a}{b-a} & a \leq x \leq b \\ \frac{c-x}{c-b} & b \leq x \leq c \\ 0 & c \leq x \end{bmatrix} \tag{1}$$

Each input variables of fuzzy sets for every road segments are classified into three descriptors, i.e. poor, fair and good as shown in the Table 1.

Fig. 1 Structure of fuzzy logic system evaluating road user perception

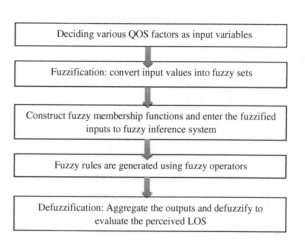

Deciding various QOS factors as input variables

Fuzzification: convert input values into fuzzy sets

Construct fuzzy membership functions and enter the fuzzified inputs to fuzzy inference system

Fuzzy rules are generated using fuzzy operators

Defuzzification: Aggregate the outputs and defuzzify to evaluate the perceived LOS

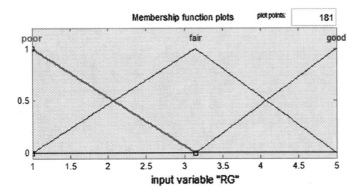

Fig. 2 Triangular fuzzy membership function for roadway geometry of Vanivihar–Rupali sq. segment of Bhubaneswar city

Table 1 Controlling parameters of triangular fuzzy membership function for roadway geometry of a road segment of Bhubaneswar

Controlling parameters	a	b	c
Poor	1	1	3.15
Fair	1	3.15	5
Good	3.15	5	5

The output variable (i.e. LOS) is classified into six subsets i.e. A to F, similar to HCM 2010 (Fig. 3) and each class is assigned with a linguistic term (e.g., A (very good), B (good), C (fair), D (acceptable), E (poor) and F (very poor). The fuzzy membership functions for LOS (Fig. 3) is based on index numbers from 1 to 6 (1 representing the worst and 6 representing the best condition of road service).

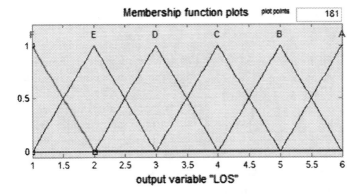

Fig. 3 LOS rating for different Fuzzy membership function

2.2 Fuzzy Inference

The fuzzified input variables were entered into the Fuzzy inference system. FIS comprises of a set of "If-Then" fuzzy rules depending upon the number of input variables.

The general form of fuzzy rule is as follows:

If $\{RG$ is $X_{RD}\}$ and $\{TF$ is $X_{TF}\}$ and $\{PC$ is $X_{PC}\}$ and $\{S$ is $X_S\}$, Then $\{LOS$ is $Y\}$

where
RG   Roadway Geometry
X_{RD} Fuzzy set for Roadway Geometry
TF   Traffic Facilities
X_{TF} Fuzzy set for Traffic Facilities
PC   Pavement condition
X_{PC} Fuzzy set for Pavement condition
S   Safety
X_S Fuzzy set for Safety, and
Y   Fuzzy set for LOS, i.e., A (very good), B (good), C (fair), D (acceptable), E (poor), F (very poor)

The number of fuzzy rules increases with increase in the number of input variables as well as the number of descriptors in membership function. There are six inputs taken into consideration in this method and each input variable has three descriptors. So total $3^6 = 729$ fuzzy rules will be generated. This is called "rule explosion problem." "Hierarchical fuzzy inference system" is the appropriate technique to solve the "rule explosion problem" and to reduce the computational complexity of number of fuzzy rules.

Two levels of hierarchical fuzzy inference system: In order to decrease the intricacy of multivariable, the fuzzy inference system (FIS) is partitioned into two subsystems: two lower level FIS and an upper level FIS, together conducing to a hierarchical fuzzy inference system. In the first lower FIS traffic operations (TO) and traffic signs (TS) are the two input variables taken and the output is traffic facilities (TF). In the second lower FIS, pavement quality (PQ) and pavement marking (PM) are taken as input variables and output is the pavement condition (PC) of the road segment. The upper level fuzzy inference system is generated to deduce and evaluate the level of service (LOS) of roadways taking in account four input variables namely roadway geometry (RG), traffic facilities (TF), pavement condition (PC) and safety (S). The output of the fuzzy inference system produces the levels of satisfaction of users as regards to the design and operating conditions of respective road segments. Figure 4 represents the reduction of the number of fuzzy rules with the help of hierarchical inference system.

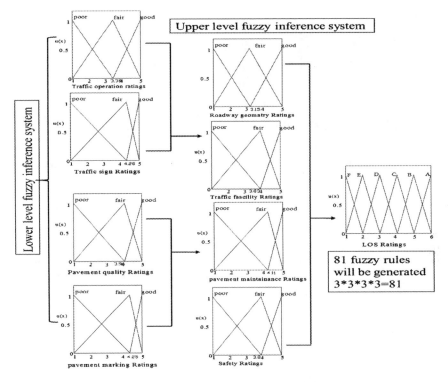

Fig. 4 Reducing number of fuzzy rules using Hierarchical fuzzy inference system

2.3 Defuzzification

It is the final step to be carried out in the fuzzy logic. Centre of gravity (or Centroid) is the commonly used method to defuzzify the continuous membership functions.

$$CG(A) = \frac{\int u_A(x).x\,dx}{u_A(x)dx} \tag{2}$$

To reflect the fuzziness of user's rating membership grades are used to assign the confidence levels. The confidence level of each LOS criteria can be found out:

$$CL_{LOS_j} = \left\{ \frac{MG_j}{\sum MG_j} \right\} * 100\,\% \tag{3}$$

where
LOS_j The jth LOS category (A–F)
CL_{LOS_j} Confidence level which LOS follows in jth category
MG_j Membership grade of jth LOS category

3 Study Area and Data Collection

To get a comprehensive method for the evaluation of LOS for various modes of transport (i.e. private as well as public) nearly about 500 responses of road users was gathered from two emerging cities of India (Fig. 5), i.e. Bhubaneswar and Vishakhapatnam, where traffic facility is portrayed by distinctive sorts of roadway features and variation in number of pedestrians, motorists and each category of vehicles. So that our survey can include every circle of transportation and people belonging to all classes can be taken into account. Hence, in this research opinions of road users regardless of age, gender, and economical class have been extracted using travelers intercept survey. This survey is a cost effective method to represent a

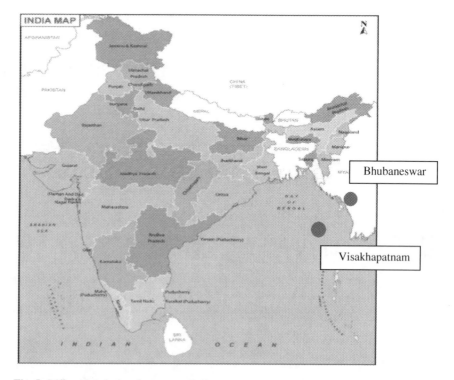

Fig. 5 Different study locations across India

wider driving population, and to collect relatively large sample size. The collected sample size is good enough to describe the discrepancy in human perception.

A number of quality of service factors have been found out which influence the satisfaction level of individual users on various transportation facilities. Based on those QOS factors a standard questionnaire was formed, which includes about 60 questions. In this survey subjects from each roadway segments from the two cities were interviewed and solicited to indicate their satisfaction level on a Likert scale ranging from 1 to 5. These satisfaction level differ from individual to individual to a great extent. This Survey carried out includes personal information such as gender, age, driving experience.

4   Results and Analysis

Case Study: Bhubaneswar (Vanivihar–Rupali square) For this segment, the values of six influencing factors are shown in Table 2. Fuzzy toolbox in MATLAB R2014a is used to build this fuzzy inference systems. In the first lower FIS TO and TS are the two input variables taken and the output is traffic facilities TF. In the second lower FIS, PQ and PM are taken as input variables and output is the PC of the road segment. The upper level fuzzy inference system is generated to deduce and evaluate the level of service (LOS) of roadways taking in account four input variables namely RG, S and the two outputs from the lower fuzzy inference systems i.e. TF and PC. The output of the fuzzy inference system i.e. LOS represents the satisfaction level of users with respect to the design and operating conditions of respective road segments.

4.1   Fuzzification

Each input variable has three descriptors (poor, fair and good) and the output variable has six descriptors namely, A, B, C, D, E and F. All the inputs RG, TF, PQM and S are fuzzified in Table 3.

Table 2 Input values of influencing factors for the road segment

Influential factors	Input values	Lower level FIS output variable	Lower level FIS output value
RG	3.15		
TO	3.38	TF	3.35
TS	4.28		
PQ	3.96	PC	3.56
PM	4.25		
S	3.8		

Table 3 Fuzzification of input variables for Vanivihar to Rupali square road segment

Input variable	Input Data	Fuzzified category	Membership grade
RG	3.15	Fair	0.72
		Good	0.28
TF	3.35	Poor	0.02
		Fair	0.98
PC	3.56	Poor	0.17
		Fair	0.83
S	3.8	Fair	0.71
		Good	0.29

4.2 Fuzzy Inference

FIS comprises of a set of "If-Then" fuzzy rules depending upon the number of input variables. The upper level FIS contains four input variables and each input having three descriptors, so the no. of rules generated is 3*3*3*3 = 81. For the fuzzified input data it is found that 29, 30, 32, 33, 38, 39, 41, 42, 56, 57, 59, 60, 65, 66, 68, and 69 numbered rules are involved in this inference system and shown in Table 4. The fuzzy inference technique using max–min composition is shown in the Table 5. After the application of Max–Min composition method, the output values for LOS A, B, C and D are found out to be 0.28, 0.71, 0.17 and 0.17 respectively.

4.3 Defuzzification

It involves center of gravity method (C.G.) which is given previously in Eq. (2). The output is defuzzified and the value of LOS rating is found out to be 4.76 for Vanivihar–Rupali square segment. This value falls largely into the category of

Table 4 Fuzzy rules using fuzzy operator

No.	If-Then Fuzzy Rules using 'and' operator
29	If (RG is Fair) and (TF is Poor) and (PC is Poor) and (S is Fair) then (LOS is Poor(E))
30	If (RG is Fair) and (TF is Poor) and (PC is Poor) and (S is Good) then (LOS is Acceptable (D))
32	If (RG is Fair) and (TF is Poor) and (PC is Fair) and (S is Fair) then (LOS is Acceptable (D))
…	…
…	…
66	If (RG is Good) and (TF is Fair) and (PC is Poor) and (S is Good) then (LOS is Fair(C))
68	If (RG is Good) and (TF is Fair) and (PC is Fair) and (S is Fair) then (LOS is Good(B))
69	If (RG is Good) and (TF is Fair) and (PC is Fair) and (S is Good) then (LOS is Very good (A))

Table 5 Fuzzy inference technique using 'Max–Min composition' method

Rule No.	Input variables				LOS	Minimum of RD, TF, PC and S
	RD	TF	PC	S		
29	F(0.72)	P(0.02)	P(0.17)	F(0.71)	E	0.02
30	F(0.72)	P(0.02)	P(0.17)	G(0.29)	D	0.02
32	F(0.72)	P(0.02)	F(0.83)	F(0.71)	D	0.02
33	F(0.72)	P(0.02)	F(0.83)	G(0.29)	C	0.02
38	F(0.72)	F(0.98)	P(0.17)	F(0.71)	D	0.17
39	F(0.72)	F(0.98)	P(0.17)	G(0.29)	C	0.17
41	F(0.72)	F(0.98)	F(0.83)	F(0.71)	B	0.71
42	F(0.72)	F(0.98)	F(0.83)	G(0.29)	B	0.29
56	G(0.28)	P(0.02)	P(0.17)	F(0.71)	E	0.02
57	G(0.28)	P(0.02)	P(0.17)	G(0.29)	D	0.02
59	G(0.28)	P(0.02)	F(0.83)	F(0.71)	D	0.02
60	G(0.28)	P(0.02)	F(0.83)	G(0.29)	C	0.02
65	G(0.28)	F(0.98)	P0.17)	F(0.71)	C	0.17
66	G(0.28)	F(0.98)	P(0.17)	G(0.29)	C	0.17
68	G(0.28)	F(0.98)	F(0.83)	F(0.71)	B	0.28
69	G(0.28)	F(0.98)	F(0.83)	G(0.29)	A	0.28

P indicates POOR, F indicates FAIR, and G indicates GOOD
Maximum of $A = 0.28$, $B = 0.71$, $C = 0.17$, $D = 0.17$

Table 6 Composite LOS for the road segments of Bhubaneswar city

Name of the Road segment	A	B	C	D	E	F	C.G.	Perceived LOS
Confidence level for each LOS category								
Vanivihar–Rupali square	21.1	53.4	12.8	12.8	0	0	4.76	B
AG square-Airport	65.8	34.2	0	0	0	0	5.02	A
Rasulgarh-Kalpana	0	19.6	19.1	19.6	21.7	20	1.33	E
Rajmahal-Bapuji nagar	32.6	67.4	0	0	0	0	4.83	B
Ram mandir-Rabindra mandap	0	2.81	45.08	13.38	25.35	13.38	3.74	C
Acharya vihar-Nicco park	2	91	5	1	1	0	4.52	B
Jayadev vihar-Vidyut marg	20.65	79.35	0	0	0	0	4.70	B
Nayapalli-CRPF NH5	43	57	0	0	0	0	4.98	B

LOS B i.e. Good. Hence, LOS B can be designated to this segment. Using Eq. (3) the confidence levels for each LOS category for this road segments are found out. The same procedure is repeated for other road segments of Bhubaneswar and Visakhapatnam city and LOS is calculated in the Tables 6 and 7.

Table 7 Composite LOS for the road segments of Visakhapatnam city

Name of the Road segment	A	B	C	D	E	F	C. G.	Perceived LOS
Confidence level for each LOS category								
VSKP station road	0	53.97	15.87	25.46	3.97	0.73	4.58	B
VSKP central mall road	0	17.28	12.57	28.27	24.08	17.8	2.69	D
Daba Garden road	9.81	84.31	1.96	1.96	1.96	0	4.7	B
KGH down road	24.24	75.76	0	0	0	0	4.8	B
RK beach road	34.83	65.17	0	0	0	0	4.98	B
RTC complex road	58.43	41.57	0	0	0	0	5.08	A
Ring road	0	7.62	63.81	9.52	9.52	9.52	3.66	C
Lower tank bund road	0	12.29	12.29	19.55	26.26	29.61	1.35	E

5   Summary and Conclusions

The basic inductions drawn from this study shows that the proposed models for homogeneous traffic flow conditions in other countries can't be applied directly to evaluate the service quality of Indian roadways due to the heterogeneity of traffic conditions and behavioural diversity of individual. Some researchers have analyzed quantitative LOS for Indian condition without considering the road user's perception, their needs and satisfaction level [10].

In this study fuzzy inference system has been implemented to figure out the complex and multiple view of road users and LOS was determination under heterogeneous traffic flow condition. Results from fuzzy inference shows that road users rate the service quality of road segments of Bhubaneswar as LOS category A, B, C and E, whereas road segments of Visakhapatnam city as A, B, C, D and F categories of LOS. This survey included every class of people regardless of age, gender, and economic background. The sample size taken is good enough to figure out behavioral diversity in human perception about the service quality provided in Indian cities.

This methodology offers unique and significant approach to define service quality provided by the transportation infrastructure and may thus conquer the constraints of conventional delay-based methods to some extent. This kind of model is new to Indian traffic condition of heterogeneity which will give fair assessment of service quality provided by the transportation infrastructure to the public. This sort of research can be further extended to evaluate comfort level of road users at grade separated interchanges, mid-block segments.

References

1. Sutaria, T.C., Haynes, J.J.: Level of service at signalized intersections. Transportation Research Record 644, TRB, National Research Council, Washington, DC, pp. 107–113 (1977)
2. Pecheux, K.K.: Quality of service and customer satisfaction on arterial streets: final report. Science Applications International Corporation, George Mason University, and Volpe National Transportation Systems Center, U.S. Department of Transportation (2003)
3. Hummer, J.E., Rouphail, N., Hughes, R.G., Fain, S.J., Toole, J.L., Patten, R.S., Schneider, R. J., Monahan, J.F., Do, A.: User perceptions of the quality of service on shared paths. Transportation Research Record: Journal of the Transportation Research Board, 1939, Washington, DC, pp. 28–36 (2005)
4. Zhang, L., Prevedouros, P.D.: Signalized intersection level of service that accounts for user perceptions. Presented at 83rd Annual Meeting of the Transportation Research Board (CD-ROM), Washington, DC (2004)
5. Flannery, A., Wochinger, K., Martin, A.: Driver assessment of service quality on urban streets. Transportation Research Record: Journal of the Transportation Research Board, 1920, Washington, DC, pp. 25–31 (2005)
6. Lee, D., Kim, T., Pietrucha, M.T.: Incorporation of transportation user perception into evaluation of service quality of signalized intersections. Transportation Research Record: Journal of the Transportation Research Board, 2027, Washington, DC, pp. 9–18 (2007)
7. Chen, X., Li, D., Ma, N., Shao, C.: Prediction of user perceptions of signalized intersection level of service based on fuzzy neural networks. Transportation Research Record: Journal of the Transportation Research Board, 2130, Washington, DC, pp. 26–34 (2009)
8. Fang, F.C., Pecheux, K.K.: Fuzzy data mining approach for quantifying signalized intersection level of services based on user perceptions. J. Transp. Eng. **135**(6), 349–358 (2009)
9. Zhang, L., Prevedouros, P.D.: User perceptions of signalised intersection level of service using fuzzy logic. Transportmetrica **7**(4), 279–296 (2011)
10. Bhuyan, P.K., Krishna Rao, K.V.: Defining level of service criteria of urban streets in Indian context. Eur. Transp. **49**, 38–52 (2011)
11. HCM "Highway Capacity Manual". Transportation Research Board, Washington, DC (2010)

Service Quality Assessment of Shared Use Road Segments: A Pedestrian Perspective

Sambit Kumar Beura, Haritha Chellapilla, Suprava Jena
and P.K. Bhuyan

Abstract At present, a suitable methodology is not available for assessment of the quality of services offered by the transportation facilities like shared use roadway segments for pedestrian use under the influence of heterogeneous traffic flow. Hence an initiative has been taken in present study to develop a new method that can be well applicable for mid-sized cities. Required data sets has been collected from a total of 60 segments of three Indian cities namely; Bhubaneswar (Odisha), Rourkela (Odisha) and Kottayam (Kerala). These segments widely vary in their traffic flow and geometric conditions. Also, both kinds of segments, i.e. with and without provision of sidewalk facilities have been considered. The quality of walking environments offered by these segments has been examined. Approximately 700 respondents differing in their age, sex, education level and driving experiences participated in perception surveys to share their satisfaction level on these segments. Ten different variables affecting the service quality significantly ($p < 0.001$) has been found out and stepwise multi-variable regression analysis has been carried out to develop a new model. Primary factors included are namely: total available width of the road, paved shoulder and sidewalk facilities in through direction; volume of pedestrians and vehicles in peak hours of traffic flow; percentage of heavy vehicles; 85th percentile speed of running motor vehicles; and interruptions created by nearby commercial activities and available public transits. Also, influences of two secondary factors namely; availability of driveways and type of separation of sidewalk facilities from the main carriageway has been examined under prevailing conditions. These influences have been reflected in the

S.K. Beura (✉) · H. Chellapilla · S. Jena · P.K. Bhuyan
Department of Civil Engineering, National Institute of Technology Rourkela, Rourkela
769008, Odisha, India
e-mail: sambit.beura@gmail.com

H. Chellapilla
e-mail: ch.haritha11@gmail.com

S. Jena
e-mail: suprava728@gmail.com

P.K. Bhuyan
e-mail: pkbtrans@gmail.com

© Springer Science+Business Media Singapore 2017
P. Deiva Sundari et al. (eds.), *Proceedings of 2nd International Conference on Intelligent Computing and Applications*, Advances in Intelligent Systems and Computing 467, DOI 10.1007/978-981-10-1645-5_55

developed model by using adjustment factors. This regression model reported high reliability in its universal application with a high correlation coefficient (R^2) of = 0.75. Also, its parameter estimates satisfied required significance criteria through several tests including t test, collinearity statistics etc. Fuzzy C-Means clustering technique has been used to define the ranges of service categories (A–F). In order to evaluate off-street pedestrian facilities, average pedestrian space (m^2/ped) has been used as the measuring parameter. The service categories predicted by the developed model for fifteen randomly selected segments has been compared with the observed ones. The high value found for exact matching (86.7 %) denotes well applicability of this developed model.

Keywords Mid-sized city · Pedestrian facility · Pedestrian level of service · Fuzzy C-Means cluster analysis

1 Introduction

Walking is one of the most essential necessities in daily life. People irrespective of their social and economic background have to walk daily to fulfill their needs. It is one of the green modes of transportation and hence, is a crucial requirement for a balanced transportation system. It is also an inevitable link in case of inter modal transfers. The significant increase in the population density of Indian cities in the last two decades has resulted in escalations of travel demand which is being leading to congestions on roads, accidents, pollution level and several other issues. Motor vehicles are being treated as a symbol of status in the rapid urbanized India. Hence, a large attention has been paid on enhancement of facilities for motorized vehicle use, while pedestrian facilities are undervalued. These facilities are improperly designed and also poorly maintained. This has resulted in huge pedestrian fatality (approximately 78 %) in last few decades. This indicates that the transportation system is not sustainable in the present context. Hence, the service quality offered by existing pedestrian facilities should be evaluated in a methodical manner to be convinced about the extent of needs for improved facilities. In this regard, this study has aimed to develop suitable methodology for the assessment of Pedestrian Level of Service (PLOS) offered by the facilities. PLOS describes the operational conditions and satisfaction level of pedestrians within the geometric and traffic flow conditions of roads.

Several researchers have developed mathematical models generally known as PLOS models for the prevailing conditions of road infrastructure in developed countries. However, those models may not be well adopted in developing countries due to differing situations. In India, the traffic flow on urban roads is highly heterogeneous which result in complex interactions among various kinds of vehicles. Hence, satisfaction level perceived by pedestrians is influenced by several types of vehicles passing them. Along with this, geometric details of existing roads are also quite different. Hence, a new model has been developed in this study that considers all major variables influencing satisfaction level of pedestrians. The model can be used to estimate the service quality (A–F) provided by the road segments of shared use by the pedestrians. Both sides of the roadway can be assessed separately. Because, the variables used in the model are only specific to one direction of the segment. In case, sidewalk facilities are not available in direction under consideration, then it is assumed that pedestrians will walk on that side, even if sidewalk facilities are available on the opposite side.

The succeeding section describes background of the study in brief. This is followed by discussions on study methodology, used data sets and procedure adopted to develop the model. The statistical significance and validation criteria of developed regression model has been tested which are discussed in details. Its performance has also been tested under field conditions. The ranges of six service categories (A–F) has been defined by using Fuzzy C-Means (FCM) clustering.

2　Literature Review

Several PLOS models have been developed for the prevailing conditions of road infrastructure in developed countries as found in the literatures. PLOS criteria have been defined both for urban segments and off-street facilities. Variables significantly affecting the perceived walking quality have been identified and have been included in the developed model. These influencing variables are summarized in Table 1. The most important variables includes: capacity of sidewalk, width of outside lane of the road in through direction and buffer area, volume of pedestrians and motorized vehicles, average speed of traffic flow, flow rate of pedestrians etc. Influences of these variables have also been examined in the current study. Additionally, behavior of pedestrians under highly mixed traffic flow conditions have been examined. Influences of those variables which are not accessed properly in previous research works (such as: presence of non-motorized vehicles and type of separation of sidewalk facilities by means of separated grade, handrails, and/or presence of buffer area) have also been studied. This study considers the evaluation of both kinds of segments, i.e. with and without provision of sidewalk facilities.

3　Methodology

This section briefly describes the criteria adopted to determine peak one hour of traffic flow, and to calculate volume of non-motorized vehicles and 85th percentile speeds of motorized vehicles. These are followed by discussions on methodology followed in regression analysis and FCM cluster analysis.

Table 1 List of variables influencing quality of walking

Ref. [...]	Variables influencing quality of walking
[1]	Walk space available to each pedestrian, speed and volume of traffic flow in peak hours
[2]	Walk space available, average walking speed, density of pedestrians in peak hours
[3]	1. Design factors (i.e. width of facilities, quality of surface, obstructions to pedestrians, available facilities for crossing the road, and support facilities); 2. Location factors (i.e. connectivity, walking environment and probable vehicular conflicts); 3. User factors (i.e. pedestrian volume, mix of road users and offered level of security)
[4]	Width of outside lane, shoulder, bicycle lane, buffer area, sidewalk, number of through lanes, traffic volume, speed of traffic flow and occupancy of on-street parking area
[5]	Width of sidewalk and its lateral separation from roadway, volume of pedestrians and their interaction with bicycles and other obstacles
[6]	Walk space available, buffer area, number of lanes, type of roadside development, presence of trees and medians, on-street parking, volume of pedestrians, motorized vehicles including mopeds, volume of bicycles, and average speed of traffic flow
[7]	Buffer area, volume of pedestrians, motorized vehicles and bicycles, and amount of driveway access
[8]	Width of sidewalk, buffer area, shoulder, bicycle lane and outside travel lane, occupancy of on-street parking area, number of lanes, volume and average speed of traffic flow
[9]	Walk space available, average walking speed, flow rate and volume to capacity ratio of pedestrians in peak hours, number of lanes, width of outside lane, bicycle lane, shoulder, sidewalk and buffer area, occupancy of on-street parking area, mid-segment demand flow rate, and average speed of traffic flow
[10]	Flow rate of pedestrians, width of sidewalk, obstructions to pedestrians, buffer area, volume of traffic in the outside lane
[11]	Walk space available, flow rate of pedestrians, average walking speed and volume to capacity ratio

3.1 Principles Followed in Collection of Required Traffic Parameters

Traffic flow on roads has been recorded using a HD video camera for two hours during morning or evening peak periods of traffic flow. Running average method has been used to estimate peak one hour. Volume of motorized vehicles (Passenger Car Units per hour or PCUs/h) has been calculated by using the equivalent PCU values suggested by IRC [12]. Volume of non-motorized vehicles has been calculated in Nos./h by multiplying four with number all non-motorized vehicles except bicycles and then adding the outcome with number of bicycles. IRC recommends equivalent PCU values for the non-motorized modes like cycle rickshaw, trolley and animal drawn vehicles etc. as approximately four times that of a bicycle. Hence four has been multiplied to equate the interruptions caused by these kind of vehicles relative to those by bicycles. Spot speed data has been collected by using radar guns. Here it can be mentioned that, these equipment should be set to approaching mode and be placed at an angle of nearly 135° to the direction of traffic flow for better accuracy in collected speeds of vehicles. Speed of at least 130 motor vehicles are collected from every segment. 85th percentile speed has been calculated from graphs plotted for cumulative percentage frequency and mid-value of spot speed ranges. Various geometric measurements has been carried out by using measuring tapes. Perception surveys have been conducted (1) to determine the overall PLOS scores of segments, and (2) to estimate adjustment factors. These are discussed later in details.

Average space available for pedestrian use inside sidewalk facilities has been observed by measuring a sample area of the sidewalk facility in the field and then determining the maximum number of pedestrians in that particular areata a given time instant. Mathematical expressions used for this purpose are shown in Eqs. (1–3).

$$A_{Ped} = 60 * (WS/V_{Ped}) \qquad (1)$$

$$V_{Ped} = V_P/(SW_E * 60) \qquad (2)$$

$$SW_{Eff} = SW_{Tot} - SW_0 \qquad (3)$$

where: A_{Ped} = average space available for pedestrians (m²/ped), WS = average walking speed (m/s), V_{Ped} = pedestrian flow divided by width of sidewalk (ped/m/min), V_P = flow of pedestrians/h during peak hours (ped/h), SW_{Eff} = effective width of sidewalk facility (m), SW_{Tot} = total width of sidewalk facility (m), and SW_0 = summation effective widths (m) occupied by fixed-objects plus shy distances plus the average width occupied by street vendors as explained in HCM 2010 [9].

3.2 Step-Wise Regression Analysis

The regression model has developed in four different steps such as: (1) identification of the variables significantly ($p < 0.001$) influencing the quality of walking or overall PLOS scores, (2) development of numerous combined or transformed forms of these variables and subsequently statistical test of their coefficient estimates, (3) selection of the most suitable model which fulfills the required significance criteria, and (4) development of adjustment factors.

3.3 Fuzzy C-Means (FCM) Cluster Algorithm

Fuzzy C-means (FCM) clustering method is developed by Dunn [13] and further improved by Bezdek [14]. It assigns every data point with a membership function whose value ranges between 0 and 1. The algorithm for this method introduced by Bezdek [14] issued in this study because of its high popularity in the field of classification problems. It is considered as an efficient and accurate algorithm used in clustering techniques [15]. Onem × nmatrix symbolized as U = $[\mu_{ik}]$ signifies the fuzzy partitions, with three boundary conditions shown in Eqs. (4a–4c).

$$\mu_{ik}\epsilon[0,1], for\ 1 \leq i \leq m\ and\ 1 \leq k \leq n \tag{4a}$$

$$\sum_{k=1}^{n} \mu_{ik} = 1,\ for\ 1 \leq i \leq m \tag{4b}$$

$$0 < \sum_{i=1}^{m} \mu_{ik} < m,\ for\ 1 \leq k \leq n \tag{4c}$$

where: 'm' is the number of observations; 'n' is the number of clusters.

This cluster algorithm is based on the minimization of an objective function or *C-means* functional. This is defined by Dunn [13] as shown in Eq. (5).

$$J(X; U, V) = \sum_{i=1}^{n} \sum_{k=1}^{m} \mu_{ik}^{w} \|X_k - V_i\|_A^2 \tag{5}$$

where: $V = [V_1, V_2, V_3, ..., V_n], V_i \epsilon R^m$ is a vector of cluster centers, which can be determined.

Where: 'X' represents the data set; 'U' represents the partition matrix; 'V' represents the vector of cluster centers, 'w' represents weight exponent which determines the fuzziness of clusters, whose defaulting value is 'two'; V_i is the mean for those data points over cluster i.

4 Study Area

The essential criteria followed in selecting study sites is to include all possible category of roadway segments which widely vary from excellent to worst in their quality of service offered for pedestrian use. Required data sets has been collected from three Indian midsized cities namely, Bhubaneswar (Capital of Odisha), Rourkela (Steel city of Odisha) and Kottayam (a city in Kerala). The traffic flow on roads of these cities are highly heterogeneous. A total of 60 segments has been considered including 29 segments from Bhubaneswar, 19 segments from Rourkela and 12 segments from Kottayam. These segments well represent the desired variability in geometric and traffic flow conditions of roadway segments in mid-sized cities. Substantial variations in geometric and operational characteristics of road segments (in peak one hour) considered are as follows: (1) number of through lanes: 1–4 (Nos.), (2) width of roadway in the direction under consideration: 3–14 m, (3) width of sidewalk facility: 1–3 m, (4) width of buffer area: 0–4.5 m, (5) separation of sidewalk facility from the main carriageway: by provision of separated grade, buffer area and/or handrails, (6) volume of pedestrians: 35–490 (Nos./h), (7) traffic volume: 361-5741 (PCUs/h), (8) 85th percentile speed of traffic flow: 23–50 km/h, (9) percentage age of heavy vehicles: 0.1–7 %, (10) extent of nearby commercial activities: very high—negligible, and (11) interruption occurring by public transits: very high—negligible.

5 Perception Survey

This section describes about the perception surveys those have been conducted in two phases as per the requirement.

5.1 Phase 1

Perceived or observed PLOS scores of segments has been determined by conducting perception surveys. This includes both roadside interviews and video surveys. The survey participants represents a good cross-section of age, sex, education and driving experiences. Children (below age of 14 years) have been excluded. The distribution sin age of participants are as: 5 % for age of below 20 years, 38 % for 21–30 years, 26 % for 31–40 years, 17 % for 40–50 years, 10 % for 51–60 years and 4 % for age above 60 years. Gender split of partisans also good which is approximately 44 % female and 56 % male.

For conducting roadside interviews, appropriate questionnaire are prepared. This questionnaire includes questions on several major factors those affect walking quality of pedestrians. These major attributes are: overall width of roadway, width of sidewalk facilities, separation of sidewalk facilities from main carriageway, perceived safety, usual congestion on the road, and dis-comfort caused due to presence of pedestrians, motorized and non-motorized vehicles, presence of public transits, nearby commercial activities, on-street parking, driveways access and presence of auto stop(s) or bus stop(s). Participants are asked to give an overall ratings on each factor according to their own satisfaction level. This data are collected on a 5-point scale, which varies from 1 (very much satisfied/agree with) to 5 (very much unsatisfied/disagree with). Approximately 10 participants per each segment have been participated in these interviews which has resulted in approximately 600 responses. To access the real-time response of another 100 users, video survey has been conducted in three sessions. In this survey several persons including students, faculty members and co-workers of National Institute of Technology Rourkela are participated. Representative video clip of each segment with an average time duration of 30–45 s has been shown to the survey participants by a professional video projector and on a wide screen. While showing the video clips, the volume of connected audio boxed is adjusted to imitate the sound generally made by real traffic flow. Questionnaire are prepared to access the response of participants which include several influencing factors those are visible in the video clips. These factors includes: overall width of roadway and sidewalk facilities, separation of sidewalk facilities from the main carriageway, perceived comfort, usual congestion on the road, and dis-comfort caused due to presence of pedestrians, motorized and non-motorized vehicles, presence of public transits and on-street parking area. Finally, overall ratings or perceived PLOS scores of segments are calculated using these data collected from above discussed surveys. These rating are obtained on a 5-point liker scale where, 1 indicates excellent PLOS and 5 indicates worst PLOS. In this way, approximately 110 scores are obtained for every segment and subsequently perceived PLOS score is calculated by doing the average of all ratings.

5.2   Phase 2

Type-2 surveys are required for estimating Adjustment Factors (AF) those are discussed in detail. The survey participants stated their opinion on two issues: (1) with how many numbers driveways (each with at least 200 vehicles/h ingress-egress in peak hours) per kilometer length of a segment they start to feel extreme discomfort, and (2) what should be the minimum height of handrails and minimum of width of the buffer area so that they will feel better comfort.

6   Analyses and Model Development

In the first step of statistical analyses, all sixty roadway segments are sorted with respect to their overall PLOS scores. From this data base, 45 (75 % of total) segments are selected randomly. These data have been used to develop the PLOS model. Remaining data are used for other purposes including model validation, model comparison and its performance test. This regression model is more instinctive in practice. Non-modelers also easily understand the sensitivity of it. The variables accommodated by the model can also be understood by the public officials, engineers and transportation planners easily. Also, new variables can be included in it further according to the changes in operational characteristics may occur over time period.

Pearson correlation analysis has been carried out on the variables listed in part 'A' of Table 2 with respect to overall PLOS scores to identify the relevant variables (with $p < 0.001$). From the extensive array of geometric and traffic flow variables collected from the study sites, ten variables have been found to be significant.

Attempt has been made to develop a suitable PLOS model using these variables. Statistical significance of its coefficient estimates has been tested in the developed model using t test. As linear regression model failed to accommodate all variables with significant coefficient estimates, several transformations (e.g. square, square root, inverse, exponential, logarithmic etc.) and/or combinations of the variables have been developed and step-wise regression has been carried out. In this process several models are rejected which could not satisfy required significance criteria and the best-fit PLOS model has been reported. In the best fit PLOS model, following four forms of the variables has been found with significant ($p < 0.05$) coefficient estimates (t-statistics): (1) $V_1 = ln(PHMV/RW)$, (2) $V_2 = ln(SW_{Eff} + SH)$, (3) $V_3 = ECA * |(35 - S_{85}) + 1|$, and (4) $V_4 = IBPT * ln[(1 + PV + PHNV) * (1 + \%HV)]$. These nomenclatures are shown in Table 2; and the developed PLOS model is shown in Table 3.

Satisfaction levels of road users decrease with increase in congestions on the roads. Under congested traffic flow conditions, 85th percentile speed (S_{85}) of motorized traffic flow achieves an extremely lower value. This kind of congested environment degrades the walking quality of pedestrians. PLOS increases with increase in S_{85} under such situations. But after a certain limit, further increase in speed creates the feelings of an unsafe environment. So to find this limit in the whole range of S_{85} values (varying from 23 to 50 km/h in study corridors) up to which PLOS increases with increase in S_{85} and then decreases with further increase in S_{85}, two graphs are plotted as shown in Fig. 1. From these two graphs this limit has been fixed as 35 km/h. Hence, a new variable has been developed as $|(35 - SPD)|$.

In the data analysis process it has been observed that, driveway access and separation of sidewalk from the main carriageway have significant influence on satisfaction level of pedestrians. Hence, perception surveys have been conducted to assess the influences of these two factors as described below. The sum of these influences have been termed as Adjustment Factor (AF) of the developed PLOS model.

1. *Adjustment factors for frequency of driveways (AF_D)* The ingress and/or egress of vehicles to driveways adversely affect the perceived satisfaction levels of pedestrians. From phase 2 of perception survey it has been observed that, pedestrians start to feel extreme dis-comfort if frequency of drives is at least three, provided each driveway has at least 200 vehicles/h ingress-egress in the peak hours. The frequency of driveways stands for average number of this facility per kilometer length of a segment. Also, it has been further observed that the averaged overall PLOS scores of road segments with provision of three such driveways is larger than PLOS scores of other segments with similar geometric and traffic conditions and provision of no driveways by an approximate value of 0.45. Values of F_D corresponding to one and two driveway facilities are shown in Table 3.

2. *Adjustment factors for separation of sidewalk facilities from main carriageway (AF_S)* Physical separation of sidewalk facilities from the main carriageway adds to the perceived satisfaction levels of pedestrians. Numerical values of these influences varies with the type of separation. These values have been derived from phase 2 perception surveys as represented in Table 3.

Table 2 List of variables collected from study corridors

Var. no.	Variable	Abbreviation	Units/values	Remark
Part A				
1	Pedestrian level of service score	$PLOS_{Score}$	5-point (1–5) scaling system	1 represents Yes, and 0 represents No
2	Number of traffic lanes in the direction under consideration	–	Number(s)	
3	Width of the roadway in the direction under consideration	RW	Meters	
4	Presence of sidewalk facilities	–	1 or 0	
5	Sidewalk facilities if separated	–	1 or 0	
6	Effective width of sidewalk facilities (if present)	SW_{Eff}	Meters	
7	Height of handrails used in sidewalk facilities	–	Meters	
8	Width of shoulder (if present) excluding average widths occupied by on-street vendors, illegal parking and other obstructions	WS	Meters	
9	Presence of gutter pan	–	1 or 0	
10	Width of gutter pan (if present)	–	Meters	
11	Presence of kerb	–	1 or 0	
12	Width of available buffer area	–	Meters	
13	Extent of commercial activities nearby	ECA	2 (high), 1.5 (medium), and 1 (minimal)	
14	Number of driveways	–	Number(s)	
15	Presence of on-street parking area	–	1 or 0	
16	Peak hour motorized vehicular volume	PHMV	PCUs/h	
17	Peak hour non-motorized vehicular volume	PHNV	Number(s)/h	
18	Peak hour pedestrian volume	PV	Number(s)/h	
19	85th percentile speed of motorized traffic flow	S_{85}	km/h	

(continued)

Table 2 (continued)

Var. no.	Variable	Abbreviation	Units/values	Remark
20	Percentage of heavy vehicles	%HV	%	
21	Interruptions by public transits	IBPT	2 (presence of auto/bus stop plus frequent stopping public transits), 1.5 (one from the criteria exists), 1 (none of these criteria exists)	
Part B				
22	Number of driveways per kilometer length of segment (each with at least 200 vehicles/h ingress-egress in peak hours)	–	Number(s)	
23	Separation of sidewalk facilities from main carriageway	–	Different grade, buffer area and/or handrails	

Table 3 Equation box

$$PLOS_{Score} = -0.650 + 0.5436 * \ln(PHMV/RW) - 0.7024 * \ln(SW_{Eff} + WS) + 0.0196 * ECA * |(35 - SPD)| + 1)$$
$$+ 0.0664 * IBPT * \ln[(1 + PV + PHNV)(1 + \%HV)] + AF$$

$NMVV = Number\ of\ bicycles + (4 * Number\ of\ non - motorized\ vehicles\ other\ than\ bicycles)$

$AF = AF_D - AF_S$

Where, PLOS_{Score}, PHMV, PHMV. RW, SW_{Eff}, WS, ECA, S_{85}, IBPT, PHNV, PV, and %HV represents the variables as explained in Table 2	AF = adjustment factors, AF_D = adjustment factors for frequency of driveways, and AF_S = adjustment factors for separation of sidewalk facilities from main carriageway

Adjustment factors

Frequency of driveways those have at least 200 numbers of vehicles/h ingress-egress in peak hours of traffic flow	Factor 'AF_D'	Separation of sidewalk facilities from main carriageway	Factor 'AF_S'
≥3	0.45	1. Different grades	0.15
2	0.30	2. Handrails of minimum 0.5 m	0.35
1	0.15	height (grade separation may also	0.5
		be present)	0.6
		3. At grade sidewalk facilities	0.7
		separated by provision of a buffer	
		area (minimum 1 m width); but	
		without provision of handrails	
		4. Wide buffer area and separate	
		grade; but no provision of	
		handrails	
		5. Provision of a wide buffer area,	
		handrails and separate grades	

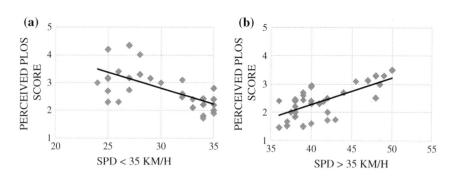

Fig. 1 Plotted graphs for observed PLOS scored and 85th percentile speeds of motorized traffic flow (S_{85}), **a** $S_{85} < 35$ km/h, **b** $S_{85} > 35$

The PLOS scores predicted by developed model has been classified into six categories by using FCM cluster technique. MATLAB has been used for executing this cluster program. Numbers of service categories has been chosen as six, i.e. 'A' through 'F' to remain consistent with the previous research works. Figure 2 shows

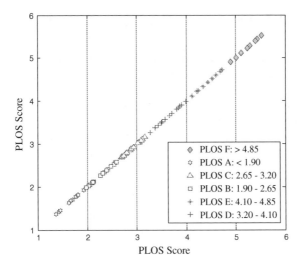

Fig. 2 Ranges of PLOS categories (A–F) defined by using FCM cluster analysis

Table 4 Ranges of PLOS Categories

PLOS score	PLOS scores corresponding to average space available for pedestrians (m²/ped)					
	>16.53	>13.06–16.53	>9.91–13.06	>7.25–9.91	>4.48–7.25	≤4.48
≤1.90	A	B	C	D	E	F
>1.90–2.65	B	B	C	D	E	F
>2.65–3.20	C	C	C	D	E	F
>3.20–4.10	D	D	D	D	E	F
>4.10–4.85	E	E	E	E	E	F
>4.85	F	F	F	F	F	F

the plot of classified scores as obtained by using the cluster analysis. Average silhouette width (ASW) has been evaluated to recognize the strength of cluster structure. It can be noted here that, an average ASW value in the range of 0.71–1, 0.51–0.70, 0.26–0.50 and 0.25 or less respectively indicates a strong, reasonable, weak and no substantial structure [16]. FCM cluster analysis on PLOS scores in present study has produced a strong structure with an average ASW value of 0.765.

In order to evaluate off-street pedestrian facilities, average pedestrian space (m²/ped) has been used as the measuring parameter. Ranges of PLOS defined in [11] for midsized cities has been adopted as shown in 2nd row of Table 4. The first column shows the ranges defined by using FCM clustering. In the absence of sidewalk facilities only this first column shall be used to determine ranges of PLOS as pedestrian space concept does not apply.

7 Statistical Significance of Developed PLOS Model

The developed PLOS model satisfied several statistical significance criteria with the data sets using which it has been developed. The input variables in any model should not be highly correlated. This criteria has been checked by using Pearson correlation analysis. In this test, the highest correlation (r) value has been found as 0.59. Hence the input variables (V_1, V_2, V_3, and V_4) in developed PLOS model are not highly correlated. This model has produced a high correlation coefficient (R^2) of 0.75 with average observations. The model has also produced an adjusted R^2 value of 0.73. This small difference (0.03 = 3 %) denotes that, if this developed model has been derived from whole population instead of a sample of 60 segments, then it would have accounted for only three percent of less variance in its prediction. The value of F-ratio (i.e. 29.871) has been determined by using Anova test. This ratio has been found to be significant at $p < 0.001$ level. This indicates that the PLOS model has been fitted significantly to the data set. Table 5 shows the coefficient estimates of input variables and other statistics. It can be observed that the coefficients are associated with minor standard errors. The t-statistics are also significant at $p < 0.05$ level. Here it can be mentioned that, a partial correlation of 0.5 or more indicates high correlations among input variables of a regression model. The input variables of developed PLOS model have very good correlations with observed PLOS scores. Collinearity statistics including Tolerance and Variance Inflation Factor (VIF) are the measures for whether there exist collinearity in the data. A tolerance value (<0.2) designates a potential problem [17]. If the largest value of VIF is greater than 10 then there is a reason for apprehension [18]. Fortunately, all tolerance values are far above 0.2 and the maximum VIF (1.261) is far below 10 for the inputs of developed model. The regression analysis is not biased which has been convinced by the average value of VIFs (1.2) which is not largely greater than 1. So collinearity is not a problem in this study.

Table 5 Coefficient estimates of input variables and other statistics

	Coefficient estimates	Standard error (SE)	t-statistics	Significance (p value)	Partial correlation	Collinearity statistics	
						Tolerance	VIF
Constant	−0.6502	0.5841	–	–	–	–	–
V_1	0.5436	0.1120	4.853	< 0.001	0.6087	0.798	1.253
V_2	−0.7024	0.1006	−6.982	< 0.001	−0.7411	0.842	1.188
V_3	0.0196	0.0085	2.306	0.026	0.3424	0.793	1.261
V_4	0.0664	0.0195	3.412	<0.001	0.47474	0.813	1.231

8 Validation and Comparison of Developed PLOS Model with Existing Models

Traffic flow and geometric data of fifteen randomly selected segments have been utilized for the validation check of developed PLOS model. Expected PLOS scores of these segments have been estimated by using the developed model. The trend line with a slope of 45.2° obtained in the graph plotted for observed and these predicted scores indicates that the model is well validated. Attempt has been made to examine the performance of this developed model with compared to other existing Pedestrian LOS models followed in developed countries. Pedestrian LOS models developed by HCM 2010 [9] and Landis et al. [4] have been chosen for this comparison purposes. These three models have been employed to predict the service categories of these fifteen roads. Predicted PLOS have been compared with observed ones. In this test, the exact match for developed PLOS model, HCM (2010) PLOS model and Landis et al. PLOS model are found as 86.70, 60.00 and 33.33 % respectively. Therefore, it is well convinced that the developed model is capable of producing better results than other models under heterogeneous traffic flows.

9 Summary and Conclusions

The traffic flow on urban roads in developing countries is highly heterogeneous. At present, a suitable methodology is not available for assessment of the quality of services offered by the transportation facilities like shared use roadway segments for pedestrian use under the influence of this highly mixed type of traffic flows. Hence a new PLOS model has been developed in this study that can be used for this purpose under the prevailing conditions of roads in midsized cities. Lower PLOS score signifies better quality of service which is an assumption in this study. The primary factors affecting walking quality of pedestrians such as: total available width of the road, paved shoulder and sidewalk facilities in through direction; volume of pedestrians and vehicles in peak hours of traffic flow; percentage of heavy vehicles; 85th percentile speed of running motor vehicles; and interruptions created by nearby commercial activities and available public transits. Influences of two other supplementary factors such as: frequency of driveways and separation of sidewalk facilities from main carriage has also been found to have considerable influence satisfaction levels of pedestrians. The former supplementary factor has increased overall PLOS scores by a maximum value of roughly 0.45; while the later factor has decreased the scores by roughly 0.7. It has been observed that, under mixed traffic flow conditions satisfaction levels of pedestrians in peak hours increases with the increase in 85th percentile speed of motorized traffic up to 35 km/h. But after this limit, further increase in speed adversely affect the satisfaction levels. This phenomenon occurs due to the fact that extreme lower speed of traffic flow causes congestion on the segment which leads to dis-satisfaction of pedestrians. On contrary, higher speed of traffic flow results a psychology of an unsafe walking environment.

The PLOS model developed by using stepwise multi-variable regression analysis has reported its reliability in universal applications. The model has got a high correlation coefficient ($R^2 = 0.75$) with averaged observations. The trend line with a slope of 45.20 obtained in the graph plotted for observed and predicted PLOS scores indicates that the model is well validated. PLOS scores predicted by the model for each segment has been classified into six categories (A–F) by using Fuzzy C-Means (FCM) clustering technique. Service categories predicted by the developed model for fifteen randomly selected segments has been compared with the observed ones. The high value found for exact matching (86.7 %) denotes well applicability of this developed model. Hence with these statistical attestations, it has been concluded that the developed PLOS model is well applicable for road segments in midsized cities under the influence of highly mixed traffic flow conditions. This model evaluates the existing roadway facilities with respect to pedestrians' perceived satisfaction levels. This method may help the transportation planners, engineers, and other policy makers in recognizing the need for further improvements of existing pedestrian facilities. Hence decisions may be taken about provision of better facilities on existing road infrastructure from a pedestrian's perspective.

References

1. Fruin, J.J.: Pedestrian Planning and Design. Metropolitan Associations of Urban Designers and Environmental Planners, New York (1971)
2. Tanaboriboon, Y., Guyano, J.A.: Level-of-service standards for pedestrian facilities in Bangkok: a case study. ITE J. **59**(11), 39–41 (1989)
3. Gallin, N.: Pedestrian friendliness—guidelines for assessing pedestrian level of service. In: Paper Presented at the Australia: Walking the 21st Century Conference, Perth, Australia (2001)
4. Landis, B.W., Vattikuti, V.R., Ottenberg, R.M., McLeod, D.S., Guttenplan, M.: Modeling the Roadside Walking Environment: A Pedestrian Level of Service. Transportation Research Record, pp. 82–88, No. 1773(1), Washington, DC (2001)
5. Muraleetharan, T., Adachi, T., Hagiwara, T., Kagaya, S.: Method to determine overall level-of-service of pedestrian walkways based on total utility value. J. Infrastruct. Plan. Manag. Jpn. Soc. Civ. Eng. (JSCE) **22**(3), 685–693 (2005)
6. Jensen, S.U.: Pedestrian and bicycle level of service on roadway segments. Journal of the Transportation Research Board 2031, pp. 43–51, Washington, DC (2007)
7. Tan, D., Wang, W., Lu, J., Bian, Y.: Research on methods of assessing pedestrian level of service for sidewalk. J. Transp. Syst. Eng. Inf. Technol. **7**(5), 74–79 (2007)
8. FDOT: Quality/Level of Service. Department of Transportation, Tallahassee, FL (2009)
9. Highway Capacity Manual. Transportation Research Board, Washington, DC (2010)
10. Vedagiri, P., Anithottam, D.D.: Pedestrian flow characteristics and level of service for sidewalks in mixed traffic condition. In: Presented at the Annual Meeting of Transportation Research Board, Washington, DC (2012)

11. Sahani, R., Bhuyan, P.K.: Pedestrian level of service criteria for urban off-street facilities in mid-sized cities. Transport (2014). doi:10.3846/16484142.2014.944210
12. Guidelines for Capacity of Urban Roads in Plain Areas. Indian Road Congress, IRC: 106, New Delhi (1990)
13. Dunn, J.C.: A fuzzy relative of the ISODATA process and its use in detecting compact well-separated clusters. J. Cybern. **3**(3), 32–57 (1974)
14. Bezdek, J.C.: Pattern Recognition with Fuzzy Objective Function Algorithm. Plenum, New York (1981)
15. Jain, A.K., Dubes, R.C.: Algorithms for Clustering Data. Prentice Hall, Englewood Cliffs (1988)
16. Rousseeuw, P.J.: Silhouettes: a graphical aid to the interpretation and validation of cluster analysis. J. Comput. Appl. Math. **20**, 53–65 (1987)
17. Menard, S.: Applied Logistic Regression Analysis. Sage University paper series on quantitative applications in the social sciences, no. 07-106. Thousand Oaks, CA (1995)
18. Bowerman, B.L., O'Connell, R.T.: Linear Statistical Models: An Applied Approach, 2nd edn. Brooks/Cole, Belmont, CA (1990)

Single Axis Tracking to Enhance Power from Solar Photovoltaic Panel

K. Saravanan and C. Sharmeela

Abstract As the electrical power is inevitable for all kind of work, the society concentrates on generation of electrical power from various modes. Power generation from fossil fuel is the traditional technology and it should be obsolete since it has many disadvantages like more cost, carbon emission and transport cost. Due to the above limitations the end user pays more cost as the generation cost per unit become more. Another biggest danger of using fossil fuel is raising heat of earth day by day and using fossil fuel is environmentally not viable. So our ruthless shunt should be of what the alternate for fossil fuel is and how to make lesser the unit cost of electricity. The solution is encouraging renewable energy [1–5]. Among the renewable energy, solar is most prominent one. It is increasingly urgent to find energy alternatives that are sustainable [6–8] as well as safe for environment and humanity. It is evident for utilizing the available solar energy is inevitable at this stage. Hence we are using Single axis tracking for utilizing the maximum available power from solar panel. The restructuring of the electric power is a serious move at present. Stake holders, facilitating their wide contribution in the manufacture distribution, how effectively uses the generated power. The restructuring is facilitating integration of internet technology in various aspects. As everyone aware the importance of shortage of power, people mover towards saving of power and generation of power for their need using solar panel. As the government also encourages by giving subsidy, more people come forward to install solar panels for their need. However as the position of the sun changes from time to time, it is advisable to implement a single axis tracker to minimize the cosine loss. In this way enhancing of 15 % power is possible and experimentally proven also.

Keywords Tracking · Solar PV · Cosine angle

K. Saravanan (✉)
SRM University, Chennai, India
e-mail: Saravanank96@gmail.com

C. Sharmeela
Anna University, Chennai, India
e-mail: Sharmeela20@yahoo.com

© Springer Science+Business Media Singapore 2017
P. Deiva Sundari et al. (eds.), *Proceedings of 2nd International Conference on Intelligent Computing and Applications*, Advances in Intelligent Systems and Computing 467, DOI 10.1007/978-981-10-1645-5_56

1 Introduction

The smart grid has further offered alternatives to participants looking to enhance the
reliability, sustainability, and capability to customer choices in energy schemes.
The power evacuation is one of the major issue due to lack of transmission system.
Because power will be produced in one area, the end user is in another area. When
you transmit the power, one problem is the transmission loss and the other problem
is the transmission system is not supporting the generated power to evacuate. To
address this problem it is mandatory to introduce smart grid concept. The solar
power generation supports for smart grid as the investment is smaller compared
with the other renewable resources. Though the smart grid and solar duo to each
other it is advisable to place a single axis mechanical tracking [9–13] to mitigate the
losses. In the single axis tracking the investment [14–16] also less. It keep tracks the
position of the sun from morning to evening. Using this mechanical tracking, the
cosine angle maintains 90°. So that the loss becomes zero as for as the cosine angle
is concerned. But still there would be few losses like dust loss, heat loss etc.

2 Solar Energy

As discussed in the introduction, the solar energy is the most promising one for the
generation of electrical power. In the incoming solar radiation as a average of 6 h
daily. If you assumes that 800 W per meter square. The total power generated is
4.8 kW. The Fig. 1 explains clearly about the position of the sun and the location
with timing.

According to International Energy Agency, solar is the biggest source of electricity than others. By reducing the use of fossil fuels, the emission of carbon to the
environment become lessor. Electrical engineers trying to utilize the incoming solar

Fig. 1 Solar illusion angle

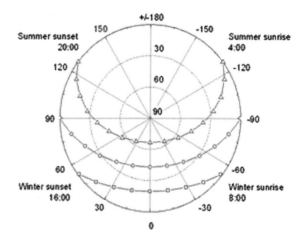

Fig. 2 Green house
emission from electricity
production

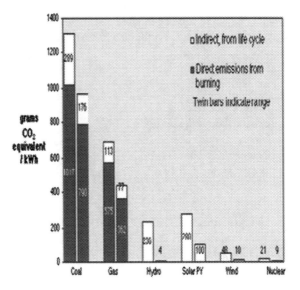

insolation as electrical power and the mechanical engineers trying to utilizes the
incoming solar radiation as a thermal power.

In general when the production is more and the cost will be less. In the same
way, recent years the production of solar panel is becoming high, in turns the cost
of the panel coming closer to lesser. The diagram (Fig. 2) shows clearly the
emission of carbon to the environment when we uses the fossil fuels. The country is
gearing up for a solar revolution as the India raised the target of solar electricity
generation from 20,000 to 100,000 MW, to be achieved by 2022. Set a cost of $100
billion, this ambitious project aims at encouraging individual and companies to
switch over to renewable energy sources. While many home owners have now
understood the benefits of solar technology, and have installed solar panel based
system in their homes, experts say that unless the government doesn't encourage
the commercial sector and developers in particular, individual efforts don't really
serve the purpose.

The altitude, azimuth angles were calculated from various values of declination
angle (δ), hour angle (ω), local solar time (LST) and equation of time (EOT) using
Eqs. (1) and (2).

$$Cos\,\theta = Sin\,\emptyset \; * \; Sin\,\delta + Cos\,\emptyset \; * \; Cos\,\delta \; * \; Cos\,\omega \qquad (1)$$

$$Cos\,A = (Sin(\delta) \; * \; Cos(\emptyset)) - (Cos(\delta) \; * \; Cos(\omega) \; * \; Sin(\emptyset))/(Cos(90 - \theta)) \quad (2)$$

In order to track the sun's ray to improve energy yield from solar panel, location
of sun at the particular period at the particular location is needed exactly. Sun in sky
found calculating various and these angles are programmed in a microcontroller
(ATMEL) which provides the location of sun for aligning PV facing in both axes.

In order to achieve this, a mounting structure for PV panel is designed which rotates around its axis compensating azimuth angle and also tilts the panel compensating altitude angle of the sun.

The function of tracker can be divided into

- Techniques of Tracker Mount
- Methods of Drives
- Sensor Controller
- Motor Controller
- Tracker Algorithm
- Data Gaining Card

3 Methods of Tracker Mount

3.1 Construction of Tracker

The Solar tracker is constructed for 100 Wp (watt peak) solar panel which is of dimension 655 × 600 mm. The four supporting legs as shown in Fig. 5 are given to the tracker to withstand the weight of the system. The solar panel is bolted at 1.2 m height from the ground to ensure safe loading and unloading of the panel into the tracker. The mounting structure is fabricated using mild steel to withstand the PV panel weight (5.3 kg) as well as the admissible wind load. The two gear box of ratio 30:1 and 15:1 each one is coupled with motor so that the speed of both the axis is reduced in the ratio of 450:1. The mounting structure is controlled in both the axis by 0.5 to 1 degree of accuracy. The permanent magnet Wiper motor is selected as DC motor drive since it is readily available and also easily suitable for solar tracking purposes. The time and date are set exactly and left unchanged before initializing the tracker program. Microcontroller first gets time from RTC and computes sun angle at that time when it is a daytime. It then checks the tilt and yaw angle of the panel from IMU. Microcontroller then activates both the motors until they tilt and yaw angle is equal to altitude and azimuth angle respectively.

These are as follows:

(A) Mechanical system design;
(B) Electrical circuit design.

4 Hardware Description

(A) Mechanical System Design. Assembling the mechanical system was the most challenging part of this system because the objective was to make an energy efficient solar tracking system which demanded intelligent operations of the tracking

motors. Generally one of these motors is used for daily tracking (east-west motion) and the daily tracking motor operates continuously based on preprogrammed microprocessor with configuration of arduino where the relay system which start motion tracking motor operates only a few times over the year. So for design and implementation process the whole mechanical system is mainly divided into three parts as follows:

(1) Permanent magnet DC motor drive
(2) Panel carrier;
(3) Panel carrier rotator.

(1) *Permanent magnet DC motor drive: The vertical motion* used for creating the seasonal angle of the sun. In this tracking system linear actuator consists of one PMDC motor, screw thread, bolt, bearing, circular rod, and Metal plate support. Experimentally it is found that this mechanical structure has a special feature of high weight lifting using allow power stepper motor. Linear motion in vertical axis (upward and downward) and is connected to one end of panel carrier and attached to the metal frame. There are some bolts and these are tied with seven 15-inch long circular rods of 2 mm diameter. There is also a 13-inch long screw thread and its diameter is 6 mm. A bolt is attached in the middle of the metal frame and this bolt is also tied with the screw thread. Four circular rods are also mortised through the wooden frame. The wooden frame moves up and down along with the bolt and the single rod hook. It works in such a way that the wooden frame does not let the bolt move along with the thread screw rather when the thread screw moves then the four circular rods mortised into the wooden frame cause the bolt to move up or down. Now when the single rod hook moves upward or downward it moves along with the panel carrier. The two ends of screw thread are placed in two bearings which helps it to rotate smoothly. These bearings are mortised into the roof and floor. One gear is also placed at the bottom of the screw thread and this gear is connected to the stepper motor gear.

Working

(2) *Panel Carrier.* Panel carrier is basically a rectangular frame made of aluminum which holds the solar panel with the help of a circular rod. One end of the horizontal base of the panel Carrier is attached with the single rod hook of linear actuator and other with the panel carrier rotator. Figure 3 shows the design and implementation of panel carrier. A PMDC motor with a gear is placed on the body of the aluminum frame.

When the PMDC motor rotates along with its gear then the panel rotates from east to west by tracking sun's daily motion actively at the two ends of solar panel. Again the rectangular aluminum frame has a rectangular mortise in its horizontal base. Single circular rod hook from linear actuator goes through this mortise. Thus it helps to lift the panel carrier in a semi-circular path to get sun's tilt angle caused

Fig. 3 Construction of motor

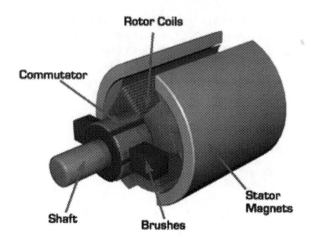

by seasonal/annual motion. While the linear actuator lifts one end of panel carrier the other end needs to be fixed with a panel carrier rotator to get the perfect circular motion.

(3) *Panel Carrier Rotator.* Panel carrier rotator is used to hold one end of the horizontal base of the solar panel carrier. One screw thread, gear, and position sensors are used in this panel carrier rotator to give a circular movement to the panel carrier. Its base is fixed on a wooden floor. Figure 5 shows the design and implementation of panel carrier rotator and experimental setup of the single axis solar tracker.

5 Worm Gear

Worm gears one which gears in the range of 20:1 as shown in Fig.4. It is the wonderful logic particularly in the single axis tracking.

Electrical Circuit Design The whole electrical system is mainly divided into three units. These are sensor unit, control unit, and movement adjustment unit. Sensor unit senses three different parameters (light, time, and position) and converts it to appropriate electrical signals. Then the electrical signals from sensor unit are sent to control unit. Control unit determines the direction of the movement of the motors both in the horizontal and vertical axes. Finally the movement adjustment unit adjusts the position of the solar module by receiving signal from the control unit. This adjustment is done by using two geared PMDC motors. Where the following configuration with the motor to sync the connectivity towards the panel are:

(A) Real time clock
(B) Position sensor

Fig. 4 Worm Gear

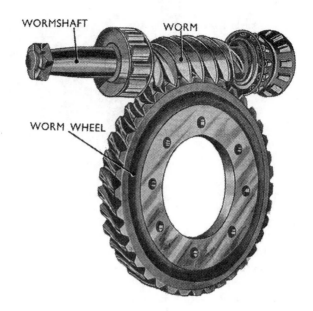

(A) Real Time Clock: Real time clock is a clock device that keeps track of the current time. There are different types of real time clock (RTC) device; among them DS1307 is used here. Microcontroller takes the month and hour values from the RTC device to track the sun's annual motion and the darkness of night to take the solar panel at its initial position. Address and data from RTC chip are transferred serially through an I2C, bidirectional bus.

(B) Position Sensor: Position sensor detects the sun's annual motion. A variable resistor is used here as position sensor. Position sensor is placed in the panel carrier rotator. When linear actuator moves linearly then panel carrier rotator rotates a semicircular path which causes the position sensor to change its voltage level. The panel carrier rotator rotates 50° in a semicircular path with respect to the horizontal axis as in the experimental location sun's latitude angle changes in between this 50°. The panel carrier rotator can rotate 75° in both sides which may also be applicable in other locations. In that case the sensor has tobe calibrated accurately. For 12 months different 12 values of sun's latitude angle are predetermined and set in the microcontroller and with respect to these values microcontroller decides how much to move the linear actuator. Panel carrier rotator rotates due to the linear actuator's linearly upward and downward motion with the panel carrier. A gear is placed with the panel carrier rotator which also rotates with it. This gear rotation causes the variable resistor's gear to rotate and this is how the resistivity of the variable resistor changes. Thus the signal is changing from the position sensor.

(1) Control Unit: Microcontroller is the main control unit of this whole system. The output from the sensor unit comes to the input of the microcontroller which determines the direction of the movement of the motors both in the horizontal and vertical axes. For this research ATmega32 microcontroller. This is from the Atmel AVR family.

Fig. 5 Single axis hardware setup

(2) *Movement Adjustment Unit*: Movement adjustment unit consists of two geared unipolar stepper motors along with their motor driver device. The output from microcontroller is sent to the motor driver. To run the unipolar stepper motor in full drive or half drive mode ULN2803 is used as motor driver IC. This driver is an array of eight Darlington transistors. Darlington pair is a single transistor with a high current gain. Thus the current gain is required for motor drive and it reduces the circuit space and complexity (Fig. 5).

The Performance and developed system was experimented and compared with both the static and continuous single axis solar tracking system. This work demonstrates that hybrid tracker system can assure higher power generation compared to static panel as well as less power consumption compared to continuous dual axis solar tracking system.

Altitude and azimuth angle from the astronomical calculation is verified for Chennai (13.08°N, 80.27°E), India by obtaining the result for every half an hour on winter solstice (December 21) and comparing it with the sun chart for the same location obtained from university of Oregon website [17]. The observation proves that the obtained result from the algorithm is the same with the sun chart diagram. The verification is shown in Fig. 6 for 8.00 a.m. as obtained from microcontroller output with graph. In addition, the astronomical calculation is also verified manually using a magnetic compass. The total maximum movement of the solar panel in a day is of 180° for the altitude movement (including movement of home position during night time). Therefore altitude motor is operating for 6 min per day. Similarly for azimuthal movement, the total maximum movement will be 360°. So the azimuth motor is operating for 12 min per day. The power consumed by the altitude motor is 10 W and azimuth motor is 8 W. The energy consumed by two DC motor is given in Eq. (3).

$$\text{Parasitic energy consumption} = [10 * 0.1] + [8 * 0.2] = 0.26\,\text{W h} \qquad (3)$$

Thus the maximum energy consumed by both the motor per day is 0.26 W h.

Fig. 6 Graphical Representation of fixed and single axis

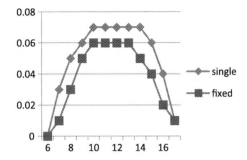

This amount of power saving will have a significant effect in large systems like heliostat powerplants where a lot of trackers are required and power saved by all the systems will show a big amount of power. Other than this the designed tracking system can also be implemented for the solar thermal systems. Finally the proposed design is achieved with low power consumption, high accuracy, and low cost (Tables 1 and 2).

Table 1 Fixed panel output

Time (h)	Power (kW h)	Voltage (V)	Current (A)
8	0.0293	14.88	2.3
9	0.0421	14.06	3
10	0.0481	13.96	3.45
11	0.0557	13.6	4.1
12	0.0545	13.25	4.12
13	0.0526	13.13	4.01
14	0.0448	13.75	3.26
15	0.0355	13.69	2.6
16	0.0158	14.13	2.45

Table 2 Single axis panel output

Time (h)	P (W h)	I (A)	V (V)
8	44.86	2.45	18.31
9	57.015	3.15	18.1
10	66.42	3.75	17.71
11	67.18	3.84	17.49
12	67.53	3.9	17.31
13	65.89	3.75	17.57
14	64.6	3.54	18.2
15	55.72	3.03	18.36
16	39.15	2.13	18.3

References

1. Deb, G., Roy, A.B.: Use of solar tracking system for extracting solar energy. Int. J. Comput. Electr. Eng. **4**(1), 42–46 (2012)
2. Tudorache, T., Kreindler, L.: Design of a solar tracker system for PV power plants. Acta Polytech. Hung. **7**(1), 23–39 (2010)
3. Shen, C.-L., Tsai, C.-T.: Double-linear approximation algorithm to achieve maximum-power-point tracking for photovoltaic arrays. Energies **5**(6), 1982–1997 (2012)
4. Liu, K.: Dynamic characteristics and graphic monitoring design of photovoltaic energy conversion system. WSEAS Trans. Syst. **10**(8), 239–248 (2011)
5. Tudorache, T., Oancea, C.D., Kreindler, L.: Performance evaluation of a solar tracking PV panel. U.P.B. Sci. Bull. Ser. C Electr. Eng. **74**(1), 3–10 (2012)
6. Mousazadeh, H., Keyhani, A., Javadi, A., Mobli, H., Abrinia, K., Sharifi, A.: A review of principle and sun-tracking methods for maximizing solar systems output. Renew. Sustain. Energy Rev. **13**(8), 1800–1818 (2009)
7. Benghanem, M.: Optimization of tilt angle for solar panel: case study for Madinah, Saudi Arabia. Appl. Energy **88**(4), 1427–1433 (2011)
8. Praveen, C.: Design of automatic dual-axis solar tracker using microcontroller. In: Proceedings of the International Conference on Computing and Control Engineering (ICCCE '12) (2012)
9. Sefa, I., Demirtas, M., Colak, I.: Application of one-axis sun tracking system. Energy Convers. Manag. **50**, 2709–2718 (2009)
10. Bakos, G.C.: Design and construction of a two-axis Sun tracking system for parabolic trough collector (PTC) efficiency improvement. Renew. Energy **31**, 2411–2421 (2006)
11. Mousazadeh, H., Keyhani, A., Javadi, A., Mobli, H.: A review of principle and sun-tracking methods for maximizing solar systems output. Int. J. Renew. Sustain. Energy Rev. **13**, 1800–1818 (2009)
12. Helwa, N.H., Bahgat, A.B.G., Shafee, A.M.R.E., Shenawy, E.T.E.: Computation of the solar energy captured by different solar tracking systems. Energy Sources **22**, 35–44 (2000)
13. Díaz-Dorado, E., Suárez-García, A.: Optimal distribution for photovoltaic solar trackers to minimize power losses caused by shadows. Renew. Energy **36**(6), 1826–1835 (2011)
14. Kelly, A., Gibson, L.: Improved photovoltaic energy output for cloudy conditions with a solar tracking system. Int. J. Solar Energy **83**, 2092–2102 (2009)
15. Al-Naima, F., Yaghobian, N.: Design and construction of solar tracking system. Int. J. Solar Wind Energy **7**, 611–617 (1990)
16. Neville, R.C.: Solar energy collector orientation and tracking model. Sol. Energy **20**, 7–11 (1978)
17. Fam, D.F., Koh, S.P., Tiong, S.K., Chong, K.H.: Qualitative analysis of stochastic operations in dual axis solar tracking environment. Res. J. Recent Sci. **1**(9), 74–78 (2012)

Author Biographies

K. Saravanan received his B.E. degree in Electrical and Electronics Engineering from Alagappa Chettiar College of Engineering and Technology, Madurai Kamaraj University, Karaikudi, Tamilnadu, India in 1998. He received his M.E. degree in Power Electronics and Drives Engineering from Anna University, Guidy Campus, Chennai, Tamilnadu, India in 2001. At present, he holds the post of Assistant Professor (Sr.Gr.) in EEE at SRM University. He has published several technical papers in national and international proceedings. His current research interests include the Renewable Energy Resources. He is a member of the IEEE, IACSIT, Life-member in ISTE.

C. Sharmeela holds a B.E. in Electrical and Electronics Engineering, M.E. in Power Systems Engineering from Annamalai University and a Ph.D. in Electrical Engineering from Anna University. At present, she holds the post of Assistant Professor (Sr.Gr.) in EEE, A.C.Tech., Anna University, and Chennai. She has done a number of consultancies for Power quality measurements and design of compensators for industries. Her areas of interest include Power Quality, Power Electronics applications to Power Systems and Renewable Energy Systems. She is a Life-member of the Institution of Engineers (India), ISTE and SSI.

Grid Interactive Level Multiplying Cascaded Multilevel Inverter for Photovoltaic MPPT

K. Saravanan and C. Sharmeela

Abstract For high voltage applications, H-bridge Cascaded MLI topology is one amid popular methods. It uses many series connected H-Bridge to produce AC voltage. The main benefits of MLI is better performance and low harmonics. Traditional MLI requires 'n' DC voltage bases to gain ($2n + 1$) stages. A new MLI by a level repetition system takes the method of a half-bridge inverter nearly twin the levels. The topology practises H-bridge MLI but offers a comparable presentation of an unequal topology in terms of voltage. Also, it keeps the virtue of unvarying stocking of the distinct cell for a MLI. The method is executed by linking a three-arm H-bridge (only two switches per phase) with the complete three-phase inverter to double the number of levels. Thus, it meaningfully recovers the power quality, lessens the switching frequency and lessens the cost and size of the power filter. The simulation results tested by the hardware setup.

Keywords AC energies · Converter topology · Equal repetition system · MLI · Control excellence

1 Introduction

For increasing control demand, the growth of Renewable energy resources is inevitable. Solar and wind energy are the most predominating resources among the other renewable energy resources [1]. Compared with wind energy, solar energy is having its inherent advantage of available in abundance and available daily [2]. In the advent of PV power generation MLI plays a vital role as it suitable to work. MLI consume recognized additional thought aimed at capability high-power,

K. Saravanan (✉)
SRM University, Chennai, India
e-mail: saravanank96@gmail.com

C. Sharmeela
Anna University, Chennai, India
e-mail: sharmeela20@gmail.com

© Springer Science+Business Media Singapore 2017
P. Deiva Sundari et al. (eds.), *Proceedings of 2nd International Conference on Intelligent Computing and Applications*, Advances in Intelligent Systems and Computing 467, DOI 10.1007/978-981-10-1645-5_57

683

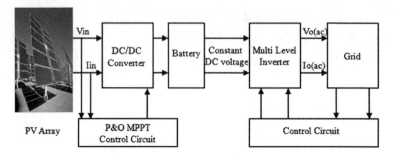

Fig. 1 Illustration of the scheme

medium voltage action, low switching loss and flexibility for operation [3–6]. The MLI uses the PV panels as dc sources and it produces like AC waveforms [7–12]. This paper explains the novel method the amount of stages in MLI, by addition merely two switches per phase [13]. The planned idea for single-phase is shown in Fig. 1. In this plan by tallying a further half bridge linked to a capacitor that preserves half the voltage of other bridges by a self-adjusting mechanism. For a three-phase system, three half-bridges (i.e., one half-bridge per phase) in equivalent are required. Thus, successfully, a three-phase full-bridge wants to be connected. Note that the dc buses of these half-bridges not consuming control. As it brings an assumed quantity of control in the first half cycle, it absorbs the similar quantity of control in the subsequent partial cycle.

In this paper, Sect. 1 explains the literature review of the present techniques places onward the drive of the effort the quantity of levels. Section 2 explains the system with block diagram. Section 3 explains the multilevel inverter (MLI). Section 4 describes the simulation results. Sections 5 deals with hardware setup and Sect. 6 concludes the work.

2 System Explanation

As solar cell is a basic component for a solar panel, the net production voltage of a PV is very little for both parallel and series or in both ways, to meet practical demands. For mathematical modelling of PV array, the fundamental equations are derived from the Fig. 2 shown.

Current source connected in equivalent with a diode gives solar cell equivalent. Say I be the current flowing out of the parallel circuit comprising a current source and diode and it is assumed by the calculation (1) (Fig. 3)

$$I = I_{pv} - I_d \tag{1}$$

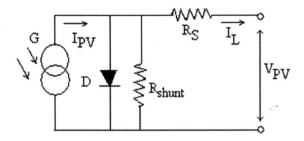

Fig. 2 Comparable network of the practical PV cell

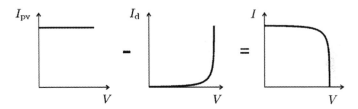

Fig. 3 Photo voltaic IV characteristic

For simulating the characteristics of the Photo Voltaic, modeling is necessary. From the equation given below the net current of the solar cell is the modification amid photocurrent (I_{PV}) and the standard diode current (I_d):

$$I = I_{pv} - I_0 \left(e^{\frac{q(V + IR_S)}{nKT}} - 1 \right) \tag{2}$$

This model depends on the cell temperature and the saturation current of the diode I_o.

$$I_{pv} = I_{pv}(T_1) + K_0(T - T_1) \tag{3}$$

Photon generated current:

$$I_{pv}(T_1) = I_{SC}(T_{1,nom}) \frac{G}{G_{nom}} \tag{4}$$

Rise in I_{sc} for unit rise in temperature:

$$K_0 = \frac{I_{SC}(T_2) - I_{SC}(T_1)}{(T_2 - T_1)} \tag{5}$$

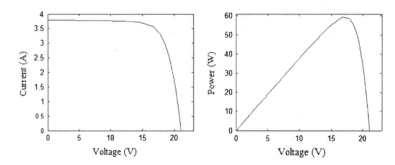

Fig. 4 Model of I-V and P-V characteristics

Diode Inundation Current at a specified temperature

$$I_0 = I_0(T_1) \times \left(\frac{T}{T_1}\right)^{\frac{3}{n}} e^{\frac{qV_0(T_1)}{nk}\left(\frac{1}{T} + \frac{1}{T_1}\right)}$$ (6)

Diode Barrage Current at standard hotness (Fig. 4)

$$I_0(T_1) = \frac{I_{SC}(T_1)}{\left(e^{\frac{qV_{oc}(T_1)}{nkT_1}} - 1\right)}$$ (7)

The buck and boost converter is unique kind of chopper, in the production voltage is either greater or less compared with the input voltage. It is nothing but a SMPS circuit topology for Boost and Buck converter. The duty cycle decides the production voltage. The rudimentary opinion of the buck and boost converter is honestly unassuming (Fig. 5).

During the On-state, the participation voltage basis is directly linked to the inductor (L). The outcome is accumulated in L. In this phase, the capacitor deliveries liveliness to the productivity freight. During the Off-state, the inductor is associated to the productivity freight and capacitor, so energy is transported from L to C and R.

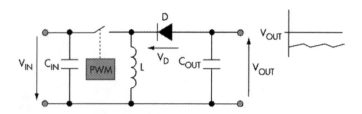

Fig. 5 Buck-boost converter circuit

3 Multi Level Inverter

The cascaded MLI is the combination of a amount of single-phase H shaped bridge inverters and is confidential hooked on equal and unequal collections created scheduled the scale of dc power sources. In the symmetric types, the amounts of the dc voltage bases of all H-bridges are equivalent while in the asymmetric types, the values of the dc voltage sources of all H-bridges are different.

Many kinds of switch methods have been obtainable for cascaded MLI [14–17]. In [18–21], different symmetric cascaded multilevel inverters have been existing. The key benefit of arrangements low change of dc voltage foundations, it is one of the utmost chief features in shaping the price of the inverter.

The main gain of this method and its processes is linked capability produce significant amount of production voltage stages (Fig. 6).

PV cell gives the required voltage for process of the network. Normally, this CHB (MLI) is able to produce **2N + 1** output voltage levels without the inclusion of Level Doubling Network (LDN). When the LDN originates into process, additional N stages of output voltage in the positive half cycle due to addition of **VPV/2**, which is the voltage of half bridge. Similarly, we get another N levels of production voltage in the undesirable partial series.

Result in a total of **4N + 1** levels of output voltage when the LDN is included. Also (LDN) remains the same. Thus, no power is consumed by the dc bus at any

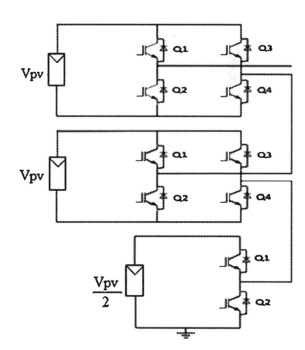

Fig. 6 Method of nine equal MLI

Table 1 Swapping board aimed at nine level MLI

	Cell 1				Cell 2				Cell 4	
	Q1	Q2	Q3	Q4	Q1	Q2	Q3	Q4	Q1	Q2
−2 VPV	0	1	1	0	0	1	1	0	0	1
−1.5 VPV	0	1	1	0	0	1	1	0	1	0
−VPV	0	1	1	0	0	1	0	1	0	1
−0.5 VPV	0	1	1	0	0	1	0	1	1	0
0	0	1	0	1	0	1	0	1	0	1
0.5 VPV	1	0	0	1	0	1	1	0	1	0
VPV	1	0	0	1	0	1	0	1	0	1
1.5 VPV	1	0	0	1	0	1	0	1	1	0
2 VPV	1	0	0	1	1	0	0	1	0	1

Fig. 7 Modes of operation of multilevel inverter

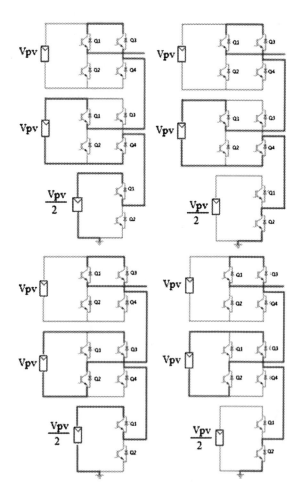

power factor of the circuit. The operation of this topology is divided into 4 modes according to the switching sequence assumed in the Table 1.

The topology can be considered as a combination of three PV Cells. Each H-bridge acts as a cell having 4 switches. The two main H-Bridges in the method can be named as Cell 1 and Cell 2.

The LDN can be named as Cell 4. If we want to expand the topology for a three phase operation then the additional H-Bridge added for each phase will be named as Cell 3. The inverter operates in four different modes as shown in Fig. 7.

Mode 1—Switches Q2 and Q3 conducts for both cell 1 and cell 2 and a negative voltage appears across the H Bridge. Since LDN is operating, switch Q1 of cell 4 conducts. This produces a negative odd level output voltage.

Mode 2—Switches Q2 and Q3 conducts for both cell 1 and cell 2 and a negative voltage appears across the H Bridge. Since LDN is not operating, switch Q2 of the cell 4 conducts and it will be bypassed. This produces a negative even level output voltage (Table 2).

Mode 3—Switches Q1 and Q4 conducts for both cell 1 and cell 2 and a positive voltage appears across the H Bridge. Since LDN is operating, switch Q1 of cell 4 conducts and the optimistic LDN added to positive. This produces a positive even level output voltage.

Mode 4—Switches Q1 and Q3 conducts for both cell1 and cell 2 and a positive voltage appears across the H Bridge. Since LDN is not operating, switch Q2 of the cell 4 conducts and it will be bypassed. This produces a positive odd level output voltage (Fig. 8).

The proposed nine level MLI is synchronized to with help zero crossing detector. The multi level inverter circuit is isolated from the grid by means of a

Table 2 Switching table for thirteen level MLI

Opt V (V)	Cell 1 (2 V)				Cell 2 (2 V)				Cell 3 (2 V)				Cell 4 (V)	
	Q1	Q2	Q3	Q4	Q1	Q2	Q3	Q4	Q1	Q2	Q3	Q4	Q1	Q2
−6 V	0	1	1	0	0	1	1	0	0	1	1	0	0	1
−5 V	0	1	1	0	0	1	1	0	0	1	1	0	1	0
−4 V	0	1	1	0	0	1	1	0	0	1	0	1	0	1
−3 V	0	1	1	0	0	1	1	0	0	1	0	1	1	0
−2 V	0	1	0	1	0	1	0	1	0	1	1	0	0	1
−1 V	0	1	0	1	0	1	0	1	0	1	1	0	1	0
0 V	0	1	0	1	0	1	0	1	0	1	0	1	0	1
1 V	0	1	0	1	0	1	0	1	0	1	0	1	1	0
2 V	0	1	0	1	0	1	0	1	1	0	0	1	0	1
3 V	0	1	0	1	0	1	0	1	1	0	0	1	1	0
4 V	1	0	0	1	1	0	0	1	0	1	0	1	0	1
5 V	1	0	0	1	1	0	0	1	0	1	0	1	1	0
6 V	1	0	0	1	1	0	0	1	1	0	0	1	0	1

Fig. 8 Topology of thirteen level multilevel inverter

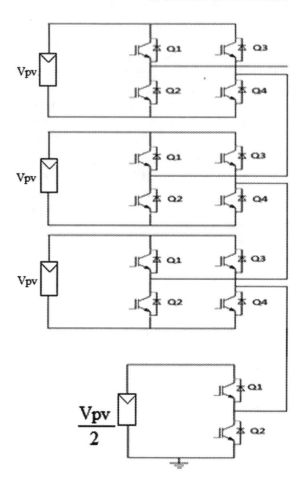

switch. The grid voltage is step down to the value of output voltage of MLI by means of step down transformer. The zero crossing detector circuit will make the waveform of grid voltage starting from the zero point. After checking the polarity of waveform, if both waveform starting from the same point the switch will close (ON position).

4 Reproduction Outcomes

The presentation circuit is assessed by pretending the topology. A Simulink model of the single phase multilevel inverter with level doubling network is developed and the output voltage waveforms are observed through the scope. The firing pulses for

Fig. 9 Pulses to the H-bridge inverter

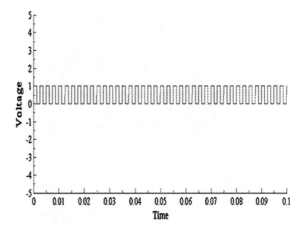

Fig. 10 Production electrical energy of nine equal MLI

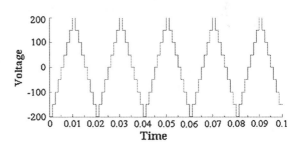

the H-Bridges in the model are generated by Sine Pulse Width Modulation (SPWM) in the Embedded Matlab Function block with appropriate coding. A universal bridge is used for including the H-Bridges for the topology.

Pulses given to the H-Bridges of the MLI circuit as shown in Fig. 9. These pulses are generated by Sine Pulse Width Modulation (SPWM).

Figure 10 shows the output voltage waveform generated by the Multilevel Inverter. The number of levels of voltage in the output waveform is more than those formed by the conventional topologies with the same number of H-Bridges.

5 Experimental Results

The hardware implementation of spilled H-bridge MLI structure with level doubling network is discussed. This topology is used for finding 9-level since the gains over the other two designs as conferred before. We also present the portraits of the waveforms. Getting stepped waveform the desired number sources (Photovoltaic cells) in series across the load at a particular instant through fast switching devices.

Fig. 11 Complete hardware setup

The next instant we need a lesser voltage level, one of the sources is eliminated from the circuitry using the switching devices. Hence we can make use of these fast switching devices to connect or disconnect a particular source across the load (Fig. 11).

Idea behind governing the production voltage magnitude by using microcontroller method is changing the width of pulses by varying a amount of reference wave. dsPIC 30F2010 is used as controller with 10 MHz clock frequency. TLP 250 is used as driver IC and IRF 840 Mosfets are used as the switching device due to cost optimization for hardware prototype.

6   Conclusion

This work concludes that Multilevel inverter is a favorable expertise in the power industry. In this project, the benefits and claims of Multilevel inverters are declared and a detailed description of different multilevel inverter topologies is offered. MLI with Level Doubling Network structure functioning is realized virtually using MATLAB SIMULINK. A detailed multilevel inverter is presented from which we concluded that the harmonic content is greatly reduced in multilevel inverter.

References

1. Nema, P., Nema, R., Rangnekar, S.: A current and future state of art development of hybrid energy system using wind and PV-solar: a review. Renew. Sustain. Energy Rev. **13**(8), 2096–2103 (2009)
2. Bakos, G.: Distributed power generation: a case study of small scale PV power plant in Greece. Appl. Energy **86**(9), 1757–1766 (2009)
3. Babaei, E., Hosseini, S.H.: Charge balance control methods for asymmetrical cascade multilevel converters. In: Proceedings of ICEMS, Seoul, Korea, pp. 74–79 (2007)
4. Wang, K., Li, Y., Zheng, Z., Xu, L.: Voltage balancing and fluctuation-suppression methods of floating capacitors in a new modular multilevel converter. IEEE Trans. Ind. Electron. **60**(5), 1943–1954 (2013)
5. Ebrahimi, J., Babaei, E., Gharehpetian, G.B.: A new topology of cascaded multilevel converters with reduced number of components for high-voltage applications. IEEE Trans. Power Electron. **26**(11), 3109–3118 (2011)
6. Kouro, S., Malinowski, M., Gopakumar, K., Pou, J., Franquelo, L., Wu, B., Rodriguez, J., Perez, M., Leon, J.: Recent advances and industrial applications of multilevel converters. IEEE Trans. Ind. Electron. **57**(8), 2553–2580 (2010)
7. Rodriguez, J., Lai, J.-S., ZhengPeng, F.: Multilevel inverters: a survey of topologies, controls, applications. IEEE Trans. Ind. Electron. **49**(4), 724–738 (2002)
8. De, S., Banerjee, D., Siva Kumar, K., Gopakumar, K., Ramchand, R., Patel, C.: Multilevel inverters for low-power application. IET Power Electron. **4**(4), 384–392 (2011)
9. Rodriguez, J., Bernet, S., Wu, B., Pontt, J.O., Kouro, S.: Multilevel voltage-source-converter topologies for industrial medium-voltage drives. IEEE Trans. Ind. Electron. **54**(6), 2930–2945 (2007)
10. Malinowski, M., Gopakumar, K., Rodriguez, J., Pérez, M.A.: A survey on cascaded multilevel inverters. IEEE Trans. Ind. Electron. **57**(7), 2197–2206 (2010)
11. Tolbert, L.M., Peng, F.Z.: Multilevel converters as a utility interface for renewable energy systems. In: Proceedings of IEEE Power Engineering Society Summer Meeting, vol. 2, pp. 1271–1274 (2000)
12. Ebrahimi, J., Babaei, E., Gharehpetian, G.B.: A new multilevel converter topology with reduced number of power electronic components. IEEE Trans. Ind. Electron. **59**(2), 655–667 (2012)
13. Chattopadhyay, S.K., Chakraborty, C.: A new multilevel inverter topology with self-balancing level doubling network. IEEE Trans. Ind. Electron. **61**(9), 4622–4631 (2014)
14. Park, S.-J., Kang, F.-S., Lee, M.H., Kim, C.-U.: A new single-phase five-level PWM inverter employing a deadbeat control scheme. IEEE Trans. Power Electron. **18**(3), 831–843 (2003)
15. Hinago, Y., Koizumi, H.: A switched-capacitor inverter using series/parallel conversion with inductive load. IEEE Trans. Ind. Electron. **59**(2), 878–887 (2012)
16. Su, G.-J.: Multilevel DC-link inverter. IEEE Trans. Ind. Appl. **41**(3), 848–854 (2005)
17. Najafi, E., Yatim, A.H.M.: Design and implementation of a new multilevel inverter topology. IEEE Trans. Ind. Electron. **59**(11), 4148–4154 (2012)
18. Pereda, J., Dixon, J.: High-frequency link: a solution for using only one DC source in asymmetric cascaded multilevel inverters. IEEE Trans. Ind. Electron. **58**(9), 3884–3892 (2011)
19. Rahim, N.A., Elias, M.F.M., Wooi, P.H.: Transistor-clamped H-bridge based cascaded multilevel inverter with new method of capacitor voltage balancing. IEEE Trans. Ind. Electron. **60**(8), 2943–2956 (2013)
20. Khazraei, M., Sepahvand, H., Corzine, K.A., Ferdowsi, M.: Active capacitor voltage balancing in single-phase flying-capacitor multilevel power converters. IEEE Trans. Ind. Electron. **59**(2), 769–778 (2012)
21. Najafi, E., Yatim, A.H.M.: Design and implementation of a new multilevel inverter topology. IEEE Trans. Ind. Electron. **59**(11), 4148–4154 (2012)

Author Biographies

K. Saravanan received his B.E. degree in Electrical and Electronics Engineering from Alagappa Chettiar College of Engineering and Technology, Madurai Kamaraj University, Karaikudi, Tamilnadu, India in 1998. He received his M.E. degree in Power Electronics and Drives Engineering from Anna University, Guidy Campus, Chennai, Tamilnadu, India in 2001. At present, he holds the post of Assistant Professor (Sr.Gr.) in EEE at SRM University. He has published several technical papers in national and international proceedings. His current research interests include the Renewable Energy Resources. He is a member of the IEEE, IACSIT, Life-member in ISTE.

C. Sharmeela holds a B.E. in Electrical and Electronics Engineering, M.E. in Power Systems Engineering from Annamalai University and a Ph.D. in Electrical Engineering from Anna University. At present, she holds the post of Assistant Professor (Sr.Gr.) in EEE, A.C.Tech., Anna University, and Chennai. She has done a number of consultancies for Power quality measurements and design of compensators for industries. Her areas of interest include Power Quality, Power Electronics applications to Power Systems and Renewable Energy Systems. She is a Life-member of the Institution of Engineers (India), ISTE and SSI.

Retraction Note to: Coronary Heart Disease Detection from Variation of Speech and Voice

Suman Mishra, S. Balakrishnan and M. Babitha

Retraction Note to:
Chapter "Coronary Heart Disease Detection from Variation of Speech and Voice" in: P. Deiva Sundari et al. (eds.),
Proceedings of 2nd International Conference on Intelligent Computing and Applications, **Advances in Intelligent Systems and Computing 467,**
https://doi.org/10.1007/978-981-10-1645-5_49

The chapter published in the book 'Proceedings of 2nd International Conference on Intelligent Computing and Applications 467, pages 583–594, https://doi.org/10.1007/978-981-10-1645-5_49 has been retracted because it contains significant parts plagiarizing another publication: 'Ms. Vishakha Pareek in the IEEE Conference—SCEECS 2016'.

The retracted online version of this chapter can be found at
https://doi.org/10.1007/978-981-10-1645-5_49

© Springer Science+Business Media Singapore 2017
P. Deiva Sundari et al. (eds.), *Proceedings of 2nd International Conference on Intelligent Computing and Applications,* Advances in Intelligent Systems and Computing 467, https://doi.org/10.1007/978-981-10-1645-5_58

Printed in the United States
By Bookmasters